manufacturing processes

$$N\left[T_1\left(L_1 + H_1\right) - T_2\left(L_2 + H_2\right)\right] = I\left(\frac{1}{N} + A + B + D + M\right) + \left(Y_2 - Y_1\right)$$

N = NUMBER

T = TIME / PIECE

L = LABOR

H = OH

I = NEW TOOLS

N = YEARS

A + B + D + M = CAPITOL %

$Y_2 - Y_1$ = EXTRA SETUP OF NEW

$$TC = NC_{OLD} + Y_{OLD} = NC_{NEW} + \left[Y_{NEW} + I\left(\frac{1}{N} + A + B + D + M\right)\right]$$

$$C = COST/PART = t(L+H) \text{ in } \$/PART$$

manufacturing processes

Arthur D. Roberts

**Formerly Department Chairman
Manufacturing Engineering Technology
Norwalk State Technical College
Norwalk, Connecticut**

Samuel C. Lapidge

**Assistant Professor
Manufacturing Engineering Technology
Norwalk State Technical College
Norwalk, Connecticut**

**Gregg Division
McGraw-Hill Book Company**

New York	Mexico
St. Louis	Montreal
Dallas	New Delhi
San Francisco	Panama
Auckland	Paris
Bogotá	São Paulo
Düsseldorf	Singapore
Johannesburg	Sydney
London	Tokyo
Madrid	Toronto

Library of Congress Cataloging in Publication Data
Roberts, Arthur D.
   Manufacturing processes.

   1. Manufacturing processes.   I.   Lapidge, Samuel C.,
joint author.   II.   Title
TS183.R6   670   76-45751
ISBN 0-07-053151-X

Cover illustration drawn by Tracy A. Glasner.

Manufacturing Processes

567890 WCWC 832

The editors for this book were Don Hepler and Alice V.
Manning, the designer was Tracy A. Glasner, the art
supervisor was George T. Resch, and the production
supervisor was Regina R. Malone. It was set in Palatino by
Black Dot, Inc.

contents

viii   Contents

A basic knowledge of the machines and processes by which material is formed into useful products is vital to anyone engaged in mechanical, manufacturing, industrial, electromechanical, and related engineering fields.

To economically design a product, the machine designer must have a concept of how it can be made. The manufacturing engineer and plant layout, production, or quality control engineers deal with the processes of manufacturing daily. Electromechanical, optical, metallurgical, and other engineers often find need for this knowledge.

This book can be completely studied in a two-term course, or the instructor can, as is necessary in many two-year engineering colleges, use selected chapters for a strong, basic one-term course. The only prerequisite is a knowledge of algebra. No one- or two-term course will make the student an expert in the many phases of manufacturing processes because each machine and process has many variations. There is not sufficient time, and really no need, for a student to learn all the complexities of the equipment used in industry today.

Most graduates find that fairly early in their careers they tend to specialize to a considerable degree. At that time they begin to study more deeply the topics which are most valuable in their field of work. Then the references given in this book, manufacturers' catalogs, and personal contacts will add the details as they are needed.

The authors have included only those processes which they find are quite extensively used or which seem to be rapidly growing in use. Some of the growth areas are in items such as new cutting tools, fine blanking, EDM, ECM, new grinding materials, and a new emphasis on certain welding processes. To the extent possible, the space given to each machine or process is related to the frequency of its overall use in industry. Different localities may, however, have quite a different balance. The instructor can then make use of the references, personal experience, or local industry to amplify topics to suit the situation.

Even though this book is basically for students in the classroom, the writing is designed so that self-study is facilitated. All new words and terms are defined as they are introduced. In many places both the shop and technical terminology are given so the student will not be confused by the sometimes different vocabularies. Clear headings and subheadings make it easy to review a chapter or to locate specific information. Review questions and up-to-date references are included at the end of each chapter.

Each type of machine or process is covered in enough detail so that the student can get the feel of what the equipment can do, its range of size and power, and its limitations. We hope that none of this data will be memorized but that the breadth of the capabilities of the equipment will be realized.

Illustrations, insofar as possible, are fairly simple, and furnished with labels and explanations so that the function of each machine can be easily understood. Pictures of complex machines, which only a specialist could understand, and the extremes of sizes (which are less often used) have been omitted from most chapters.

Many of the line drawings were prepared to help students grasp the fundamentals quickly, and they have been classroom tested.

The mathematical theory of cutting, grinding, forging, and similar work is not covered extensively in this book. While this is a splendid field of work, it is highly specialized and it is not necessary at this level of study. The equations and formulas that are included are those which people in manufacturing need to do their work.

Numerical control is now in use in small and large plants all over the world. The use of N/C machines has changed many concepts and made possible many economies in manufacturing. Thus, N/C machines are mentioned as simply another type of lathe, drill press, etc. These N/C machines should always be kept in mind today as one thinks of how to make things. The basic N/C ideas are presented in Chap. 5 so that the student can understand the later references.

The *arrangement of chapters* has been influenced by possible laboratory experience which, if feasible, should accompany the study of this subject. The lathe, drilling, milling, and grinding processes are basic and are the most frequently available in college machine labs; thus they are described first.

Laboratory experience on a variety of machine tools is, even for the design engineers, highly desirable if the time can be arranged. For engineering technicians it is a real necessity. Students have a much different idea of the meaning of ± 0.001 in. (0.025 mm) if they have tried to achieve it on a lathe, milling machine, or OD grinder. Laboratory sessions do not need to be long. The authors have found that even a 2-hour period used in performing typical machining operations can be of considerable value to students.

The *metric system,* as well as the U.S. system, is shown throughout the book. Today practically the whole world is using the metric system, and many products in the United States are being made to this system. Thus, this book is "dual dimensioned," as are many drawings in England and the United States.

The changeover to metric is not difficult, though the size of the numbers associated with some specifications (such as microfinish) are quite different. Most problems in the book are solved using both the customary and metric systems. "Equivalent" metric sizes are usually rounded off to even numbers rather than the exact equivalent because when we begin to "think metric," we will use 3 mm and 6 mm instead of the exact 3.175 and 6.35 mm equivalents of ⅛ or ¼ in.

The study of manufacturing processes should be a continuing one, as new methods, refinements, and totally new developments are happening all the time. The authors believe that this book will give a good basic start on what can be a challenging, exciting, constantly changing field.

Acknowledgments: This book could not have been written without the help of nearly 300 companies which have generously provided catalogs, technical literature, suggestions, and illustrations. The names under the illustrations identify some of these companies. The trade associations have also been an important source of information and help, and students should contact them when they need assistance.

Our sincere thanks to many who reviewed, or assisted with, specific chapters. Among these are Prof. Carl German, who wrote much of Chap. 1, and Prof. Marie Kiss, who helped with Chap. 30, Plastics. Thanks also to Mrs. Margaret Roberts for spending many hours typing the text and the numerous letters necessary in this project.

Arthur D. Roberts
Samuel C. Lapidge

properties and basic metallurgy of metals

How a metal will respond to a particular manufacturing process or how it will meet a specific design requirement is of vital concern from both a safety and a cost viewpoint. Therefore, such mechanical properties as hardness, tensile strength, ductility, and toughness must be known for each material.

The subject is large; there are many excellent books available and many tables of the physical properties and metallurgy of metals, so no attempt will be made here to cover the complete subject. However, some terms are used quite frequently in all manufacturing processes today. Developing an understanding of these terms and their basic relationships to the simple metallurgical structure and the machinability of metals is the purpose of this chapter.

HARDNESS

Hardness, in a practical engineering sense, can be defined as resistance to penetration. It will be shown later in this chapter that hardness is a function of the microstructure of the metal, but first one must understand the scales by which it is measured. All six scales are arbitrary, and several of them are closely related.

Mohs' scale for hardness is a *scratch* test. It merely compares natural minerals, each of which will scratch the one with the next lower number. The scale is:

10—Diamond	7—Quartz	3—Calcite
9—Corundum or	6—Orthoclase	2—Gypsum
sapphire	5—Apatite	1—Talc
8—Topaz	4—Fluorite	

This method is almost never used in industry, but it is listed here because it is referred to occasionally in the literature.

For general comparison:

Tungsten carbide	About 9.2 Mohs' scale

[1]This chapter written with Prof. Carl. German, Department Chairman, Materials Technology, Norwalk State Technical College, Norwalk, Conn.

Hardened tool steel	About 7.5
Low-carbon Steel	About 4.3
Aluminum	About 2.3

The *metric system* is used in all five of the following hardness testing systems. All loads are in grams or kilograms, and all measurements are in millimetres. Thus, we have no conversion problem as these have been standard in the United States and Europe for many years. The first four are indentation tests.

Rockwell Hardness Testers

For most tests these use a cone-shaped diamond point (Fig. 1-1). An initial preload (minor load) of 10 kg is applied so that the diamond is well seated on the material and then the full load of 15 to 150 kg is applied and released. A dial gives the hardness reading directly. Thus, this is a fast, accurate way to test hardness.

There are about 15 different scales which can be used. This is accomplished by varying the preload (3 or 10 kg), the major load (15 to 150 kg), and the penetrators. However, the three scales used most often are:

Rockwell C scale (R_c or ISO designation HRC) uses a 150-kg load, diamond penetrator. This is for metals harder than cold-rolled steel, up to very hard steels, and carbides.

Approximate readings would be:

Hardened tool steel, R_c 50–65

A file is about R_c 50

Carbides, R_c 80–90

Cold-rolled steel, R_c 22

Rockwell A scale (R_A or HRA) uses the diamond penetrator with a 60-kg load. This is used for brittle materials, such as tungsten and titanium carbide tool

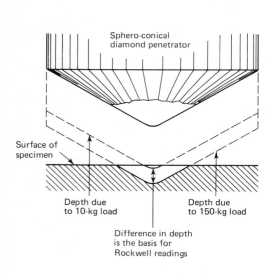

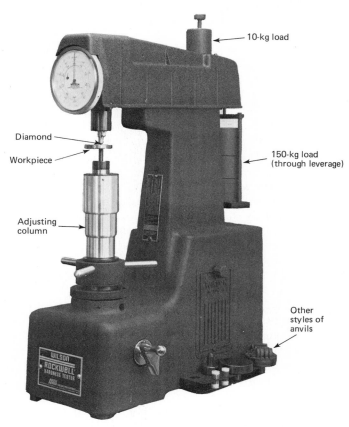

Fig. 1-1   A Rockwell hardness tester equipped for normal and superficial hardness testing. (*ACCO, Wilson Instrument Division.*)

bits, which will crack under the 150-kg load. Carbide tools = R_A 87–96 approx.

Rockwell B scale (R_B or HRB) uses a $\frac{1}{16}$-in.-diameter (1.59-mm) steel ball with 100-kg major load. This is used for softer materials, such as soft steel, brass, and aluminum.

Approximate readings would be:

Cold-rolled steel (R_c 22), HRB 100

Brass, HRB 50

Aluminum, HRB 25

Brinell Hardness Test

Brinell hardness testers (Fig. 1-2) use a steel or carbide ball of 10-mm diameter for most tests. The load is usually 3000 kg (about 3 tons), though on soft metals a 5-mm ball and a 500-kg load may be used. The *diameter* of the indentation made by the ball is measured under a low-power microscope. The hardness values are then obtained from a table based on this formula: BHN = load divided by area of indentation. ISO designation is HB.

The *disadvantages* of the Brinell scale are:

1.   The 10-mm ball makes quite a large dent on soft materials.

2.   The hardened steel or carbide indenters cannot be used for testing materials above BHN 750 (about R_c 75).

3.   The diameter of the indentation must be hand measured, in millimetres, to determine the hardness number.

The *advantages* of the Brinell scale are:

1.   A continuous scale of numbers from below soft aluminum, BHN 30, up to hardened steel, BHN 750.

2.   The indentation leaves a permanent record of the test.

3.   A minor point, but useful, is that the BHN divided by 10 is approximately equal to the R_c value.

4.   The rather large indentation made tends to "average" the hardness of multiphase materials.

5.   There is a fairly close relationship between the BHN and the tensile strength of some steels. TS ≈ BHN × 450.

6.   Direct-reading models are also available.

Brinell hardness testers are made by several companies. Costs start at about $600.

Microhardness Testers

Both the Vickers and the Knoop microhardness tests may be performed with the same testing machine, changing only the indenter. Several makes are available, one of which is shown in Fig. 1-3. They cost upwards of $3000.

These machines must be able to smoothly apply loads from 1 g to at least 1 kg and must have a built-in microscope with up to at least 500 × magnification.

Vickers hardness testing uses a diamond penetrator as shown in Fig. 1-4a. The loads used are quite light, usually from 1 to 50 kg, which makes it easy to test fragile or thin materials. A single scale is used for all materials from soft to very hard. To get the hardness reading, the diagonal (usually 0.2 to 1.5 mm) of the square indentation is measured under a microscope. The hardness value is obtained from tables or from the formula shown in Fig. 1-4.

The reading is given as DPH (Diamond Pyramid Hardness), VPN (Vickers Pyramid Number), or HV (ISO designation, Hardness Vickers). The DPH number for a material is very nearly the same as the

Fig. 1-2 An automatic, hydraulically operated, direct-reading Brinell hardness tester. (*Ametek, Inc.*)

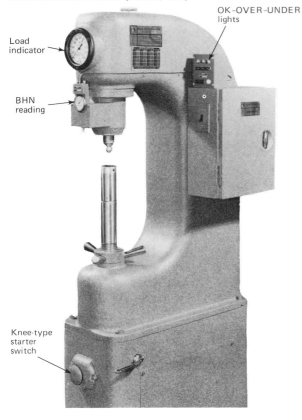

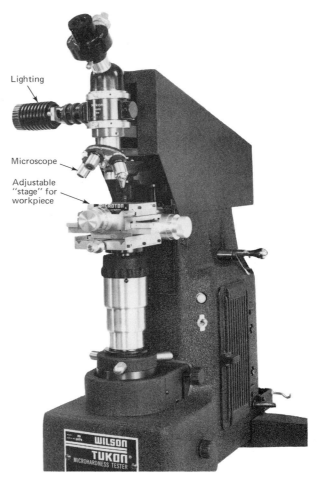

Fig. 1-3 A microhardness tester for use with Vickers or Knoop diamond indenters. (*ACCO, Wilson Instrument Division.*)

BHN reading: about 8 percent higher above 300 on the scale.

Knoop microhardness testing also uses a continuous scale. It uses a diamond penetrator which has been ground to an elongated diamond shape (Fig. 1-4b), and it uses very light loads: 1 g to 1 kg. Thus, it is especially useful in laboratory work on very small sections of a sample and on very thin materials such as foils and thin (0.025-mm) coatings.

The long diagonal of the impression is measured in millimetres under a microscope since the diagonal may be as small as 0.002 mm (0.0008 in.). The hardness values are obtained from a table or can be computed. The Knoop hardness number, KHN or KN, using a 500-g load, is very close to the BHN and DPH.

Shore Scleroscope

This is a *rebound* method of measuring hardness (Fig. 1-5). A diamond-tipped hammer (2.3 g standard

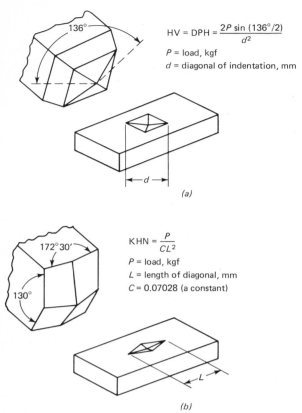

$$HV = DPH = \frac{2P \sin (136°/2)}{d^2}$$

P = load, kgf
d = diagonal of indentation, mm

(a)

$$KHN = \frac{P}{CL^2}$$

P = load, kgf
L = length of diagonal, mm
C = 0.07028 (a constant)

(b)

Fig. 1-4   Diamond indenters and formulas for Vickers and Knoop microhardness testers.

weight) is dropped from a fixed height of 9.89 in. (251 mm) in the standard model and 0.704 in. (18 mm) in the direct-reading model. The diamond makes a minute indentation in the material and rebounds in the glass tube. As some of the energy is absorbed by the material, the amount of rebound will vary according to the hardness of the material: the harder the metal, the higher the rebound. The operator gages, by eye, the maximum height of the rebound on an arbitrary scale. This is the scleroscope reading. The scale was made so that quenched tool steel of highest hardness reads 100. A direct-dial-reading model is also available.

Advantages of the Shore scleroscope are:

1.   It is very simple to operate and does not mar the workpiece.

2.   It is frequently used without a bench stand, as an easily portable hardness tester, for use on both very large and small parts.

3.   It has a continuous scale and can be used on almost any material.

4.   It is quite inexpensive, with complete outfits from $500.

General Precautions when Testing Hardness

The thickness of the piece being tested can make quite a difference in the selection of the choice of hardness-testing equipment. Thin sections of 0.010 in. (0.25 mm) to 0.030 in. (0.75 mm) thick require the use of light loads. The Vickers method is reliable down to 0.005-in. (0.125-mm) thickness, and the Knoop method can be used for foils of less than 0.001 in. (0.025 mm).

CAUTION: Do not specify hardness values of too close tolerance. Specification of R_c 65–66 or BHN 300–305 are rather unrealistic. In tests run on a single piece of steel within an area of 1 in.² (625 mm²), the Rockwell hardness readings can vary by two or more points. Most materials are *not* homogeneous, so they have a "built-in" variation of up to 5 percent.

Parts with a diameter of less than 1 in. (25.4 mm) will not give accurate readings, though they may be off by only 1 or 2 percent. Correction tables are available from most manufacturers.

A *comparison chart* (Table 1-1) gives the approximate relationships among the five methods described. *Do*

Fig. 1-5   The Shore scleroscope shown used on a large roll. (*Shore Instrument & Mfg. Co., Inc.*)

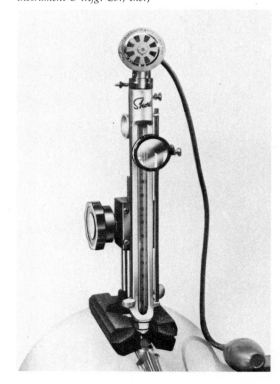

TABLE 1-1 APPROXIMATE EQUIVALENCIES OF SEVERAL HARDNESS TESTS (FOR STEEL ONLY)

Rockwell R_C and R_B	Brinell, BHN 3000-kg Load	Vickers, DPH	Knoop, KHN	Shore Sceleroscope	Examples
95 R_C					
90	Cannot	2800	2300	120	Carbide tools
80	test	1800	1300	115	
70	850*	1100	830	106	Hardened tool steel
60	653*	700	620	81	High-speed steel
50	485	525	540	67	Steel files
40	375	390	410	52	
30	285	290	315	42	Tool steel—annealed
20	223	230	245	33	Cold-rolled steel
95 R_B	207	207	230	32	
90	183	190	205	28	Gray iron—soft
80	146	154	170	22	
70	121	126	140	19	
60	105	100	128	15	
50	83†	93	100	12	Brass
40	75†	82	90	10	
30	67†	75	80	7	Aluminum
20	61†	68	73		

NOTE: Vickers, Knoop, and Brinell readings are quite similar.
BHN $\div$ 10 $\cong$ R_C readings
*Carbide ball.
†500-kg load.

not regard these as exact ratios because too many variables enter into all hardness tests and different tables of "equivalents" give different figures.

THE BASIC STRUCTURES OF METALS

Every manufacturing process which alters the shape of a solid metallic object is affected by the arrangement of the atoms in the material. The physical properties of metals (such as high density, heat conductivity, electrical conductivity) as well as the mechanical properties (such as tensile strength, hardness, ductility, etc.) are related to the bonding forces holding like and/or unlike atoms together as well as to the chemical mix of atoms.

The structure of a metal may be analyzed at three levels:

1. Atomic (or submicroscopic)
2. Microscopic (up to 2500× magnification—optical microscope)
3. Macroscopic (up to 10× magnification)

At the atomic level, which will be discussed in this chapter, the smallest building block in a metal (element or alloy) is called the *unit cell*. The three most common types are shown in Fig. 1-6.

Crystals

Metallic materials are described as being crystalline. A single crystal is made up of millions of unit cells (or, if it is an alloy, a combination of unit cells) with a repetitive pattern in all directions within the crystal. Single crystals of various sizes are now available commercially.

Most materials are polycrystalline, as is evident if a sheet of dip-galvanized steel is examined. The individual crystals or grains of zinc have different orientations and meet one another at their grain boundaries. Another example is shown in Fig. 1-7, a sketch of the microstructure of pure iron. Keep in mind that Fig. 1-7 is a flat plane section of the metal which cuts through the individual crystals.

CAUTION: Any attempt to make general statements in classifying the thousands of materials available by referring to their crystalline structure can be very misleading.

For example, brass, which is FCC, and magnesium, which is HCP, are both very easy to machine. However, some stainless steels are also FCC and some titanium alloys are HCP, and they are both quite difficult to machine. Much of this

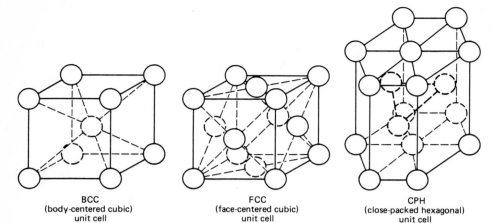

BCC
(body-centered cubic)
unit cell

FCC
(face-centered cubic)
unit cell

CPH
(close-packed hexagonal)
unit cell

Fig. 1-6   The three most common types of unit cell structure found in metals.

difficulty is due to their rapid work-hardening rate, as explained later in this chapter.

The *bonding forces* within the crystals are called *metallic* (or electron cloud), *covalent, ionic,* and *van der Waals.* These will not be discussed in this book. For more information, see references at the end of this chapter. The atoms in one crystal do not often perfectly align with the atoms in other crystals when they meet at the grain boundaries. Thus, the bonding forces at the grain boundaries (intercrystalline forces) will be less than the intracrystalline bonding forces. Therefore, the properties of a metal will be affected by the size, shape, chemical makeup, and orientation of the grains (crystals) in its structure.

The atomic bonding forces mentioned above are responsible for the strength of the individual crystals. Similar forces at the grain boundaries help hold the many crystals together. The measure and behavior of these attractive forces explain the behavior of the materials we cut, shape, and bend.

Mechanical Forces between Crystals

A very important example of the action of the forces between crystals is the standard *tensile test.* We will assume that we are testing a strong copper alloy since the steel curve has an extra "bump" in it.

A sample has been machined to the shape and size shown in Fig. 1-8. This sample is fastened between the jaws of a testing machine like the one shown in Fig. 1-9. These machines are made in sizes which can exert pulling (tensile) or pushing (compression) forces of from 10 to 100 tons (910 to 9100 kgf or 90 to 900 kN).

As the testing machine is pulling the sample, the "stretching" of the metal between two accurately marked points 2 in. (50 mm) apart is being accurately measured by a gage similar to that shown in Fig. 1-10.

The "stretch" and the load are measured at regular intervals and used as the basis for the curve shown in Fig. 1-13.

Before discussing the changes which the metal undergoes during the test, a few definitions must be understood.

Load P is the reading on the dial of the testing machine, in pounds, kilograms-force, or newtons.

Elongation is the reading of the extensometer gage, in inches or millimetres, between two points on the test piece. It is read in 0.001-in. or 0.02-mm units.

Stress is force per unit area, in pounds per square inch (psi), newtons per square metre (N/m^2), or kilograms-force per square millimetre (kgf/mm^2).

Note: N/m^2 is officially given the name pascal (Pa), but in deriving equations it is preferable to use the basic units.

$$\text{Stress} = S = \frac{\text{load}}{\text{area}} = \frac{P}{A}$$

Fig. 1-7   Crystalline arrangement of the grain structure of typical composition of pure iron (*not* cast iron).

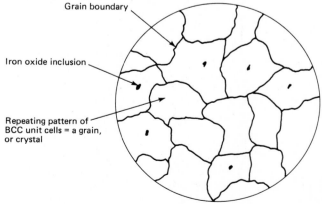

Grain boundary

Iron oxide inclusion

Repeating pattern of BCC unit cells = a grain, or crystal

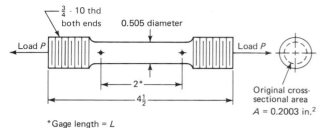

Fig. 1-8 Standard machined tensile test specimen.

where A = cross-sectional area, perpendicular to the axis of the load.

Strain e is the amount of elongation per unit of length, ΔL. In this test the strain is in inches per inch, or millimetres per millimetre over a 2-in. (50-mm) gage length; therefore, $e = \Delta L/2$ or $e = \Delta L/50$ metric.

Percent elongation is the maximum increase in length ΔL (between 2 in. or other gage lengths) at the *breaking* point, divided by the original gage length L.

$$\text{Percent elongation} = \frac{\Delta L}{L} \times 100$$

Modulus of elasticity E (Young's modulus) is stress divided by strain, at a common point in time, before the yield point is reached.

$$E = \frac{\text{lb/in.}^2}{\text{in./in.}} = \text{lb/in.}^2 \text{ (psi)}$$

Fig. 1-10 Electronic extensometer shown connected to a standard test specimen. Read-out of the strain is made by connecting to a chart recorder. (*Tinius Olsen Testing Machine Co., Inc.*)

Fig. 1-9 (a) A typical 100,000-lb. (445kN) hydraulically operated tensile tester, using a standard test piece of mild steel.(b) Triple-reading scale which is now available. (*Tinius Olsen Testing Machine Co., Inc.*)

(a)

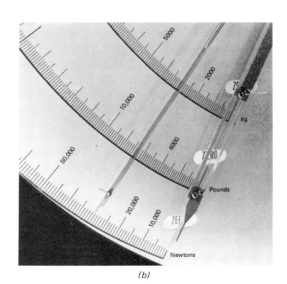

(b)

or

$$E = \frac{N/m^2}{mm/mm} = N/m^2 \text{ or pascals (Pa)}$$

EXAMPLE 1-1

30×10^6 psi $= 207 \times 10^6$ kPa $= 207$ GPa (gigapascals).

Tensile and shear forces are exerted on the test sample in Fig. 1-8. A highly simplified diagram of what happens is shown in Fig. 1-11. Notice that the crystal structure (made up of atoms) is deformed. It increases in length and decreases in width. This same thing happening to millions of crystals in the metal accounts (at least basically) for the "necking" which occurs before a test bar is broken by tensile forces.

The shear strength of most ductile metals is less than the straight tensile strength, so the actual break will be the "cup and cone" shape shown in Fig. 1-12a. Very brittle materials such as grey cast iron will break in pure tension as shown in Fig. 1-12b.

Of course, an actual piece of metal will have many different grain sizes, impurities, "vacancies" in the crystals, etc., all of which will change the properties of the material from the perfect models shown.

stress-strain curve

The results of the data collected from a tensile test will often, for a ductile metal, show a curve like that in Fig. 1-13. This curve, plotted for a metal, can give considerable information about the metal. Referring to Fig. 1-13,

OA—is a straight line, showing a constant change in length per unit change in load. This change is *elastic*, that is, if at stress S, the load was removed, the test sample would return to its original length, just like an elastic band. The strain OA' is usually from 0.003 to 0.010 in./in. or mm/mm.

AC—shows much greater strain (stretch) with very little change in load. This is a *plastic* change. The material will *not* return to its original length. Ductile materials will exhibit "necking" as shown in the sketches in Fig. 1-13.

elastic vs. plastic strain

The atoms in the material while under the *elastic* portion of the curve OA move slightly *within* the crystals and at the grain boundaries, but do not part from their immediate neighboring atoms as shown (somewhat oversimplified) in Fig. 1-14a.

The *plastic* portion of the curve AC shows permanent deformation because planes of atoms within the crystals *slip* with respect to one another as shown in Fig. 1-14b. At the same time, entire grains are rearranging themselves where the grain boundary forces of attraction are overcome.

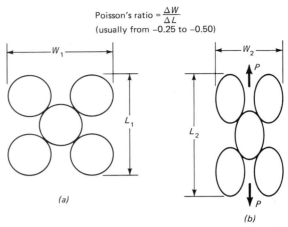

Poisson's ratio $= \frac{\Delta W}{\Delta L}$
(usually from −0.25 to −0.50)

Fig. 1-11 Simplified diagram of the effect of a tensile load on atoms in a metal. (a) Atoms in unstressed condition; (b) atoms deformed by tensile stress.

Tensile strength (Engineering TS) is taken as the value at point B on the curve. The stress at point C, the breaking point, or the value at point G (computed by dividing the load by the reduced cross section as the part necks in) are of no practical value. The progress to failure, from the maximum stress B to breaking at C, is quite rapid.

Yield strength, elastic limit, or yield point A on Fig. 1-13, is the greatest stress the material can stand and still return to its original gage length. The stress used in designing a piece of equipment should be less than this value for satisfactory service performance. This is sometimes called the *maximum design stress*.

Modulus of elasticity, or Young's modulus, E is the *slope* of the line OA.

$$E = \frac{\text{stress } S}{\text{strain } e} = \frac{\Delta Y}{\Delta X} = \frac{AA'}{OA'}$$

Fig. 1-12 Two types of fracture resulting from a metal failing under tensile stress. (a) "Cup and cone" fracture of ductile metals; (b) brittle fracture—less than 6 percent elongation.

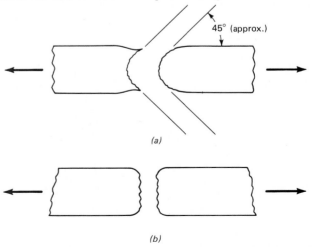

45° (approx.)

(a)

(b)

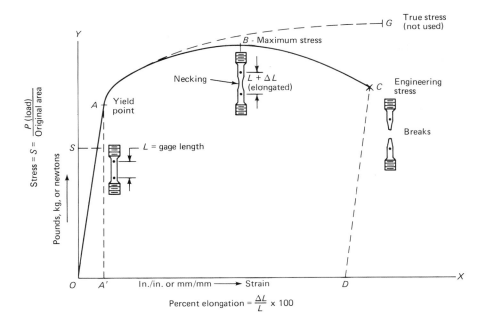

Fig. 1-13 A stress-strain curve for a ductile metal such as copper.

or

$$e = \frac{S}{E}$$

Thus, the higher the value of E, the less a material will distort, or bend, under a given load. The material with the higher modulus of elasticity is a *stiffer* material.

CAUTION: Changing the hardness or the alloy of a material such as steel or aluminum
Does not appreciably change the modulus of elasticity
Does change the yield point (YP), the ultimate tensile strength (UTS), and the percent elongation

This may seem odd until one realizes that the elastic portion of the curve occurs *before* the atoms have left their original "neighbors." Thus, the principal forces at work are the *inter*atomic forces, which vary only with the basic elements such as iron, aluminum, or copper.

Once the stress is high enough to start planes of atoms *slipping* past one another, plastic flow begins,

and the various physical conditions of the grain boundaries and grain composition have their effects. A few values of modulus of elasticity are:

All steels, $E = 29$ to 30×10^6 psi (200 to 207×10^6 kPa)*

All aluminum alloys, $E = 10$ to 11×10^6 psi (69 to 76×10^6 kPa)

Cast iron, due to the presence of free carbon in different forms and percentages, has a wide variation.

Cast iron, $E = 12$ to 20×10^6 psi (83 to 138×10^6 kPa)

The information in the last few paragraphs is summarized in Fig. 1-15. The quantities shown are only approximate, but the relationships are basically correct. From this graph, percent elongation would be the strain $e \times 100$. For example, the aluminum, at breaking, shows 0.40 in./in. Thus, $0.40 \times 100 = 40$ percent elongation.

Note that the YP for 0.20 C (0.20 percent carbon) is 45,000 psi and for 0.50 C is 70,000 psi, but the stiffness (Young's modulus) is the same for both. Note also that the slope of the curve for aluminum is less (not as

Fig. 1-14 Simplified drawing of what happens during elastic and plastic tensile deformation (strain). (*a*) Elastic strain; (*b*) plastic (permanent) strain—the atoms have new "neighbors."

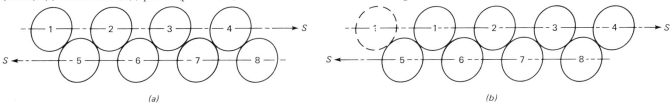

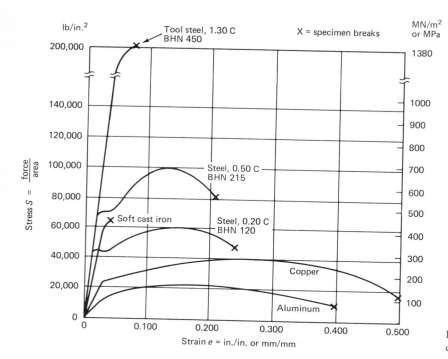

Fig. 1-15 Approximate stress-strain curves for comparison of several metals.

steep) than that of either copper or steel; therefore, it has a smaller modulus of elasticity and will bend more under the same load.

strain hardening or work hardening

An interesting condition exists in Fig. 1-13 on the curve from A to B. The force required to stretch the bar is *increasing*, even though the bar is "necking" down, causing the cross-sectional area to *decrease* and the true stress to rapidly increase.

This strengthening phenomenon is called *work hardening*. It happens whenever metallic materials are plastically changed by any manufacturing process (such as rolling, bending, or forming) without added heat. This also causes considerable increase in hardness.

A simple example of this is the use of the 10-mm ball on a Brinell hardness tester on annealed brass as shown in Fig. 1-16. The lefthand view shows the large grain structure of the soft annealed brass. Figure 1-16b shows what happens under 3000 kg (6600 lb) of pressure. Since the material under the indenter is being plastically deformed, it work-hardens and, of course, resists further penetration.

The many parallel lines within the individual grains in Fig. 1-16b are called *slip lines* and were produced because sets of planes of atoms in the crystal are displaced with respect to one another (as in Fig. 1-14) by the externally applied force of the hardness tester. A similar microstructure results from rolling or forging operations, and it also occurs in the vicinity of a *cutting tool* removing chips.

Recrystallization-anneal simply means that the warped and stresed crystals (as in Fig. 1-16b) are returned to a normal, relaxed condition as in Fig. 1-16c. This is done by heating the metal to a high enough temperature (well below its melting temperature) for a period of time sufficient for *new* stress-free crystals to grow. If enough time at temperature is maintained, the grains will increase again to the size shown in Fig. 1-16a. This is called *grain growth*, and it occurs because the smaller grains are simply absorbed by adjacent grains, which then become larger.

HOT AND COLD WORKING

This recrystallization phenomenon is utilized in making *hot-rolled steel* (HRS) and hot forgings. The necessary plastic deformation of the material (rolling down to size or hammering to shape) is done while the material is held at a temperature *above* its recrystallization temperature.

The growth of new grains during a hot-working operation continuously offsets the work-hardening effect. Thus, hot working avoids the higher strength and hardness which would result if the material was worked at lower temperatures.

Cold working as done on cold-rolled steel (CRS) is performed well below the metal's recrystallization temperature, often close to room temperature. If the cold work is extensive (such as forming a deep cup in a punch-press operation), this may require periodic interruptions for a recrystallization-annealing heat treatment.

In spite of their somewhat higher cost, cold-worked materials are used in many applications because of their improved tensile strength, hardness, and smooth, bright surfaces. Moreover, close dimensional tolerances are (in CRS round and square bars, for example) standard manufacturing practice. A 1-in.-diameter CRS bar will be held to 0.997- to 1.000-in. diameter.

MACHINABILITY VERSUS MICROSTRUCTURE

Machinability is a term used to classify materials according to the relative ease with which they may be cut. The measure of machinability which is often used is the cutting speed (in feet per minute or metres per minute) at which a cutting tool will last 60 min.

Actual testing is more often done in terms of cubic inches of material removed before the cutting tool wears a standard amount, i.e., *tool life*. The factors involved in the testing are shown in Fig. 1-17.

An excellent example of the work-hardening (cold-working) effect which machining has on a metal's microstructure is shown in Fig. 1-18. Notice especially that the finished surface shown at the extreme right has been affected to a depth of about 0.008 in.

(0.2 mm). The original material was annealed (soft) commercial brass, 70 percent copper, 30 percent zinc.

Of course, besides microstructure (atomic crystalline arrangement), the metal-removal mechanism is an interplay of work-hardening rates, temperature in the vicinity of the cut, friction on the tool-chip interface, orientation of slip planes, and other factors.

With the above in mind, the correlation between microstructure, as altered by heat treatments, and the chemistry of the material will be shown. One material (plain carbon steel), one machine (a lathe), and one tool (a carbide turning tool) will be discussed. The student can, through use of reference books, follow the same procedure for other materials.

Iron-Carbon Diagram

Before charting microstructure, it is necessary to know the terminology used. The metallurgist uses an *equilibrium phase diagram* to simplify the classification of potentially available microstructures. The binary (two-element) equilibrium diagram (or phase diagram or constitutional diagram) for the alloying possibilities of *iron* and *carbon* to form plain, unalloyed, carbon steel is shown in Fig. 1-19.

Steel is iron plus up to approximately 2.0 percent

Fig. 1-16   Example of the effect of strain hardening and annealing on the crystal structure of alpha brass. (*a*) Annealed brass; (*b*) during test; (*c*) after annealing. (*Microphotographs by Bridgeport Brass Co., a division of National Distillers and Chemical Corp.*)

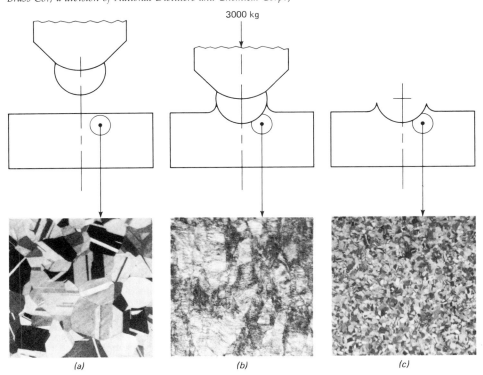

3000 kg

(a)   (b)   (c)

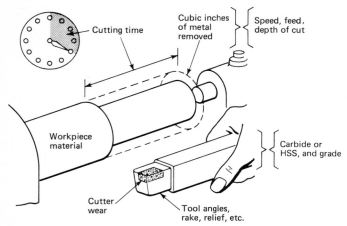

Fig. 1-17 Some of the factors involved in determining the machinability of metals and plastics.

carbon, though most common steels have up to 1.5 percent carbon plus other elements. Iron with more than 2 percent carbon becomes the material called *cast iron*, which uses up to about 3.5 percent carbon. There are many types of plain carbon steels, which range from 0.08 percent (weight) carbon (the balance is iron) to approximately 1.5 percent (weight) carbon and have various degrees of hardness, strength, and toughness due to combinations of heat treatments, cold working, and hot working. These are possible because iron is an allotropic element. *Allotrophy* simply means that iron (like 15 other elements) may exist in more than one crystal system as a solid. This behavior is shown to be a function of temperature as shown on the 0 percent carbon (or 100 percent iron) vertical lefthand axis of Fig. 1-19. From absolute zero to 1670°F (910°C), iron exhibits a BCC unit cell arrangement. From 1670 to 2550°F (1400°C), it is FCC (see Fig. 1-6), then BCC again up to its melting point at 2800°F (1538°C).

solid-state diffusion (iron-carbon)

Before moving to the right on the horizontal axis of Fig. 1-19, another property of metals must be explained. In an alloy system, it is possible for one element to be considered dissolved in the other in the *solid* state.

In the case of iron and carbon, the carbon atoms occupy the spaces between the iron atoms when these spaces are large enough. Since wrought (not cast) steel is being studied, only that part of the diagram below 2066°F (1130°C) will be used. Since this is well below the melting points of the alloys which will be discussed, all microstructural changes due to compositions and heat treatments will be the result of the carbon atoms moving about (or diffusing) through the iron atoms in the *solid* state, or, *solid-state diffusion*.

austenite

If an 0.8 percent (weight) steel sample were viewed through a microscope fitted with a hot stage (small oven with windows) at 1450°F (788°C), the microstructure shown in Fig. 1-20a would be visible and represents the condition at point 1 on the Fe-C diagram. This phase is called *austenite*. It is an FCC crystal, and *all* the carbon atoms are squeezed into the spaces between the iron atoms and thus cannot be seen with the microscope. The view through the microscope will remain the same while the temperature is dropped to 1333°F (723°C). Below this temperature, with 0.8 percent carbon, austenite cannot exist.

pearlite

At 1333°F (723°C), the crystal system changes to BCC and there is no longer room in the spaces for the carbon to remain in solid solution. *Cooling slowly* (furnace cool or equilibrium cool) gives the carbon atoms time to move out of the spaces, and the microstruc-

Chip. Fine recrystallized grain size due to cold work

Crystal distortion below the final surface

Original 70-30 brass

Unaffected material

Fig. 1-18 An example of work hardening during cutting of brass. The piece was heat-treated to emphasize relative grain sizes. (*Sketched from S. Ramalingham and J.T. Black, Paper No. 71 - WA/Prod - 22, ASME, 1971.*)

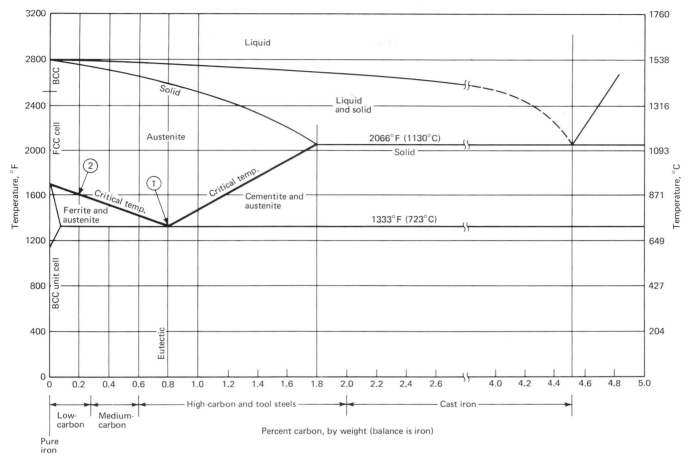

Fig. 1-19   Simplified iron-carbon equilibrium phase diagram.

ture shown in Fig. 1-20b results. This mixture of two phases is called *pearlite*. Notice the parallel-plate arrangement alternating white and black. The white strips are practically pure soft iron, or *ferrite*, which is BCC. The "squeezed out" carbon atoms have combined with iron to form the hardest *micro*constituent found in plain carbon steel, *cementite* or *iron carbide* (Fe_3C), in the form of the "black" plates. There is no further change in microstructure upon cooling to lower temperatures.

The amount of pearlite present will decrease as the percentage of carbon decreases as shown in Fig. 1-20c to e.

martensite

If, instead of cooling slowly from point 1 in Fig. 1-19, the steel specimen is quickly plunged into an iced, 10 percent salt brine solution, the carbon atoms will be trapped in the spaces between the iron atoms. When the FCC crystal tries to change, it forms a body-centered tetragonal crystal which is called *martensite*, the hardest *structure* in plain carbon steel (over 600 BHN). A martensitic steel is shown in Fig. 1-20f.

This hardness is due to the high internal stresses caused by the trapped carbon atoms; therefore, martensite is quite brittle, susceptible to cracking, and never used commercially in the "as-quenched" condition.

If the martensitic specimen is reheated to a temperature of 800 to 1000°F long enough, the internal stress level will be reduced because some of the trapped carbon atoms will be allowed to squeeze out and form Fe_2C and Fe_3C. Figure 1-20g shows the lower hardness readings and the microstructure resulting from the heat treatment called *tempering*.

Thus, tempering means to heat treat so as to lower the internal stresses and hardness to a level required by a part's final use. The longer the tempering time or the higher the tempering temperature, the softer the material will be.

spheroidizing

If the tempering treatment is continued long enough for all the carbon atoms to form Fe_3C, the microstructure shown in Fig. 1-20h would result, with the cementite in the form of spheroids. This is called

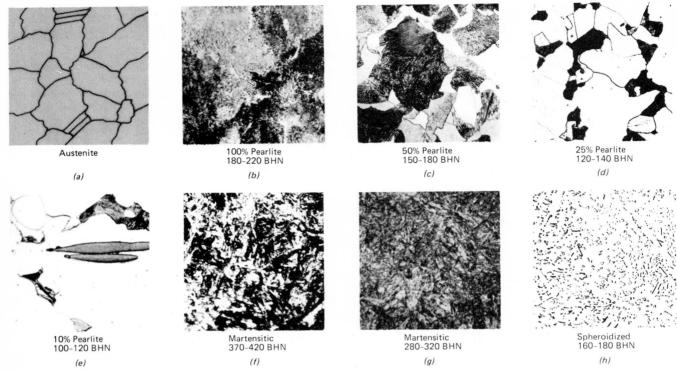

Fig. 1-20   Photomicrographs of steel in several metallurgical structures.

spheroidized steel, and it is the softest condition in which any carbon steel can exist. This microstructure may also be produced by taking pearlite (Fig. 1-20b) and holding it at a temperature just below 1333°F long enough for the cementite plates to form spheroids.

This heat treatment is time consuming, and therefore expensive, so it is used mostly on steels having 0.5 percent or more carbon (such as tool steels) which would otherwise be difficult to machine. These steels are often rehardened after machining.

partially pearlitic structures

The microstructures shown in Fig. 1-20d and e are produced by analyzing the changes occurring when a 0.20 percent carbon steel specimen (AISI-1020) is cooled slowly from point 2 on Fig. 1-19. Initially it is held at a high enough temperature (1650°F, 900°C) so that it is FCC. As it cools from 1580°F (860°C) to 1333°F (723°C), it descends through a two-phase region of the equilibrium diagram (ferrite and austenite). During this time, as part of the austenite is changing to ferrite, the carbon atoms are being absorbed by the remaining austenite. Finally, at 1333°F (723°C), the *remaining* austenite now analyzes to 0.8 percent carbon and, on cooling, produces small colonies of pearlite, the dark areas of Fig. 1-20d and e.

Microstructure Determines Tool Life

Tool life was measured in this series of tests by running the test until the edge of the carbide tool was worn 0.015 in. (0.375 mm). At least twice this amount of wear is often allowable in production. The 0.015-in. limit was used merely to save time and material.

Tool wear is the removal of tool surface material by abrasion from the material being cut. As will be shown, hardness tests do not always predict the abrasiveness of the material. Sometimes the limit of tool wear is dictated by the surface finish required since worn cutting tools produce rough finishes.

The graph in Fig. 1-21 correlates two of the factors determining machinability with the microstructures just discussed. All other machining variables are held constant. The vertical axis labeled "tool life, in.³" refers to the amount of material which can be removed in a lathe-turning operation before 0.015 in. (0.375 mm) wear is found on the carbide tool bit.

Notice that as the carbon content goes up, the percent of pearlite increases. Since the tool bit must break more of the hard cementite plates, the tool life decreases. If the shape of the cementite is changed from brittle plates into balls (spheroids), which roll or flow out of the way of the advancing cutting edge, the tool life increases. Notice the same hardness reading in Fig. 1-21c and d even though the machinability is different.

The *martensitic* material in Figs. 1-21f and g can be so hard that carbide tools or grinding are needed for removing material.

free-machining steels

Notice that Fig. 1-21a is not a plain carbon steel. Extra sulfur has been deliberately added to form the brittle manganese sulfide stringers shown. This improves the machinability as compared with a plain steel, which has 0.05 percent sulfur maximum, because the hard MnS particles act as "chipbreakers" so that the steel will shear ahead of the cutter at a much lower stress.

Sometimes a small amount of *lead* is added to the steel. This lead does not alloy with the steel but is found as lead globules at the grain boundaries. The lead's low tensile and shear strengths "lubricate"

Fig. 1-21   The approximate relationship between the metallurgical structure of steel and the tool life of a carbide tool bit used in turning it on a lathe. (*Photomicrographs from Buehler, Ltd.*)

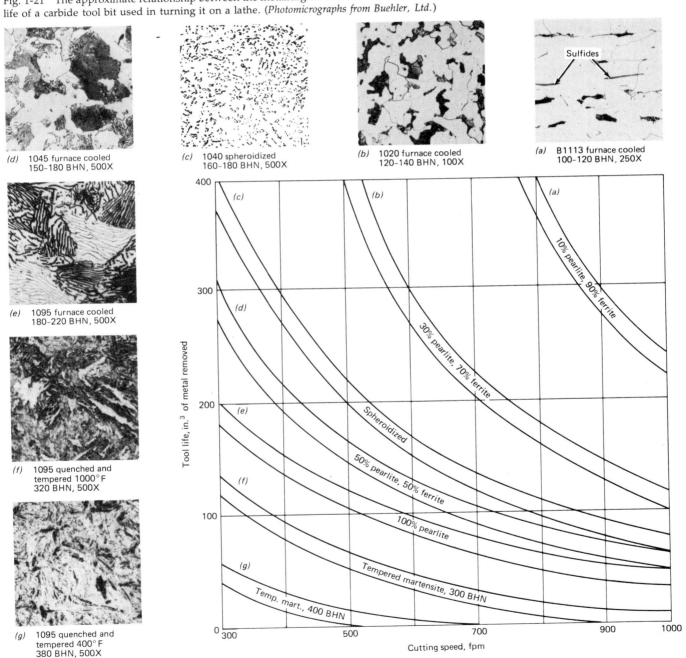

(d) 1045 furnace cooled
150-180 BHN, 500X

(c) 1040 spheroidized
160-180 BHN, 500X

(b) 1020 furnace cooled
120-140 BHN, 100X

(a) B1113 furnace cooled
100-120 BHN, 250X

(e) 1095 furnace cooled
180-220 BHN, 500X

(f) 1095 quenched and
tempered 1000°F
320 BHN, 500X

(g) 1095 quenched and
tempered 400°F
380 BHN, 500X

grain boundary movement. This greatly improves the machinability of the steel.

Other Terms Defined

There are several terms used in describing the heat treatment of metals which describe factors that affect machinability but which were not easily included in the previous discussion.

Normalizing means to heat steel for long periods of time at temperatures about 100°F (38°C) above the critical temperature followed by cooling in air. This is done so that the elements in the microstructure can "migrate" until the entire workpiece is more nearly uniform or homogeneous in its grain structure.

This will improve machinability somewhat but, more importantly, it ensures that any further heat treatments will give uniform properties throughout the entire workpiece.

Annealing can be applied to most metals. It consists of heating the metal above its critical or recrystallization temperature and cooling it very slowly (often in the furnace). This removes the stresses which were introduced by previous operations so that the part is less likely to warp during machining. These stresses are created by welding, casting, or any form of cold working.

Note: Recrystallization-anneal, previously mentioned, is a process used only for nonferrous metals.

PRECIPITATION (AGE) HARDENING

Not all alloys harden as dramatically or in the same manner as steel when they are quenched from an elevated temperature. For example, a 95.5 percent aluminum–4.5 percent copper alloy, when annealed [held at a temperature of 130°C (266°F) long enough to produce its softest condition], has a BHN of 45 and a tensile strength of 25,000 psi. If this same alloy is held at 550°C (1022°F) for 12 h (hours) and quenched suddenly in water, its hardness reading may go up to BHN 60, with a tensile strength of 35,000 psi.

This hardness may subsequently be raised to BHN 135 and a tensile strength of 60,000 psi by heating it again to 130°C (266°F) for 3 or 4 h. However, holding it too long at this temperature will cause it to soften again to its original BHN 45.

Those alloys which respond to this sequence of thermal treatments are said to be *age-hardening* or *precipitation-hardening alloys*. These property changes are *not* due to a crystal change as in steel.

Let us now describe approximately what happens. The annealed (softest) microstructure of this Al-Cu alloy consists of two visibly different phases as shown in Fig. 1-22a. One phase, the white area, is practically pure Al. The dark areas contain nearly all the Cu and are almost 50-50 Al and Cu, or $CuAl_2$. Holding this composition at 550°C allows the copper atoms to diffuse *uniformly* throughout the aluminum, displacing aluminum atoms in random sites in the FCC aluminum unit cells. It is now a single-phase metal (analyzing to 4.5 percent copper) and remains so on quenching as indicated in Fig. 1-22b. If this material is now heated at 130°C for a few hours, the copper atoms *start* to move toward each other in an *attempt* to form the $CuAl_2$. If the heating is stopped just before the $CuAl_2$ forms, the material is said to be age-hardened.

The material is much harder and stronger because this *attempt* to form $CuAl_2$ causes a stressed condition at the atomic level, since the copper atoms have distorted the uniform crystalline structure of the aluminum. This condition is visible only under an electron microscope. This same process will take place if the metal is left at room temperature for a long time, maybe 3 to 6 months, that is, it is "aged" naturally. However, this is neither as quick nor as reliable as the artificial aging just described.

The age-hardening process takes place on other alloys of aluminum, some stainless steels, and some other metals. For example, an annealed 90 percent copper–2 percent beryllium alloy with 35,000 psi tensile strength can be age-hardened to 195,000 psi—over a 500 percent increase.

To change the properties of a material, three

Fig. 1-22 Representation of precipitation (age) hardening of 95.5 percent aluminum, 4.5 percent copper alloy. (*a*) Annealed structure (two-phase); (*b*) age-hardened (single-phase).

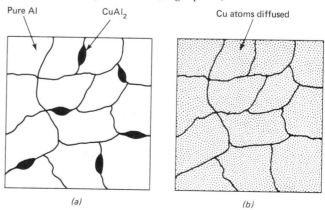

Pure Al $CuAl_2$ Cu atoms diffused

(a)   (b)

methods may be used separately or in combination. They are:

1. Alloying the material, such as increasing the carbon in iron to make steels

2. Heat treating, to soften or harden or strengthen as described in this chapter

3. Work hardening, as in cold-rolling steel or rolling sheets of aluminum or copper

This chapter has indicated a few of the ways these processes are used. It is a big, complex, interesting subject, as browsing through the references at the end of this chapter will quickly indicate.

Review Questions and Problems

1-1. Which method and testing machine would you use to check the hardness of:
   a. A large, machined iron casting.
   b. A large, machined aluminum casting.
   c. A 0.020-in.-thick (0.50-mm) piece of cold-rolled copper strip.
   d. A HSS tool bit $\frac{1}{2} \times \frac{1}{2} \times 3$ in. (13× 13 × 75 mm).

1-2. Which would bend the least under a 5000-lb (22.2-kN) load, a material with a Young's modulus of 10×10^6 psi or one with a modulus of 20×10^6 psi?

1-3. In the graphs of Fig. 1-15,
   a. Which is the strongest metal shown?
   b. What is the percent elongation of the 0.50 C steel?
   c. Which is the most brittle material shown, and how can you tell?

1-4. According to Fig. 1-15, for 0.20 C steel, the yield point is about what percent of the maximum (engineering) tensile stress?

1-5. Will the crystal structure of a metal give any indication of machinability? Explain your answer.

1-6. Is it possible to classify a material as brittle or ductile by the type of break in a tensile test? Explain your answer.

1-7. A shaft in a machine bends too much.
   a. Can the problem be solved by heat treating to a harder condition? Why or why not?
   b. What properties of the shaft will be changed by the hardening?

1-8. Explain why repeated bending of a piece of soft copper wire (at one spot) will eventually cause it to break.

1-9. Explain why it would be unwise to weld a part which has been
   a. Strain hardened.
   b. Heat treated to HRC 45.

1-10. What advantage is there is specifying cold-rolled steel instead of hot-rolled steel when designing a machine?

1-11. Differentiate between steel and cast iron based on their carbon content.

1-12. Why does the machinability of steel tend to decrease as its carbon content increases?

1-13. Which would be the least expensive steel to machine, a 100 percent pearlitic microstructure or a 100 percent spherodized microstructure? Explain your answer.

1-14. AISI 1112 steel is resulfurized (extra sulfur added), and 12L14 steel has lead added to it. Is it worthwhile paying 10 percent extra for these steels?

1-15. When a large casting is made, the outside hardens first, causing internal stresses. What heat treatment would you recommend to remove these stresses?

References

1. Avner, Sidney H.: *Introduction to Physical Metallurgy*, McGraw-Hill, New York, 1964.
2. Chisholm, A. W. J., and Redford: *The Assessment of Machinability*, Society of Manufacturing Engineers, Detroit, Mich., MR72-163, 1972.
3. Enberg, E. H.: *A Guide to Selecting the Best Indentation Hardness Test*, Testing Equipment Systems Div., Ametek, Inc., Lansdale, Pa.
4. Guy, Albert G.: *Physical Metallurgy for Engineers*, Addison-Wesley, Reading, Mass., 1962.
5. Kehl, George L.: *The Principles of Metallographic Laboratory Practice*, McGraw-Hill, New York, 1949.
6. U.S. Industrial Planning Div., Air Material Command, *Machinability Research Program*, vol. 2, Curtis-Wright Corp., 1951.
7. Van Vlack, Lawrence H.: *Elements of Materials Science*, 2d ed., Addison-Wesley, Reading, Mass., 1964.

measuring equipment

Much of our present way of life depends on our ability to measure accurately the products produced by our sophisticated manufacturing equipment. Parts made at plants hundreds of miles apart must fit together properly. Replacement parts for equipment several years old must be made so that they fit accurately. Pieces made in one plant, but on many different machines, must all fit together smoothly at assembly. The demand for more accurate machines can only be met as we develop more repeatably accurate measuring instruments.

In this chapter several of the basic measuring instruments are described. There are hundreds of variations and combinations of these basic tools. However, an engineer who has good background knowledge of the fundamental instruments should easily be able to understand and use these various "specials."

THE METRIC SYSTEM

The international system of measurements which is being developed by the International Standards Organization (ISO) is already used in many companies in the United States. NASA started doing some metric design work in 1971. Thus, a knowledge of this system is becoming more important.

The units and decisions of the ISO are officially referred to as SI, from the French *Système International d'Unités*. So, when you see information labeled ISO or SI, it refers to the decisions or recommendations of this international group. These decisions are not binding on any country, but many of them are being used. In the United States, the American National Standards Institute (ANSI) is the final source of our standards. ANSI works with the ISO and with industry when writing these standards.

In this book, both U.S. and metric figures are

given. Many of the metric equivalents have been rounded off to even millimetres, or one- or two-place decimals, because this is quite likely the way we will use them as we "Go Metric." The ISO in many cases has "preferred" series of sizes (such as in taps and drills) which they hope will become world standards. The actual details are worked out with the national groups involved in manufacturing the products being considered.

In converting inches into millimetres, closely comparable accuracy is possible if one less decimal place is used with millimetres than with inches. There are more detailed suggestions in the metric guides which are available.

Appendixes J and K show the conversion tables, and Table 2-1 gives some "baseline" conversions which can be helpful.

DEFINITIONS

Precision is the closeness with which a measurement can be read directly from a measuring instrument. It is the smallest marked increment on the instrument, and it is usually marked on the scale or dial as $\frac{1}{64}$, 0.001, 0.02 mm, etc. In practice one can often "guesstimate" a number between these divisions though not accurately.

Accuracy is a measure of how close the reading is to

TABLE 2-1 METRIC TO ENGLISH, QUICK REFERENCE CONVERSION

1 inch (in.) = 25.4 millimetre (mm) exactly	
1 micrometre (μm) = 1 micron = 0.001 mm = 0.000 040 in. approx.	
1 metre = 39.370 008 in.	0.5 in. = 12.7 mm exactly
10 mm = 0.393 70 in.	0.250 in. = 6.35 mm exactly
1 mm = 0.040 in. approx.	0.005 in. = 0.127 mm exactly
0.5 mm = 0.020 in. approx.	0.001 in. = 0.025 mm approx.
0.02 mm = 0.0008 in. approx.	$\frac{1}{32}$ in. = 0.79 mm approx.

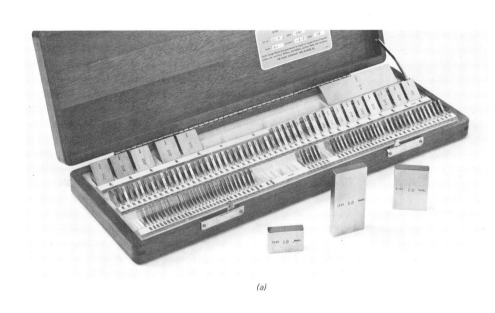

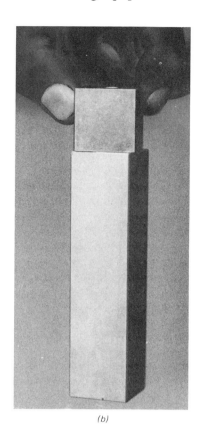

Fig. 2-1   (a) A 121-piece set of gage blocks. Six fraction blocks in the center of the front row. (b) Two gage blocks wrung together. (*DoAll Company.*)

the true size of the part being measured. This involves temperature, the mechanics of the measuring instrument, the pressure exerted by the instrument, and, sometimes, the skill of the inspector. Thus an instrument or machine may have a precision of 0.001 in. with a repeatable accuracy of 0.0005 in.

The Basis of Length Measurements

The international standards of length are now defined as the *metre*, which is 1,650,763.73 wavelengths of a rare gas, krypton 86, under specific conditions, and the *inch*, which is 25.4 mm exactly or 41,929.399 wavelengths of krypton 86.

GAGE BLOCKS

To provide a practical standard for use in industry, gage blocks (Fig. 2-1), sometimes called *Jo blocks* or *Hoke blocks*, are produced by a number of companies. These are square or rectangular blocks of special tool steel, sometimes chrome plated, or solid carbide which have been lapped to a finish of 1 μin. (microinch) (0.000 001 in., 0.025 μm) or less.

TABLE 2-2   GAGE BLOCK TOLERANCES

	Tolerances in Inches (1-in. Blocks)		
Class	Size, in.	Flatness, in.	Parallel, in.
AA	±0.000 002	0.000 003	0.000 003
A	+0.000 006	0.000 004	0.000 004
	−0.000 002		
B	+0.000 010	0.000 006	0.000 005
	−0.000 006		

	Tolerances in Millimetres (25-mm Blocks)		
Class	Size, mm	Flatness, mm	Parallel, mm
AA	±0.000 05	0.000 08	0.000 08
A	+0.000 15	0.000 10	0.000 10
	−0.000 05		
B	+0.000 20	0.000 10	0.000 10
	−0.000 10		

NOTE: 0.000 05 mm is very close to 0.000 002 in.; thus the two standards are very similar.

The accuracy of these blocks varies according to certain classifications. Different makers use slight variations from the values shown in Table 2-2, and "Master" and A+ classes are used by some.

Gage blocks are guaranteed at 68°F or 20°C. Any variation in temperature will change the length. For example, at 78°F a 1.000 000-in. block will be approximately 1.000 065 in. long.

All workpieces are, of course, subject to the same expansion. Thus, parts made in the factory at 65 to 75°F (18 to 24°C) will not measure the same in a temperature-controlled inspection room. However, it is not often that accuracy to 0.000 050 in. (0.001 27 mm) is required.

The gage blocks come in sets of selected sizes, often from 0.100 025 to 4.000 000 in. One manufacturer claims that its 84-piece set will make half a million different lengths.

Gage blocks can be "wrung" together so that it takes an appreciable effort to pull them apart. To do this, two blocks are carefully cleaned, and then pressed firmly together as one is slid over the other. The finish is so nearly perfect that all air is forced out from between them. They are held together somewhat by atmospheric pressure and also by intermolecular forces.

CAUTION: Keep your fingers off the finished surfaces, and clean the blocks before putting them away so that they will not rust.

In building up to a dimension, always satisfy the righthand digit first. For example, if a height of 1.5724 is needed,

First use 0.1004, which leaves 1.4720

Second use 0.1220, which leaves 1.3500

Third use 0.3500, which leaves 1.0000

Finally use 1.0000

This is necessary because, for example, only the 0.101 through 0.149 blocks are in sizes varying by 0.001. The larger blocks are in steps of 0.050 in.

Gage blocks can be purchased up to 24 in. long and can be "built up" to 48 in. Metric gage blocks come in comparable sets and lengths.

These blocks can be used to check all other measuring instruments. The class AA set is used only by the inspection department, in a temperature-controlled room.

HAND-HELD MEASURING INSTRUMENTS

Steel Rules

This very accurately divided relative of the "ruler" everyone used in school is still widely used for quick approximate checking of dimensions, for layout work when $\frac{1}{64}$-in. accuracy is sufficient, and for quick checking of hole sizes, stock sizes, and cutter diameters. Almost every machinist (and most engineers) keeps a 6-in. rule (15-cm) handy all the time (Fig. 2-2).

Lengths of steel rules range from 1 to 48 in. and up to 144 in. in some cases. Metric rules are made up to 1 m (39.37 in.) long, with divisions of 0.5 or 1.0 mm.

Scales may be on two edges or four edges, using both sides of the rule. They may be in fractions or decimal inches or in combinations of these and millimetres.

Using the steel rule is a simple matter of accurately reading the divisions. However, remember that the ends do get worn, so that accurate measurements are often made "from inch to inch."

It is preferable to use the largest gradations you can, as it is much easier to make an error in reading a sixty-fourth scale than in reading the thirty-seconds scale. With a hand magnifying glass, readings can be made to $\frac{1}{128}$ or 0.005 in. (half a hundreth), though for this accuracy a vernier-type measuring instrument would be preferable. Metric scales can be read to 0.5 mm (0.020 in.)

Combination Squares

The steel rule, heavier and wider, fitted to a "head" with 90° and 45° edges becomes a very useful tool for checking or laying out work (Fig. 2-3). The addition of the "center head" and protractor creates a versatile, compact instrument for a variety of work when specified tolerances and quantity of work are suitable.

The "blades" are made in lengths from 4 to 24 in. long, with English divisions to $\frac{1}{64}$ or $\frac{1}{100}$ in. and metric divisions of 0.5 and 1 mm. A protractor head for setting angles is also shown.

Calipers—Spring Type

In the early days of this century a skilled machinist would machine an excellent running fit between the shaft and its bearing in, for example, a steamboat engine by skillful use of simple spring-type calipers (Fig. 2-4).

Actual measurements made with these are only as accurate as one's ability to read the steel rule since calipers merely serve to *transfer* a part dimension to a measuring instrument, or the reverse.

Outside calipers are used for measuring outside diameters, lengths, thickness, etc., and will reach into places where a steel rule cannot be used, such as the diameter of a shaft mounted between centers on a lathe or machined grooves (Fig. 2-5).

Inside calipers are used for measuring inside diameters, widths or lengths of keyways and slots, lengths and widths of milled pockets, and similar dimensions (Fig. 2-5). Both calipers are made in lengths from 2 to

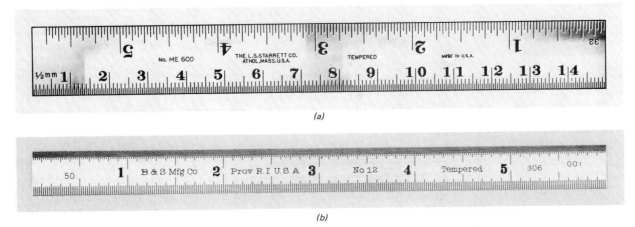

Fig. 2-2   6-in. (15-cm) steel scales. (*a*) Scales $\frac{1}{32}$ in. and 0.5 mm. (*The L. S. Starrett Company.*) (*b*) Scales $\frac{1}{100}$ and $\frac{1}{50}$ in. (*Brown & Sharpe Mfg. Co.*)

12 in. A slightly different type, the "firm joint" caliper, is made in lengths up to 36 in.

Using the outside or inside spring caliper requires that one get the "feel" of the proper tightness of adjustment. If the instrument just very lightly drags along the surface being measured, it is properly set. If tightened too much on the work, these instruments will spring quite easily and give a very inaccurate measurement. The calipers must also be aligned on the work, or they will be measuring a diagonal (and, therefore, larger) distance.

Notice that the scale reading must be at a *tangent* to the ends of the calipers. These ends are not in flat contact with the work but are shaped to give point contact only (Fig. 2-6).

Dividers (Fig. 2-4*c*) are used mostly for transferring a measurment to a workpiece. The dividers are set to the desired dimension by using a steel rule. Their sharp points are then used to scratch into a "blued" surface, a circle, or an arc, or to lay out a distance along a line. Dividers are made in lengths up to 12 in. in both spring and firm joint styles.

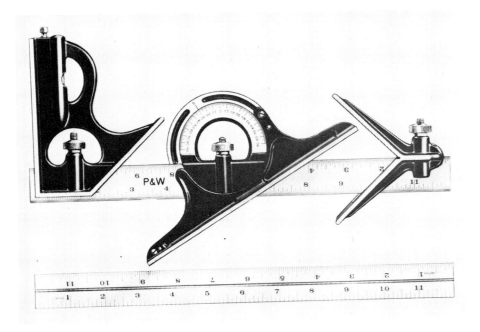

Fig. 2-3   A combination square with center head and protractor attachments. (*Colt Industries, Inc., Pratt & Whitney Machine Tool Division.*)

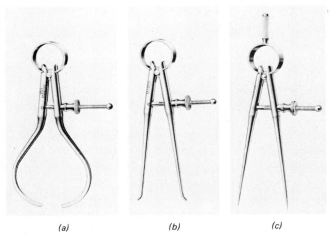

(a)   (b)   (c)

Fig. 2-4   Spring-type hand-measuring instruments. (a) Outside calipers. (b) Inside calipers. (c) Dividers.

PRECISION HAND-HELD MEASURING INSTRUMENTS

The Micrometer

This instrument (Fig. 2-7) is sometimes called a *micrometer caliper*, and in its various sizes and styles is one of the most used measuring instruments. The measuring ability of a micrometer depends on a very accurately ground nut and thread of 40 threads per inch. One-fortieth of an inch is 0.025 in. Thus, each complete turn of the thimble moves the spindle 0.025 in. Four times 0.025 is 0.100 in., and this is the major division marked off on the barrel or stationary part of the micrometer.

Referring to Fig. 2-7, note that the barrel is marked off in tenths of an inch from 0 to 9, with divisions of 0.025 each in between. The final "thousandths" readings are marked from 0 to 24 around the circumference of the thimble. Thus, to read a micrometer one mentally (or at first on paper) adds up three numbers as shown in Fig. 2-8a.

metric micrometers

When the metric system is used, all mechanical engineering dimensions are given in millimetres, even on long dimensions when centimetres could be used. This is to avoid confusion, just as in the English system a lathe is specified as 100-in. capacity, not 8 ft 4 in.

In the English system, we think in terms of 0.001 and 0.0001 in. The metric system uses 0.01 mm (about 0.0004 in.) or $\frac{1}{50}$ mm (0.02 mm = about 0.0008 in.) as common units. Thus, the 0.02 mm is quite close to the familiar 0.001 in. Our "tenths" (0.0001 in.) is close to 0.0025 mm.

The metric micrometer (Fig. 2-8b) has a spindle screw thread with a pitch of 0.5 mm. The sleeve is divided into 0.5-mm markings, but arranged so that the 1-mm markings are on top, or longer, so that they can be read easily. Each 5 mm, up to 25 mm, is numbered on the sleeve. The readings are made as shown in Fig. 2-8b.

how to handle a micrometer

When measuring small parts, the part is (for a right-handed person) held in the left hand, and the micrometer in the right hand (Fig. 2-9a). Notice that the "mic" is held so that the thumb and index fingers are free to rotate the thimble.

If the work is larger, or being held in a machine, or if one is using large micrometers, it is easier to hold the mic with both hands. The right hand does the adjusting, and the left steadies the instrument as shown in Fig. 2-9b.

Notice that even 24- and 30-in. micrometers have the same 1-in., or 25-mm, measuring capacity. Thus, larger mics are listed as, for example, 5 to 6 in. or 29 to 30 in. (125 to 150 mm or 725 to 750 mm). To enable one to set these larger mics to zero, the manufacturer often furnishes a very accurately ground bar the length of the smaller distance between spindle and anvil, or gage blocks are used.

Fig. 2-5   Typical uses of outside and inside calipers.

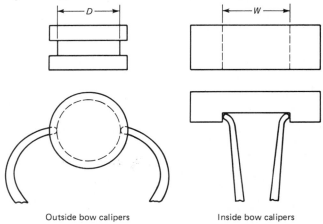

Outside bow calipers   Inside bow calipers

Fig. 2-6   How outside and inside calipers are read on a steel scale.

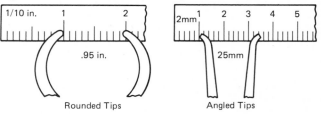

Rounded Tips   Angled Tips

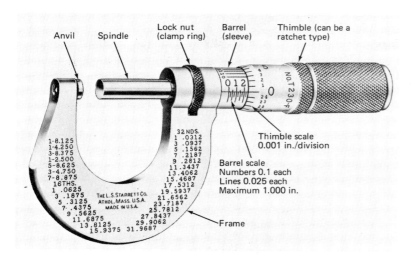

Anvil   Spindle   Lock nut (clamp ring)   Barrel (sleeve)   Thimble (can be a ratchet type)

32 NDS.

Thimble scale 0.001 in./division

Barrel scale
Numbers 0.1 each
Lines 0.025 each
Maximum 1.000 in.

Frame

Fig. 2-7   A 0 to 1-in. micrometer, showing a reading of 0.250 in. (*The L. S. Starrett Company.*)

accuracy of a micrometer

Reading a micrometer is relatively simple, but one soon finds that different people may get different readings with the same mic on the same piece. This is because of the differences in "feel" or tightness to which each one turns the thimble. You can "squeeze" a mic up to 0.001 in. (0.025 mm). To correct this, many micrometers have ratchet or friction thimbles which are preset to a light pressure.

The micrometer itself is manufactured to 0.0001 in. or better tolerance. However, due to errors in "feel," or to errors in aligning the mic squarely with the work, or to possible dirt, oil, or metal dust on the

work, it is usually agreed that the final reading is accurate to within 0.0002 to 0.0005 in. (0.002 to 0.005 mm) at best.

The 0.0001 (tenths) can be estimated between the 0.001 readings or, on some mics, can be read from a vernier scale on the barrel. In the same way, 0.001 mm may be estimated or read from a vernier.

There are many variations on the standard micrometer. Some have interchangeable anvils. There are screwthread mics, depth mics, special mics for measuring tubing walls and paper thickness. Others are thin blade mics (for measuring recesses and keyways), deep throat mics (for sheet metal), and many others. However, the measuring end of practically all of these is the same as that of the micrometer described here.

The Vernier Caliper

If you wanted to check a part which measured 0.950 × 2.355 × 6.718 in. using micrometers, you would need three mics: 0 to 1 in., 2 to 3 in., and 6 to 7 in. Or you could use one 12-in. vernier caliper to make all the measurements with a single instrument. However, the vernier caliper (Fig. 2-10) can be read only to the nearest 0.001 in. ±0.0005 in. or 0.02 ± 0.01 mm.

The vernier caliper is made up of a beam or bar marked in inches and hundredths (0.1), with small divisions of either 0.025 or 0.050 in. The beams are from 6 to 48 in. (150 to 600 mm) long. Metric vernier calipers are divided into centimetres (10 mm) and millimetres.

The vernier scale is part of the sliding jaw and is graduated in 0.001 in. or 0.02 mm (Fig. 2-11).

The vernier principle is used on a variety of instruments and is basically quite simple. Between the

Fig. 2-8   Example of readings on (*a*) inch and (*b*) metric micrometers.

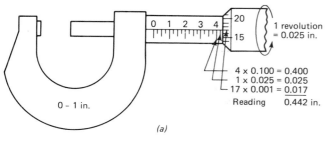

0 – 1 in.

1 revolution = 0.025 in.

4 × 0.100 = 0.400
1 × 0.025 = 0.025
17 × 0.001 = 0.017
Reading   0.442 in.

(a)

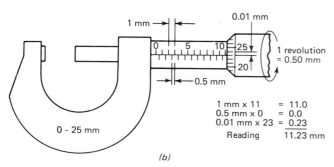

1 mm

0.01 mm

0.5 mm

1 revolution = 0.50 mm

0 – 25 mm

1 mm × 11   = 11.0
0.5 mm × 0   = 0.0
0.01 mm × 23 = 0.23
Reading   11.23 mm

(b)

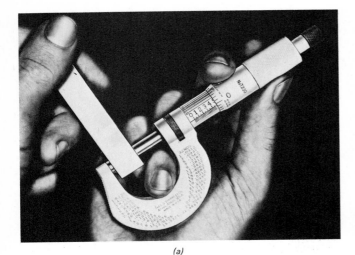

Fig. 2-9 How to hold a micrometer. (*a*) When measuring small work; notice that this micrometer has a "tenths" vernier. (*b*) When measuring large work. (*The L. S. Starrett Company.*)

inch marks on the beam, the graduations are the same as on the barrel of a micrometer; the smallest division is 0.025 in. If the *vernier* scale is graduated in 25 equal parts, the total of which is equal to only 24 of the 0.025-in. divisions, then the vernier scale is 0.025 in. short, or 0.001 in. short per division (Fig. 2-11). Thus, if the vernier is moved to the right so that its seventeenth division aligns with a marking on the beam, the vernier has moved 17 × 0.001 = 0.017 in.

To read an inch vernier requires adding *four* figures: the inches, tenths, and 0.025-in. (or 0.050-in.) divisions, and the 0.001 in. from the vernier. The last three are read the same as a micrometer.

metric vernier caliper

The metric vernier caliper is constructed in the same way as the English model. However, the main scale is marked off in centimetres (10 mm) divisions, with 1- and 0.5-mm divisions in between. The slider is graduated in $\frac{1}{50}$- (0.02) mm divisions (Fig. 2-11).

The readings shown in Fig. 2-12 are:

Fig. 2-12*a*		Fig. 2-12*b*	
Inches	= 1.000	4 cm × 10	= 40.00 mm
0.1 in. × 4	= 0.400	1 mm	= 1.00
0.25 in. × 1	= 0.025	0.5 mm	= 0.50
Vernier	= 0.011	Vernier 9 × 0.02	= 0.18
Reading	1.436 in.		41.68 mm

suggestions for reading a vernier caliper

1. Always read the vernier caliper at the *zero* mark on the vernier scale, *not* at the edge of the sliding jaw.

2. Look for a pattern of alignment like that shown in Fig. 2-11*a*. It is possible to find a pattern like that in Fig. 2-11*b* (which is actually a 0.0005-in. reading) and various stages in between. Thus, two people can read the vernier 0.001 in. differently. A magnifying glass is sometimes helpful.

3. Many vernier calipers have both *inside* and *outside* scales due to the width of the measuring points. The inside scale is usually on the top side of the beam.

4. Alignment in all directions is very critical.

These same vernier scales are used on depth gages, height gages, and a few other instruments. There are also vernier calipers with dial readings, and some which can be used as depth gages. Some use separate sets of jaws for inside and outside measurements, which eliminates the special "inside" scale.

Special-Purpose Hand-held Measuring Instruments

Besides the special styles of micrometers and vernier calipers, there are many special-purpose measuring instruments. Two of these are described here. For others, see the catalogs of the toolmakers.

small hole gages

Holes, slots, grooves, and recesses from 0.125 to 0.500 in. (3.175 to 12.70 mm) are frequently difficult to measure accurately.

The small hole gages (Fig. 2-13) can be placed in shallow or deep openings and expanded, by means of the knurled knob at the top, until they just drag slightly. The distance across the contacts can then be measured with a micrometer.

As these, like the bow calipers, are *transfer-*

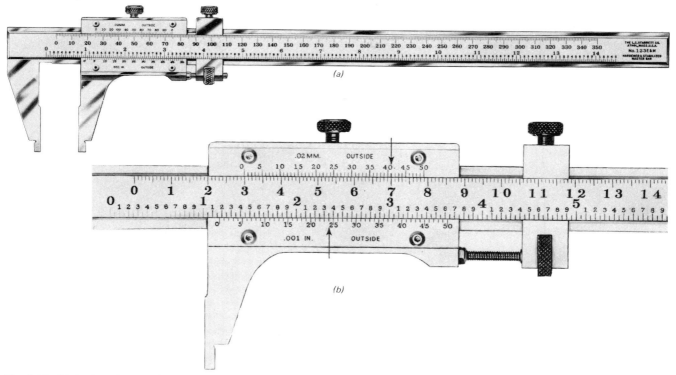

Fig. 2-10 (a) A 12-in. (350-mm)vernier caliper with both inch and metric scales. (b) Closeup of the scales. The reading is 1.174 in., 29.82 mm. This scale is in centimetres. Newer model at top is marked in millimetres. (*The L. S. Starrett Company.*)

Fig. 2-11 The vernier scale. (a) The basic principle. (b) How the actual readings will look. (c) An ambiguous reading.

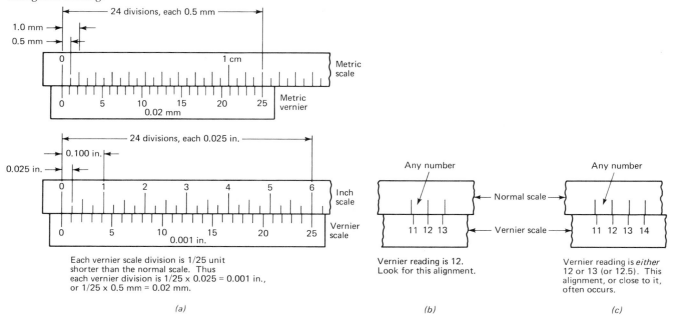

Each vernier scale division is 1/25 unit shorter than the normal scale. Thus each vernier division is 1/25 x 0.025 = 0.001 in., or 1/25 x 0.5 mm = 0.02 mm.

(a)

Any number

Normal scale

Vernier scale

11 12 13

Vernier reading is 12. Look for this alignment.

(b)

Any number

11 12 13 14

Vernier reading is *either* 12 or 13 (or 12.5). This alignment, or close to it, often occurs.

(c)

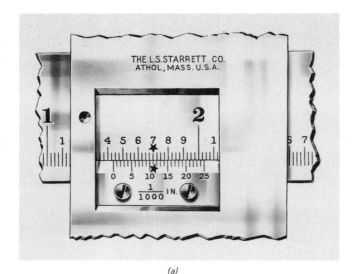

(a)

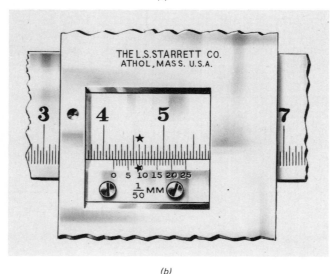

(b)

Fig. 2-12 Typical vernier caliper readings. (*a*) With 25 divisions, reading is 1.436. (*b*) Metric, with 25 divisions of 0.02 mm each, reading is 41.70 mm. (*The L. S. Starrett Company.*)

measuring instruments, they can be used for either the metric or the English system. The accuracy of the measurement will not be as close as if it was made by a direct-reading instrument.

telescoping gages

For larger holes, slots, keyways, etc., $\frac{1}{2}$- to 6-in. (12.7- to 150-mm) telescoping gages (Fig. 2-14) are quick and quite accurate to use. Either one (or both) of the contact plungers telescopes into the body and is spring-loaded. The gage is inserted in the opening and the plunger allowed to expand. After the proper "feel" is obtained, the contacts are locked in position by a turn of the knurled screw in the end of the handle. The gage is then removed from the opening.

The size of the opening is determined by measuring across the two contacts with a micrometer or vernier caliper. Use gentle pressure, as the plunger can be moved by too much force. Since this is a transfer-measuring instrument it can be used with the metric system.

CAUTION: It is wise to remove the gage, after locking it, and try it again in the opening. It is very easy to measure a chord instead of a diameter, or to tilt the gage slightly in one or more planes, thus giving a shorter or longer reading.

For *direct reading* of openings from 2 to 32 in. (50 to 800 mm), the *inside micrometer sets* (not shown) would be used or special gages would be made in the tool-room.

INSPECTION BENCH MEASURING INSTRUMENTS

The following group of instruments are used principally for inspection, layout, and setup in the inspection department. To use them, a very flat, stable surface is necessary as a reference plane.

Surface Plates

Surface plates (Fig. 2-15) are very accurately ground or scraped cast iron or granite slabs. They are made in sizes from about 9 × 12 in. up to 48 × 144 in. The smaller plates are made in accuracies (guaranteed flatness) of ±0.000 025 to ±0.0001 in. (0.000 64 to 0.0025 mm), and the largest are guaranteed accurate to ±0.0003 to 0.0012 in. (0.0076 to 0.030 mm), depending on the quality of the plate.

There is some preference today for the granite plates as they do not rust, they are very hard, and if something is dropped on granite, it will chip but will not raise a burr as steel plates will. However, the granite plates are very heavy and must be adequately supported on a stand. The smaller cast-iron plates are lighter and portable and, due to their ribbed construction, will maintain their accuracy without additional support.

Vernier Height Gage

If you were asked to check the center-to-center distances between the holes shown in Fig. 2-16 to ±0.001 in. (0.025 mm), you might be able to use a vernier caliper if the radius of the nibs was less than the radius of the holes. However, a much easier and more accurate method is to use the vernier height

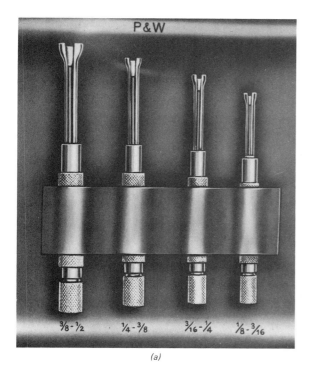

(a)

(b)

Fig. 2-13 A set of small hole gages and an example of one use of them. (*Cold Industries, Inc., Pratt & Whitney Machine Tool Division.*)

gage (Fig. 2-17) on a surface plate, with a *dial indicator* as shown in Fig. 2-18.

The vernier height gage is a vernier caliper which has been set vertically in a firm, flat baseplate. It also has, instead of the nibs, a projection to which various accessories may be clamped. For measuring, a small-diameter dial indicator is often used.

The vernier height gage is made with inch and/or millimetre scales and heights from 6 to 24 in. (150 to 600 mm).

Even though the vernier height gage is a direct-reading instrument, it does not usually directly give the desired dimension. This is because both the dial indicator and its contact point are adjustable. Thus, to

Fig. 2-14 A set of telescoping gages, $\frac{5}{16}$ to 6 in., and an example of their use. (*The L. S. Starrett Company.*)

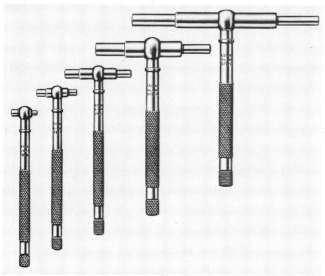

(a)

(b)

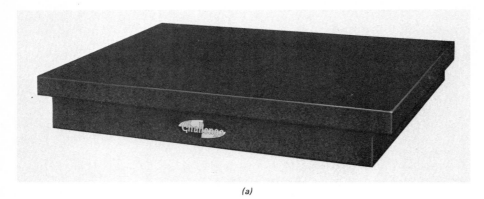

(a)

(b)

Fig. 2-15   Surface plates. (a) Black granite. (b) Steel. (*The Challenge Machinery Co.*)

measure actual height above the surface plate, it would have to be "zeroed in" with a gage block. This is seldom necessary.

The *use of the vernier height gage* for measuring is illustrated in Fig. 2-18. If the vernier height gage when set up without zeroing in reads 1.086 in. when set on hole A (Y dimension, Fig. 2-16) and 2.839 in. at hole B and both holes are the same diameter, then the distance A-B (vertically) is 2.839 − 1.086 = 1.753 in., which is within the tolerance. However, remember that the vernier scale measurement is accurate to ±0.0005 and the hole diameters may vary by 0.001, so the 1.753 measurement itself is probably ±0.001 in accuracy.

Height Micrometers

Several Companies manufacture height micrometers (Fig. 2-19) under various names. They are made in heights from 12 to 48 in. and are often accurate to 0.000 050 or 0.000 010 in. (0.001 27 to 0.000 25 mm).

These height micrometers are quick and easy to read because the inches and 0.100 increments are read from the column and the scale at the top of the column is in 0.025-in. divisions, the same as a micrometer. The extra large thimble is divided into widely spaced 0.001-in. divisions and 0.0001-in. divisions are read on the vernier.

As shown in Fig. 2-19, the vernier height gage can now be used as strictly a transfer instrument, and the readings can be made on the height micrometer faster and more accurately. In fact the "vernier height gage" can now be purchased as a simple column, without any divisions marked on it as shown Fig. 2-19.

Electronic Comparators

At the right of Fig. 2-19 is shown an electronic comparator which can be used to eliminate most errors caused by a dial indicator. These instruments, equipped with very free-acting, sensitive probes, can measure deviations as close as 0.000 010 in., though the range at this magnification is only 0.0003 in. The most useful scale for checking dimensions is probably the one with a range of 0.030 in. and scale divisions of 0.001 in.

The electronic comparator is used almost entirely as a transfer instrument or, with somewhat different gaging heads, as a comparator. The setup shown in Fig. 2-19 will cost close to $1500.

Coordinate Measuring Machines

To carefully check the "first piece" from a $150,000 numerical control machine can, if it is at all complex,

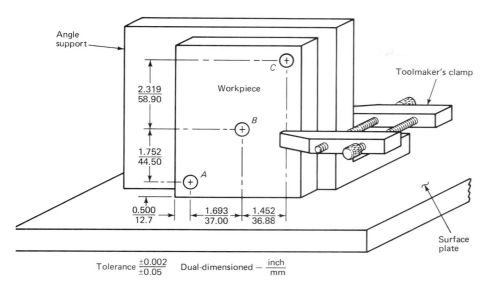

Angle support

Workpiece

Toolmaker's clamp

C

$\frac{2.319}{58.90}$

B

$\frac{1.752}{44.50}$

A

$\frac{0.500}{12.7}$ $\frac{1.693}{37.00}$ $\frac{1.452}{36.88}$

Surface plate

Tolerance $\frac{\pm 0.002}{\pm 0.05}$ Dual-dimensioned — $\frac{inch}{mm}$

Fig. 2-16   A simple example of a hole pattern which could be inspected with a vernier height gage equipped with a small dial indicator.

take 60 to 180 min, using any of the previously described methods. This is too long a time to have an idle numerical control (N/C) machine. Thus, coordinate measuring machines (Fig. 2-20) have been developed by several companies.

These machines consist of a very accurate surface on which to rest the work, and a stylus which can be equipped with different shape and diameter tips. The stylus runs on X- and Y-axis supports which are equipped to activate, in lighted numbers on the cabinet, the amount of movement in each axis to

±0.0005 or 0.0001 in. (0.0125 or 0.0025 mm). The read-out can be reset to zero at any time; thus, either coordinate dimensions (from a common zero) or incremental dimensions can be read quickly and checked against the drawing.

A printout of all dimensions is available (at an extra cost) on most machines. The Z (vertical) axis can also be checked, by means of attachments, and a Z-axis read-out and printout is often available. These machines can inspect a workpiece in one-tenth the time it previously took, and more accurately.

Fig. 2-17   A Vernier height gage being used to scribe a locating line. Note the granite surface plate used to insure accuracy. (*The L.S. Starrett Company*)

Fig. 2-18   Use of a vernier height gage to check hole locations on a workpiece.

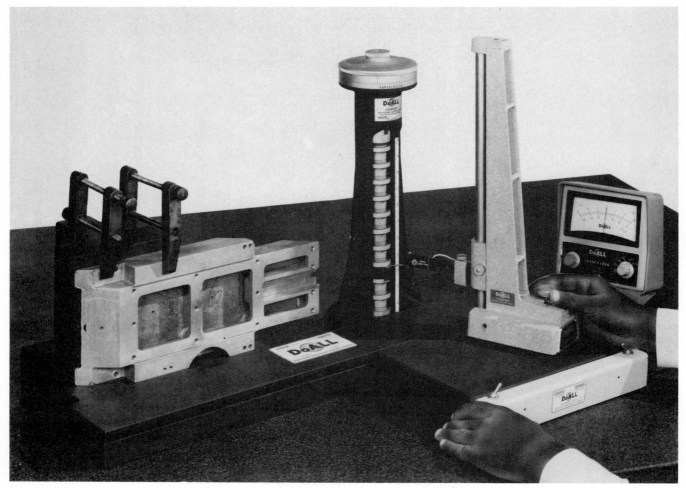

Fig. 2-19   Examples of the use of a modern height micrometer and electronic indicator. (*DoAll Company.*)

Sine Bars and Sine Plates

If it is necessary to measure an angle closer than 5 min (which is possible with the vernier protractor), the sine plate or sine bar (Fig. 2-21) is frequently used. The working surfaces of the sine plate are from about 3½ × 6 to 6 × 12 in., so that a part may be mounted on them. The sine bar is used for laying out angles, especially in toolmaking. It is about 1 × 6 or 1 × 12 in. in size (25 × 150 or 25 × 300 mm).

The center-to-center distance of the two round bars under the plate is 5 or 10 in. ±0.000 05 to ±0.0003 in. (0.001 25 to 0.0075 mm), depending on the size and quality. The top surface is parallel with the bottom of the two cylinders within about 0.0001 in. (0.0025 mm). To be accurate, they must be used on a surface plate. Metric sine plates and bars are made in 100- and 200-mm sizes.

To use a sine plate (Fig. 2-22), the plate is raised at one end by means of gage blocks until the dial indicator shows that the top surface of the part is parallel

with the surface plate. This is a trial and error procedure and somewhat slow.

Once this vertical distance is known, the sine of the angle can be computed:

$$\text{Sine } A = \frac{\text{gage block height}}{\text{sine bar length}} = \frac{1.4723}{5.000} = 0.294\,46$$

Looking this up in a five-place sine table and interpolating,

Angle A = 17°7′30″

If the measurements are carefully made, this could be accurate to within ±6 seconds.

Dial Indicators

A dial indicator (Fig 2-23) was used with the vernier height gage. This is a small 1- to 1½-in. indicator in order to keep within space and weight requirements. Larger indicators are made in several sizes from 1¹¹⁄₁₆- to 4-in. diameter to assist when using the sine bar and in many other shop measurements. The most used sizes are from 2- to 2¾-in. diameter. Dial indicators are used to align work on lathes and other machines, to check concentricity of shafts, to measure deflection, and for dozens of other jobs in manufacturing. Dial indicators are made to be attached by several types of mounting brackets and in a variety of positions.

The finest graduations on the dial are usually 0.001, 0.0005, or 0.0001 in. or 0.01 mm (0.0004 in.), 0.005 mm (0.0002 in.), and 0.002 mm (0.000 08 in.). The value of the distance between two adjacent lines on the dial is always printed on the face of the dial.

Only about 100 graduations can be clearly marked around a 2¼-in. dial, and most indicators can make only 2½ revolutions of the hand. Thus, if 0.0001-in. graduations are needed, the total *range* of measurement will be only 0.025 in. If 0.001 in. graduations are close enough for a particular job, the range will be 0.250 in. Of course, you can estimate partial spaces and read closer than the markings shown, but not very accurately.

The entire dial can usually be rotated, so that the zero may be located anywhere around the dial, and locked in place by the thumb screw shown in Fig. 2-23.

There are *long-range dial indicators* which, with the aid of a counting dial inside the larger one, will measure from 1 in. (shown) up to 12 in.

Accuracy of dial indicators is plus or minus one division or better. Many of these are made to American Gage Design (AGD) standards. The cost for AGD dial indicators is about $25 to $50 each.

PRODUCTION GAGES

Even though the inspection department has the final responsibility for checking work, it is wise, wherever possible, to have the person at the machine check his or her own work. The machinist will not necessarily inspect every piece, but often has time to inspect every second or fifth part or even to stop the machine to check critical dimensions such as bored holes or critical shaft diameters. To do this, micrometers or

Fig. 2-20 A coordinate measuring machine in use on a large sheet metal part. *X* and *Y* axes shown. *Z* axis can be added. (*Bendix Automation and Measurement Division.*)

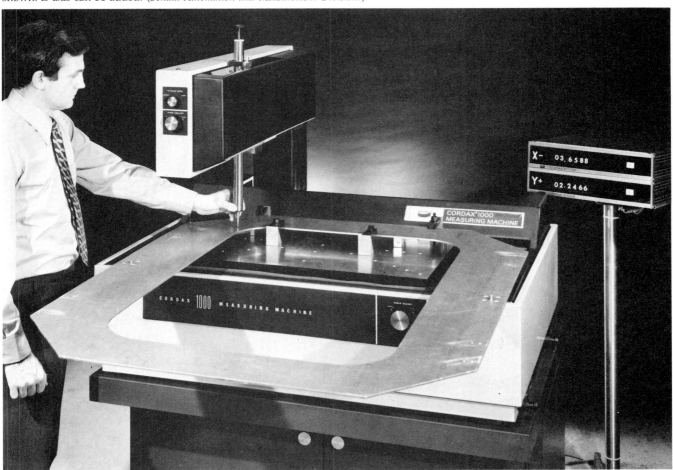

Fig. 2-21 Two sine plates (5 and 10 in.) and a 5-in. sine bar shown on a black granite surface plate. Tapped holes on working surface and edges are for clamps. (*DoAll Company.*)

Fig. 2-22 An example of the use of a sine plate.

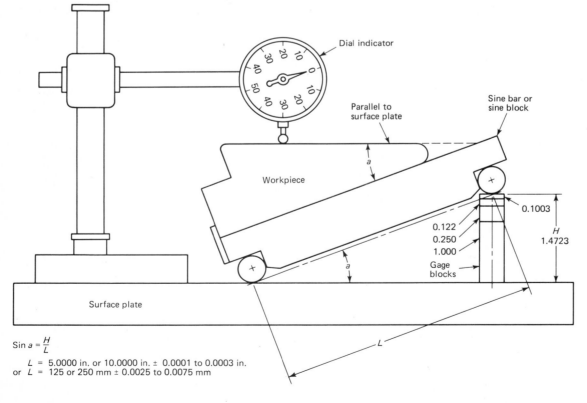

$$\text{Sin } a = \frac{H}{L}$$

L = 5.0000 in. or 10.0000 in. ± 0.0001 to 0.0003 in.
or L = 125 or 250 mm ± 0.0025 to 0.0075 mm

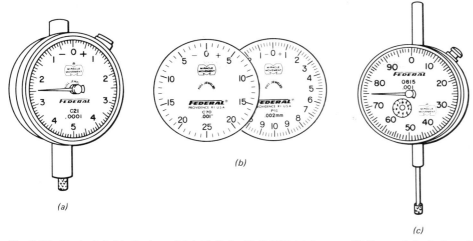

Fig. 2-23 Typical dial indicators. (*a*) A 2¾ dial with 0.0025 total range. (*b*) Some of the inch and metric dials which are available. (*c*) An indicator with a 1.000-in. range. (*Federal Products Corp., a division of Esterline Corp.*)

vernier calipers may be used, but quicker, sufficiently accurate methods are available.

Plug Gages

The plug gage (Fig. 2-24) is a quick, quite accurate way of checking that a hole is within the specified tolerance. These gages are made in sizes from 0.005 to 8.010 in., or the equivalent metric sizes, in classes XX, X, Y, Z. Class XX is the highest quality, with tolerances of 0.000 02 on gages up to about 1.500 in. diameter.

Most plug gages are made with a *go* end (often longer) and a *no-go* end (often shorter); smaller diameters are often simply rods fastened into holders.

Use of plug gages is illustrated in Fig. 2-25. The desired size is 0.763 to 0.765 in., a tolerance of 0.002; thus a class Z gage (0.0001) would be satisfactory. Practically all "gagemaker's tolerance" (usually 10 percent of the allowable part tolerance) will be applied *plus* on the *go* member and *minus* on the *no-go* member of the plug gages.

Of course, this means that a few "good" parts will be rejected. However, the go member tends to wear down, so it is on the safe side. The no-go member lasts longer than the go because it seldom enters the hole.

In many shops, the gage handles are color coded, and only certain colors may be used each month, while the previously used gages are being checked by the inspection department.

Ring Gages

For quick checking of outside diameters, ring gages (Fig. 2-26) are used, as are snap gages, which are discussed next. Tolerances are the same as for the plug gages, and sizes made range from 0.059 to 8.510 in. (1.5 to 216 mm) and over on special order.

The go and no-go ring gages are separate, but the no-go has an annular groove in the periphery, as shown in Fig. 2-26.

The gagemaker's tolerance is applied *minus* on the *go* member and *plus* on the *no-go* member. This is because the go gage must slide over the part; thus, any wear will tend to increase this measurement. The no-go should not slide onto the work. If it does, the part is undersize.

Snap Gages

For very fast checking of diameters or flat features on a part, snap gages (Fig. 2-27) can also be used. They most frequently have one large fixed anvil and two adjustable anvils which are either round or square.

Snap gages commonly come in sizes up to 6 in. (150

Fig. 2-24 Plug gages. (*a*), (*b*) Plain cylindrical. (*c*) Progressive cylindrical. (*d*) Reversible cylindrical, notches on no-go ends. (*Dearborn Gage Co.*)

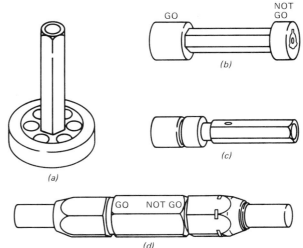

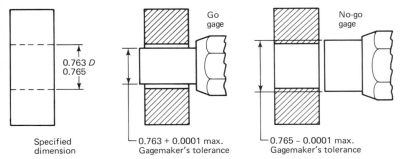

Go
gage

No-go
gage

0.763 D
0.765

Specified
dimension

0.763 + 0.0001 max.
Gagemaker's tolerance

0.765 – 0.0001 max.
Gagemaker's tolerance

This is a *limit* gage. It does not *measure* the hole.

Fig. 2-25   Typical use of a plug gage.

mm) and are listed up to 24 in. (600 mm) with tubular frames. The possible adjustment is about 0.250 in. (6.4 mm) though greater for the larger sizes.

These gages are set by hand, using gage blocks as standards. Thus, their accuracy depends on the "feel" and skill of the person setting them. The front anvil is the go (largest dimension). In use, an inspector merely slides the snap gage onto the work. If it is within tolerance, the gage will slide on until the part touches the no-go anvil and then stop.

CAUTION: Do not force the gage over the part, it *will* spring a few "tenths."

Thread Gages

These gages are made like ring gages for outside (male) threads and like plug gages for inside (female) threads (Fig. 2-28). They are made for machine (straight) threads in class 2 and class 3 threads, coarse, fine, and extra fine series. They are also made for NPT (National Pipe Taper) and NPTF (Dryseal) pipe threads and for metric threads.

The go gages represent the maximum metal condition, that is, on a male thread, the maximum permissible pitch diameter, major diameter, and minor diameter. The go gage should turn freely onto the

Fig. 2-26   A variety of ring gages. No-go gages have a ring around the center. (*Dearborn Gage Co.*)

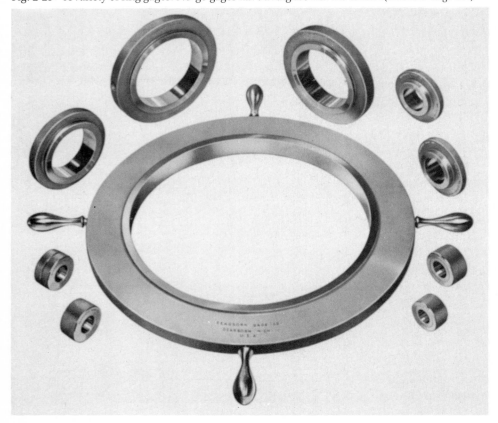

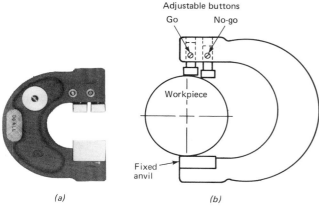

(a) (b)

Fig. 2-27 A snap gage. (a) As received, the anvils are set in the inspection department, and the size is stamped on the disk shown at upper left. (*DoAll Company.*) (b) How the snap gage is used for fast inspection.

complete length of the thread. The no-go gage should not thread on more than $1\frac{1}{2}$ turns. The thread plug gages should operate similarly. Using gages for taper pipe threads is somewhat more complex and will not be described here.

Snap thread gages are also available for fast checking of male threads. They have go and no-go rolls similar to the plain snap gages previously described.

Precision Bore Gages

A plug gage only checks that a hole is not undersized for the length of the go plug and is not oversized at the surface, the only point checked by the no-go plug. If the actual hole diameter and possible variations along the depth of the hole must be known, the precision bore gage (Fig. 2-29) will determine this quickly and accurately.

The diameter is not read directly, but the dial will show the amount above or below the original zero

setting, and the actual size can easily be calculated. This is a comparator-type measuring instrument, and the gage is "zeroed in" or set to a master ring gage before it is used.

These gages consist of a dial indicator mechanically connected to various sizes of gage heads. These heads, up to about 1.600 in., consist of spring steel nibs with the actuating rod in the center. Larger sizes require more complex arrangements.

The *range* of each head is limited. The very small (0.0185 to 0.380 in.) sizes have only 0.0025- to 0.0045-in. range. Thus, several gaging heads are needed according to the size of the work done in each shop. Sizes are available up to 14.000 in. These gages can easily be converted for metric use.

CAUTION: These gages read to 0.0001 in. (0.0025 mm); thus, dirt or rough hole finishes will greatly decrease their accuracy. Be sure to "rock" the gage head slightly in the hole to get the minimum reading, when the head is at 90° to the hole axis.

Air Gages

These are very versatile comparator-type gages and are made in several styles (Fig. 2-30). The readings may be from a dial, or a column of colored liquid or a "float" in a tube. They are made with amplifications of up to 50,000× for very precise work. Of course, at this high amplification, the total range of measurement is only 0.0003 in., so it would be used only for extremely precise work. Greater ranges are available.

As the scale charts for various ranges are easily changed, air gages can be converted to metric scales and the "gain" knobs quickly adjusted to suit.

The gaging element is typically a cylinder with two or more holes (jets) in it. These jets allow air to flow out. When the gaging element is placed in a hole, the

Fig. 2-28 Thread gages, ring and plug types. Short end of the plug gages is no-go. (*DoAll Company.*)

(a)

(b)

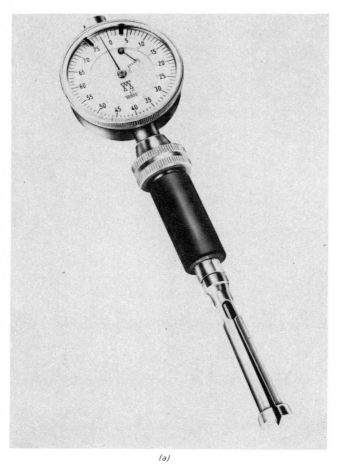

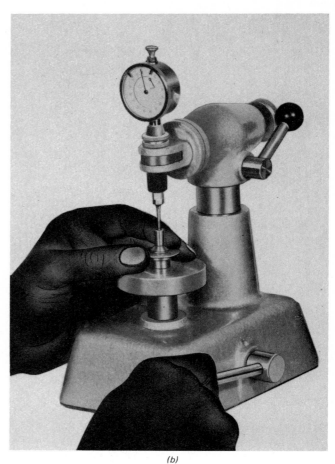

(a)   *(b)*

Fig. 2-29   A precision bore gage and a typical use. The indicators on the dial show the allowable limits to aid in fast inspection. (*Foster Supplies Company.*)

flow of air is restricted, and this is registered on the dial or column and compared with the previously set limits. The limit indicators are positioned at setup by the use of master gages (Fig. 2-30a). The operator can very easily see if the part is within tolerance.

Some *advantages* of air gages are:

1.   High amplification
2.   No metal-to-metal contact needed
3.   Easy readability, and therefore low operator error
4.   No mechanical moving parts
5.   Adjustable amplifications and easy scale changes
6.   Multiple gaging columns easily used

Some *disadvantages* of air gages are:

1.   Quite small range
2.   Sensitive to surface finish of the hole
3.   Expensive master gages needed for each size measured

4.   Fairly frequent adjustment needed

SPECIAL GAGING AND MEASURING EQUIPMENT

Optical Comparator

For checking shapes which are difficult or impossible to check or measure by ordinary means, the optical comparator (Fig. 2-31) is frequently used. It is also valuable in checking and measuring small parts, tool setting, gears, thread forms, and odd-shaped holes.

The workpiece is placed between a light and a lens-mirror combination. This shows a magnified image of the piece on a frosted glass screen. The magnification is commonly 10× to 40×, though over 100× is available.

The screen sizes are most frequently from 10 to 30 in. (250 to 750 mm). High magnification on a small screen will limit the maximum size of the part being measured. For instance, if 40× was used on a 10-in.

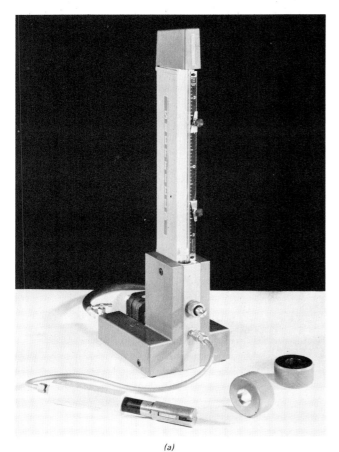

(a)

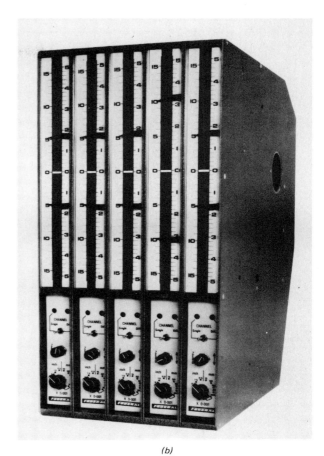

(b)

Fig. 2-30   Air gages. (*a*) An air gage and the go and no-go master setting plugs. (*Bendix, Automation & Measurement Division.*) (*b*) A solid-state electronic column gage, inch/metric switch, uses electronic gage heads. (*Federal Products Corporation, a division of Esterline Corp.*)

diameter screen, the largest object which could be completely seen would be $\frac{10}{40}$ or $\frac{1}{4}$ in. (6.4 mm).

Frequently, the desired shape is very carefully drawn at, say, 20× on a plastic sheet which is fitted over the comparator screen. The part is then placed on the "stage" and moved into position with the micrometer screws, and the shape is easily checked. At 20:1 magnification, 0.001 in. becomes 0.020 in. on the screen, which is easily seen. Drawings and scales can easily be used with metric dimensions.

Distances can also be measured in either the X or the Y axis, or both, by use of the large barrel micrometers which read directly to 0.0001 in. Metric micrometers could also be installed on the optical comparator.

The parts to be measured may be clamped on the movable stage, mounted between centers, or mounted on a great variety of special fixtures.

Angle measurement can be done on the optical comparator in two ways. First, a template may be accurately drawn, with the desired angle and even the permissible tolerance laid out. Second, the entire ring around the viewing screen can be rotated.

Optical Flats

A simple, inexpensive, rugged optical instrument is the quickest and most accurate way to check on the flatness of a surface if the surface is smooth enough to reflect light. This is the optical flat (Fig. 2-32), a round or square piece of tempered glass or quartz which has been ground and polished flat within 0.000 004 in. (0.0001 mm) or closer.

These are made in diameters of from 1 to 10 in., with the 2- to 4-in. diameter being the most used size. They may be finished on one or both surfaces and may be "coated" with titanium dioxide to give brighter contrast on the band patterns. A set of four flats (1- to 3-in. diameter) and a monochromatic light costs from $300 to $600.

To use these, it is necessary to use a *monochromatic*

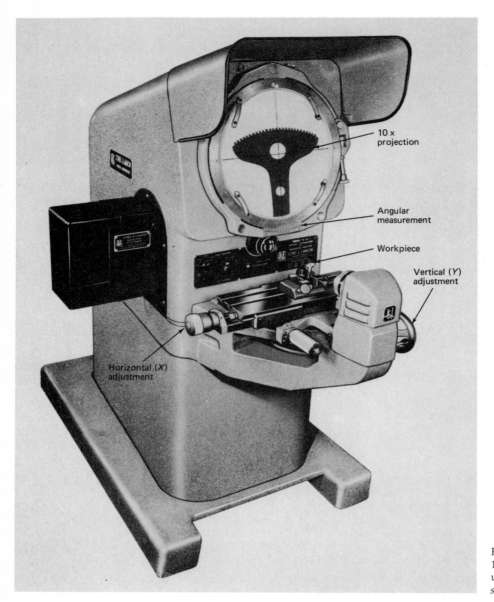

10 x
projection

Angular
measurement

Workpiece

Vertical (Y)
adjustment

Horizontal (X)
adjustment

Fig. 2-31 An optical comparator with a
14-in.-diameter (355-mm) screen, being
used at 10:1 enlargement. (*Jones & Lamson Division of Waterbury Farrell.*)

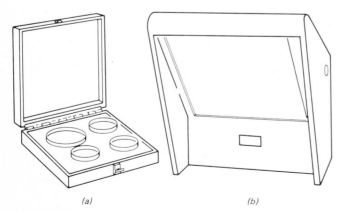

(a)   (b)

Fig. 2-32 (*a*) A set of optical flats. (*b*) A monochromatic light
source. (*Van Keuren Co.*)

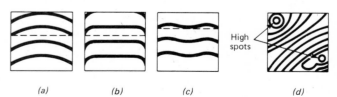

High
spots

(a)   (b)   (c)   (d)

Fig. 2-33 Some typical patterns showing deviation from perfect
flatness. (*a*) Convex, edges one-half wavelength 5.8 μin. low; (*b*)
Nearly flat, sides 5.8 μin. low, (*c*) Hollow in the center, low at
edges about 3 μin.; (*c*) Two high spots, with valley 69.6 μin. deep.
(*Van Keuren Co.*)

(single wavelength) light source. This is standardized as helium gas with a wavelength of 0.000 023 1 in. (0.000 588 mm). When this light shines through the optical flat onto the work, a series of dark and light bands appear on the surface of the flat, unless they have been wrung together. These bands are caused by the reinforcement and interference of the light "rays" reflected from the work and the top surface of the optical flat.

In its simplest and most frequent use, inspecting flatness, the pattern will show the condition of the surface as shown in Fig. 2-33. The bands may be closely spaced (if the two surfaces are nearly in contact) or widely spaced (if there is a wider angle between the surfaces).

A Modern Measuring Instrument

The *laser* (Light Amplification by Stimulated Emission of Radiation) is finding an expanding use in precise measurement today. It is relatively simple to operate, requires no contact, and is accurate, even over distances of 100 ft (30,000 mm) or more. As shown in Fig. 2-34, the laser-beam interferometer combination can be mounted conveniently on the machine frame, and one portion of the beam is aimed at the specially shaped target prism mounted on the machine carriage.

The laser interferometer is connected to a computerized control and read-out console which will visibly display the measured dimension to 3, 10, or 100 μin. (0.000 08, 0.000 25, or 0.0025 mm) least count. The accuracy is about 0.000 010 in./ft or 0.001 in./100 ft.

The laser used is a helium-neon gas type with a half-wavelength of about 12.4571 millionths of an inch. The logic circuitry will measure in either direction with a resolution of about 3.1143 millionths of an inch.

The laser beam, about 10-mm. (0.4-in.) diameter, is visible and laser interferometer compensation for ambient air temperature, material temperature, coefficient of expansion, and atmospheric pressure is provided.

In use as a measuring instrument (Fig. 2-35), the movement of the reflector target causes interference bands to move across a reference point in the interferometer. A count of the number of bands provides a measure of the target movement.

Fig. 2-34 A laser interferometer in use checking the accuracy of a machine. (*Perkin-Elmer Corp.*)

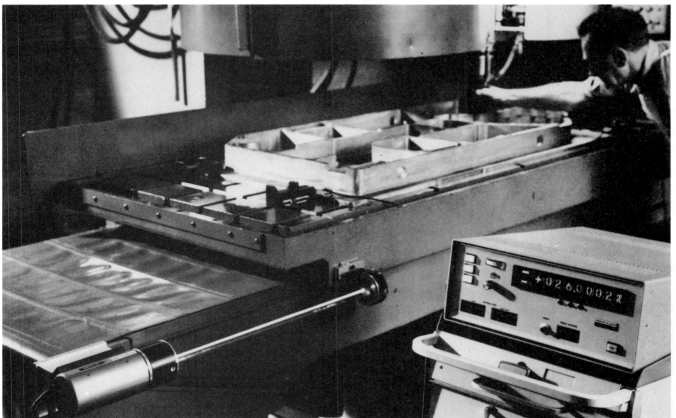

Advantages of the laser system are:

1. No pressure on machine or parts due to the measuring equipment
2. Absolute measurement regardless of lead screw accuracy over distances of 100 ft or more

Disadvantages of the system are:

1. Interruption of the beam by chips or any object passing across it will result in an incorrect measurement.
2. Too rapid machine motion may go faster than the counting speed capabilities of the electronic logic.

SURFACE ROUGHNESS MEASUREMENT

In writing of the use of some of the gaging equipment, the errors due to "rough" surfaces were mentioned. This raises the problem of what is "rough" or "smooth," and how this can be measured.

Not too many years ago, many design engineers used a series of letters such as *f, ff, g,* etc., to try to describe the desired finish. Today, a clearly defined numerical system is used. This system is based on the microinch (μin.) (0.000 001 in.) or, in the metric system, micrometre (μm) (0.000 001 m).

This is *not* the total distance between peaks and valleys, but an *average* amount of irregularity above or below an assumed centerline (see Fig. 2-36). All of this is "computed" by equipment such as that shown in Fig. 2-38.

For quick, fairly close estimates of surface finish, plates such as those shown in Fig. 2-37 are very often used. If the fingernail is run across the representative finish on the plate and then across the machined surface, a close estimate of the microinch finish can be made.

The notations in Fig. 2-37*a*, such as M, G, S, T, etc., merely identify the "lay" or pattern of scratches made by milling, grinding, shaping, or turning, etc.

The appearance is quite different, but the *finish* in microinches or micrometres is the same.

CAUTION: Do not depend on just looking. Do use the fingernail test. Appearances are very deceptive as you can see from Fig. 2-37.

The *electronic surface indicator* (Fig. 2-38) is a much more accurate method of checking finishes. The dial, with a range from 1 to 1000 μin. (0.025 to 25 μm), indicates readings obtained from either a hand-held or bench-mounted probe which has a 0.0005-in.-diameter (0.015-mm) diamond point.

The chart will record, with selectable scales, the "picture" of the surface, which will look something like the top line of Fig. 2-36.

Roughness width cutoff is controlled by a selector switch. In most measurements the 0.030-in. (0.75-mm) cutoff is used. This means that the electronic circuit records all the peaks and valleys over a distance of 0.030 in. and then starts over again. This width records the major irregularities. If the 0.010 or 0.003 width is used, these in effect measure the roughness of one-third or one-tenth of the sides of the slopes of the 0.030 profile and then start over again. The 0.030-in. (0.75-mm) cutoff is the specified standard.

Waviness refers to irregularities from a mean line which are of greater spacing than the roughness. The *waviness height* is the peak-to-valley height. This could be measured with an electron height gage and a surface plate.

Rms or AA and, in England and sometimes in the United States, CLA should follow any number which specifies surface finish. The two represent mathematical methods of averaging the heights of peaks and valleys. They mean **root mean square**, **arithmetic average**, and **centerline average**, which is the same as AA. The method of computing each of these is illus-

Fig. 2-35   A simplified diagram of the light paths of a laser interferometer.

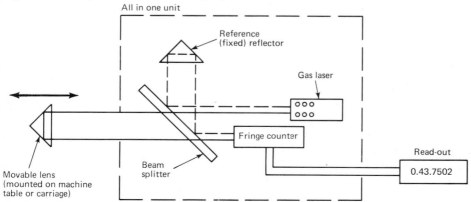

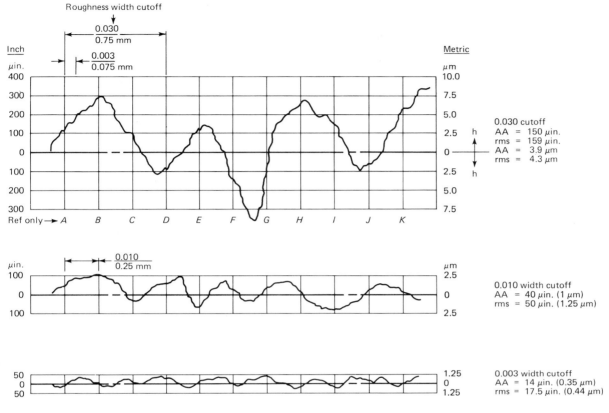

Computations for 0.030 cutoff width—roughness
readings
Readings taken at vertical, lettered lines

	h	h^2
A	125	15,625
B	290	84,100
C	100	10,000
D	90	8,100
E	125	15,625
F	150	22,500
G	100	10,000
H	250	62,500
I	140	1,960
J	80	6,400
K	200	40,000
$n = 11$	1650	276,810

$$\text{AA} = \frac{\Sigma h}{n}$$

$$= \frac{1650}{11} = 150 \ \mu\text{in.}$$

$$\text{rms} = \sqrt{\frac{\Sigma h^2}{n}}$$

$$= \sqrt{\frac{276,810}{11}} = 159 \ \mu\text{in.}$$

Note: 150 μin. = 3.75 μm
159 μin. = 3.98 μm

Fig. 2-36   Representation of an actual trace of the surface of a machined surface and the readings
resulting.

trated in Fig. 2-36. Of course, this is done electronically in the actual surface-measuring instrument.

Rms has long been the standard. However, the official standard is now AA, which reads about 8 to 11 percent *less* than rms on most surfaces. Most instruments today can be switched to either system. Using the "scratch" test panels, it is not possible to discriminate within 11 percent, so it does not matter too much which it is called.

Interestingly, surfaces produced by EDM (electrical discharge machining) and sand or shot blasting methods do not give accurate readings with the electronic instrument. This is because these processes produce "circular indentations" instead of the "scratches" produced by conventional metal-cutting processes.

The surface finish which can be expected from some machining processes is shown in Fig. 2-39. Do not specify any finer finish than is really needed. Finer finishes can be very expensive.

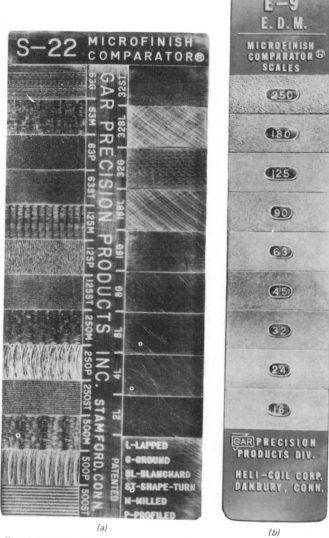

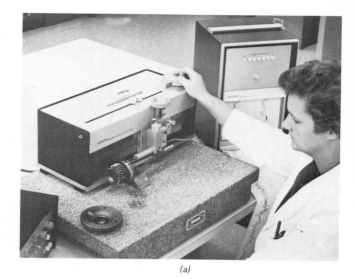

(a)

Fig. 2-37 Microfinish comparator surface finish scales. The one shown in (a) is used for most machining operations. (*Gar Electroforming Div., Mite Corp.*)

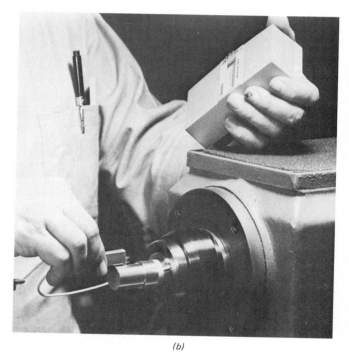

(b)

Fig. 2-38   Surface roughness measuring equipment. (a) An inspection room setup. Note the chart tracing at the right. (b) A portable hand-held instrument which can be used at the machine. (*Clevite Corporation Gaging and Control Division.*)

Measuring on the Machine, Digital Read-out

Accurate measurement, using the hand dials on machines, depends to quite an extent on the skill of the operator. Even the operator can forget the count of the number of turns of the feeding wheel.

To eliminate errors and speed up the work of both setup and cutting, several types of *digital read-out* devices are available. The one shown (Fig. 2-40) is electronic, shows + or − from any desired "zero" point, and can be switched to read in inches or millimetres.

This read-out's accuracy of about ±0.0001 in. (0.002 mm) is independent of the accuracy of the machine's

lead screws, so more accurate work may be done on even an older machine. Costs range from $500 to $1000 per axis, and they can be attached to almost any machine tool.

WHAT TOLERANCE SHOULD YOU USE?

Do not let the talk and writing about sophisticated measurements in the millionths inflence your com-

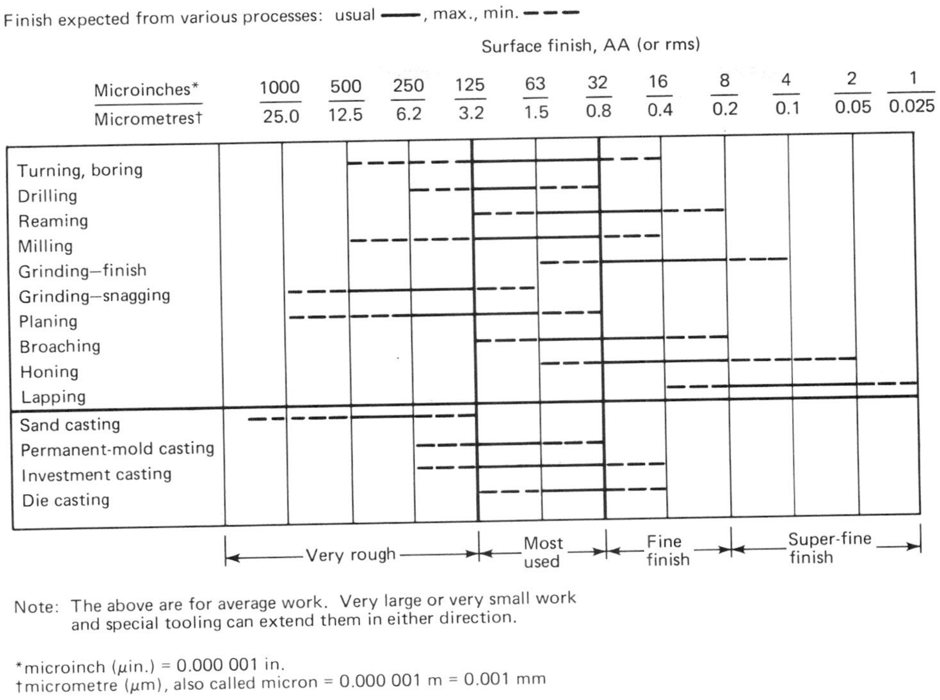

Finish expected from various processes: usual ——— , max., min. — — —

	Surface finish, AA (or rms)										
Microinches*	1000	500	250	125	63	32	16	8	4	2	1
Micrometres†	25.0	12.5	6.2	3.2	1.5	0.8	0.4	0.2	0.1	0.05	0.025
Turning, boring											
Drilling											
Reaming											
Milling											
Grinding—finish											
Grinding—snagging											
Planing											
Broaching											
Honing											
Lapping											
Sand casting											
Permanent-mold casting											
Investment casting											
Die casting											

←———Very rough———→ ←Most used→ ←Fine finish→ ←Super-fine finish→

Note: The above are for average work. Very large or very small work and special tooling can extend them in either direction.

*microinch (μin.) = 0.000 001 in.
†micrometre (μm), also called micron = 0.000 001 m = 0.001 mm

Fig. 2-39   Finish expected from various processes.

mon sense. There are still thousands of parts made which only require tolerances of 0.015, 0.010, or 0.005 in. (0.4, 0.25, or 0.125 mm). These are easily checked with micrometers, with the production gages mentioned, or with special gages made to check a specific workpiece.

It takes experience and good judgment to dimension parts for ease and economy in manufacturing without sacrificing any really needed quality. The next chapter contains some additional information on some ways this is done.

Fig. 2-40   One type of digital read-out (DRO) shown attached to a milling machine. (*Micro-Line, Jamestown, N.Y.*)

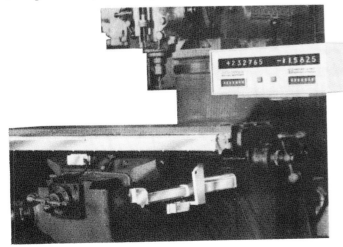

Reveiw Questions and Problems

2-1. What are the international standards of length and how are they established?

2-2. Describe several uses for gage blocks as they were mentioned throughout this chapter.

2-3. What gage blocks would you use to make the dimensions 2.6756 in. and 3.0702 in.?

2-4. How accurate is a steel rule, and why should you not always use the smallest possible graduations?

2-5. Describe some uses for a combination square.

2-6. You are reducing the diameter of a 2-in. bar to $1\frac{1}{4}$ in. in several cuts. Should your outside caliper be set to $1\frac{1}{4}$? How should it be used?

2-7. Could you use inside spring calipers to measure a groove down the inside of a part?

2-8. What steps should be taken when using spring calipers to ensure accurate readings? How accurate are these readings?

2-9. What are dividers used for?

2-10. Make a sketch and label the principal parts of a micrometer caliper (or "mike," "mic," or micrometer).

2-11. a. In what units is an inch micrometer marked on the barrel and on the thimble?
    b. In what units is a metric micrometer marked on the barrel and on the thimble?

2-12. How is a micrometer held when taking a measurement?

2-13.   What is the "built-in" accuracy of a micrometer?

2-14.   Describe a vernier caliper and the principle on which it is constructed.

2-15.   What are the divisions on these calipers?
  a.   In inches.
  b.   In millimetres.

2-16.   What is the smallest increment to which a vernier caliper can be read?

2-17.   Would you use small hole gages or a vernier caliper to measure inside diameters of: 0.194, 0.763, and 4.892 in.? Explain your choice.

2-18.   Is a telescoping gage or a vernier caliper more accurate in measuring a 3.893-in.-diameter hole? Why?

2-19.   What is a surface plate and for what is it used?

2-20.   Why is the height micrometer rapidly replacing the vernier height gage?

2-21.   a.   Describe a coordinate measurement machine.
  b.   Why is this machine worth $15,000 or more?

2-22.   Draw a sketch of a sine bar illustrating its use.

2-23.   Describe several situations in which dial indicators would be useful. Would it be any advantage to use the larger diameter dials?

2-24.   Would you have your inspection department check a batch of work using plug and ring gages? Explain your answer.

2-25.   Are snap gages faster to use than ring gages? Why?

2-26.   When are precision-bore gages used? Why not just use plug gages?

2-27.   What are the advantages and disadvantages of an air gage? How closely can they be read?

2-28.   a.   For what is the optical comparator used?
  b.   What limits the size of the work which can be measured or compared on the optical comparator?

2-29.   What are optical flats and how are they used?

2-30.   How is a measurement of a machine's table movement made with the laser interferometer?

2-31.   What are the advantages and disadvantages of the laser interferometer?

2-32.   How are standard surface finish plates used?

2-33.   In what units are standard finishes expressed in a. U.S. units? b. Metric units?

2-34.   To what microinch or micrometre finish are most parts in industry finished?

2-35.   List all the internal measuring devices in this chapter and the uses and limits of each.

2-36.   Do the same for external measuring equipment.

References

1.   Busch, Ted: *Fundamentals of Dimensional Metrology*, 3d ed., Delmar Publishers, Inc., Albany, N.Y., 1964.

2.   *Handbook of Industrial Metrology*, Society of Manufacturing Engineers, Detroit, Mich.

3.   *Metric Practice Guide*, American Society for Testing and Materials, E 380-72, 1973.

4.   Michelon, Leno C.: *Industrial Inspection Methods*, rev. ed., Harper & Row, New York, 1957.

5.   Roth, E.S.: *Functional Gaging of Positionally Toleranced Parts*, Society of Manufacturing Engineers, Detroit, Mich., 1964.

Note: Manufacturers' catalogs and descriptive literature are a splendid source of useful information.

quality control, tolerancing, and dimensioning

The basic purpose of manufacturing is to produce parts to a specified shape and size. The specifications for the shapes and sizes are furnished to the shop by part drawings or manufacturing drawings. These drawings contain all the information necessary for the complete manufacture of the desired part.

Interchangeability

In present-day manufacturing, large quantities of parts are produced in large production lots. It is necessary that these parts be made to definite size and shape to ensure that each part, or assembly of parts, going into the final product is completely interchangeable. If interchangeability is not achieved, selective assembly will be required, that is, each part must be selected to fit its mating part. Selective assembly is costly and should be avoided wherever possible.

Inspection

It has often been stated that no two things can ever be exactly alike. This also holds true with manufactured parts. There are several factors which cause variation in parts manufactured in the same way, on the same machine, and by the same operator. A few of these factors are variations in raw materials used, inaccuracies in the machines used, incorrect or worn tooling, improper method of manufacture, and human errors.

Some procedure must be set up to ensure that faulty parts are rejected and that the manufacture of faulty parts does not go uncorrected. This is the job of the inspection department.

There are three basic areas of inspection: (1) receiving inspection, inspection of all incoming material and purchased parts; (2) in-process inspection, in-spection of parts as they are being produced; and (3) final inspection, inspection of the finished product before shipment to the customer.

QUALITY CONTROL

It is the function of quality control to set up standards and inspection procedures to ensure quality in the finished product. To be absolutely sure that each and every part manufactured meets the requirements of the part drawing, 100 percent inspection would be required. Since inspection is expensive and does not directly affect the value of the finished product, inspection should be kept to a minimum.

With the use of automatic and semiautomatic machines which can produce large quantities of parts with little variation in size, methods of inspection have been devised which eliminate the need for 100 percent inspection. The procedures used are called *statistical quality control*. Note that the word "quality" as used in manufacturing does not necessarily mean the "best" but implies "the best for the money."

Statistical Quality Control

As far back as the eighteenth century, it was noted that the variation of measurements made on a sample of pieces manufactured under the same conditions would display a surprising degree of regularity. The variations in measurements would form a pattern which would closely approximate a distribution which is called the **normal distribution curve**. It was also noted that the larger the number of measurements, the closer the data obtained would approach a "normal" distribution.

One important and interesting fact about the normal distribution curve is that it can be completely

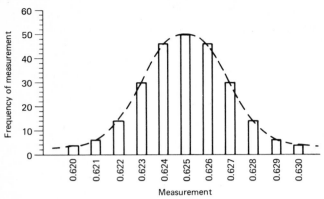

Fig. 3-1   Frequency distribution of sample of 250 pieces measured.

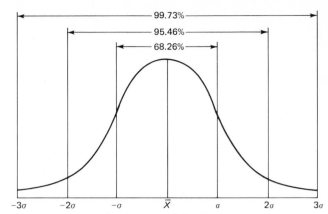

Fig. 3-2   Areas under the normal distribution curve.

defined by its mean and standard deviation. If we know the mean and standard deviation, we can determine the area under the curve between any two points on the horizontal scale. In statistical quality control, this would be the number of measurements which would be expected to fall within any given area.

The mean (measure of central tendency) is the arithmetic average of the sum Σ of all data recorded and is called $\overline{X}$ (read X bar). The normal distribution curve is symmetrical about this value. The **standard deviation** (measure of variation) is the square root of the average of the squares of variations (difference between the measurement and the mean) and is called σ (Greek letter sigma). In mathematical form

$$\overline{X} = \frac{\Sigma X_i}{n}$$

$$\sigma = \sqrt{\frac{\Sigma(X_i - \overline{X})^2}{n}}$$

where n = the number of measurements
X_i = value of each measurement

As an example, consider a lot of 250 pieces machined to a finished diameter of 0.625 in. (15.875 mm). After machining, the pieces are measured and the results plotted on a bar graph as shown in Fig. 3-1. The height of each bar represents the number of pieces yielding the dimensions shown on the horizontal scale. Notice how the measurements tend to group about the center of the graph. A curve drawn connecting the top center of each bar represents the smooth curve of the readings and will approach a normal distribution curve.

Figure 3-2 shows a curve of normal distribution. The horizontal scale is in standard deviations σ above and below the mean $\overline{X}$. The area under the curve between several points is shown. If the plot of measurements taken approaches the normal distribution

curve, 99.73 percent of all measured values will fall within the $\overline{X} \pm 3\sigma$ points, 95.46 percent between $\overline{X} \pm 2\sigma$, and 68.26 percent between $\overline{X} \pm 1\sigma$. Other values of area for various values of standard deviations from 0.01 to 4.0σ are given in tables in textbooks on statistics. (See the bibliography at the end of this chapter.) Based on this principle, many of the statistical quality control methods are formulated.

Referring again to the sample lot plotted in Fig. 3-1, if the values obtained are used in the formulas for $\overline{X}$ and σ, the mean $\overline{X}$ is 0.625 in. (15.875 mm) and the standard deviation is 0.002 in. (0.051 mm). If the design engineer specifies that the part can vary $\pm$ 0.006 in. (0.152 mm) representing 3σ, then from the normal distribution curve, 99.73 percent of all parts manufactured will be expected to fall within this area. This represents a rejection rate of only 0.27 percent.

If, however, the engineer specifies a variation of only $\pm$ 0.002 in. (0.051 mm), or 1σ, then only 68.26 percent of all manufactured parts would be expected to fall within this area. Under these conditions, the expected rejection rate would be 31.74 percent. If the $\pm$ 0.002 in (0.051 mm) must be held, the process used to manufacture the parts shown in the example will have to be improved or a more accurate method of manufacture used.

When a manufacturing process results in a large percentage of rejects, a reduction in the rejection rate by changing or improving the process will undoubtedly increase the manufacturing cost. This increased cost will be compensated for by the reduced cost of rejected material. There is a point, the point of diminishing returns, where the increase in cost for an additional reduction in rejects will be greater than the value of the rejected material that can be saved. This point falls somewhere between a 5 and 10 percent rejection rate. Therefore, most manufacturing processes are considered economical when the rejection rate is below 10 percent.

process control charts

Process control charts are commonly used in quality control to maintain a continuous evaluation of the manufacturing process. The control chart shows the distribution of the averages of the observed measurements of each sample lot taken during the actual manufacture of the parts. In most cases, only sampling inspection is used to reduce cost. The sample size usually runs from three to six pieces and is taken at predetermined intervals of time or quantity.

Two such charts are the chart for average values, the $\overline{X}$ chart, and the chart for ranges, the R chart. Figure 3-3 shows an example of these two charts based on the machining operation used in the previous example covering a three-day period.

A sample is taken, in this case five pieces, and the diameter of each is measured. The average of these five measurements is then plotted on the $\overline{X}$ chart. Average values are plotted instead of individual readings because it is known that sample averages tend to approach the normal distribution curve more closely than do individual readings. The central line, or mean, $\overline{\overline{X}}$ shown on the $\overline{X}$ chart is the average of the averages of each sample of five readings. Expressed mathematically,

$$\overline{\overline{X}} = \frac{\Sigma \overline{X}}{K}$$

where $\overline{X}$ is the average value of each sample group and K is the number of groups of five measurements each.

The control chart has limit lines called *control limits*, which are set at three standard deviations (3σ) of the sample averages from the mean $\overline{\overline{X}}$. These are the upper control limit (UCL $= \overline{\overline{X}} + 3\sigma$) and the lower control limit (LCL $= \overline{\overline{X}} - 3\sigma$). In mathematical form

$$\sigma_x = \sqrt{\frac{\Sigma(\overline{X} - \overline{\overline{X}})^2}{K}}$$

It can be shown theoretically that the chances of a point falling above or below the limit lines are less than 3 out of 1000, provided that the variations are caused by normal manufacturing causes and that no change has been made in the manufacturing process.

The chart for ranges, the R chart (Fig. 3-3), is obtained from the same sample groups that were used to determine the values of $\overline{X}$. The central line, or mean, of the R chart is the average of the ranges of each sample group. The control limits on the range chart depend on the number of units in each group and are determined by multiplying the average range $\overline{R}$ by a factor which is given in textbooks on statistical control. (See the bibliography at the end of this chapter.)

Once the control charts have been established, all values of $\overline{X}$ and R will fall within the control limits if the variations are caused only by chance variations present in the manufacturing process. If either the $\overline{X}$ or R values fall outside of the control limits, some corrective action is required.

The $\overline{X}$ chart of Fig. 3-3 indicates that the manufacturing process was under control until 9 o'clock on Wednesday. At this time the value of $\overline{X}$ was above the UCL. Investigations into the cause showed that the tool had become worn. Replacement of the tool brought the process back into control.

It should be remembered that the control limits of the process control charts do not represent the limits of the manufacturing process nor do they represent the specification limits of the manufacturing drawing. Points which fall outside of the control limits do not necessarily represent rejected material but only signal that some corrective action is required to prevent manufacturing faulty parts.

The scope of this book does not allow further discussion of the various procedures used in quality control, and the reader is referred to any of the many texts on statistical quality control.

Inspection by Sampling

Inspection operations are costly and should be limited to only the amount of inspection necessary to maintain the required quality. When a small number of rejects can be tolerated, sampling inspection may

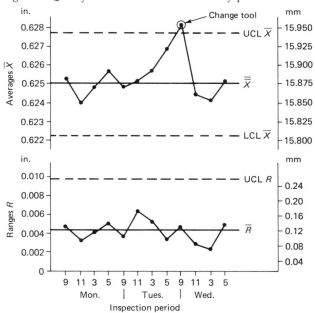

Fig. 3-3   Quality control charts for a three-day period.

be used to advantage. The cost of 100 percent inspection, in most cases, far exceeds the value of the small number of rejected parts which would be expected when a suitable sampling inspection procedure is used. Any reject material would be discarded at final assembly.

Sampling inspection is used widely on large quantities of purchased parts. In the sampling procedure, a specified number of pieces are selected at random from the entire lot and inspected. According to previously determined standards, the entire lot is either accepted or rejected on the basis of the inspection of the sample lot.

Various plans for sampling inspection are discussed in any of the many textbooks on quality control. One such plan might require a sample of 120 pieces to be inspected from a lot of 1200 pieces. If three or fewer of the 120-piece sample were found defective, the entire lot would be accepted. If more than three were defective, the entire lot would be rejected and returned to the vendor.

In the above example, the objection is sometimes raised to accepting the lot when some defective parts appear in the sample. It should be emphasized that if the sample showed no defective material, this would be no guarantee that the entire lot did not include some defective material not present in the sample tested. In fact, it has been found that for a desired amount of protection against accepting lots which contain more than a predetermined percentage of defects, sampling plans using a large sample size and a relatively large number of allowable defects in the sample are far superior to those plans using small samples with zero allowable defects. Plans using large sample sizes can discriminate better between satisfactory and unsatisfactory lots.

Before the widespread use of sampling inspection for quality control, it was commonly believed that the sample inspected should be some fixed percentage of the entire lot. It is now known that the size of the sample, not the percentage, is more important. The scope of this book does not allow for discussion of the proof of these statements.

When using any sampling inspection plan, one must be aware of the risks of accepting defective materials or rejecting acceptable material. A plan should be selected which will satisfy the requirements of the consumer of the material and be acceptable to the manufacturer so that each party will be aware of the risks involved.

TOLERANCE

Note: The drawings in the balance of this chapter are dual dimensioned in inches and millimetres. The millimetre equivalents are carried to two decimal

places since that is as close as a metric micrometer can be read directly. Table 3-1 shows how close these metric equivalents are.

When only metric dimensioning is used, there is a tendency to use multiples of 0.02 mm since this equals 0.000 79 in., which is just slightly less than our 0.001 in., and 0.02 mm is the unit reading on most metric dial indicators.

It would be desirable to manufacture all parts to the exact size specified on the manufacturing drawings. As pointed out previously, this is impossible. It is necessary to allow some variation from the desired dimensions. This variation is called the *manufacturing tolerance* or *tolerance* for short. It is the design engineer's duty to specify tolerance on all dimensions tight enough to maintain the necessary quality and loose enough to provide for economical manufacture. The smaller the tolerance, the more expensive the part, other things being equal.

There are three basic ways of specifying tolerances: (1) bilateral tolerance, (2) unilateral tolerance, and (3) limiting dimensions or manufacturing limits. Examples of each of the three methods are shown in Fig. 3-4. Note that each of the three methods results in parts made to the same tolerance and having the same limiting dimensions.

Bilateral tolerances (Fig. 3-4a) are used where the parts may vary in either direction from the desired or nominal size. The sizes of parts manufactured will tend to fall in a normal distribution centered about the nominal dimension.

Unilateral tolerances (Fig. 3-4b) are used where it is important for the dimension to vary in only one direction. Parts manufactured will tend to fall close to the desired dimension but can vary in only one direction. Unilateral dimensioning is also used in those manufacturing operations where the variation is most likely to fall in only one direction. An example is the drilled hole. Since the drill is made close to the nomi-

TABLE 3-1 MILLIMETRE EQUIVALENTS USED IN THE DRAWINGS OF CHAPTER 3, SHOWING THE VARIATION FROM EXACT EQUIVALENCY

in.	Approx. Metric, mm	Exact Inch Equivalent	Deviation from Exact Inch Equiv.
0.750	19.05	0.7499	−0.0001
0.500	12.7	0.500	Exact
0.020	0.51	0.0201	+0.0001
0.015	0.38	0.014 96	−0.000 04
0.010	0.25	0.009 84	−0.000 16
0.005	0.13	0.005 11	+0.000 11
0.003	0.08	0.003 15	+0.000 15
0.002	0.05	0.001 97	−0.000 03
0.001	0.03	0.001 18	+0.000 18
1.000	25.4	1.000	Exact

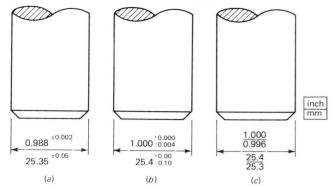

Fig. 3-4   Dimensioning of tolerances. (a) Bilateral; (b) unilateral; (c) limiting dimensions.

nal hole size, it is seldom possible to drill a hole undersize. All drilled holes should carry only a plus tolerance.

Specifying tolerance by using *limiting dimensions* (Fig. 3-4c) is the most frequently used method. It is especially useful when the total tolerance is small or in large-scale production work where gages are extensively employed.

When dimensioning using limits, all external dimensions (such as shaft diameters) usually have the *maximum* limit shown first or on top. All the internal dimensions (such as bearing bores) have the *minimum* size shown first or on top. This arrangement is for the convenience of the machine operator. The operator, in the case of a shaft diameter, will turn down the diameter to the dimension shown above the line. When the operator overshoots this dimension, the shaft will still be within tolerance. The opposite will hold true when enlarging a hole.

Where there is no tolerance specified on a dimension, a "standard" tolerance will apply. For fractional dimensions, the tolerance is often $\pm \frac{1}{64}$ in. (0.40 mm) and for decimal dimensions, the tolerance is often $\pm$ 0.005 or 0.010 in. (0.125 or 0.25 mm). These values will vary with different manufacturers and are always specified in a special note in the title block.

Allowance

The amount of free space between a shaft and the bearing must be limited so that the shaft will be located accurately but will neither bind nor set up excessive vibrations which could damage the shaft or bearings. The amount of free space is called **clearance**.

In some cases where parts are to be assembled permanently, they may be pressed together. Where a press fit is desired, a negative clearance or **interference** is deliberately dimensioned into the mating parts to ensure a tight fit. The amount of minimum clear-

ance or maximum interference designed into mating parts is called the **allowance**.

Fits

The best method of determining allowances and tolerances for mating parts is to use the fits recommended by the American National Standards Institute in tables thay have developed for the purpose (ANSI B4.1-1967). These tables specify the preferred limits and fits for cylindrical parts divided into five basic types of fits. Each of the basic five types is again subdivided into classes.

Running and sliding fits (RC) are intended for parts which are free to rotate or slide when the proper lubrication is applied. They are subdivided into nine classes (RC 1 through RC 9).

Locational clearance fits (LC) are used for stationary parts which must be located accurately but must be freely assembled and disassembled. These fits are subdivided into eleven classes (LC 1 through LC 11).

Transition locational fits (LT) are used where accuracy of location is important, but where a small amount of clearance or interference can be tolerated. They are subdivided into seven classes (LT 1 through LT 7).

Locational interference fits (LN) are used where accuracy of location is of prime importance, along with rigidity of alignment. There are only two subclasses (LN 2 and LN 3) and both give interference fits which must be assembled under light pressure.

Force or shrink fits (FN) have larger amounts of interference and are subdivided into five classes. The classes run from a light drive fit (FN 1), to a force or shrink fit (FN 5), requiring a heavy pressing force or shrink fit for permanent assembly.

Figure 3-5 shows two examples, RC 2 and FN 2, of mating parts dimensioned for specific fits. Notice that "standard hole" practice is used. In standard hole practice, the minimum size of the hole is taken as the base dimension (in this example 1.000-in. diameter) and the tolerance and allowance are figured from this basic hole size. This method is preferred to making the shaft the basic size because the hole can then often be made with standard tools.

selective assembly

Where fits are desired which are extremely close and the tolerances are small, selective assembly may be used, with parts dimensioned for a transitional fit. Every part is measured and marked as small, medium, and large or within a certain numerical tol-

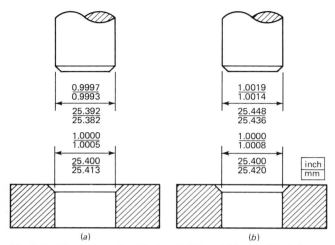

Fig. 3-5   Dimensions for (a) class RC 2 and (b) class FN 2 fits.

erance. At assembly, a small shaft would be used with a small bearing hole, medium with medium, and large with large. This will produce approximately the same allowance on all assemblies without having any interference fits. While this is expensive, it is the only practical way to assemble some ball bearings, compressor cylinders, and similar items.

DIMENSIONING

Intelligent placement of dimensions requires good engineering judgment and a thorough knowledge of manufacturing processes. The fact that there will be tolerance buildup must never be forgotten, and the results of such buildup must be taken into consideration.

There are three approaches to determining the correct location of dimensions. They are: (1) dimensions to indicate the engineering priorities, (2) dimensions for ease of manufacture, and (3) dimensions to simplify inspection. Since inspection does not add directly to the function or quality of the parts manufactured, it is seldom considered. The best approach

is to dimension for ease in manufacture while maintaining the requirements of the engineer.

All features of the part to be manufactured must be dimensioned, but care must be used not to overdimension. Figure 3-6 shows an example of a simple part which is overdimensioned. In this example, the width of the slot is given, also the distance from each side of the slot to the corresponding edge. These three dimensions are added to give the overall dimensions also given. Remember that the tolerance on each dimension must be considered. In this example, the tolerance of each of the three feature dimensions must be added, giving an overall dimension of 2.500 in. + 0.005 in., − 0.010 in. (In metric this would be 63.5 mm + 0.13, − 0.26 mm.) Notice that this tolerance exceeds the allowable tolerance given on the overall dimension ± 0.005 in. (0.13 mm). When making this part, the full tolerance on each feature cannot be used if the overall dimension is to be held.

Figure 3-7 shows how this piece can be dimensioned to eliminate the conflict of tolerance buildup. In this example, the width of the slot has been left off the drawing. Notice that the depth of the slot has been given along with the overall height of the part.

When calculating tolerance buildup and when dimensions are added as in the example in Fig. 3-6, the plus (+) tolerances of each dimension are added to arrive at the total plus tolerance, and the negative (−) tolerances of each dimension are added to arrive at the total negative tolerance.

When dimensions are to be subtracted, the effect the tolerances will have on the result is sometimes confusing. A simple approach is to think not of subtracting, but of adding a negative number. In the example in Fig. 3-7, to calculate the width of the slot, first add the two distances the slot is from each edge. Then add this result as a *negative* number to the overall dimension. To convert to a negative number, multiply by minus one (−1) and then add to the overall dimension. Remember, the tolerances must also be multiplied by minus one.

inch

0.750 $^{+0.000}_{-0.005}$

0.750 $^{+0.005}_{-0.000}$

1.000 $^{+0.000}_{-0.005}$

2.500 $^{+0.005}_{-0.010}$

mm

19.05 $^{+0.00}_{-0.13}$

19.05 $^{+0.13}_{-0.00}$

25.40 $^{+0.00}_{-0.13}$

63.50 $^{+0.13}_{-0.26}$

Overall length by sum of feature dimensions

Fig. 3-6   Example of overdimensioning.

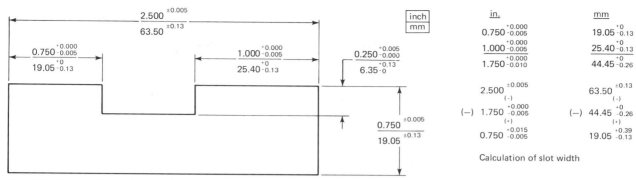

Fig. 3-7   Correct dimensioning of piece where slot width is relatively unimportant.

In manufacturing the part in Fig. 3-7, due to the method of dimensioning, it will be necessary to mill the slot in two cuts because of the possible variation in the width of the slot. The cutter would be positioned flush with the top surface and moved to position to mill one side of the slot in relation to one side of the piece. A second cut will be required to machine the other side of the slot in relation to the other side of the part.

The cutter may have to be positioned differently for each piece made because of the possible variations in the width of each piece, and two cuts are required.

The cost of machining this part can be reduced by redimensioning the part drawing. Figure 3-8 shows the same part where the width of the slot is given and the bottom of the slot is dimensioned from the bottom of the piece. A standard cutter can now be used to cut the width of the slot.

Base-Line Dimensioning

Where parts are to be made on numerical control (N/C) machines, *base-line dimensioning* is often used (see Fig. 3-9). One advantage of this method of dimensioning is the elimination of tolerance buildup. A second advantage of base-line dimensioning, when working with N/C machines, is that the programmer would use the intersection of the two base lines as the

zero reference point, and all dimensions used in writing the program can be read directly from the drawing.

Tolerance buildup is most noticed in the dimensioning of a series of holes. Figure 3-10*a* shows a piece requiring five equally spaced holes dimensioned in a conventional way. Notice that the tolerance between any pair of adjacent holes is the same, but the tolerance between the two end holes, due to tolerance buildup, is ±0.020 in. (0.52 mm). Figure 3-10*b* shows the same piece dimensioned using base-line dimensions. Here the tolerance between any pair of holes, including the two end holes, is the same ±0.010 in (0.26 mm) with no tolerance buildup.

When pieces are designed to fit together, tolerance buildup is of extreme importance. Figure 3-11 shows an example of two pieces, consisting of a tongue and groove, mating together. Calculations show that there will be clearance of 0.005 in. (0.125 mm) to 0.015 in. (0.04 mm) at the bottom of the tongue, and the overall length of the two pieces will vary from 1.995 in. (50.67 mm) to 2.020 in. (51.32 mm).

In Fig. 3-12, the same two pieces are redimensioned using the same basic dimensions used in Fig. 3-11. Note that in this case the clearance at the end of the tongue can vary from *minus* 0.010 in. (0.25 mm) to *plus* 0.020 in. (0.51 mm). The −0.010 in. (0.25 mm) indicates that under this condition there will be a

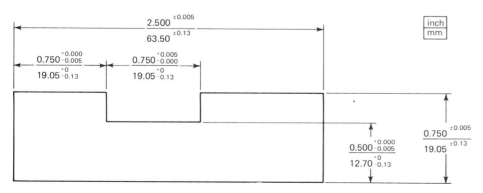

Fig. 3-8   Dimensioning of piece where slot width is important.

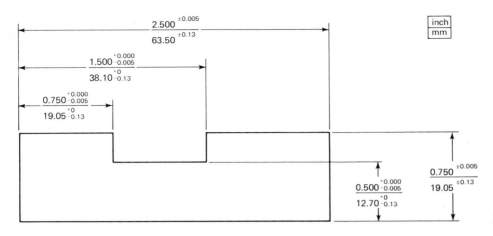

Fig. 3-9   Base-line dimensioning

0.010-in. (0.25-mm) maximum space between the shoulder of piece A and the top of piece B. In a condition such as this, there is no simple way to calculate the height of the two assembled pieces. For practice, the student should verify the overall height shown on the figure.

Geometric Dimensioning

So far the discussion on dimensioning has covered only size and locating dimensioning with no mention of the variation in the geometry of the part. Geometric dimensioning establishes the amount of variation in *form* which will be acceptable. If no geometric tolerances are specified, all parts meeting size and finish requirements would be acceptable regardless of the amount of variation in form. Among the important geometric features are: straightness, flatness, squareness, parallelism, and concentricity. Tolerances on these features are specified by standard notes and symbols included on the drawings.

Straightness, most used on round shafts and pins, specifies the amount the part may deviate from true straightness (Fig. 3-13).

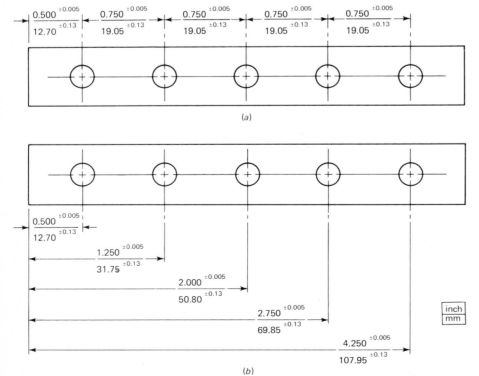

Fig. 3-10   Dimensioning for reduction in tolerance buildup. (*a*) Tolerance buildup; (*b*) reduced tolerance buildup.

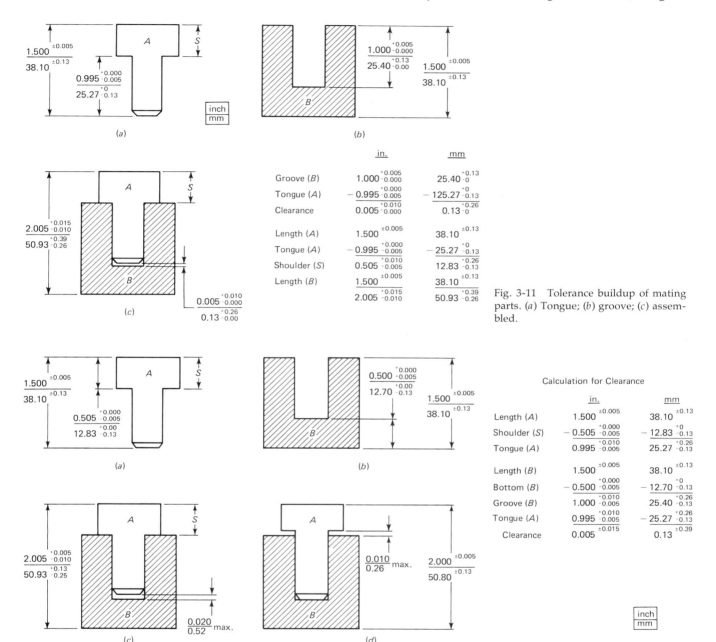

Fig. 3-11 Tolerance buildup of mating parts. (a) Tongue; (b) groove; (c) assembled.

Fig. 3-12 Excessive buildup of tolerance on mating parts. (a) Tongue; (b) groove; (c) bottom clearance; (d) top clearance.

Flatness specifies the amount of variation or waviness allowed on any flat surface (Fig. 3-14).

Squareness specifies the amount of variation from perpendicularity one surface may have with reference to another (Fig. 3-15).

Parallelism specifies the amount one surface may vary from another in regard to being parallel (Fig. 3-16).

Concentricity specifies the amount of eccentricity one diameter can have with another, expressed in total indicator readings. *Eccentricity,* by definition, is the amount the center of one circle is displaced from the center of a second circle (Fig. 3-17).

When measuring concentricity between two diameters, the part is located on one diameter and slowly rotated. The eccentricity is measured by using a dial indicator on the second diameter (Fig. 3-18). When the piece is rotated so that the eccentricity is maximum in the vertical direction, as shown in Fig. 3-18a, the dial indicator will read maximum. If the

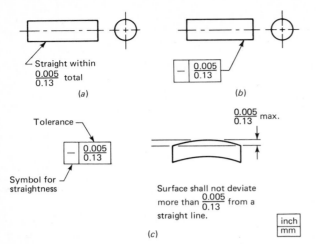

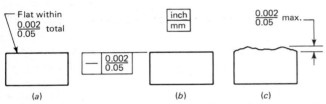

Fig. 3-13 Straightness specification regardless of feature size (RFS). (*a*) Note; (*b*) symbol; (*c*) meaning.

Fig. 3-14 Flatness specification, RFS. (*a*) Note; (*b*) symbol; (*c*) meaning.

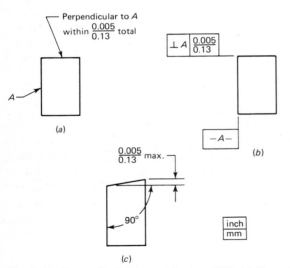

Fig. 3-15 Perpendicularity specification, RFS. (*a*) Note; (*b*) symbol; (*c*) meaning.

part is then slowly rotated 180°, the dial indicator will read a minimum as shown in Fig. 3-18*b*.

The amount of variation in reading between the highest and lowest reading will be twice the amount of eccentricity present. This variation is called *total*

indicator reading (TIR) or sometimes *full indicator reading* (FIR). When more than two diameters are involved, concentricity is often specified on all diameters. Fig. 3-19 is an example of a shaft involving three diameters, each held concentric to the other two.

True-Position Tolerancing

True-position tolerancing denotes the variation from the basic or theoretically exact position for any feature. (Feature means a hole, slot, boss, or surface.)

In the conventional method of expressing tolernaces, rectangular coordinates are used to locate the feature, such as a hole, and each coordinate carries its

Fig. 3-16 Specification for parallelism, RFS. (*a*) Note; (*b*) symbol; (*c*) meaning.

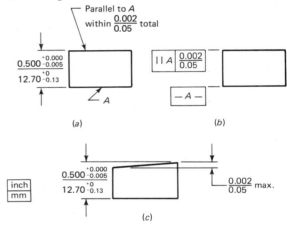

Fig. 3-17 Concentricity specification, RFS. (*a*) Note; (*b*) symbol; (*c*) meaning.

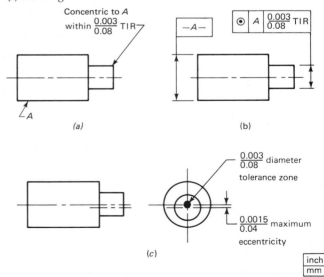

own tolerance. This produces a square or rectangular tolerance zone (Fig. 3-20). Since the centerlines of the feature may fall anywhere within the tolerance zone, maximum variation from the true position will (in a square zone) fall at a position 45° from the horizontal, or vertical, at one of the four corners of the tolerance zone. The actual variation is then 1.4 times the tolerance specified on each of the coordinates (±0.005 in. × 1.4 = ±0.007 in. max. variation) (±0.125 mm × 1.4 = ±0.175 mm).

In true-position tolerancing, the tolerance zone for cylindrical features, such as holes, is cylindrical. The true-position tolerance is the diameter of the toler-

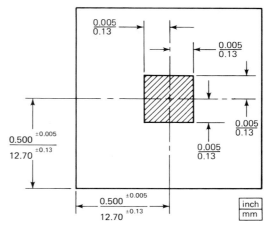

Fig. 3-20   Square tolerance zone resulting from tolerance applied to each coordinate dimension of a feature.

Fig. 3-18   Measuring concentricity or eccentricity by total indicator reading. TIR equals high reading minus low reading. (*a*) High indicator reading; (*b*) low indicator reading.

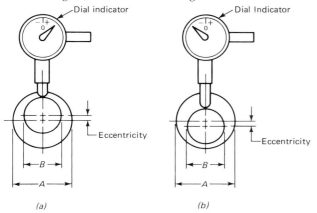

Fig. 3-19   Three diameters held concentric to one another. (*a*) Note; (*b*) symbol.

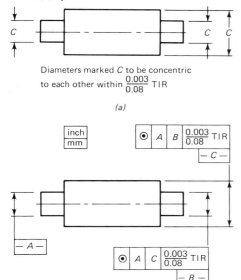

ance zone, with the center of the zone located at the true position according to the dimensions (Fig. 3-21). In this type of dimensioning, the maximum variation from true position allowed is the same regardless of the direction in which the variation occurs.

When true-position tolerancing is used, dimensions locating the true position of the feature are without tolerance. They are identified by a note, or the word BASIC (abbreviated BSC). Figure 3-22 shows examples of a part dimensioned using true-position tolerancing.

It is interesting to note that when no specification is given limiting the extent that a hole can deviate from being perpendicular, the tolerance zone specified in the true-positioning tolerance will define the limit of perpendicularity. The axis of the hole must fall completely within the cylindrical tolerance zone (Fig. 3-23).

maximum material condition (MMC)

Unless otherwise specified, true-position tolerances on location of features apply *only* at maximum material condition. *Maximum material condition* (MMC) is that condition where the feature of a part contains the maximum amount of material, holes are at minimum allowable size, and shafts are at maximum allowable size. The tolerance specified on the drawing applies *only* at MMC, and the tolerance is increased as the feature departs from the MMC by the amount of such departure.

In many cases the limits of location may be exceeded, as in the case of the center distance between two clearance holes, if the holes are not at maximum material limit of size. The increase in hole size from MMC will increase the amount of clearance with the mating part.

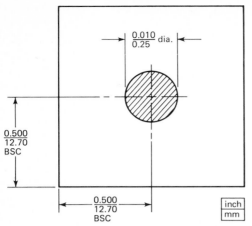

Fig. 3-21 Circular tolerance zone used with true-position tolerances.

In cases where the limit of location must be held regardless of feature size, a note "Regardless of Feature Size (RFS)" will be incorporated on the drawing.

All tolerances on the geometry of a part apply regardless of feature size unless otherwise specified. Figure 3-24 is an example of the specifications of a pin designed to fit in a hole. At MMC, a 1.000-in.-diameter (25.4-mm) pin with a straightness specification of 0.005 in. (0.13 mm) will fit into a hole of 1.005-in. (25.53-mm) diameter. If the pin is made smaller in diameter, the tolerance on straightness may be exceeded by the amount the pin deviates from MMC.

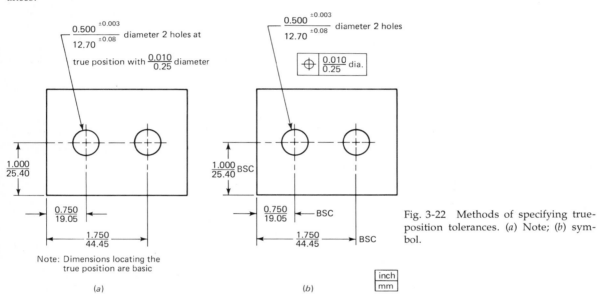

Fig. 3-22 Methods of specifying true-position tolerances. (a) Note; (b) symbol.

(a)

(b)

Fig. 3-23 Perpendicularity of a hole when perpendicularity is not specified.

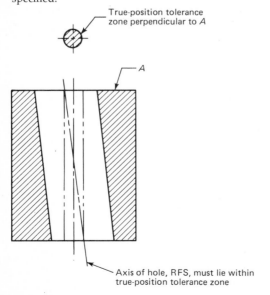

Fig. 3-24 Variation in tolerance when MMC is specified. (a) Note; (b) symbol; (c) meaning.

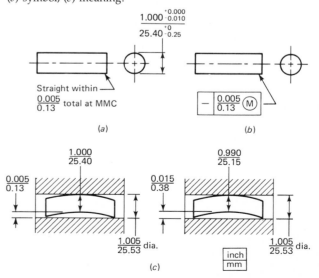

56

For additional information on dimensioning and tolerancing, refer to the bibliography at the end of this chapter.

Review Questions and Problems

3-1. Why is inspection of manufactured parts necessary, and what is the primary responsibility of the inspection department?

3-2. What is the difference between inspection, quality control, and statistical quality control?

3-3. If the variation of a dimension on parts manufactured using a particular process follows a normal distribution curve, what percentage of parts would have a dimension falling between plus and minus one standard deviation from the mean? Plus and minus two standard deviations? Plus and minus three standard deviations?

3-4. The diameter of a shaft was measured using five sample lots of five samples each taken at predetermined times during manufacture. The measurements obtained are listed below.

Sample				
1	2	3	4	5
0.624	0.625	0.625	0.626	0.625
0.625	0.623	0.626	0.626	0.625
0.625	0.625	0.623	0.625	0.626
0.624	0.625	0.625	0.625	0.622
0.626	0.627	0.624	0.625	0.623

    a. From the above data compute the value of $\overline{\overline{X}}$.

    b. Draw $\overline{X}$ and R charts. Show UCL and LCL. (UCL = 2.114 R for R chart.)

3-5. Using the $\overline{X}$ and R charts constructed for Problem 3-4 above, plot the average and range values for each of the following samples. Is the process in control?

Sample				
1	2	3	4	5
0.625	0.625	0.625	0.620	0.622
0.624	0.626	0.626	0.624	0.622
0.621	0.626	0.626	0.623	0.627
0.623	0.626	0.624	0.627	0.627
0.627	0.627	0.624	0.625	0.624

3-6. Why are average values ($\overline{X}$) used on a control chart instead of the individual (X) values?

3-7. Why is it necessary to use both a chart for averages and a chart for ranges?

3-8. How can the risks involved in using a particular sampling plan be determined?

3-9. Define tolerance, allowance, and clearance.

3-10. Define positional tolerancing and give its advantages.

3-11. What is meant by selective assembly and what are the advantages and disadvantages?

3-12. What is meant by overdimensioning and what is the disadvantage?

3-13. Describe geometric dimensioning.

3-14. List the important geometric features which are covered by geometric dimensioning.

References

1. French, T. E., and C. J. Vierck: *A Manual of Engineering Drawing for Students and Draftsmen,* 10th ed. McGraw-Hill, New York, 1966.

2. Freud, John E.: *Modern Elementary Statistics,* 3d ed., Prentice-Hall, Englewood Cliffs, N.J., 1967.

3. Grant, E. L., and R. S. Leavenworth: *Statistical Quality Control,* 4th ed., McGraw-Hill, New York, 1972.

4. *International Standards,* ISO 1000, International Organization, American National Standards Institute, New York, 1973.

5. Military Standard, MIL-STD-8, *Dimensioning and Tolerancing,* Department of Defense, Defense Supply Agency, Washington, D.C., 1963.

basic cutting information

MACHINABILITY

All the following information on forces, angles, tool materials, and machines is for one purpose: to help decide how to remove the maximum amount of metal at the lowest cost, consistent with the required tolerance and finish. The comparison of the possible rate of metal removal (in cubic inches per minute, in.³/min, also CIM, or cubic centimetres per minute, cm³/min) for different materials, keeping cost, tolerance, and finish consistent, is the most practical measure of machinability.

To compare, one must have a standard. The national standard is B1112 hot-rolled steel = 100 percent machinability. Thus, if a stainless steel is rated at 50 percent, and an aluminum alloy is rated at 500 percent, this means that only half as many cubic inches per minute of stainless steel and five times as many cubic inches per minute of aluminum can be removed with the same tool life, compared with B1112 HRS. This may be accomplished by decreasing the cutting speed of the stainless steel and cutting the aluminum at five times the speed which would be used for the B1112 steel.

The following list includes the most used measures of machinability.

1. Allowable cutting speed (for 60-min tool life)
2. Cubic inches per minute of stock removed, with comparable tool life
3. Cutting tool life, in minutes, between sharpenings
4. Tool forces, energy, or power required
5. Temperature of the cutting tool or chips

There are also charts showing a straight-line relationship between hardness (in BHN) and machinability. While this relationship is, in a general way, correct, it must be used with caution, as other factors such as strain hardening and microstructure can drastically change the machinability.

What Affects Machinability?

The answer to this is a list of 15 variables in the *Tool Engineer's Handbook.* These include:

1. Tool geometry and material
2. Cutting fluids
3. Rigidity of cutting tool, machine, and workpiece
4. Microstructure of the material

The entire study must finally result in decisions about the feeds, speeds, and depths of cut to be used when machining the materials. These three factors will accurately determine the *time* (which is basically *cost*) for machining, for removing a given amount of material.

The following points can act as *guides,* but use them with care. *Machinability tends to decrease as:*

1. Hardness increases
2. Tensile/yield strength increases
3. Strain-hardening tendencies increase
4. Carbon content of steels increases
5. Grain size decreases
6. Hard oxide, carbide, or silicate inclusions increase

THE PROCESS OF CUTTING

A metal tool does *not* slide through metal as a jacknife does through wood, nor does the tool "split" the metal or plastic material as an axe does a log. Actually, the metal is *forced* off the workpiece by being compressed, shearing off, and sliding along the face of the cutting tool. Figure 4-1 shows this action in a "slow motion" series of pictures using two tool bits, each with a different cutting angle.

Types of Chips

Three types of chips are referred to in the industry:

Type 1, a **segmented,** broken, or discontinuous chip, which always results from cutting brittle materials such as cast iron. Figure 4-1*b* shows how each chip breaks by shearing off as the cutter moves forward. This type of chip is easy to dispose of and falls freely from the cut, so cutters such as drills and end mills do not get clogged up easily.

A **Type 2, continuous** chip is formed by the majority of metals since they are more ductile than cast iron and, thus, do not break off but are merely displaced. Typical steel chips are shown in Fig. 4-2.

These chips can, as will be shown later, be directed away from the work and the operator. However, they can also cause a tangled mess. These snarled chips are sharp and hot. They are hard to get off the tool post and the work, and do not pack readily in containers for future disposal. Sometimes this continuous chip is tightly coiled and not too messy, as shown in Fig. 4-2*d*. It is still hard to pack, so that usually an effort is made to avoid this type of chip. To avoid

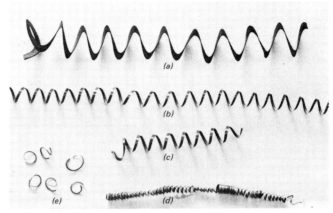

Fig. 4-2 Types of chips, lathe cuts. (*a*) A heavy continuous chip. (*b*), (*c*) Lighter continuous chips. (*d*) Tightly curled continuous chips. (*e*) Short, broken "figure 9" chips. (*Photograph by Samuel Lapidge.*)

these snarled and curled chips, several steps may be taken as described in Chap. 8.

The ideal chip is a short, broken one in the shape of a figure 9 (or 6) (see Fig. 4-2*e*). This falls quickly into the chip pan and is easily packed for disposal.

A **Type 3** chip is a continuous chip which, due to the pressure and heat against the tool, causes a small particle of the material to temporarily stick to the end of the tool. This is called a **builtup edge** *(BUE)*. An exaggerated view of this is shown in Fig. 4-3. Actually the BUE is often only about 0.020 in. long and 0.010 in. thick (0.5 × 0.25 mm) or less. It can be felt more easily than it can be seen.

This BUE increases in size in front of the tool bit, and when it becomes fairly large, a piece of the BUE breaks off and goes off with the chip as in Fig. 4-4*c*. Notice also in Fig. 4-3 that it is the BUE which is actually doing the cutting over at least part of the tool-bit surface. This roughness on the cutting edge can spoil the finish of the work, decrease tool life, and use additional power. The use of the proper cutting fluid and higher speeds can frequently eliminate the BUE, and it should be avoided if possible.

The photomicrographs in Fig. 4-4 show how the metal in chips has actually flowed due to the forces exerted in cutting. Figure 4-4*a* is AISI 8640 steel cut at high speed, making a continuous chip. The slope of the grains shows the line on which the metal was shearing during the cut. Figure 4-4*b* shows the type of continuous chip which results when very heavy lathe cuts are taken. The smooth side is against the tool bit and the inside is corrugated. Figure 4-4*c* shows a piece of a BUE being carried off with the chip. This is about 75 times larger than actual size.

Fig. 4-1 Cutting action of two styles of cutting tools. Test (*a*) High rake angle; (*b*) low rake angle. (*Cincinnati Milacron.*)

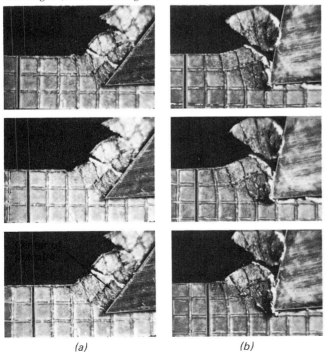

(a) (b)

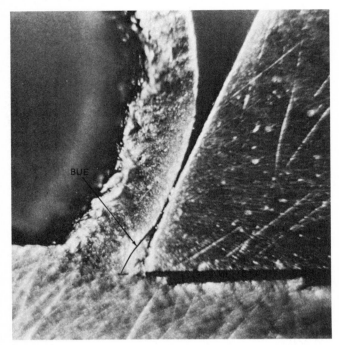

Fig. 4-3   A builtup-edge (BUE) chip. (*Cincinnati Milacron.*)

Figure 4-4*d* shows the deep fractures which occur when cutting brittle materials. These chips break up into segmented, discontinuous small pieces.

The Cutting Theory

The theory of what actually happens when metals are cut has been, and is still, the subject of many investigations in colleges and laboratories throughout the world. The working engineer is not usually interested in the details, though it is a good field to enter if one is interested. The following will give some basis for understanding the causes of wear on cutting tools.

Two types of cuts, orthogonal and oblique, are made by cutting tools. *Orthogonal cutting* (Fig. 4-5*a*) is the simplest, as the cutting edge goes in a straight line through the material. Sawing, broaching, a slotting cutter, and a lathe cutoff tool are examples of this.

Oblique cutting (Fig. 4-5*b*) is typical of many lathe tools, most milling cutters, and drills. The force diagram for these tools is quite complex to draw though basically the same as for the orthogonal. Thus, the following description is for orthogonal (straight-line) cutting.

Forces on the Chip

Figure 4-6 shows one way of representing the forces acting on a chip. The tool presses on the chip with force *R*, which can be divided into a friction force *F* and a normal force *N*. At the shear plane, the work exerts an equal and opposite force *R'*, which can be broken up into a force *S* along the shear plane and a force *N'* normal to this plane.

The *shear angle* varies with different materials and with the tool geometry. As the shear angle increases, the shear plane becomes shorter and, therefore, the total shearing force becomes relatively less. This would indicate preference for high-positive-rake-angle (cutting angle) cutters. However, this is not always preferable, as will be mentioned later.

Fig. 4-4   Photomicrographs of several types of chips. (*The Monarch Machine Tool Co.*)

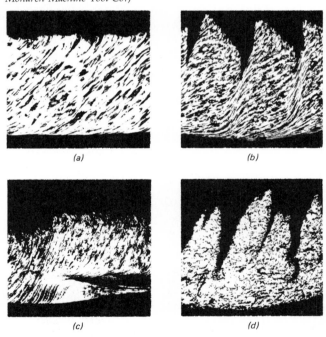

(a)

(b)

(c)

(d)

Fig. 4-5   Sketches showing (*a*) orthogonal and (*b*) oblique cutting.

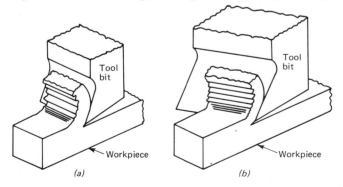

(a)

(b)

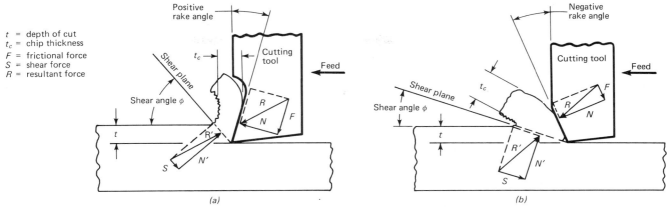

t = depth of cut
t_c = chip thickness
F = frictional force
S = shear force
R = resultant force

Fig. 4-6 Force diagram on the chip, orthogonal cutting. (*a*) With positive cutting (rake) angle. (*b*) With negative cutting angle.

Forces on the Cutting Tool

The three forces on a lathe cutting tool, as shown in Fig. 4-7, can be measured by a dynamometer. These forces can be quite large. For example, in one report, cutting a high nickel alloy

$$F_t = 655 \text{ lb } (2.91 \text{ kN})$$
$$F_L = 345 \text{ lb } (1.54 \text{ kN})$$
$$F_R = 180 \text{ lb } (0.80 \text{ kN})$$

which makes a resultant force of 710 lb (3.16 kN).

Or, in another actual cutting situation: a lathe was using about 20 of its available 30 hp, turning a 6-in.-diameter shaft at 230 fpm.

Then

$$F = \frac{20 \text{ hp} \times 33,000 \text{ ft} - \text{lb/min/hp}}{230 \text{ fpm}} = 2870 \text{ lb}$$

$$(12.8 \text{ kN})$$

force on the tool bit. The chips on this job came off the work a dark purple color and would sometimes fly 4 ft up and 4 ft away. Not a very safe condition, even though it was a short-run job.

The largest of the three forces is always the tangential (downward when turning). The axial (longitudinal) force, which is due to the feed motion, is about 35 to 55 percent as large as the tangential. The radial force, which tends to push the tool back out of the work, is about 25 to 30 percent as large as the tangential.

The radial force, when too large, can cause "chatter," which gives a poor finish. All these forces are affected by the feed, speed, and depth of cut (which will be studied next) and by the cutting-tool angles discussed in Chap. 8.

Cutting Speed

Cutting speed is the rate at which the chip goes past the cutting edge of the tool. The *workpiece* may be rotating, as when turning on a lathe, or the *cutter* may be rotating, as when drilling a hole with a drill press.

Cutting speed is the most important factor in tool life. Excessive speed can "burn" a cutter. This means heating it to a temperature at which it becomes softer than the metal which it is cutting. The result is sometimes startling; for instance, in 3 or 4 s (seconds) the end of a drill may "flow" to one-half its original diameter and become a deep purple color. Too slow a speed

Fig. 4-7 The three forces on a lathe cutting tool bit when doing a turning operation.

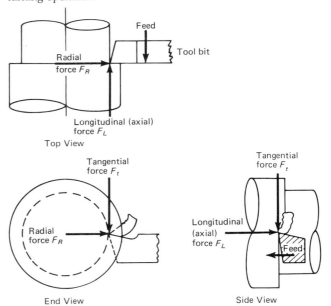

is, of course, an expensive way to do business. It also tends to encourage the BUE, which is undesirable.

When starting to specify the cutting conditions for a machining operation, the engineer looks up in a handbook, or in the company's specifications, the cutting speed recommended for the material being used. This speed has been determined by experiment (empirically). No one has yet found a way to calculate the speed from the properties of the material.

Using the basic figure from the handbook, the engineer may increase or decrease it according to the conditions of the cutter, workpiece, and machine. There is no "magic" in any table of cutting speeds. In fact, it is rather confusing when speeds specified by different reliable sources are compared. Suggested cutting speeds are given in the following chapters.

Calculation of cutting speed is quite easy, though charts and tables are frequently used. Cutting speed is specified in *feet per minute* (fpm) or *metres per minute* (m/min). For quick figuring m/min = fpm/3.

If the diameter D, *in inches*, and the speed in revolutions per minute (rpm) are known, then the cutting speed can easily be computed. To calculate cutting speed, refer to Fig. 4-8. Always use the *largest* diameter, as that is the fastest.

Circumference (*in inches*) = πD = distance traveled in one revolution

Circumference (in feet) = $\pi D/12$ ft/rev
Distance (in feet) traveled in 1 min

$$= \frac{\pi D}{12} \frac{ft}{rev} \times \frac{rev}{min} = fpm$$

This computes to

$$fpm = \frac{rpm \ (D)}{3.82}$$

If we assume that $\pi/12 = 1/4$, then

$$fpm = \frac{rpm \ (D)}{4} \qquad (4\text{-}1)$$

This approximation is perfectly adequate, as the exactly correct fpm certainly is not known within the 5 percent error which is introduced.

If, as is more often the case in actual practice, the *diameter* and the *cutting speed* are known, then the above equation becomes

$$rpm = \frac{4(fpm)}{D} \qquad (4\text{-}2)$$

The resulting rpm is 5 percent *over* the theoretically correct value, in this case, which is not important because frequently the machine being used has only a limited number of speeds available. Usually the

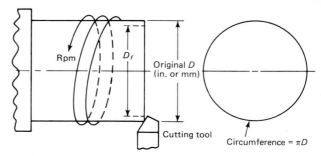

D_f = final diameter, after cutting. It is *not* used to compute cutting speed

Fig. 4-8   The terminology used in computing rpm and cutting speeds.

closest speed *lower* than the computed number is used.

to compute cutting speed in metric units
Diameter Dm = diameter-metric and is *always* in millimetres (mm).

$$Circumference = \pi Dm = \frac{\pi Dm}{1000} \ m/rev$$

And cutting speed in metres per minute is

$$m/min = \frac{\pi Dm}{1000} \times \frac{rev}{min} = \frac{(rpm)Dm}{318}$$

If we assume that $\pi/1000 = 1/300$, the result will be 6 percent *less* than the theoretical value. This is not usually significant in this type of application.

Thus, cutting speed in metric units is

$$m/min = \frac{(rpm)Dm}{300} \qquad (4\text{-}3)$$

Or, if cutting speed in metres per minute, and diameter in millimetres are known (as is usually the case),

$$rpm = \frac{(300) \ m/min}{Dm} \qquad (4\text{-}4)$$

When first using these formulas, there is sometimes a question as to which diameter should be used. Figure 4-9 shows a few examples of the D to use.

Two examples will show why it is necessary to use these formulas when specifying speeds for use in the shop.

EXAMPLE   4-1
Drilling aluminum at 300 fpm, using a ⅛-in. HSS (high-speed steel) drill and using Eq. (4-2),

$$rpm = \frac{4(300)}{0.125} = 9600$$

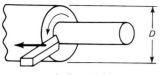

Lathe—turning

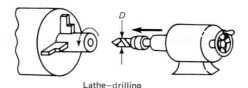

Lathe—drilling

Boring

Turning

Vertical
boring machine
(two tools cutting
at the same time)

Workpiece

D'

Machine table

D   (use)

Use *D*. *D'* will,
of necessity, run
at slower cutting
speed

Fig. 4-9   Examples defining the diameter *D* to use in cutting speed formulas with different tools.

As no standard machine in the shop will go this fast, use the fastest available rpm, which will probably be 4000 rpm or less.

The metric equivalent will be a cutting speed of 100 m/min, drill 3.2-mm diameter. Using Eq. (4-4),

$$rpm = \frac{300(100)}{3.2} = 9400 \ rpm$$

(Use maximum speed available.)

EXAMPLE   4-2

Turning the O.D. of a 5-ft stainless steel casting at 45 fpm and using Eq. (4-2),

$$rpm = \frac{4(45)}{(5)(12)} = 3 \ rpm$$

Note the change from feet to inches. Large machines for this type of work often have speeds below 1 rpm.

The metric equivalent will be a cutting speed of 15 m/min and a diameter of 1500 mm. Using Eq. (4-4) gives

$$rpm = \frac{(300)(15)}{1500} = 3 \ rpm$$

Figure 4-10 shows a graphical solution of the above problems which is often close enough, and many tables are published for both English and metric systems.

Feedrate

This is the rate at which the cutter advances along the work (as in turning in a lathe) or through the work (as in drilling). Feedrate is basically expressed in *inches per revolution* (in./rev, or sometimes ipr) or in millimetres per revolution (mm/rev). Figure 4-11 illustrates the meaning of feedrate.

Once again, as in cutting speed, the figure used for feedrate is selected from tables or on the basis of past experience. The amount of feedrate can be from 0.001 in. (0.025 mm) or less to 0.050 in. (1.25 mm) or more per revolution.

Inches per minute (ipm) feedrate (or millimetres per minute, mm/min) is a specification which is frequently used today, especially since numerical control machines often use this figure for setting feeds, either manually or on the tape. Some conventional milling machines also use this feedrate. The maximum feedrate usually available is 30 to 40 ipm (760 to 1000 mm/min).

Computation for lathe cuts, drills, reamers, etc., is simple. Feedrate in inches per minute is

$$ipm = \frac{inches}{rev} \times \frac{rev}{min} = (ipr)(rpm)$$

Feedrate for multiple-blade milling cutters is slightly different and will be discussed in Chap. 10.

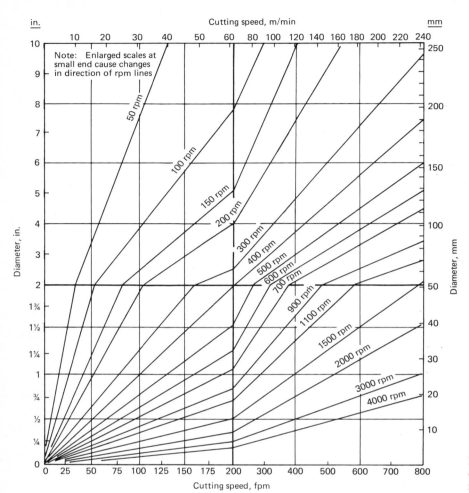

Fig. 4-10   Graphical solution of the required rpm, using either English or metric information.

EXAMPLE 4-3

Some feedrate calculations are: A lathe, turning at 350 rpm, with a feed of 0.008 ipr,

$$ipm = 350 \times 0.008 = 2.8 \text{ ipm}$$

Or, in metric, using 350 rpm and feedrate 0.20 mm/rev,

$$mm/min = 350 \times 0.20 = 70 \text{ mm/min (2.76 ipm)}$$

Depth of Cut

This term refers to the distance, in inches or millimetres, that the point of the cutter is below the surface of the work while it is cutting, as shown in Fig. 4-12.

Notice that in Fig. 4-12a the material removed from the diameter is twice the depth of cut. Some lathes have dials which read in "reduction of diameter" instead of depth of cut, which makes it "failsafe" in use.

For axial cutting tools such as drills and reamers, depth of cut as defined above does not apply. However, *depth* is used to describe the *total depth* which the drill, etc., has penetrated into or through the work.

Depth of cut in lathe turning and milling cuts may vary from 0.001 in (0.025 mm) to 0.500 in. (12.7 mm) and over in large machines. This will be discussed more fully in the chapters dealing with each of those machines.

CUTTING TOOL MATERIALS

There are six materials from which cutting tools of all kinds are made, but two, high-speed steel and carbides, are used for about 90 percent of all cutting.

Any material which is used for cutting tools should have certain characteristics:

1. Hardness, to penetrate the workpiece material
2. Toughness, to withstand shock
3. Resistance to softening at the high temperatures often involved when cutting metals
4. Relative ease of sharpening by grinding with standard grinding wheels

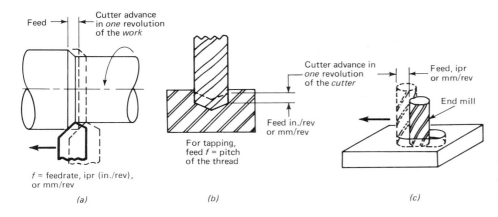

Fig. 4-11 Illustrations of the meaning of feedrate for different cutting tools. (a) Turning—lathe; (b) drilling; (c) milling a slot.

Tool Steel

This was the original cutting material. Sometimes referred to as "high-carbon steel," it is often a W-1, W-2, or 0-2 tool steel with 0.08 to 1.4 percent carbon and sometimes with small amounts of vanadium, manganese, and chromium added.

While these steels can be hardened to HRC 55 to 65, they lose their hardness rapidly at temperatures above 400°F (204°C). Thus, they are good only for woodworking or for low-speed short-run jobs on mild steel or nonferrous alloys. They are fairly easy to shape, so they can be used to make form tools for aluminum and wood. Plain high-carbon tool steel is almost never used for production tools in a metalworking shop.

High-Speed Steel

High-speed steel (HSS) is made from one of two special classes of tool steel, the tungsten-alloyed (T series) and the molybdenum-alloyed (M series). Table 4-1 shows the composition and characteristics of some of the more widely used HSS alloys. Today the M1 and the M2 HSS are the most widely used. These were developed as alloys to replace T1, which uses a high percentage of tungsten (W), a metal which was very scarce in this country during World War II. The added molybdenum (Mo) gives it properties which

equal the T1, and the resulting tools are less expensive.

Manufacturers sometimes specify that their cutting tools are "cobalt steel." As can be seen from Table 4-1, a number of tool steels contain cobalt. The M30 and M40 series are often used because they do have better tool life, yet are not too difficult to sharpen.

In general, the higher-numbered tool steels are harder and/or tougher, more difficult to grind for sharpening, and more expensive. Thus, drills and lathe tools made from alloys such as T15 and M42 are used principally for cutting the "problem" metals which have low machinability.

special high-speed steels

Especially since 1968, the metallurgy of steels for cutting tools has shown remarkable progress. Several special, higher priced, high-speed steels have been introduced.

CPM Rex HSS by Crucible Specialty Metals, a Division of Colt Industries, is a hot isostatic pressed HSS made by powder metallurgy (Chap. 22). This is available in several grades: M2, M7, M3, T15, and others. These have very high impact strength and good grindability. This material has given two to four times the tool life of other HSS materials when used for milling cutters and lathe tools on tough steels, titanium, and superalloys.

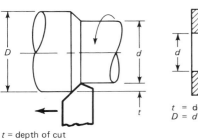

t = depth of cut
d = D - 2t

(a)

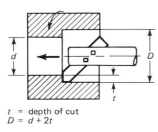

t = depth of cut
D = d + 2t

(b)

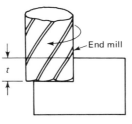

t = depth of cut

(c)

Fig. 4-12 Illustrations of the meaning of depth of cut. (a) Lathe—turning; (b) lathe—boring; (c) end milling—step.

TABLE 4-1   COMMONLY USED HIGH-SPEED STEEL ALLOYS

AISI Type	Approximate Alloying Composition, Percent							Grindability Index G*
	C	W	Mo	Co	Cr	V	Cost	
Tungsten High-Speed Steels								
T1	0.75	18.0	---	---	4.00	1.00	Med.	6.5
T2	0.80	18.0	---	---	4.00	2.00	Med.	4.5
T4	0.75	18.0	---	5.00	4.00	1.00	Med.	
T5	0.80	18.0	---	8.00	4.00	2.00	Med.	
T6	0.80	20.0	---	12.00	4.50	1.50	High	3.0
T8	0.75	14.0	---	5.00	4.00	2.00	Med.	
T15	1.50	12.0	---	5.00	4.00	5.00	High	0.8
Molybdenum High-Speed Steels								
M1	0.80	1.50	8.00	---	4.00	1.00	Low	7.0
M2	0.85	6.00	5.00	---	4.00	2.00	Low	4.5
M3 Class 1	1.05	6.00	5.00	---	4.00	2.40	Med.	1.0
M3 Class 2	1.20	6.00	5.00	---	4.00	3.00	Med.	0.9
M4	1.30	5.50	4.50	---	4.00	4.00	Med.	
M7	1.00	1.75	8.75	---	4.00	2.00	Low	3.5
M10	0.85	---	8.00	---	4.00	2.00	Low	3.3
M30	0.80	2.00	8.00	5.00	4.00	1.25	Med.	
M33	0.90	1.50	9.50	8.00	4.00	1.15	High	
M42	1.10	1.50	9.50	8.00	3.75	1.15	High	4.0

*Grindability index = $G = \dfrac{\text{in.}^3 \text{ metal removed}}{\text{in.}^3 \text{ of grinding wheel wear}}$ $\begin{cases} 12\text{--}100 = \text{easy} \\ 3\text{--}12 = \text{average} \\ 0\text{--}3 = \text{difficult} \end{cases}$

Some of the new grinding wheel materials are raising this G ratio considerably.

UCON, from Union Carbide Company (in 1972), is an alloy of columbium 50 percent, titanium 30 percent, tungsten 20 percent, with a surface hardening treatment giving HRA 92. It has no carbides, yet made as replaceable inserts, it easily competes with C5-C7 carbides. This material requires less horsepower. It is especially useful when cutting up to 1500 fpm (450 m/min) on low-alloy steels, with deep cuts and feeds up to 0.020 in./rev (0.50 mm/rev). It is not recommended for cast iron, stainless steel, titanium, or the superalloys.

The M40 series (M41 through M47) is a new high-carbon–cobalt HSS which can cut high-strength steels, titanium, and high-temperature alloys but has a better grindability index than T15 (see Table 4-1).

Cast Alloys

Cast alloys are used mostly for lathe tools and specials. There are four brands available. The approxi-

TABLE 4-2   APPROXIMATE COMPOSITION OF SOME CAST-ALLOY CUTTING TOOL MATERIALS

Alloy	Co, %	Cr, %	W, %	C, %	Hardness, R_c
Blackalloy*	42–44	24	20–22	2.0	63–67
Crobalt†	40–48	30–33	14–20	2.0–3.0	59–64
Stellite‡	38–53	30–32.5	10.5–17.5	1.8–2.5	55–63
Tantung§	45–47	28–30	15–18	3.0	60–64.5

*Blackalloy Co. of America.
†Crobalt, Inc.
‡Stellite Div. Cabot Corp.
§VR/Wesson Co.

mate compositions are given in Table 4-2. Notice that this is *not a steel*. It is principally cobalt (Co), chromium (Cr), and tungsten (W).

These cutting tools are melted in electric furnaces and cast to shape in "chill molds." This gives the material good hardness (about R_c 63) and over 300,000-psi (2070-MPa) transverse rupture strength. It will stand temperatures up to 1500°F (816°C). It is suggested for use when cutting speeds are too high for HSS but too slow for efficient use of carbides.

Cast alloys can be cast to form for specially shaped tools and can be relatively easily ground with aluminum oxide wheels (diamond wheels are not needed). They are made as solid tool bits, cutoff blades, and standard throwaway inserts.

Carbides

The term *carbides* includes a growing list of cutting materials which are basically tungsten or titanium carbides or combinations of these with a binder of cobalt or nickel.

These are very hard materials—85 to 95 R_A—with good transverse rupture strength—150,000 to 350,000 psi (1030 to 2410 MPa). They can be run at temperatures up to 1800°F (982°C). The modulus of elasticity is almost three times that of steel. Thus, solid carbide tools (such as small drills and small boring bars) are much "stiffer" than HSS tools; that is, they will bend or deflect only about one-third as much as HSS tools.

Carbide tools are made by powder metallurgy (see Chap. 18). The powder is made by sintering tungsten, titanium, etc., with the required percent of carbon in large furnaces. This makes a very hard large carbide ball. This is crushed and sifted, and then combined with cobalt and other powder and pressed into shape and sintered. The resulting "blanks" or "inserts" may be used as is or ground to size with diamond wheels.

TABLE 4-3 GRADES OF CARBIDES USED IN CUTTING TOOLS

Grade	Approximate Composition, %				Hardness, HRA	Use
	Wc Average	Co	Ti	Ta		
Used for Machining Cast Iron, Nonferrous Metals, Titanium, 300-Series Stainless Steels, etc.						
C-1	90	6–12	---	0–3	89	Roughing cuts
C-2	93	5–6	---	0–2	92	General use
C-3	93	4–6	---	0–2	92.5	Finishing cuts
C-4	97	3–4	---	---	93.5	Precision cuts
Used for Machining Steels (TiC added)						
C-5	73	8–10	6–8	10–12	91	Roughing cuts
C-6	74	5–9	7–12	10–12	92	General use
C-7	76	4–6	10–14	5–10	93	Finishing cuts
C-8	75	4–6	10–20	4–8	94	Precision cuts
C-9, C-10, C-11	Special for shock cuts and durability					
C-12, C-13, C-14	Special for impact and interrupted cuts					
C-15–C-19	Special purpose (Rockbits, hot flash removal, etc.)					

NOTE: As the percent of cobalt decreases, toughness decreases and allowable surface speed increases.

grades of carbides

There are eight principal grades and several special grades of carbide, all shown in Table 4-3. The very hard grades are also very brittle due to the lower percent of cobalt. Thus, they cannot be used for interrupted cuts, but they are especially good where long tool life with light cuts is desired.

Many companies make carbide tools and each one uses a different numbering system for the different grades. Moreover, each manufacturer uses slightly different alloys and processing, so these comparisons are *not* accurate. Each alloy must be tried on the type of work in a specific plant.

The grade classification will, however, guide the buyer to a limited few alloys for trial. Certainly one would not try a class 1 alloy for roughing cuts in steel.

coated carbide inserts

In 1970 a new process was introduced. It consists of metallurgically bonding 0.0002 to 0.0003 in. (0.0051 to 0.0076 mm) or more of titanium carbide (TiC) onto a tough grade of tungsten carbide (WC) which, of course, contains the cobalt binder.

This thin coating, under proper conditions, has a remarkable effect on tool life. Friction forces are reduced up to 25 percent at the tool-chip contact zone, which reduces the cutting temperatures about 150°F (65.6°C) and may give up to four times the life of regular carbides under proper conditions.

Applications of these coated tools are mostly in finish or mild roughing cuts on metals such as steels which tend to alloy with, or weld to, the regular carbide grades. They are not at present recommended for use on titanium or the nickel-based alloys.

Limitations of the TiC-coated tools are:

1. They cannot be reground, as this would remove the coating.
2. They should not be used where sand inclusions or heavy scale would quickly wear off the coating.

A coating of titanium nitride (TiN) is also being used with very similar results.

titanium carbides

Titanium carbide combined with tungsten carbide is a standard cutting tool alloy. However, TiC is now being combined with nickel and molybdenum binders to create new, much lighter weight cutting tools with remarkable properties, though they are more expensive.

This alloy, introduced about 1969, is as hard as, and nearly as strong as, tungsten carbide. These cutting tools, used principally as inserts, can be run at cutting speeds of from 100 to over 1000 fpm (305 m/min), with depth of cut to 0.060 in. (1.5 mm), and heavy feeds. Finishes to 10 μin. (0.25 μm) can be attained.

The principal reasons for these results are:

1. Ability to stand chip erosion at high speeds
2. Resistance to chip-to-tool welding, with decreased friction
3. Higher thermal conductivity than tungsten carbide and higher hot hardness.

These tools are comparable to ceramics in their resistance to edge buildup and cratering; thus they hold their size and give excellent surface finishes. TiC may be used on steel, cast iron, most stainless steels, and some high-temperature materials.

other carbides

RAMET L, from VR/Wesson Company, a submicron grain size, dispersion-strengthened carbide, was introduced in 1969. It is as hard as regular tungsten carbide but much stronger. Thus, it enables a carbide tool to make heavy, low-speed cuts (even interrupted cuts) in cast iron, titanium, and high-temperature alloys. It is *not* recommended for steels or high cutting speeds.

Carboloy 545, available in 1973 from General Electric Co., is an aluminum oxide–coated carbide tool. It can cut at very high speeds, like ceramic tools, but is much less likely to break.

Cast carbide, introduced by Teledyne Firth Sterling,

has no cobalt binder. It is a mixture of titanium, tungsten, and carbides. This material seems especially adapted to cuts of 0.100 to 1.5 in. (2.5 to 38.0 mm) deep and feeds to 0.010 to 0.180 in./rev (0.25 to 4.6 mm/rev) on steel and stainless-steel castings and forgings, even with heavy scale. It is *not* recommended for cast iron, titanium, or the high-nickel alloys, or for light cuts.

Other materials will doubtless be developed, and other companies will make similar products in the future.

> **CAUTION:** The capabilities of some of the new cutting tools actually exceed the capacity of many lathes being made today. With capabilities of 1000 fpm (305 m/min), a 2-in.-diameter (50-mm) bar would be run at 2000 rpm. How long will the lathe bearings last at this speed and with heavy chip loads?
>
> One company found that "the chips come off like bullets" and that with "spindles turning as fast as 3000 rpm, the chucks would not stay closed." Thus, these speeds and feeds require more horsepower (kilowatts), better bearings, and heavy, rigid machines.

Ceramic Tools

Ceramic cutting tools are basically aluminum oxide (Al_2O_3) with small additions of other oxides such as magnesium oxide (MgO), titanium oxide (TiO), and nickel oxide (NiO). Because of their chemical composition, they are sometimes referred to as *sintered oxides*. The melting point is about 3725°F (2050°C), and the hardness on the Mohs' scale is 9 compared with diamonds, 10.

These tools, usually triangular or square inserts, are made by powder metallurgy (see Chap. 22). They are compacted to shape, sintered, and then diamond ground to size.

The very high hardness, low friction, and ability to cut well even at 2000°F (1100°C) make these ceramic tools sound very attractive. However, their extreme brittleness and low tensile strength make them very easy to chip. Thus, they should be used in toolholders giving maximum support and in machines which are rigid, free of vibration, and of sufficient power to handle high material-removal rates.

Ceramic inserts are best used with negative rake angles for relatively light cuts (for good finish) at high speeds on cast iron and hardened steels. They are used on the softer steels only if very high cutting speeds can be used and if good surface finish is required. They do not work well on aluminum,

titanium, the 300-series stainless steels, and nickel alloys.

Diamonds

A natural or synthetic diamond is the hardest known substance. However, it is also a low-strength, very brittle substance, and expensive to reduce to a particular shape and size. A typical diamond-pointed turning tool is shown in Fig. 4-13.

Thus, diamonds are used only for special operations such as shaping carbide dies, turning glass rods, and some boring operations where very fine feeds and high speeds can be used, such as jewelry work.

To use diamonds for cutting,

1. Cuts must be light and at very high speeds.
2. Toolholder and machine must be rigid.
3. Small tool angles, especially clearances, must be used.
4. Diamonds should never be used on interrupted cuts.

Of course, diamonds are used for dressing and truing grinding wheels, and, when crushed, they are used in grinding wheels for cutting carbides (see Chaps. 14 and 15).

diamond lathe and milling tools

Man-Made artificial diamonds are being made by General Electric in industrial quantities. In 1973 G.E. introduced the Compax diamond tool blank. This uses a formed diamond point on a carbide base for machining nonferrous materials such as brass and aluminum, and abrasive, nonmetallic materials such as glass-filled plastic.

The Man-Made diamonds are more resistant to chipping, sharper, and more consistent than natural diamonds.

Megadiamond

In 1968 a new concept in diamond tools was introduced. The name could come from the fact that "mega" is one million and that is the number of pounds per square inch it takes to make these diamond cutting tools. These Megadiamond cutters are made in round, square, and triangular shapes, much the same as carbide inserts. See Fig. 4-14 for some of the cutting tools.

Since these cutters are small, the company makes holders or simple adapters so that these diamond cutters can be used in standard carbide throwaway toolholders. The cutters can be indexed to use several cutting edges.

The advantages of these shaped Megadiamonds

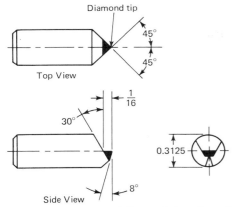

Fig. 4-13 One type of diamond-tipped turning tool.

Fig. 4-14 Some typical Megadiamond tools for lathe and milling work. (*Megadiamond Corporation.*)

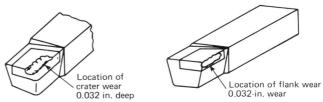

Fig. 4-15 Illustration of flank and crater wear on a carbide tool bit. It appears the same on HSS.

are that they are practically as hard as natural diamonds but have $1\frac{1}{2}$ times the transverse rupture strength and two to three times the realizable compressive strength, as well as being in an easily usable shape.

Like natural diamonds, Megadiamond cutters are not usually used for cutting steel. However, for difficult to machine nonferrous materials (such as the high-silicon aluminums and fiber glass–reinforced plastics) and for the very hard carbons used in jet-engine rotary seals, the Megadiamond cutter may last many times longer than carbides.

TOOL LIFE

How many minutes or hours should a tool be used before it is removed from the machine to be sharpened or replaced? The answer may be from 5 min to 8 h (hours) or more.

One report states that on an interrupted cut on a nickel alloy, one carbide tool lasted from 5 to 20 min. A change in carbide alloy gave a tool life of 2 h, which was considered excellent. Or, machining aluminum with the proper grade carbide, a cutter may last a full 8-h shift or more. Frequently in comparing materials, feeds, speeds, etc., a tool life of 30 to 60 min is the basic time used.

How does one know when a tool has failed? When has its "life" ended? In a small shop, a cutter is usually removed or sharpened because the finish of the work has become too rough, the chips are blue, chatter is starting, the operator noticed that the cutter looked dull, or the size was not holding. If it is a carbide tool, maybe it chipped and would not cut at all.

Technically, in controlled test conditions, the "life" is how long it takes before a certain amount of wear occurs on the flank or cratering occurs on the top face of the tool. A $\frac{1}{32}$-in. or 0.032-in. (0.80-mm) wear limit

is frequently used, as shown in Fig. 4-15. To save time and material during testing, smaller wear areas are sometimes used.

Effect of Temperature

High temperature in cutting is a *result* of other factors, but it is also a limiting factor in tool life. In general, the following maximum safe operating temperatures apply:

Plain high-carbon steel	400°F (200°C)
High-speed steel	1000°F (540°C)
Cast cobalt alloys	1400°F (760°C)
Carbides	1600°F (870°C)
Ceramics	2000°F (1100°C)
Diamonds	2000°F (1100°C)

Most of the heat generated in cutting is carried away by the chips—maybe up to 75 percent. The rest is absorbed by the cutting tool and the workpiece or partially carried away by cutting fluid if it is used. Anything that affects the temperature will affect tool life. In general, higher temperatures cause shorter tool life.

Effect of Cutting Speed

This is the factor which, for any given cutting material, will make the greatest difference in tool life.

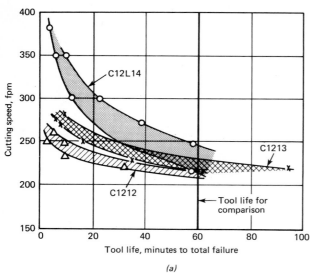
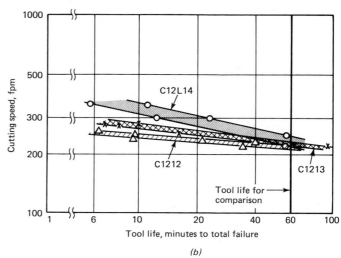

Fig. 4-16   Tool life at various cutting speeds, hot-rolled steels. (*a*) Drawn with cartesian coordinates. (*b*) Drawn with log-log coordinates. (*Bethlehem Steel Corp.*)

Figure 4-16*a* and *b* are two ways of showing how cutting speed affects tool life. Figure 4-16*a* is graphed using standard cartesian coordinates, and it indicates that the effect of cutting speed varies considerably with the alloy being cut. The leaded C12L14 steel, while it can be cut at over 350 fpm (107 m/min), must be cut at about 230 fpm (70 m/min) to get a 60-min tool life. The C1212 steel, which is very similar to the B1112 free-machining steel, increases tool life from 20 to 60 min with a change in cutting speed of only about 20 fpm (6 m/min). These tests were run to total failure of the HSS tool bit, which should never be done in industry.

Figure 4-16*b* shows the same information plotted on a log-log scale, changing the curves to straight lines. Most of the graphs of tool life, finish, etc., are plotted this way for some reason. This log-log graph is good, but this writer believes that it takes most people longer to get significant information from it. Thus, most of the succeeding graphs have been replotted from logarithmic to cartesian coordinates.

These graphs, from the Bethlehem Steel Company Homer Research Laboratories, are excellent because they show the *spread* which can be expected when running tests on metals. They comment "tool-life data can only be (truly) represented by *bands*, as a result of systematic variations in chemistry and properties from the outside of the section of the steel bar to about mid-radius." These same variations plague production in the shops, especially with castings.

The *60-min tool life* (*V*60) is only one basis of comparison. It is *not* to be interpreted as an "ideal" figure. In a simple operation, with short tool-changing time, it is often more economical to use high speeds to in-

crease production even if tool life is only 15 to 30 min because the tool expense is low. On a job with several tools at one station, a 2-h tool life might be more economical.

Effect of Feeds and Depths of Cut on Tool Life

An increase in feed and depth of cut will shorten tool life but not nearly as much as an increase in cutting speed.

An empirical formula which includes the effect of speed, feed, and depth of cut is

For steel

$$V_{60} = \frac{CA}{d^{0.14} f^{0.42}} \qquad (4\text{-}6)$$

For cast iron

$$V_{60} = \frac{CA}{d^{0.10} f^{0.30}} \qquad (4\text{-}7)$$

Where CA = constant. For steel, with carbide, CA = 67; with HSS, CA = 20. For cast iron, with carbide, CA = 9; with HSS, CA = 2.6

d = depth of cut, in.
f = feed, in./rev

Although the above values are approximate and were worked out quite a few years ago, they do indicate the relative importance of the two factors.

Depth has a small exponent, indicating that a large increase in depth will require changing the cutting speed only slightly, in order to maintain the same tool life. In fact, doubling the depth of cut requires only about a 10 percent decrease in V_{60}.

Feedrate, with a somewhat higher coefficient, has more effect, but according to these equations, doubling the feedrate requires only about a 25 percent reduction in cutting speed to maintain a 60-min tool life.

Thus to remove the maximum number of cubic inches (or cubic centimetres) per minute with the same tool life, the rule is

1. Increase depth of cut
2. Increase feedrate
3. Decrease the cutting speed

CAUTION: When you follow the above rule, the *power* needed will increase and the *finish* will get considerably worse as the feedrate is increased.

Actually, the depth of cut is often limited by the actual amount of stock to be removed or by the strength of the workpiece. The feedrate, as shown later, definitely determines the finish if all other factors remain constant.

POWER REQUIRED FOR CUTTING

It is very seldom that a metal-cutting machine is overloaded. The supervisors, the better machine operators, and the methods engineers soon learn the capabilities of each machine just by observation. More often the machines are working at only 25 to 50 percent of their capacity.

However, with the need to increase production with faster metal removal and multiple cuts, a supervisor or methods engineer can put too much load on a machine—too much for the available horsepower, bearings, and gears. Thus, it is advisable to know what happens when speeds, feeds, and depths are increased or decreased.

CAUTION: The power used is *not* necessarily directly related to either the tool life or the finish on the work.

Unit Horsepower

It is known that as speed, feed, and depth of cut are increased, the total power needed is also increased. This is not actually in a direct ratio. A rough approximation is that the ratio is about 0.8:1 for depth, 0.9:1 for feed, and about 0.7:1 for speed. However, a useful approximation of needed horsepower, or kilowatts, can be computed on a volume basis.

The product of feed f and depth d is the rectangular area of the chip being formed. When this is multiplied by the cutting speed in *inches* (or centimetres) per minute, the product is the cubic inches (or cubic centimetres) per minute of metal being removed (in.³/min or often CIM, or cm³/min).

Volume of metal removed per minute is found by

$$\text{in.}^3/\text{min (CIM)} = f \times d \times (\text{fpm})(12) \qquad (4\text{-}8)$$

where f = feed, in./rev
  d = depth of cut, in.
  fpm = cutting speed, ft/min

Or in metric, $\qquad\qquad\qquad\qquad\qquad (4\text{-}9)$

$$\text{cm}^3/\text{min} = \frac{f}{10} \times \frac{d}{10} \times (\text{m/min} \times \cancel{100}) = f \times d \times \text{m/min}$$

where f = feed, mm/rev
  d = depth, mm
  m/min = cutting speed, m/min

These basic formulas are directly applicable to lathe work and are correct for all work, though they take a slightly different form for drilling and milling. Those formulas will be explained in the chapters on drilling and milling.

With the large number of variables which can influence the power needed for a job, there is no accurate method of computing it. If a dynamometer setup is available, the forces can be measured for particular situations. This has been done many hundreds of times, and Table 4-4 shows some representative figures for the unit horsepower. **Unit horsepower** is the horsepower needed to remove *one cubic inch per minute, or one cubic centimetre per minute,* of a specified material on a specified machine at 100 percent efficiency with sharp tools. This is often abbreviated as hp CIM or hp/cm³/min or kW/cm³/min (1 hp = 746 W, 1 in³ = 16.39 cm³)

CAUTION: The figures in Table 4-4 are *not* to be considered precise. Some sources list figures at 80 percent efficiency with partially dull cutters, other sources give quite different figures.

To use Table 4-4 to estimate needed power

$$\text{hp} = \text{volume per minute} \times C$$

where C = hp CIM or hp/cm³/min or kW/cm³/min $\qquad (4\text{-}10)$

An *efficiency* factor must be applied as no machine works at 100 percent efficiency. The basic equation is

hp available for work = hp of motor on machine × efficiency

or $\qquad\qquad\qquad\qquad\qquad\qquad\qquad (4\text{-}11)$

$$\text{hp}_w = \text{hp}_m \times E$$

TABLE 4-4   UNIT POWER REQUIREMENTS* AT 100 PERCENT EFFICIENCY

Material	Horsepower per Cubic Inch per Minute			Kilowatts per Cubic Centimetre per Minute		
	Drilling[1]	Milling[2]	Turning[3]	Drilling[1]	Milling[2]	Turning[3]
Mild steel, to 25 R_c	1.0	1.0	0.9	0.0456	0.0456	0.041
Medium steel, 25-35 R_c	1.6 .	1.8	1.3	0.073	0.082	0.059
Hard steel, 35-50 R_c	1.9	2.1	1.5	0.0866	0.0956	0.0683
Soft cast iron	0.8	0.7	0.5	0.0365	0.032	0.0228
Hard cast iron	0.9	1.1	1.0	0.041	0.050	0.0456
Aluminum	0.35	0.4	0.3	0.016	0.0183	0.0137
Brass	0.5	0.6	0.4	0.0228	0.0274	0.0183
Bronze	0.6	0.8	0.7	0.0274	0.0365	0.032
Stainless steel: 400 series 300[4] series	1.3 1.6	1.3 1.8	1.1 1.7	0.059 0.073	0.059 0.082	0.05 0.0775
Titanium	1.0	1.0	1.1	0.0456	0.0456	0.05
Nickel alloys	1.6	1.6	1.5	0.073	0.073	0.0683

*Values given here are approximately the average of values from several sources. Individual values from reliable sources may vary from these values up to 30 percent.
[1]With feeds 0.002 to 0.008 in./rev (0.05–0.2 mm/rev)
[2]With feeds 0.002 to 0.012 in./tooth (0.05–0.3 mm/tooth)
[3]With feeds 0.005 to 0.015 in./rev (0.125–0.375 mm/rev)
NOTE: Heavier feeds require less unit horsepower.
Double these feeds will need about 20 percent less power.
[4]Free-machining alloys 30–40 percent less. Conversion factors: 1 in.³ = 16.39 cm³, 1 hp = 0.746 kW

where E is expressed as a decimal figure of percentage.

Thus, the needed horsepower is

$$hp_m = \frac{hp_w}{E} \qquad (4\text{-}12)$$

EXAMPLE   4-4
   Rough turning a 3-in.-diameter (76-mm) aluminum bar at 500 fpm (153 m/min), feed = 0.015 in./rev (0.38 mm/rev), depth = 0.200 in, (5.1 mm), at 75 percent efficiency.
Volume:
Using Eq. (4-8),
   CIM = (0.015)(0.200)(500)(12) = 18 in.³/min
or using Eq. (4-9),
   cm³/min = (0.38)(5.1)(153) = 296 cm³/min
Horsepower:
Using Eqs. (4-10) and (4-12),

$$hp = \frac{18(0.30)}{0.75} = 7.2\text{-hp motor needed}$$

In metric (kilowatts power):
Using Eqs. (4-10) and (4-12),

$$hp = \frac{296(0.0137)}{0.75} = 5.41 \text{ kW needed}$$

Comparison:
   Since 1 hp = 0.746 kW, (7.2 hp)(0.746) = 5.37-kW motor needed

EFFECT OF SPEED, FEED, AND DEPTH ON FINISH

Often the required finish on the workpiece is the limiting factor in deciding what quantities to use for speed, feed, and depth of cut. Of course, holding the tolerance is of equal or greater importance. However, a ± 0.0001-in. (0.0025-mm) tolerance is meaningless if the finish is only 125 μin., which is ±0.000 125 in. [3.1 microns (μm) or 0.0031 mm]. Thus, these two important factors are closely allied.

Finish versus Speed

In general, finish improves slightly as cutting speed increases. At higher speeds more of the generated heat goes off with the chip, and as speed increases, the BUE becomes smaller and finally does not form at all. At some "critical speed," the BUE is no longer formed and the finish stays the same even with a further increase in cutting speed. This critical speed will vary with different materials, but it is usually above 100 fpm (30 m/min).

   Slow speeds are especially undesirable when using carbide tools since they waste time (money), and the tools may actually wear out faster at the slow speeds.

Finish versus Depth of Cut

When a large chip is removed, the finish is often poor. However, the increase in roughness due to deeper cuts is quite small. Thus, using a deeper cut is a good way to increase the rate of removing metal.

Finish versus Feedrate

Feedrate is the most critical quantity when finish is being considered. Its effect is modified by the *nose radius* of the tool bit (see Chap. 8). Figure 4-17 illustrates how the surface finish (depth of irregularities) changes when the nose radius is changed. The depth of this "groove" d in the material can be calculated:

$$d = R\sqrt{R^2 - (f/2)^2} \qquad (4\text{-}13)$$

where R = nose radius of cutter, in. or mm
f = feedrate, in./rev, or mm/rev

Using this equation, the finish can be calculated easily on the computer for a wide variety of conditions. Since finish is in *average* roughness, $d/2$ will correspond to the microinch reading.

CAUTION: The Machinability Development Service of the General Electric Company found that with the same feed, speed, and nose radius different materials gave quite different finishes. Thus, Eq. (4-13) is only approximately correct in actual use.

For best finishes, use low feeds, 0.001 to 0.010 in./rev (0.025 to 0.250 mm/rev), high speeds, and proper nose radius, with small depths of cut. The combination of low feedrate and shallow depth results in a

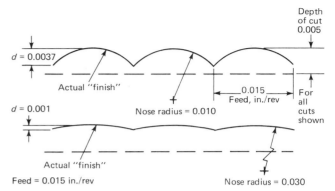

Fig. 4-17   The geometry of the effect of tool nose radius on workpiece finish.

chip with a very small area. Thus, speeds can be at least twice those used for roughing without adversely affecting the tool life.

A summary of some of the ideas discussed in the last few sections is shown in Table 4-5.

Computer Data

The question of the optimum (best considering all factors) speeds, feeds, and depths of cut can now be solved by a computer program. Enough solid data have been accumulated so that tool life under given conditions can be determined.

The Metallurgical Products Department of General Electric Company has a "Hi-E" hand calculator, and the European EXAPT program has a section which can be used. The latest is the NCECO (N/C Economics) computer program. In 1973 the Machinability Data Center, Metcut Research Associates, Inc., offered this program to industry at under $500.

TABLE 4-5   SUMMARY OF THE EFFECTS OF SPEED, DEPTH OF CUT, AND FEEDRATE ON SOME FACTORS

	Finish	Tool Life	Forces on Cutting Tool	Total Stock Removed
Cutting speed	Increase in speed tends to improve the finish. Use 1½ to 2 times normal speed for light finish cuts.	More speed means more heat; thus tool life decreases with increased speed. NOTE: Carbide's life decreases also if speed is too low.	More speed means less friction. Actual forces increase only slightly with higher speeds. With negative rake cutters, and carbides, very high speeds (600 fpm or over) actually decrease forces.	Increases, 1:1 with speed. However, this can be offset by the shorter tool life.
Depth of cut	Greater depth makes slightly poorer finish in most cases.	Only a small decrease in tool life.	Forces increase 0.8:1 ratio with increase in depth of cut.	Increases 1:1 with depth. A good way to increase production.
Feedrate	Finish is much poorer as feedrate increases, as tool marks show on the work.	The smallest decrease, in relation to the increased production.	Force increases but less than that due to increase in depth.	Increases 1:1. Best method within the limits of allowable finish.

The input is the data on the workpiece material, tool material, etc., and the output is the total costs and total operation times for each operation investigated. The program can be run on a small computer and should easily save its low cost, especially for long-run jobs, though the program works well with any given lot sizes.

CUTTING FLUIDS

Some materials such as cast iron can be very efficiently cut "dry." Most materials can be cut faster, and with better finish, if some kind of fluid is used during the cutting.

Cutting fluids are used for one or more of the following reasons:

1.   Cooling the work and the cutting tool
2.   Lubricating the tool and chip surfaces
3.   Forming a protective chemical film on the tool surface—"antiwelding" agent
4.   Flushing chips away from the work area
5.   Protecting the tool, work, and machine from rust
6.   Providing better finish on the work because of items 1 to 5

The cutting fluid must fulfill certain requirements. It must

1.   Be nontoxic, so the operator's skin will not be damaged
2.   Not chemically attack the painted or machined surfaces of the machine, work, or cutter
3.   Create a minimum of smoking or fogging when being used
4.   Be stable in use and storage, not decompose and, thus, lose effectiveness and create undesirable odors

Most commercial products today are well qualified in regard to these factors, so the engineer seldom has to worry about them. However, not all cutting fluids are equally well qualified in each criterion, so these must be kept in mind.

In a large percentage of machining, the cooling of the work is the most important factor. In jobs such as threading and broaching, lubrication is more important. Ideally, both should be done at the same time. However, this is not possible, so compromises have been worked out.

Today there are hundreds of trademarks and dozens of variations on the basic formulas. However, a few basic principles and products are still important. The major divisions in cutting fluids are:

1.   *Soluble (emulsified) oils.* Water is the best cooling medium, but it does not lubricate and it causes rust. Therefore, a mixture of oil, emulsified with soap, or sulfonates is added. Then a "wetting agent" to help the water "stick" to the surfaces and germicides to control bacteria are also added.

This "soluble" oil is mixed with water in ratios of one part oil to five parts water (1:5) or 1:10, 1:20, and even occasionally 1:50 or over. This is the most widely used cutting fluid.

The mixture has a milky color, is inexpensive, gives excellent cooling and moderate lubrication, and will not stain most metals, including copper.

2.   *Straight cutting oil.* The oil used today is usually mineral oil, as it is fairly inexpensive, a good lubricant, and does not create odors, etc.

A *fatty oil*, such as lard oil, is mixed in small amounts (5 to 10 percent) with the mineral oils because it is a better lubricant than mineral oil, especially under pressure. However, it does become rancid and smell bad much too quickly if used alone.

3.   *Sulfur and chlorine.* Either one or both may be added to the above basic fluids. These provide a tougher, more stable lubricating film between the chip and the cutting tool.

There is some evidence that these also react somewhat below the surface of the metal to reduce the forces in the shear zone.

Sulfurized oils are used especially on steels, for threading and broaching, since they do not stain steel. They will, however, stain copper. Thus, the chlorinated oils or plain oils are better for use on copper.

Phosphorus is also added, the same as sulfur and chlorine. It is not as effective as these in preventing metal "welding," but it is effective in cutting down on friction and wear on the tools.

Synthetic (chemical) solutions are available from many manufacturers. These contain a variety of chemicals such as nitrates, phosphates, amines, alcohol, and many others. These are successfully competing with the "standards," especially for the more difficult jobs.

Many of these have, mixed into them, some mineral oil, sulfur, etc., and might be called *"semichemical" cutting fluids.*

Choosing a Cutting Fluid

This is not always a simple task. Theoretically, there is one best fluid for each job. Occasionally a change in cutting fluid will make a dramatic difference in the results. One source reports:

1. Drilling 410 stainless steel (45 R_c) at 75 fpm: only 5 holes per drill could be made when using 1:20 soluble oil. Using either chlorinated or sulfurized mineral oil, 15 or 16 holes could be made with one drill.

2. Reaming a tough maraging steel (321 BHN): only 35 holes could be reamed using 1:20 soluble oil and 190 holes when a heavy-duty chlorinated oil was used.

It might be nice to always use the perfect fluid but, in most manufacturing plants, the usual run of a specific part is from 10 to 50 pieces. No plant could afford to change the fluid in a lathe, milling machine, or grinder every 50 pieces. Thus, many companies use two or three general-purpose liquids for most of their machining.

Table 4-6 shows suggested cutting fluids for some materials.

Application of Cutting Fluids

Flood cooling is the most frequently used. The coolant is kept in a tank (often the base of the machine). A pump, usually with its own motor, circulates the fluid through pipes and flexible hoses up to the work. This may be at anywhere from 5 to 80 psi (34 to 552 kPa) and is often controlled by a hand valve regulated by the machine operator.

For carbide tools especially, the fluid must be continuous and applied at the cutting area. Sometimes more than one cooling nozzle is needed. From 3 to 60 gallons (11.4 to 227 litres) per minute may be used.

Mist coolant (a solution of soluble oil or synthethics in air, atomized and blown at the work) is used on some jobs, such as drilling aluminum. Sometimes the liquid, temperature, pressure, and volume are such that the liquid vaporizes and no return pump is needed.

Mist coolant gives an excellent view of the work and sometimes better tool life and finish. However, it does have quite limited cooling power and may require venting.

Handling the Cutting Fluid

The fluid in most machines is recirculated continuously from the reservoir (in the machine base or separate) to the work area and back to the reservoir. The chips and fine particles of metal must be separated from the liquid. This is often done is settling tanks and filters. Magnetic separators, centrifuges, and flotation are also used, especially where large amounts of fluid must be cleaned.

Eventually the oil becomes dirty, or odor starts,

and the oil must be replaced. This poses a problem because dumping it anywhere causes considerable pollution. Some plants find that they can "recondition" the oil by the use of chemicals and filters.

In high-speed operations such as automatic screw machines (cutting aluminum and brass) and grinders, there is a *mist* caused by the speed and the resulting atomization of the oil. In 1972 several oil companies found an additive which materially reduced this difficulty. As the federal law allows a maximum oil mist in the air of 5 mg/m³, these new oils may quickly save the few cents a gallon extra cost.

Review Questions and Problems

4-1. What is meant by the "machinability" of a material?

4-2. What material is used as a standard in measuring machinability?

4-3. What factors tend to cause a decrease in machinability?

4-4. Name the three types of chips and describe each.

4-5. Describe the ideal type of chip.

4-6. Describe a builtup edge and its effect on the workpiece and the cutting tool.

4-7. Is orthogonal or oblique cutting the most common in actual machining?

4-8. Sketch the three directional forces acting on a tool bit, and indicate the relative magnitude of each.

4-9. What is cutting speed? In what U.S. and metric units is it measured?

4-10. What are the effects on the tool bit, the work, and production of too high or too low a cutting speed?

4-11. Calculate the cutting speed in feet per minute when machining each of the following diameters at the number of revolutions specified.

	a	b	c	d	e
Diameter, in.	4	12	$\frac{1}{4}$	6	75
rpm	240	60	120	90	25

4-12. What rpm should be used to machine the following diameters at the specified cutting speeds?

	a	b	c	d	e
Diameter, in.	4	0.032	25	6	9
Cutting speed, fpm	90	100	300	70	50

4-13. Calculate the cutting speed in metres per min-

TABLE 4-6 SUGGESTED CUTTING FLUIDS

Material	Turning	End Milling	Drilling	Tapping
Low-carbon steel	Synthetic compounds or soluble oils, 1:20. Use chlorinated or sulfurized oils for tougher or harder steels. Some can be cut dry with carbide tools.	Sulfurized mineral oil or heavy duty soluble oil, 1:20. Synthetic compounds may work better with carbide tools.	Synthetic (chemical) fluids sulfo-chlorinated mineral-lard oil for HSS drills. Soluble oil 1:10 heavy duty for carbides.	Sulfurized oils. sulfo-chlorinated mineral, or lard oils. Medium to heavy duty as needed.
Cast iron	Cut dry, especially when using carbide tools. Use soluble oil 1:20 to keep dust down; it may help in finish cuts. Use heavy duty soluble oil or added sulfo-chlorinated for harder castings.	Cut dry with carbides. Use more lubricant than with turning. Heavy duty soluble oil, or sulfo-chlorinated additives.	Dry, or soluble oils. Use EP (extreme pressure) types for heavy feeds.	Dry, or synthetic (chemical) fluids, heavy duty. Soluble oils, sulfurized chlor-inated, medium to heavy duty.
Stainless steel	Sulfo-chlorinated mineral oil, light to heavy duty depending on the cut. Medium to heavy duty soluble oil, as above, 1:5. Sometimes cut dry, with carbides.	Medium and heavy duty soluble oils, 1:5. Synthetics, light duty, may be best for finish cuts.	Sulfo-chlorinated mineral or mineral-lard oil, medium or heavy duty for HSS drills. Soluble oils, with sulfur or chlorine as needed for carbide drills.	Free machining, use soluble oil medium to heavy duty. Others, use sulfo-chlorinated mineral oil, or lard oil mixture.
Aluminum alloys	Dry, or light soluble oils, 1:15 to 1:30 or kerosene preferably with oil. There are several oils and chemical (synthetic) fluids made especially for aluminum.	Same as for turning.	Soluble oil, 1:15, light duty for carbide drills, added sulfur or chlorine for HSS. Some special compounds made for aluminum.	Soluble oil, sulfo-chlorinated for better tap life. Heavy duty for higher speeds and larger taps. Sperm oil or lard oil.
Nickel-base alloys	Heavy-duty sulfo-chlorinated mineral-lard oils. Heavy-duty soluble oils and soluble sulfo-chlorinated oils.	Same as for turning.	Medium to EP (extreme pressure) sulfo-chlorinated mineral or lard oils, or combinations.	Same as drilling.
Copper and brass	Be careful not to stain. Use plain soluble oil or chlorinated soluble oil. Mineral or mineral-lard oil.	Plain soluble oil, 1:30, or mineral-lard oil.	Soluble oils, 1:20 or chlorinated. Mineral-lard oil, light.	Mineral-lard oil, light duty. Sulfurized for heavier cuts if staining is okay, or use chlorinated.

ute when machining each of the following at the number of revolutions specified.

	a	b	c	d	e
Diameter, mm	102	305	229	152	1500
rpm	240	60	120	90	25

4-14. What rpm should be used to machine each of the following diameters at the specified cutting speed?

	a	b	c	d	e
Diameter, mm	102	2.5	635	152	229
Cutting speed, m/min	27	30	91	21	15

4-15. What is feedrate? In what units (U.S. and metric) is it expressed? Is 0.008 in./rev a heavy or light feedrate?

4-16. What is the depth of cut, and in what U.S. and metric units is it measured? Is 12.7 mm a deep cut?

4-17. What characteristics should cutting tool materials have?

4-18. Name the four most used materials for making cutting tools and describe the advantages and disadvantages of each.

4-19. What is tool life and how is it expressed?

4-20. How does the operator know when a tool should be sharpened or replaced?

4-21. Describe the effects of temperature, cutting speed, depth of cut, and feedrate on tool life.

4-22. What is meant by "unit horsepower" and how is it affected by the machine's efficiency?

4-23. How do speed, feedrate, nose radius, and depth of cut affect the finish of a workpiece?

4-24. a. Why are cutting fluids used?
       b. Why is one fluid not equally effective for all kinds of work and metals?

4-25. Describe how cutting fluids are applied.

4-26. Calculate the proper rpm to drill a $\frac{1}{2}$-in.-diameter hole in the end of a 3-in.-diameter bar of mild steel at a cutting speed of 60 fpm.

4-27. a. What is the rpm required to bore a 2-ft-diameter hole at 0.015-in. depth and 0.015 in./rev feed in an aluminum casting if the allowable speed is 350 fpm?
       b. What horsepower is needed if efficiency is 75 percent?

4-28. a. What rpm should be specified to turn a bar 2500 mm in diameter if the depth of cut is 0.6 mm and the allowable cutting speed is 20 m/min? Feed is 0.30 mm/rev.
       b. What power, in watts, is needed to do this job if efficiency is 80 percent?

4-29. Calculate the rpm required to bore a 63-mm-diameter hole in an aluminum casting if the cutting speed is 90 m/min.

4-30. What is the actual cutting speed being used when drilling a $\frac{1}{8}$-in.-diameter hole in mild steel at 2000 rpm?

4-31. What is the cutting speed being used when boring a 3-ft-diameter hole in an aluminum casting at 6 rpm?

References

1. Bhattacharyya, A., and J. Ham: *Design of Cutting Tools*, Society of Manufacturing Engineers, Detroit, Mich., 1969.

2. Brown, R. H., and E. J. A. Arnarego: *The Machining of Metals*, Prentice-Hall, Englewood Cliffs, N. J., 1969.

3. *Cutting and Grinding Fluids: Selection and Application*, Society of Manufacturing Engineers, Detroit, Mich., 1967.

4. *Industrial Diamond (the)*, Industrial Diamond Association of America, 1964.

5. Kalish, Herbert S.: "Some Plain Talk About Carbide," *Manufacturing Engineering and Management*, July 1973.

6. *Machinability of Steel*, Bethlehem Steel Corporation, Bethlehem, Pa.

7. *Machining Data Handbook*, 2d ed., Machinability Data Center, Metcut Research Associates, Inc., 1972. (Also several other publications by the same company.)

8. *Producibility/Machinability of Space Age and Conventional Materials*, Society of Manufacturing Engineers, Detroit, Mich., 1968.

9. Silkmann, H. J.: *How Not to Use the New Tool Materials*, Society of Manufacturing Engineers, Detroit, Mich., MR73-204.

10. Swinehart, H. J. (ed.): *Cutting Tool Material Selection*, Society of Manufacturing Engineers, Detroit, Mich., 1968.

11. ASTME, Tool Engineers Handbook, 3rd ed., McGraw-Hill, New York, 1976.

12. Whitney, E. Dow: *Ceramic Tools*, Society of Manufacturing Engineers, Detroit, Mich., TE73-205.

numerical control

In planning production in today's manufacturing plant, the engineer must consider numerical control (N/C) machines. The machine designer can use parts which before numerical control were difficult or impossible to make. The manufacturing engineer can save up to 80 percent of the time needed for short-run jobs. The production control department can schedule more accurately because the running time is set by the tape. The draftsperson may dimension the parts so that the numerical control programmer can work more quickly.

Thus, N/C machines must be considered not as a separate class but as the most modern type of lathe, drill press, or milling machine, etc. Since they will be discussed briefly in the same chapters as the "standard" machines, it is worthwhile to consider the basis of numerical control machinery.

This chapter will not teach you how to program nor will it give you details of specific machines. There are several good books listed among the references at the end of this chapter which can be used if more detailed information is desired.

THE TAPE USED FOR CONTROL

The tape by which most N/C machines are controlled is a 1-in.-wide tape into which has been punched a series of holes. The tape may be made of paper, Mylar, other plastics, or combinations of these. The dimensions of holes and spacings are standardized in documents issued by the Electronic Industries Association (EIA) and the Aerospace Industries Association of America, Inc. (AIA).

The Coding

Because a very positive on-off signal is required, the *binary* system is used. The straight binary system is quite awkward to write and to read. For example, the

number 2048 would be written 101001110001. Thus, the binary coded decimal system (BCD) was developed (Fig. 5-1).

There is also an ASCII (pronounced Asky) code. This is the American Standard Code for Information Interchange. It also uses the binary coded decimal system and is part of an effort to achieve an international standard for all processing systems (computer, telegraph, N/C, etc.). Thus, the actual ASC II coding is somewhat different from the BCD developed for N/C machines. Some MCUs (machine control units) can interpret either code, but BCD is still the most frequently used.

Figure 5-1 shows how an eight-track system is used to control all N/C machines. The small ninth track near the center of the tape is sprocket holes, which are used to drive the tape through the mechanical tape readers or for control puposes in the optical tape readers.

Track 5 is a "parity bit" or "check" hole. The N/C machines, tape punches, and tape readers are wired so that they will only accept a command which has an *odd* number of holes. If an even number of holes appears, the equipment will stop. For example, the code for $3 = 2 + 1$ has only two holes punched. Thus, track 5 is also punched, giving the odd number of holes required.

As the BCD system has no code (except a blank tape) for zero, track 6 is read as zero.

Track 8 is only punched when the *instructions* to the machine are completed. For example, the tape might instruct the machine to move to X 10.500, Y 01.250, Z 0.625, change speed to 350 rpm, feed to 4.5 ipm, and drill a hole there with tool number 4. (This can all be easily coded on the tape.) This is called a *block* of information—a complete instruction. To show that the block is completed, an *end of block* (EOB or EB) code is signaled by a hole in track 8. Incidentally, the next command might be very short. For example, Move to *Y* 03.750. As all other conditions remain the same, they do not need to be repeated, so a very short block is all that is needed.

A typical block of information is shown in Fig. 5-2. The *TAB* code is used so that when the program is typed out on a program sheet, the numbers for *X*, *Y*, etc., are aligned one under the other.

The *TAB key* serves to align the information properly on the typewritten copy and also punches a five-hole code into the tape. Some N/C machines use this TAB code, and some just disregard it. However, it is a convenience in checking a program to have the columns aligned.

Tape Punching

The tapes for numerical control machines may be punched on a special typewriter, such as that shown in Fig. 5-3. These tape punches have a semistandard keyboard, so they are quite easy to operate.

The typist simply copies the data which has been written out by the N/C programmer. Each letter, number, or symbol which is typed appears on a sheet of paper (usually a programming form) and also punches the proper code into the paper or Mylar tape.

Using the Computer

N/C programs are sometimes done on the computer. In this case, the computer, or a special "off-line" converter, makes the tape. These tapes frequently are a few hundred feet long.

Today some N/C machines are connected directly into a computer system and no tape is used since the signals go directly from the computer to the machine control unit of the N/C machine. Due to the speed of the computer, several N/C machines may be working on different parts, all programs coming from one computer.

The Tape Reader

These tapes are fed through a tape reader (Fig. 5-4). This reader may have lights shining down so that when a hole appears in the tape, the light shines onto a photodiode. This photodiode changes the light energy into electrical energy, which is amplified and used to control the machine. These readers usually operate at 300 rows of punched holes per second, though they have capacities of from 100 to 1000 rows per second.

Some tape readers use mechanical switches which are operated by eight sprockets in the reader. These

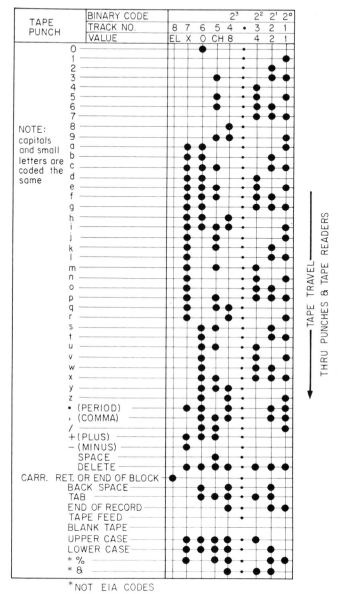

Fig. 5-1 The EIA RA-244 "Character Codes for Numerical Machine Tool Control Perforated Tape." (*Electronics Industries Association.*)

sprockets ride on the tape. When a hole is present, the sprocket drops down, thus closing a switch which sends a signal to the N/C machine.

THE MACHINE CONTROL UNIT (MCU)

These signals are sent into a machine control unit (Fig. 5-5) which contains the electric and electronic

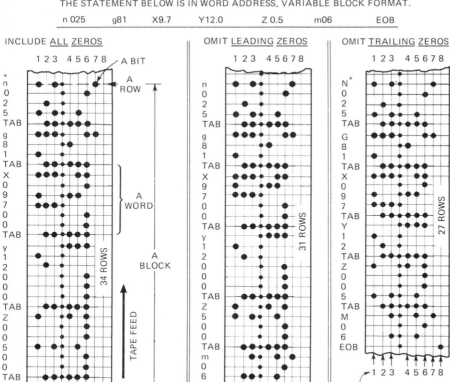

THE STATEMENT BELOW IS IN WORD ADDRESS, VARIABLE BLOCK FORMAT.

n 025   g81   X9.7   Y12.0   Z 0.5   m06   EOB

* The letters may be written as either capitals or small letters.
The punched coding is the same in BCD.

Fig. 5-2   A typical block of information coded in three slightly varying methods. Variation is due to omitting leading or trailing zeros as required by different N/C machines.

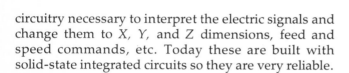

Fig. 5-3   A Flexowriter typewriter which can punch and copy N/C tapes. The punch is at the left rear; the reader on the left front of the typewriter. (*Friden Division of The Singer Company.*)

Fig. 5-4   An electromechanical tape reader which operates at up to 60 characters per second. (*Tally Corp.*)

circuitry necessary to interpret the electric signals and change them to X, Y, and Z dimensions, feed and speed commands, etc. Today these are built with solid-state integrated circuits so they are very reliable.

Feedback

The dimension commands are actually commands for

the machine to move a certain number of "pulses." These pulses may each be 0.001, 0.0002, or 0.0001 in. (or 0.01, or 0.02 mm). Most N/C machines have a "counter" on the machine. This may be a rotating or linear electric device which "counts" how far (how many pulses) the table or carriage of the machine has *actually* moved. When the actual motion count from the machine exactly matches the command count, the motion is stopped. This matching process is referred to as *feedback* (see Fig. 5-5).

Some N/C machines do not use feedback but rely on the accuracy of a synchronous motor drive.

TYPES OF N/C MACHINES

There are dozens of models and makes of N/C machines, but they are basically of three types.

Point-to-Point (NPC)

The point-to-point numerical control machines are the least expensive, costing from $12,000 to $40,000. These machines can drill, tap, ream, etc., and can do milling in a straight line parallel to either the X or Y axis. These machines do *not* control the path the spindle travels between two locations, except vertically and horizontally.

For example, Fig. 5-6 shows two locations, A and B. The spindle may travel paths 2, 3, or 4 (usually path 2), but it cannot travel on path 1 if it is a point-to-point machine. If it is only necessary to drill one hole at A and one at B, the path between does not matter.

Milling in straight lines parallel to the X and Y axes still allows a lot of milling to be done, as shown in Fig. 5-7. Notice that the inside corners will have a radius (as in all milling) but the outside corners will be sharp. Thus, the NPC machines can do a large percentage of the work done in most shops.

Point-to-Point with Circular Interpolation

The second type of N/C machine is a modification of the first, so that *circular* arcs may be cut. The MCU

has added "computer" capabilities so that if the starting and ending points of the arc and the radius are punched into the tape, the MCU will generate a series of short, straight cuts which will very closely approximate a circle. The finish is usually very good, and the "flats" are so small that the circular cut can often be used without any smoothing.

This added capability is available at a cost ranging from $300 to $1500 extra, depending on the machine. With it, an NPC machine can do almost all the work needed in many manufacturing plants.

Contouring N/C Machines (NCC)

To control the angle of cut between the two corners of the workpiece in Fig. 5-8, the MCU must control the X and Y feedrates so that the Y feedrate is 5/8.660 = 1:1,732 in relation to the X feedrate. This is called *linear interpolation*. This additional ability is what makes a *contouring* N/C machine. With this ability the NCC machine can cut any shape curve and any desired slope. These machines are expensive, costing from $75,000 to $250,000 each. However, for many parts with sloped sides or odd shapes, they pay for themselves very quickly.

Fig. 5-6   Four possible straight-line paths from point A to point B.

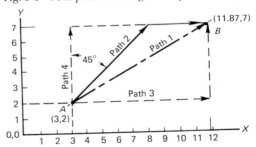

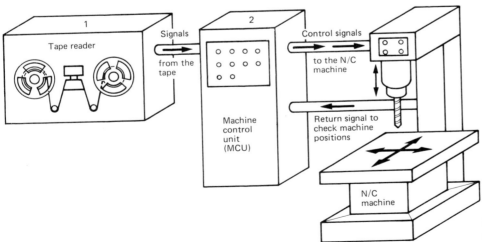

Fig. 5-5   The process of converting tape codes into machine action.

Fig. 5-7   Some milling cuts which can be done on simple point-to-point N/C machines.

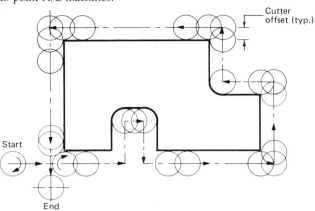

Fig. 5-8   A simple contouring cut which requires linear interpolation.

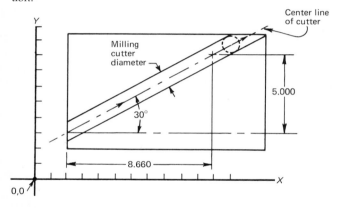

DIMENSIONING

The Zero Point for Dimensioning

Numerical control dimensioning is based on the familiar cartesian coordinate system. This system, when applied to a vertical-spindle machine, looks like the one shown in Fig. 5-9.

When the machine spindle is *horizontal*, as in Fig. 5-10, the cartesian coordinates have been rotated about the X axis. Notice that, in this case, the viewpoint is *from* the tool toward the machine table. This actually is more convenient, as the operator stands close to the spindle, so that he or she can more easily change cutters.

Quite a few N/C machines can be programmed in all four quadrants using plus and minus signs as needed and with the zero at any convenient location. However, a great many machines are built with the zero point electronically established at one corner of the machine table. Figure 5-11 shows two possible positions of this fixed zero point.

Z Dimensions

The Z axis is defined as the axis on a line through the spindle. As you can see in the previous drawings, the Z dimension is positive (+) when the spindle is moving *away* from the work.

Several of the most widely used point-to-point N/C machines, including some with circular interpolation,

Fig. 5-9   The cartesian coordinates shown on a vertical-spindle machine.

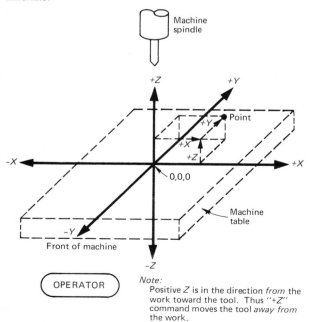

Note:
Positive *Z* is in the direction *from* the work toward the tool. Thus "+*Z*" command moves the tool *away from* the work.

Fig. 5-10   The cartesian coordinates shown on a horizontal-spindle machine.

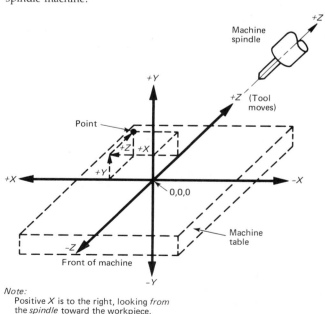

Note:
Positive *X* is to the right, looking *from* the *spindle* toward the workpiece.
Positive *Z* moves *tool* away from work.

do not code the Z dimension on the tape. The reasons for this are:

1. It requires additional electronic control in the MCU, which is expensive.

2. Computing the Z dimension is complicated by the fact that drills, reamers, and milling cutters come in many different lengths.

3. The workpiece may have features at different heights.

These three reasons do make computing the Z dimension more difficult, but it certainly can be, and is on many N/C machines, computed quite easily by an experienced programmer. However, the N/C machines which do not put the Z dimension on the tape do use cams or micrometer stops to control this travel. These are easily and quickly set by the machine operator and are accurate to within 0.002 to 0.005 in. (0.05 to 0.125 mm).

For most drilling and threading operations, the Z distance merely has to be deep enough. For milling cuts, the above tolerances are frequently close enough.

More than Three Axes

We live in a three-dimensional world, but numerical control machines are often spoken of as having four- and five-axis capabilities. Actually the added axes are created by rotating part of the N/C machine about the X, Y, or Z axis. Figure 5-12 shows an example of this situation. The a, b, and c axes rotate around the X, Y, and Z axes, respectively. These machines are more complex to program, and are usually programmed on the computer. The basic process is still the same as in the simpler machines.

THE PROGRAMMING OF N/C MACHINES

Much N/C programming is done on the computer. In fact, most of the programming for the contouring machines would be too tedious and complex for hand programming. However, the output on the tape is in the same format whether done by hand or by the computer. A great many N/C parts are done by hand programming.

Word Address Program

The EIA and AIA agreed on a fairly standard set of letters and their meanings. These are used in *word address programming*, which is the most widely used in the world today. The letters may be either capital or small letters.

A letter is used before each set of figures. The letter

serves as a switching signal (Fig. 5-13) which directs the numbers into the proper section of the electronics in the machine control unit. Thus, a Z-axis command to locate 2.550 in. from zero would read Z02550 (no decimal points are used), and it will send 2550 into the section or "address" of the MCU which controls the motion along the Z axis. Similarly, F055 will send a signal to the feedrate-controlling section of the

Fig. 5-11   Two possible positions for a "fixed" zero on an N/C machine table. (a) one is in the first and the (b) in the third quadrant.

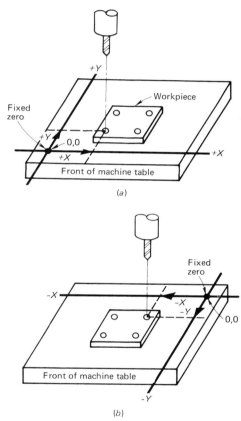

(a)

(b)

Fig. 5-12   The notation used to designate up to six axes on a machine.

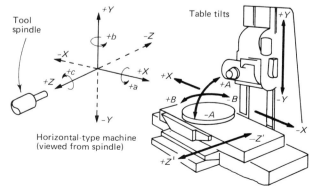

Horizontal-type machine (viewed from spindle)

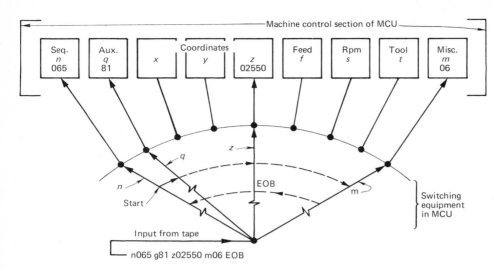

Fig. 5-13 One arrangement of the switching circuits for directing tape commands to the proper part of the N/C machine.

MCU. This will electrically set the feedrate control on the machine to 5.5 ipm.

These letters are usually arranged in a certain definite order merely for the convenience of the people using the program. However, different manufacturers may set up their programming sheets differently, but the MCU will still send the signal to the proper address.

some standard commands

Some letters such as X, Y, and Z for motion along those axes, F for feedrate, and S for spindle rpm are self-explanatory.

However, signals are also needed for stopping to change tools, rewinding the tape, etc. The letter M (for miscellaneous) is used for this group. Typical are:

M00—Stop so that the operator can inspect the work.

M02—Rewind the tape; ready for next piece.

M06—May activate an automatic tool changer or stop the machine so that the operator can change the tool.

Other M codes control coolant flow, turn the spindle on and off, etc.

Another group of signals is for "canned cycles." This callout causes a complete cycle of the machine. For example:

G81—For drilling: The spindle moves to the location specified, rapidly approaches the work, slows down to feedrate to drill through, then rapidly returns to the start position in Z axis. The next block of tape is read immediately unless M06, etc., has been programmed.

G84—For tapping: Much the same as drilling, but the spindle reverses to back the tap out of the hole.

G79—For milling: The spindle moves to the called for X and Y position, then goes rapidly down close to the work and cuts into it at feedrate. The spindle then *stays down* and cuts a slot or keyway to the next X or Y command position.

All these G codes are *modal*. They stay in operation until a new G code is programmed. Other G codes control boring, arc or slope cuts, etc.

a brief word address program

The following program does not represent any one specific N/C machine. It is an example of how some of the programming features are used. Assume a single-spindle N/C machine. The machine operator can change cutters in 15 to 20 s (seconds).

The workpiece, in position on the machine table, is shown in Fig. 5-15. The holes may be drilled in any convenient order. The order used is chosen to illustrate certain programming features. No Z dimensions will be punched into the tape.

comments on the sample program (Fig. 5-16)

The machine operator is usually given a copy of the completely typed-out program. On simple machines, such as the one we are using, the operator needs this in order to know what cutting tools to use, when to change them, and, sometimes, what depth settings to use. The setup instructions will tell where to put the workpiece on the machine table.

N—is simply a sequence number, for easy reference to any line of the program. Each line is a "block" so the EOB code will be punched after each line. This sequence number is shown in lights on the MCU panel so that the operator can follow the program.

G81—is the canned cycle for a complete drilling operation. This is modal, i.e., it stays in effect until changed by another G code.

X and *Y* dimensions—are obtained by adding the setup dimensions to the part dimensions.

S—is speed in revolutions per minute (rpm). It is either calculated or read from a table. The material is cold-rolled steel (CRS). The coded number is the actual rpm to be used.

F—is the feedrate for cutting through the work. This is in inches per minute (ipm) or millimetres per minute (mm/min). Direct ipm coding is used here. Some N/C machines use special types of coding.

M—is for the miscellaneous functions mentioned previously in this chapter.

The N/C machine used for this job has nine depth-setting cams. Each can be set for a different cutting tool or work height. They are numbered 1 through 9, and can be called into operation by using M51 through M59. The operator checks these on the program, and adjusts them all as needed before starting the job. This is part of the "setup" (getting ready) time. These M51 through M59 cam codes are modal.

ADVANTAGES AND DISADVANTAGES OF NUMERICAL CONTROL

The program shown here is very simple, yet it contains most of the basic elements involved in a large percentage of N/C programming. Various N/C machines have greater or less versatility.

The major *advantages* of numerical control machines are:

1. Lower tooling costs because the tape ensures accuracy of from ±0.001 to ±0.0001 in. (0.025 to 0.0025 mm) if the machine is in good condition.

2. Easy design changes. This is especially important in prototype work. Holes can be added, moved, or omitted simply by changing the tape or dialing in by hand.

3. Simplified inspection since the tape will produce the same accuracies part after part as long as the cutting tools are properly maintained. Thus, only occasional inspection is needed except for some critical hole diameters.

4. Reduced scrap is a result of item 3 above. The tape does not get tired or have family problems which can cause an operator to make mistakes.

5. Reduced space requirements because the tape is a small item to store. Formerly large jigs and

Fig. 5-14 A low cost N/C Pratt & Whitney Tape-O-Matic point-to-point drill, ream, tap machine, which will also do light milling. Note the digital readout and sequence number readout at the upper right. (*Photograph by Samuel Lapidge*)

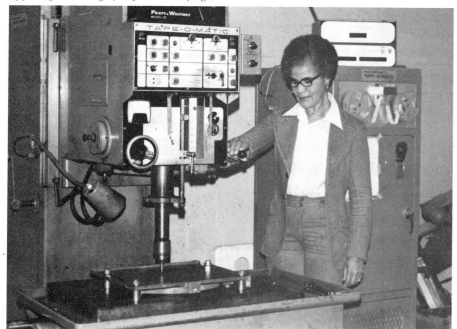

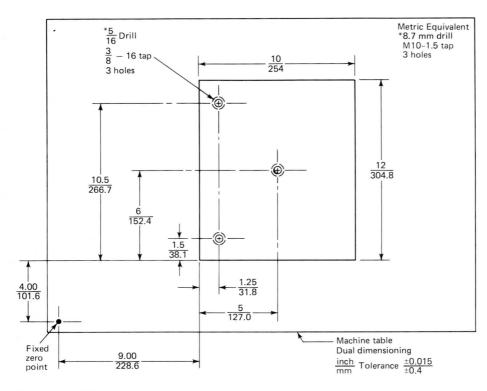

Fig. 5-15 A sample workpiece to be programmed for drilling and tapping on an N/C machine.

Seq. No.	Prep. Function	Distance from Zero, in.	Distance from Zero, in.	Speed, rpm	Feed, ipm	Misc. Function	Comments
N 000	G 00	X 00.000	Y 00.000	S 000	F 00.0	M 00	
		Set up lower lefthand corner X = 9.000, Y = 4.000					Operator instruction
N001	G81	X10250	Y05500	S713	F049	M51	5/16 drill, hole 1
N002			Y14500				Hole 3
N003		X14000	Y10000			M06	Hole 2, tool change
N004	G84	X14000	Y10000	S625	F156	M52	Tap 2, 3/8–16, new cam
N005		X10250	Y14500				Tap 3
N006			Y05500			M02	Tap 1, rewind tape, tool change

English units
(a)

Seq. No.	Prep. Function	Distance from Zero, mm	Distance from Zero, mm	Speed, rpm	Feed, mm/min	Misc. Function	Comments
N 000	G 00	X 000.00	Y 000.00	S 000	F 000.	M 00	
		Set up lower lefthand corner X = 228.6, Y = 101.6					Operator instruction
N001	G81	X26040	Y13970	S713	F124	M51	8.7-mm drill, hole 1
N002			Y36830				Hole 3
N003		X35560	Y25400			M06	Hole 2 tool change
N004	G84	X35560	Y25400	S625	F396	M52	Tap 2, M10–1.5, new cam
N005		X26040	Y36830				Tap 3
N006			Y13970			M02	Tap 1, rewind tape, tool change

Metric units
(b)

Fig. 5-16   Sample N/C programs for the workpiece in Fig. 5-15 in (a) English and (b) metric dimensions.

fixtures often had to be kept for years, and these require a great deal of storage space.

Some *disadvantages* of N/C machines are:

1. They are expensive compared with conventional drills, milling machines, and lathes. However, in most cases they have quickly amortized this cost.

2. New people must be trained:
- **a.** New electronic servicepersons
- **b.** New mechanical servicepersons
- **c.** One or more programmers
- **d.** Machine operators must learn new techniques

3. More careful planning of cutting tools and operation sequences is needed if N/C machines are to be used efficiently.

4. Drafting procedures may need to be changed to make programming quicker and more accurate.

5. Do not expect drills, milling cutters, etc., to cut any faster (except that the N/C machine may be more rugged and powerful than previous machines).

6. Any change can produce resistance from people unless it is carefully planned for, and all participants are fully informed well ahead of time. This requires planning at both management and line levels.

Small shops and giant corporations are finding that this "new" type of machine can perform difficult jobs quickly and accurately and can save hours of time on fussy small-lot jobs. Thus, it is worthwhile to become familiar with the wide range of prices, types, and capabilities of numerical control equipment.

Review Questions and Problems

5-1. How are N/C machines controlled?
5-2. What type of code is used on N/C machine tapes?
5-3. What is a "parity bit" and why is it used?
5-4. What is meant by a "block" of information?
5-5. How does the machine know when a block is completed?
5-6. What are the functions of the tab code?
5-7. How does a tape reader operate?
5-8. What is a machine control unit (MCU)?
5-9. What is "feedback"? What is its function?
5-10. What is meant by a point-to-point (NPC)

numerical control machine, and what are its advantages and disadvantages?
5-11. What is meant by circular and linear interpolation?
5-12. Define the "zero point" and its location.
5-13. How is the Z axis defined on either a horizontal- or vertical-spindle N/C machine?
5-14. Why is the Z axis not programmed on all N/C machines?
5-15. What is meant by a "canned cycle"? Why are they used on many N/C machines?
5-16. What are the major advantages of N/C machines?
5-17. What are the disadvantages of N/C machines?
5-18. How does numerical control affect the actual cutting time needed to make a part?
5-19. Plot the following points using a cartesian coordinate system.

X	Y
−1.0	3.5
−5.0	3.0
−4.0	0.5
3.0	3.0
5.0	2.0
0.8	−2.0
1.25	1.0
−3.0	−3.0
−6.5	−3.5
4.0	−2.0
2.5	−3.5

5-20. In what quadrant would each of the following fall?
- a. X positive, Y positive
- b. X negative, Y negative
- c. X negative, Y positive
- d. X positive, Y negative

References

1. Barron, Charles H.: *Numerical Control for Machine Tools*, McGraw-Hill, New York, 1971.
2. DeVries, M. A. (ed.): *NC/CAM Profits for the 70s*, Numerical Control Society, Inc., 1973.
3. Roberts, A. D., and R. C. Prentice: *Programming for Numerical Control Machines*, McGraw-Hill, New York, 1968.

Write to the manufacturers of N/C machines and equipment for information.

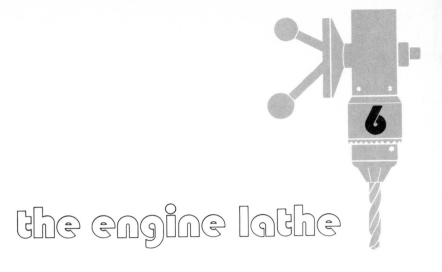

the engine lathe

The lathe, in its many forms, is one of the most widely used machine tools in all parts of the world. If you look at any piece of machinery, from the potter's wheel to jet engines, and at the lathe itself, and notice the number of rotating parts, bushings, bearings, pins, and round handles, you can see why this is so. A lathe is used basically for making round parts.

The engine lathe is the most basic, simplest form of the lathe. Figure 6-1 shows a medium-sized lathe. This model is equipped to produce both English and metric feeds and threads. A graph similar to Fig. 4-10 is at the top left on the headstock. Figure 6-2 shows one of the larger lathes.

The engine lathe, while it is remarkably versatile, is not usually used for high-production work. If more than 25 to 50 pieces are to be machined and there are more than two or three cuts to be made, the job can probably be done more quickly and economically on a turret lathe. Large quantities (from 1000 to 5000 and over) would be made on one of the automatic turning machines described in Chap. 7.

THE PARTS OF A LATHE

The **headstock** is the powered end and is always at the operator's left (Fig. 6-3). This contains the speed-changing gears and the revolving, driving spindle, to which is attached any one of several types of work holders, as described in this chapter. The center of the spindle is hollow so that long bars may be put through it to be held by the chuck (Fig. 6-15) for machining.

The **tailstock** is nonrotating but can be moved, on hardened ways to the left or right, to adjust to the length of the work. It can also be offset about 1 in. (25 mm) for cutting small-angle tapers.

The **carriage** can be moved left or right either by handwheel (hand feed) or power feed. This pro-

vides the motion along the Z axis. (The Z axis, on any machine, is on the line through the rotating spindle.) During this travel, turning cuts are made (Fig. 6-7a).

The **apron**, attached to the front of the carriage, holds most of the control levers. These include the levers which engage and reverse the feed lengthwise (Z axis) or crosswise (X axis) and the lever which engages the threading gears. The "start-stop" clutch lever is usually close beside the apron. Frequently dual levers are furnished, one on each side of the carriage, for the convenience of the operator.

The **cross slide** is mounted on the carriage and can be moved in and out (the X axis) perpendicular to the carriage motion. This is the part that moves when *facing* cuts (Fig. 6-7b) are made with power feed or at any time a cut must be made "square" with the Z axis. This, or the compound, is also used to set the depth of cut when turning. The cross slide can be moved by its handwheel or by power feed.

The **compound rest**, or *compound* for short (Fig. 6-4), is mounted on the carriage. It can be moved in and out by its handwheel for facing or for setting the depth of cut. It can also be rotated 360° and fed by its handwheel at any angle. The compound does not have any power feed, but it always moves longitudinally (left and right) with the cross slide and the carriage.

The **tool post** is mounted on the compound rest. This can be any of several varieties but in its simplest form is merely a slotted cylinder which can be used to fasten the toolholder securely. The tool post can be moved left or right in the tee slot in the compound and clamped in place. It can also be rotated so as to present the cutter to the work at whatever angle is best for the job.

The **bed** of the lathe is its "backbone." It must be rigid enough to resist deflection in any direction

Fig. 6-1   A modern engine lathe. The charts on the left end show the lever settings to use for English and metric work. Top speed is 2000 rpm. (*South Bend Lathe.*)

Fig. 6-2   A large engine lathe for turning steel mill and paper mill rolls. Note that all parts are the same as, but larger than, Fig. 6-1. (*LeBlond.*)

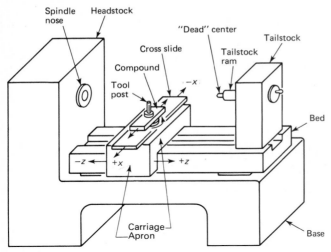

Fig. 6-3   The principal parts of an engine lathe.

under load. The bed is made of cast iron or a steel weldment, in a box or I-beam shape, and is supported on legs, cabinet, or bench.

The **ways** of the lathe (Fig. 6-5) are the flat or vee-shaped surfaces on which the *carriage* and the *tailstock* are moved left and right. Each has its separate pair of ways, often one flat surface, for stability, and one vee way for guidance in a perfectly straight line. (Dovetail slides with adjustable gibs are used on the cross slide and compound.)

These ways are hardened and scraped or ground to close tolerances. The basic accuracy of movement of the carriage depends on the ways. (Incidentally, this is true of many machine tools which have a moving table and carriage.)

The **size** of a lathe is specified by two or three figures:

1. The largest diameter workpiece which will clear the bed of the lathe. The center, of course, is the headstock spindle center.

2. The largest diameter workpiece which will clear the cross slide is sometimes also specified.

3. The longest workpiece which can be held on centers between the headstock and the tailstock.

These three dimensions are illustrated in Fig. 6-6. A 10 × 30 in. (250 × 750 mm) lathe has a swing over the bed of 10-in. diameter (250-mm), and a distance between centers of 30 in. (750 mm). The range of lathe sizes is shown in Table 6-1.

WORK DONE ON ENGINE LATHES

The terms *rough cut* and *finish cut* are often used, not only on lathe work, but on milling, planing, boring, etc., as will be described in later chapters.

Rough cut, or roughing cut, implies two conditions:

1. It is a heavy cut, removing as much metal as possible.

2. It does not give the final size or finish.

Often more than one rough cut is necessary to get close to the required size. Depths of rough cuts will be from 0.030 to 1.00 in. (0.750 to 25.4 mm) and feeds will be from 0.010 to 0.040 in. (0.250 to 1.0 mm) per revolution. Very heavy cuts may require slower speeds than normally specified.

Finish cut means that these last one or two cuts will bring the work to final size and required finish. On a lathe, these cuts are shallow, 0.005 to 0.020 in. (0.125 to 0.500 mm) deep, and the feed is from 0.001 to 0.008 in. (0.025 to 0.200 mm) per revolution. The rpm can often be double that usually specified because these light cuts remove very little metal, so they generate very little heat.

Sometimes the desired tolerance and finish cannot be machined on the lathe, so a small amount [often as little as 0.005 in. (0.125 mm)] is left by a semifinish cut on the lathe. The final finish cut may then be done on a grinding machine.

Cuts which Can Be Made on an Engine Lathe

turning (Fig. 6-7a)
Cutting so as to reduce the outside diameter is called *turning*. The tool bit travels parallel to the axis of the work (Z axis) through motion of the carriage, prefera-

Fig. 6-4   The compound rest and its relation to the carriage. The "standard" tool post is also shown.

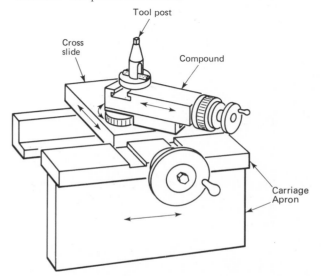

Fig. 6-5 End view of the ways of a lathe. (Photograph by Samuel Lapidge.)

bly with power feed. Turning is the most frequently used operation on the engine lathes. It is usually in a straight line, but it may follow a contour if a tracer lathe or N/C lathe is used.

facing (Fig. 6-7*b* and *c*)

Facing always decreases the length of the work or the thickness of a flange and creates a flat surface. Facing is done by feeding the cross slide or compound in or out. This operation is primarily used to smooth off a saw-cut end of a piece of bar stock or to smooth the face of a rough casting. If both ends of a workpiece are rough, it must be removed from the chuck, reversed, and secured again for the second cut. Figure 6-8 shows turning and facing cuts being done on lathes.

boring (Fig. 6-7*d*)

Boring is used to increase the inside diameter of a hole. As you can see from the illustration, a boring tool cannot possibly make a hole in solid material. The original hole is made with a drill, or it may be a cored hole in a casting or forging.

Boring on a lathe can only be done if the hole is on the Z axis, that is, in line with the center of the

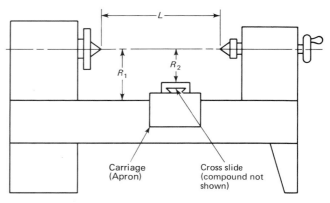

$2 \times R_1$ = swing over bed
$2 \times R_2$ = swing over cross slide
L = distance between centers, longest workpiece

Fig. 6-6 How the size of a lathe is specified.

tailstock. A workpiece may be offset on a faceplate, or fixture, so that the hole is properly aligned.

Boring achieves three things:

1. It brings the hole to the proper size and finish. A drill or reamer can only be used if the desired size is

TABLE 6-1 DATA ON ENGINE LATHES

Size	Swing (Diameter) over Bed		Length between Centers		Power		Approx. Cost
	in.	mm	in.	mm	Hp	kW	
Small	10	250	30	750	$\frac{1}{4}+$	0.2+	$800 up
Medium	20	500	60–72	1500–1800	5–15	3.7–11.2	$15,000 up
Large	48	1200	240	6000	20–60	15–45	$30,000 up

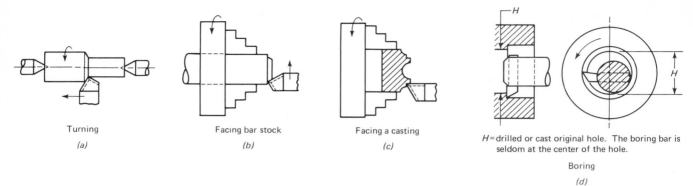

Turning
(a)

Facing bar stock
(b)

Facing a casting
(c)

H = drilled or cast original hole. The boring bar is
seldom at the center of the hole.

Boring
(d)

Fig. 6-7   Diagrams of some lathe cuts which can be made.

(a)   *(b)*

Fig. 6-8   Actual lathe cuts. (*a*) Turning. (*b*) facing. *Note:* The compound does not need to be
turned as shown. (*Clausing Corp.*)

"standard" or if special tools are ground. The boring
tool can work to any diameter, and it will give the
required finish by regulating feed, speed, and nose
radius.

2.   It will *straighten* the original drilled or cast hole.
Drills, especially the longer ones, may wander off
center and cut at a slight angle because of eccentric
forces on the drill, occasional hard spots in the material, or uneven sharpening of the drill. Cored holes in
castings are almost never really straight. The boring
tool, being moved straight along the ways with the
carriage feed, will correct these errors.

3.   It will make the hole concentric with the outside diameter within the limits of the accuracy of the

chuck or holding device. For best concentricity, the
turning of the outside diameter and the boring of the
inside diameter is done in one setup, that is, without
moving the work between operations.

CAUTION: Boring bars are subjected to radial
and tangential force, and they will *bend*. Thus the
hole, though straight and concentric, can be *tapered*. The remedies are:

1.   Use the largest diameter boring bar which
will fit into the hole.

2.   If the hole is small, a solid-carbide boring bar
can be used and will deflect much less.

3.   Take light cuts, maybe only 0.002 to 0.010 in.

(0.050 to 0.250 mm) deep for the last one or two passes.

center drilling (Fig. 6-9a)

Center drilling is most frequently done so that the workpiece can be held between centers. The "combination drill and countersink" is often used. This is held in a drill chuck placed in the tailstock. The included angle is 60°, which is standard for the points on the lathe centers.

Another use for center drilling is to obtain a well-centered location as a starting hole for a drilling operation. This will help keep the drill aligned as it cuts deeper. As the center drill is a short, stiff cutting tool, it will not bend and, therefore, will drill a quite accurately located hole.

threading (Fig. 6-9b)

This can be done on an engine lathe with a single-point tool although it is a slow method. The tool bit, for a standard unified national or metric thread, is sharpened to a 60° angle, and it is fed to depth in increments of 0.002 to 0.010 in. (0.050 to 0.250 mm) per pass by the handwheel on the compound. It will take three to ten passes to cut a 60° vee thread to full depth.

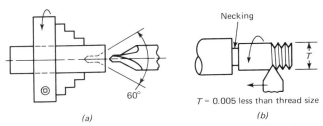

Fig. 6-9 Diagrams of more lathe cuts. (a) Center drilling; (b) outside (male) thread.

The compound is set at 29°, which makes the tool cut mostly on the leading edge. An example is shown in Fig. 6-10. By grinding the tool bit to the proper size and shape, any *style* or *shape* thread (such as acme or buttress) may be cut. The variety of *pitches*, in inches or millimetres (or threads per inch), which can be cut is limited by the gearing available on the lathe. The lathe shown in Fig. 6-1 has two extra shift levers so that it can cut both English and metric pitch threads.

Inside threads can also be cut with a single-point tool although this is also slow and somewhat difficult.

Except for small quantities of parts, special shapes, or pitch threads, there is usually a better way than the above (see especially Chap. 11).

Fig. 6-10 An unusually long thread being cut on an engine lathe. Notice the "follow rest" supporting the radial and tangential tool pressures. (*Clausing Corp.*)

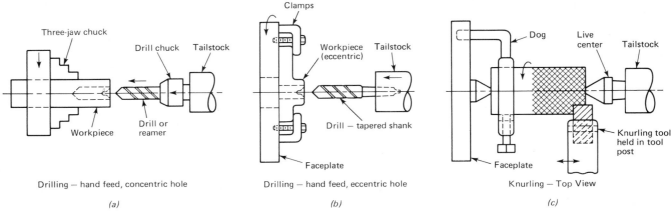

Fig. 6-11 More work which can be done on an engine lathe.

drilling and reaming (Fig. 6-11a and b)

The only way drilling and reaming can be done is if the hole is at the center of the lathe's Z axis, that is, in line with the center of the tailstock. Work may be bolted onto a faceplate in an offset position (Fig. 6-11b) so that the desired hole is properly aligned.

The drill or reamer may be held in a drill chuck in the tailstock, or it may have a tapered end which fits directly into the tailstock (Fig. 6-11b). Sometimes reamers are held in "floating" holders in the tailstock. These allow the reamer to center itself on the previously drilled hole.

CAUTION: Deep holes (over three times the diameter of the drill) tend to "run out." The reamer *will* *not* correct this condition. The hole must be bored if alignment is important.

knurling (Fig. 6-11c)

Creating a rough geometrical surface around a part is called *knurling*. This may be either straight or diamond pattern and may be a fine, medium, or coarse pattern.

This is done either for appearance or to make a surface which may be gripped securely by the hand or fingers. Several styles of knurling tools are available. The one shown in Fig. 8-19 is the type most used on engine lathes.

Knurling is done at one-third to one-half of the speed of turning but with a high feed (0.030 to 0.040 in./rev or 0.750 to 1.0 mm/rev). It will take from one to five passes of the knurling tool to create a good, sharp pattern.

CAUTION: The knurling operation increases the outside diameter of the work by from 0.010 to 0.050 in. (0.250 to 1.25 mm). This may not make any difference and may even be of advantage if a temporary press fit is needed, but this *must* be remembered when planning your work.

cutoff or parting (Fig. 6-12)

A thin blade in a special holder is used for cutoff or parting, usually to cut a specific length off a bar of stock. The sharpening and setting of the tool bit and the speeds and feeds are more critical than in turning because the cutoff blade is "buried" in the work.

This same type of tool can be shaped so that it will cut *grooves* of any width with flat or rounded bottoms. Cutting these grooves is called *necking*.

Getting exact *lengths* (along the Z axis) on an engine lathe is more difficult than getting exact diameters. Special stops with dial indicators and other measuring devices are used. As will be seen in Chap. 7, this is more easily done on the automatic or turret lathes.

taper cuts (Fig. 6-13)

Taper cuts may be made in five different ways on an engine lathe.

1. Short tapers may be made by *setting the compound* at an angle and hand feeding, using the compound lead screw (Fig. 6-13a).

2. Long, small-angle tapers may be made by *offsetting the tailstock*. A simple geometric calculation, involving the length of the taper and the angle, must be made to calculate the amount of offset (Fig. 6-13b).

Fig. 6-12 A cutoff (parting) operation on an engine lathe.

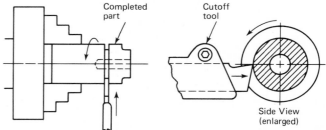

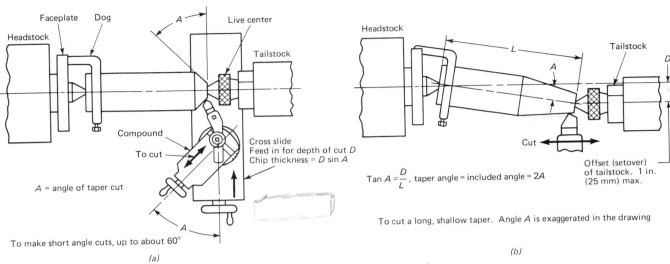

Headstock   Faceplate   Dog   A   Live center   Tailstock

Compound
To cut
Cross slide
Feed in for depth of cut D
Chip thickness = D sin A

A = angle of taper cut

A

To make short angle cuts, up to about 60°

(a)

Headstock   L   A   Tailstock   D

Cut

$\text{Tan } A = \dfrac{D}{L}$, taper angle = included angle = 2A

Offset (setover)
of tailstock. 1 in.
(25 mm) max.

To cut a long, shallow taper. Angle A is exaggerated in the drawing

(b)

Fig. 6-13   Two methods of making taper cuts on an engine lathe without special attachments.

Fig. 6-14   A small N/C lathe with a four-sided turret on the cross slide. Note that there are no handwheels. (*Clausing Corp.*)

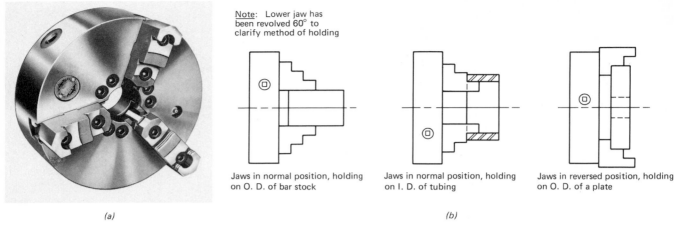

Note: Lower jaw has been revolved 60° to clarify method of holding

Jaws in normal position, holding on O. D. of bar stock

Jaws in normal position, holding on I. D. of tubing

Jaws in reversed position, holding on O. D. of a plate

(a)

(b)

Fig. 6-15   (a) A three-jaw universal chuck with reversible top jaws. (*Cushman Industries, Inc.*)
(b) Some ways of using the three-jaw universal chuck.

3. A *taper attachment* is available on many lathes. This is attached to the rear of the lathe.

4. A template may be used with a tracer attachment (Fig. 6-25) on the lathe. This type of attachment is discussed in more detail later in this chapter.

5. *Numerical control lathes* (Figs. 6-14 and 6-26) can cut a taper by simply specifying the X (offset) and Z (length) dimensions of the cut and punching these dimensions into the tape.

WORK HOLDING

As the engine lathe is often used for lot sizes of 50 or more, special clamping fixtures and specially bored chuck jaws may be worthwhile. However, a few standard items will hold 90 percent of all the lathe work.

Three-Jaw Chuck

Sometimes called a *scroll, self-centering,* or *universal chuck,* the three-jaw chuck (Fig. 6-15) is the most frequently used work-holding device. The closing force is exerted through an internal *scroll plate* which has a continuous "thread." Thus, all three jaws close uniformly and automatically center a round or hexagonal workpiece. These jaws come in matched sets of three and are curved to hold either the inside or the outside of the work (Fig. 6-15b).

Special *soft jaws* are often bought. They come unshaped and can be turned, bored, or milled to fit precisely the workpiece's diameter or shape.

CAUTION: If more than 3 to 6 in. (75 to 150 mm) of the workpiece projects from the chuck, the tangential and radial cutting forces may deflect the work. This will cause chatter, taper, and eccentricity, and occasionally even push the part out of the jaws. Long workpieces must be turned between centers.

Chucks are accurately made but are subjected to considerable wear, so they are not usually used when concentricity is critical. The size of a three-jaw chuck is its outside diameter. They are made from 5 in. (125 mm) to 48 in. (1200 mm) and larger. A 10-in. (250-mm) chuck costs about $350.

Four-Jaw Chuck

Sometimes called an independent chuck, the four-jaw chuck (Fig. 6-16a) not only has one more gripping surface but each jaw is independently adjusted. Thus, it can be used for two special purposes (Fig. 6-16b):

1. It can securely grip rectangular and irregular shapes so that drilling or boring of off-center holes can be done.

2. Eccentrically located holes on round work can be centered and, thus, easily drilled, bored, tapped, etc.

Accurately centering work on a four-jaw chuck is somewhat time-consuming as moving one jaw affects the others, so several trials are needed. Thus, if very many pieces are to be machined, special fixtures or some other method of holding the work is usually used.

Between Centers

Between centers (Fig. 6-10) is the most accurate, solid method of holding parts made from bar stock such as shafts. This requires a *dog plate,* some type of dog (Fig. 6-17), and two centers. The centers are held in place by a "locking taper" such as Morse, Brown &

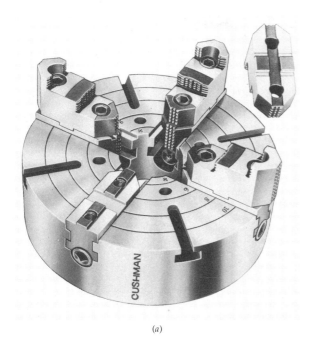

(a)

Note: Fourth jaw omitted for clarity.

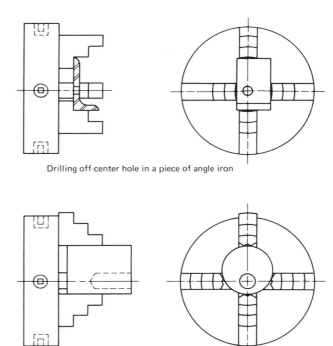

Drilling off-center hole in a piece of angle iron

Drilling off-center hole in round bar

(b)

Fig. 6-16 (a) A four-jaw independent chuck with reversible jaws. (*Cushman Industries, Inc.*) (b) Two examples of drilling or boring eccentrically located holes.

Sharp or Jarno (see Appendix D). The work must have 60° center holes drilled in both ends.

The driving dogs come in sizes from $\frac{3}{8}$- to 6-in. (9.5- to 150-mm) diameter capacity and several shapes.

CAUTION: The dog takes quite a bit of space on the end of the work. To machine this section, the workpiece must be turned end-for-end, and a second operation performed.

The "live" center in the headstock rotates with the work. It is a solid piece and it is "dead" insofar as any rotational wear is concerned. The center in the tailstock was originally also a "dead," solid, fixed piece. As the work turned on it, it had to be lubricated, often with white lead in oil. If the operator

forgot to keep applying the lubricant, the center would get hot and "burn," that is, soften to the point of losing its shape. Carbide-tipped centers are an improvement although they are brittle.

Today the live center (Fig. 6-18) is used a great deal. This center has thrust and rotating bearings so that the tapered end can rotate with the work. They are

(a)

(b)

(c)

Fig. 6-17 Dogs for use in lathe work. (a) Safety screws. (b) Heavy service. (c) Universal dog for round or square work. (*Vulcan Tool Co.*)

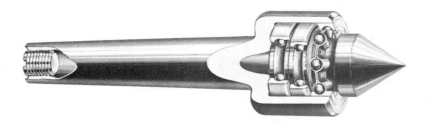

Fig. 6-18 A heavy-duty live center, showing the thrust bearings. (*Royal Products Division of Curran Mfg. Co.*)

accurate to 0.0005 to 0.0001 in. (0.0125 to 0.0025 mm) or better and are very durable. They are made in many sizes, the largest of which will take a thrust of over 10,000 lb (44.5 kN or 4.5 metric tons).

Spring Collets

These are made with round, square, hexagonal, and special-shaped openings (Fig. 6-19). The collet is threaded at the back end so that turning the drawbar pulls the collet into a tapered hole in the spindle nose which squeezes the collet against the work. This is the most accurate way of holding work on center without center drilling. Spring collets are available in sizes from $\frac{1}{16}$- to 1-in. (1.6- to 25-mm) diameter. They have only about $\frac{1}{64}$-in. (0.4-mm) adjustment, so a separate collet must be purchased for each size and shape.

Jacobs Collet Chuck

This type of chuck (Fig. 6-20) is similar to the spring collet but uses "rubber-flex" gripping collets, which distribute the load over more holding surfaces, keep the grip parallel to the work, and are less likely to mark the work than other methods. These are available to hold work from $\frac{1}{16}$- to $1\frac{3}{8}$-in. (1.6- to 35-mm) diameter.

This collet is tightened by an impact or "hammer" blow delivered by using the handwheel. Each rubber-flex collet has a $\frac{1}{8}$-in. adjustment, so no special sizes are needed for metric or decimal size work. Accuracy is excellent, 0.001-in. (0.025-mm) runout or better at the spindle nose. Cost is about $500 for a complete set.

Face Driver

A face driver (Fig. 6-21) can be used at the headstock end of work mounted between centers. This replaces the dog and faceplate. The face driver may fit into the taper in the headstock or be mounted like the chucks.

This driving method is especially valuable because it allows the entire length of a workpiece to be machined as there is no driving dog in the way (see Fig. 6-21b).

These drivers are used for many sizes of work from valves used on automobile engines up to 16,000-lb [7257-kg (mass)] rubber-mill rolls.

The basic principle is simple. The center point, which locates the part, is spring-loaded. The five or more hardened drive pins are backed by either a totally enclosed hydraulic chamber or an elastomer equalizing material. Thus, the penetration of all pins will be the same within about ±0.002 in. (0.050 mm) even though the surface may be somewhat rough. Total penetration will range from 0.010 to 0.020 in. (0.25 to 0.50 mm) per pin.

Work from 0.27- to 13.57-in. (7- to 345-mm) diameter can be held with stock-size drivers. Steel, cast iron, or nonferrous metals up to a hardness of 36

Fig. 6-19   A partial set of collets, with drawbar and nose piece for a small lathe. (*Photograph by Samuel Lapidge.*)

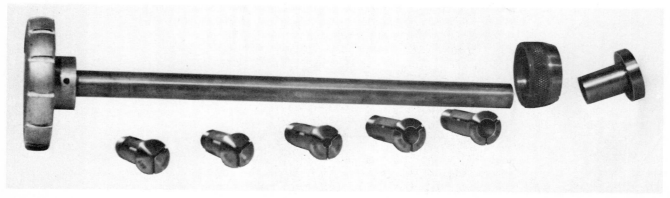

Fig. 6-20. The Jacobs spindle nose lathe collet in use on a lathe. Insert shows the multiple gripping teeth set in rubber. (*The Jacobs Manufacturing Company, a subsidiary of Chicago Pneumatic Tool Co.*)

Fig. 6-21 (*a*) A medium-sized face driver. (*b*) One way of using a face driver to save money. (*Madison Industries, Inc.*)

(a)

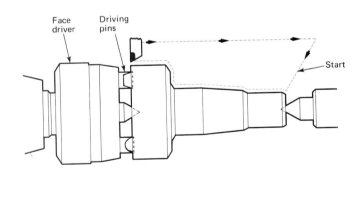

(b)

HRC can be handled. With extra care, work up to 40 to 42 HRC can be turned or ground.

The thrust force necessary to seat the driving pins into the work is from under 500 to over 5000 lb (2.2 to 22.4 kN). Up to 3000 lb (13.4 kN) thrust can be applied by exerting a hard pull on the tailstock handwheel of a medium-sized lathe. For heavier pressures, a hydraulic cylinder is incorporated in the tailstock.

The face driver costs from $270 up. The driving pins can be resharpened or new pins can be bought quite inexpensively.

Spindle Noses

Spindle noses, on which all chucks can be mounted on engine lathes, are of three types.

(a)   *(b)*   *(c)*

Fig. 6-22   Three types of lathe spindle noses. (*a*) Threaded; (*b*) camlock; (*c*) taper key.

1. *Threaded nose* (Fig. 6-22*a*) is supplied with the small- and medium-sized lathes. The spindle thread varies from 1 in. 10 threads to 2¾ in. 8 threads. Standard metric threads have 2 mm (12.7 threads per inch) to 4 mm (6.35 threads per inch) pitch and diameters close to the inch equivalents.

2. *DI camlock* spindle nose (Fig. 6-22*b*) is made in nominal sizes from 3 to 20 in. (75 to 500 mm) outside diameter. Two to six pins on the chuck project into the spindle nose and are locked into position by cams which are tightened with a standard tee-handle wrench. There is a short taper which accurately locates the chuck on center. This type of fastening is quick and easy to use, and it centers the chuck accurately.

3. *Type L taper key drive* (Fig. 6-22*c*) uses a long taper to drive it and a threaded collar to pull it up tight. The collar is tightened with a spanner wrench. These are made in sizes numbered from L00 to L3, with flange diameters from 3.5 to 10.0 in. (90 to 250 mm).

Mandrels

Sometimes mandrels are used so that the entire length of a workpiece can be machined. These hold and locate a part from its center hole (Fig. 6-23).

Taper mandrels can be purchased but are frequently made to fit one special piece of work. The very shallow taper, 0.006 to 0.008 in./ft or 0.5 to 0.67 mm/m, locks the work securely as long as the cutting tool pressure is toward the large end. The mandrels are supported on centers and driven by a dog and faceplate. The concentricity of work done on a mandrel is usually excellent. It is limited by the accuracy of the mandrel and by the alignment between the tailstock and headstock centers.

VARIATIONS OF THE BASIC ENGINE LATHE

Gap Bed Lathe

This lathe is used when large-diameter faceplate work is needed (Fig. 6-24). The lathe shown has a special extension to the left on the compound to hold the cutting tools close to the work.

Toolmaker's Lathe

A toolmaker's lathe (not illustrated) appears to be about the same as the engine lathe; however, it is more accurately made and sometimes has more feeds and speeds.

Tracer Lathe

Sometimes called a *duplicating lathe*, the tracer lathe (Fig. 6-25) is often an engine lathe with a tracer attachment. This attachment is usually fastened to the back of the lathe although a few are fastened to the front.

The lathe *compound* is permanently or temporarily replaced by one made especially for tracer work. The

Fig. 6-23   A solid tapered mandrel shown in working position.

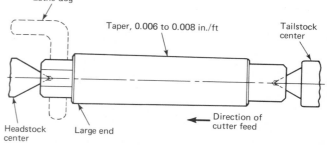

cross-slide lead screw is used for the initial and subsequent depth settings, but the in and out (X axis) movement is controlled by hydraulic or air pressure which, in turn, is controlled by the action of a stylus which moves along a template (pattern) which has the desired form.

The *template* is usually handmade from ⅛- to ⅜-in.-thick (3.0- to 10.0-mm) steel or aluminum. The shape may be rough saw cut and then filed or machined to the final size, shape, and accuracy. Some templates are actual full-size parts as shown in Fig. 6-25.

Costs of the tracer lathe attachment may be from a few hundred to a few thousand dollars. The templates are relatively inexpensive to make (from $30 to $150), and setup time is about 20 to 40 min.

Thus, a tracer lathe can be used economically on even small lots of 10 or 15 pieces if the shape needed would otherwise require several tool settings or contour tools especially ground for curves and special angles.

Accuracy of the tracer is excellent. Many will be accurate within ±0.001 in. (±0.025 mm) of the template shape, and successive parts may compare with each other even more closely.

CAUTION: 1. Shapes such as grooves with 90° sides and undercuts usually cannot be made with the tracer attachments.
2. The tool bit is usually one with a 55° or 60° point, set so that it can cut with either side.
3. Before deciding to duplicate a complex shape, the cutting positions of the tool and the actions of the combined X and Z motions must be carefully considered.

N/C Engine Lathes

The N/C lathe (Fig. 6-26), while it costs $25,000 to $150,000, can do an amazing variety of work to accuracies ±0.001 in. (0.025 mm) and often better.

The first thing one notices is the lack of any handles. They are not needed because all functions are controlled by the codes which have been punched into the tape. This includes X (cross slide) and Z (lengthwise) movements, spindle speeds, feedrates, and rotating the toolholder to several positions.

Fig. 6-24   Gap bed lathe with 12-in. (300-mm) space between face plate and carriage. Larger gaps are available. (*LeBlond.*)

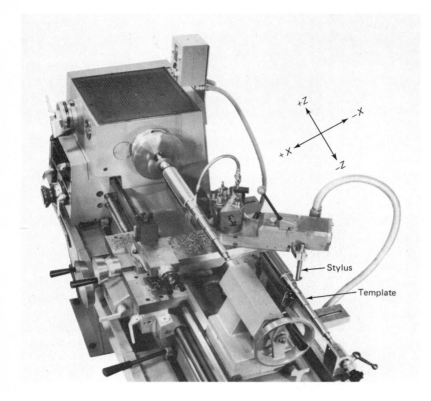

Fig. 6-25 A tracer lathe. Either a round or flat template may be used. (*Clausing Corp.*)

These lathes can cut slopes, circular arcs, and threads. They can do work between centers (such as shafting) or faceplate work (such as wheels) of many sizes and shapes.

At least four cutting tools can be mounted in the square cross-slide turret. The turret can be indexed (by tape control) to any one of 8 or 12 positions.

Cutting tools are *preset* to the length and position specified by the. N/C programmer, so setup and changing of dull cutters is fast allowing the N/C lathe to spend more of its time cutting.

Writing the "program" giving the X and Z coordinates and other commands is not difficult. Many times all calculations are done by hand, though the computer is often quicker, especially if the shape is at all complicated. Several simplified computer programs are available so that the programmer needs only to learn a sort of "pidgin English" version of regular shop words in order to write a computer program.

tracer lathe compared with N/C lathe

The tracer lathe costs considerably less than the N/C lathe, and it is simpler to set up. The cost of a template is the same or less than the cost of making an N/C tape. However, the N/C lathe can use four or more preset tools and can position them at various angles. Thus, it can make plunge cuts and undercuts, which are impossible on a tracer lathe. The N/C lathe can rough and finish turn with separate tool bits, cut radii, grooves, and threads—all in one setup without disturbing the workpiece.

The accuracy of the tracer lathe is limited by the accuracy of the template. The N/C lathe's accuracy is limited only by the condition of the ways, lead screws, etc., of the machine, and the smallest motion possible due to the tape reader and the driving motors. This is usually ± 0.001 to ± 0.0002 in. (± 0.025 to ± 0.005 mm).

For quickly and accurately duplicating relatively simple shapes, the tracer lathe is the most profitable. If the part requires other operations such as grooving or threading, the N/C lathe is probably preferable. If the tracer lathe is used, these other operations require extra setup time, loading and unloading time, etc.

ECONOMICS OF ENGINE LATHE USE

In the general family of machine tools, the engine lathe is relatively inexpensive. It is a fairly simple, easily repaired machine which can be expected to last 10 or more years with low upkeep expense. As it is very versatile, the use factor is usually quite high, and it is made in a wide variety of sizes and styles.

Setup time (that is, the time to get all the tools, chucks, etc., in place on the machine and ready to start cutting) is quite low; often 10 to 20 min is enough

Fig. 6-26 A modern slant bed N/C lathe. Notice the two turrets. These, with the tail stock, allow this machine to be used for either chucking or between centers work as shown. (*Jones and Lamson Division of Waterbury Farrel*)

except on very large lathes which require the use of a chain hoist. The short setup time makes the engine lathe especially economical when only 1 to 20 parts are needed.

However, it can, in general, make only one cut at a time, and then the cutting tool has to be changed and/or reset. Because of this, it is used principally for rather simple work in medium quantity lots or for complex work when only one or two parts are needed as in a model shop or toolroom.

Cutting time, based on allowable feeds and speeds, is as fast as on any machine.

EXAMPLE 6-1

If a stainless-steel shaft of 1.50-in. (38-mm) diameter is to be turned down and the recommended cutting speed is 40 fpm (12.2 m/min), with the feed set at 0.010 in./rev (0.250 mm/rev), the time required to turn a 3-in. (76-mm) length will be

$$\text{rpm} = \frac{4(40)}{1.5} = 107 \text{ rpm} \qquad \text{(Eq. 4-2)}$$

$$\text{Feed} = (107)(0.010) = 1.07 \text{ ipm} \qquad \text{(Eq. 4-5)}$$

$$\text{Time} = \frac{3.00}{1.07} = 2.8 \text{ min cutting time}$$

Using metric dimensions:

$$\text{rpm} = \frac{300\ (12.2)}{38} = 96 \text{ rpm} \qquad \text{(Eq. 4-4)}$$

$$\text{Feed} = (96)(0.250) = 24.1 \text{ mm/min} \qquad \text{(Eq. 4-5)}$$

$$\text{Time} = \frac{76}{24.7} = 3.15 \text{ min cutting time}$$

Note: The difference in time between metric and inch dimensions is due to the approximations used in each equation, as discussed in Chap. 4.

Time for a facing cut would be computed the same way. The *length* of the facing cut is the *radius* of the work for a solid bar. The rpm is computed by using the largest diameter. Because the diameter being cut decreases to zero at the center, the cutting speed will also decrease to zero. However, it is not usually economical to change speed during the cut.

For economical production of large quantities, automatic and semiautomatic lathes are used. These are discussed in Chap. 7.

Review Questions and Problems

6-1. On which parts of an engine lathe are the following fastened or mounted?
   a. Three-jaw chuck
   b. Tool post
   c. A live center
   d. The bed
   e. The compound
   f. Which of these can be rotated around a vertical axis?

6-2. What is the purpose of the ways on a lathe? How many are there?

6-3. a. What is the largest diameter workpiece which can be held in a 500 × 14,000 mm lathe (in millimetres and in inches)?

   b. What kind of engine lathe is made so that extra large diameter work can be machined?

6-4. Can a boring cut be made at 10 mm off center, held in a three-jaw chuck in a lathe? Why or why not?

6-5. In what ways are turning and facing alike and different?

6-6. Name and sketch three other types of cuts or work which can be done on an engine lathe.

6-7. What would you consider the best and the next best ways to make a 15° taper (included angle) cut, 10 in. long, on a 3.5-in.-diameter bar? Why?

6-8. Would you estimate that the three-jaw chuck, four-jaw chuck, or faceplate would be most frequently used on an engine lathe? Which is the least used?

6-9. What is the largest and the smallest size collet mentioned in this chapter? Why would you use them?

6-10. If you wished to turn a 150-mm-diameter shaft down to 125 mm at both ends, how would you prefer to hold this 1500-mm (60-in.) long workpiece?

6-11. If, in question 10, the ends are to be reduced in size for a distance of 200 mm each, how long would it take? Assume that the deepest roughing cut is 15 mm, roughing feed is 0.32 mm/rev, finishing feed is 0.12 mm/rev, and cutting speed is 150 m/min.

6-12. a. Convert the above to English units and refigure the time.

   b. Explain why the answers are not the same.

   c. Should both times theoretically be the same?

6-13. If you were running a small shop, doing about $50,000 per year of contouring work, would you buy a tracer lathe or a numerical control lathe? Size needed is 10 × 48 in.

Note: A Flexowriter for punching N/C tapes costs about $3500.

References

1. *How to Run a Lathe*, South Bend Lathe Co., South Bend, Ind.

2. Krar, O., Oswald, and St. Amand: *Technology of Machine Tools*, McGraw-Hill, New York, 1969.

3. Nordhoff, W. A.: *Machine-Shop Estimating*, 2d ed., McGraw-Hill, New York, 1960.

4. Roberts, A. D., and R. C. Prentice: *Programming for Numerical Control Machines*, McGraw-Hill, New York, 1968.

5. ASTME: *Tool Engineers Handbook*, 3rd ed., McGraw-Hill, New York, 1976.

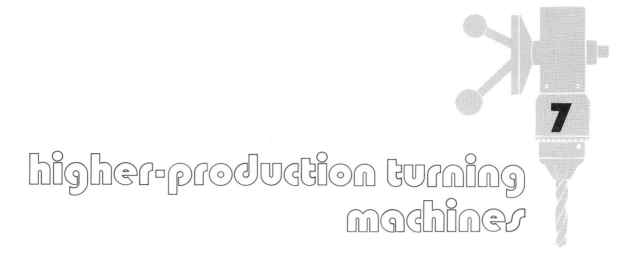

higher-production turning machines

Even though these are called "turning" machines, they do the same things an engine lathe can do, that is, boring, facing, cutoff, drilling, reaming, etc.

TURRET LATHES

A very versatile, relatively low cost, *hand-operated* lathe for medium production quantities (from 30- to 300-piece lots) is the *turret lathe*. These come in several sizes and two basic styles and cost from $23,000 to $85,000 plus tooling. Both have the basic motions shown in Fig. 7-1.

One feature which makes the turret lathe so versatile is the six-sided turret on which a wide variety of cutting tools may be mounted. Each time the operator brings the turret lathe back from the workpiece, the turret indexes clockwise, bringing a new tool (or set of tools) into position for cutting.

The second feature is the two toolholders which can be mounted on most turret lathe cross slides. The front toolholder is usually a four-sided one as shown in Figs. 7-1 and 7-2 and in detail in Fig. 7-4a. This toolholder can be indexed quickly and accurately to bring turning tools, facing tools, grooving or chamfering tools, etc., up to the work.

The rear tool post (Fig. 7-4b) usually holds a single tool, although it is not limited to this. This tool is most frequently a cutoff tool (sometimes combined with a chamfer), although forming and necking (grooving) tools can be used. Thus, the turret lathe has 11 stations which can hold one or more cutters each. Some of the most frequently used toolholders (for turret mounting) are shown and described in Figs. 7-5 through 7-7.

The dials on the cross slide are graduated in 0.001 in. or 0.02 or 0.025 mm, as desired. The lead screws are, of course, threaded in U.S. or metric to correspond to the dials. Tolerances of ±0.002 in. (0.05 mm) can be easily held on diameters. Lengths can be held to ±0.005 in. (0.125 mm). Closer tolerances can be held if conditions are right.

Ram-type Turret Lathe

This is a light- to medium-duty turret lathe (Fig. 7-1), quick and easy to handle. The turret is mounted on a ram which slides in ways machined on the saddle. The complete saddle can be positioned forward or back to bring the turret close to the work, and then it is locked in place.

These lathes are made in sizes and power as shown in Table 7-1. When handling long bars of stock fed through the headstock, collets are often used to hold the work. When operating as "chuckers" with each part being hand loaded, various chucks are used. The chucks may be the conventional hand-operated three-jaw universal type or special two- or four-jaw models. Frequently air or hydraulically operated "power" chucks are used for speed in loading and control of pressure on the workpiece.

Power feed is available for the turret (ram) and in both directions (X and Z) on the cross slide, though hand feed may also be used. Six to twelve feedrates may be available.

Speed may be from 50 to 4000 rpm depending on the size of the lathe. Larger lathes have lower speeds.

Trip stops, which stop the feeding motion of the turret at any predetermined point, are easily adjusted for each individual turret face.

Saddle-type Turret Lathe

This is a heavy-duty turret lathe in which the turret is mounted directly on the saddle (Fig. 7-2), and the saddle slides on the ways of the machine. This makes

TABLE 7-1   DATA ON TURRET LATHES

Sizes	Bar Stock Capacity*	Chuck sizes	Drive Motor	Swing over Ways	Cost (No Tooling)
			Ram Type		
1 to 5	1–2½ in. (25–64 mm)	10–15 in. (250–380 mm)	15–25 Hp (11–18.6 kW)	Up to 22 in. (550 mm)	$10,000 to $50,000
			Saddle Type		
1L–5L or 1A–5A	2½–9 in. (60–230 mm)	Up to 36 in. (910 mm)	20–75 Hp (15–56 kW)	Up to 36 in. (910 mm)	$20,000 to $75,000

*Through the spindle.

a more rigid turret, avoiding the overhang which occurs when a ram-type turret is advanced to the work and the possible sagging out of line.

The capacities and power available with saddle-type turret lathes are shown in Table 7-1. These vary with each manufacturer.

Powr feed is standard on turret and both axes of the cross slide. *Speeds* run from 20 to 1500 rpm depending on the size of the turret lathe.

These heavy-duty turret lathes use much the same type of tooling as the ram type. However, the tooling is often much larger, in proportion to the large work which these lathes can handle.

A *cross-feeding turret* is sometimes built into the

Fig. 7-1   A ram-type turret lathe. (*The Warner & Swasey Co.*)

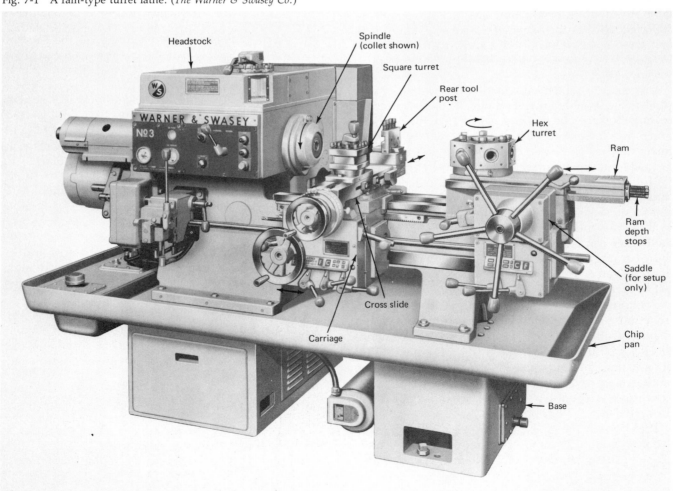

saddle. This permits facing cuts, interior grooves, and large-diameter boring to be done from the turret.

Figure 7-3 shows a turret lathe similar to Fig. 7-1, but equipped with a semiautomatic feed so that long lengths of bar stock can be machined, cut off, and fed in for the next piece.

Economics of the Turret Lathe

Using a turret lathe instead of an engine lathe can often cut machining time by 25 to 75 percent and handling time by 25 to 50 percent. Moreover, as many cuts are made with the work in one place, concentricity is more easily maintained. However, the original cost of the turret lathe is considerably higher than that of the engine lathe. Tooling is expensive [a slide tool (Fig. 7-6) can cost up to $1000], and setup time will run from 1 or 2 h on small jobs up to 8 h for a complex job on the large machines.

Machining time, as compared with an engine lathe, is *not* shortened. Feeds and speeds remain the same for all types of lathes. The time saving is principally because of the *combined cuts* possible on the turret lathe and because so much machining can be done with only one chucking of the part. For example, the cross-slide tools may be turning one diameter while the turret tools are boring and turning two or three other diameters.

Notice that this is not *all* gain, as the rpm must be set for the largest diameter being machined during the combined cut. Thus the boring operation, for example, will be done more slowly than if it were done alone.

Figure 7-8 shows a fairly simple turret lathe setup using all stations except the back tool post and the sixth turret side. Most of the stations have a single cutter doing a single job. This is quick and easy to set up, and replacing dull tools will not require much adjustment. Production was reported as 2.4 pieces per hour.

Figure 7-9 shows a quite complex tooling setup which was designed for a high production job. Only 5 of the 11 possible stations are used, which cuts the turret indexing time about in half. The special revolving center pilot allows somewhat heavier feeds, which also cuts the time. The production was increased from 6 to 9 pieces per hour when this tooling was used as compared with the previous tooling, which was similar to that in Fig. 7-8. However, the tooling cost about $1000 more, the setup time was longer, and it is somewhat more difficult to replace one cutter without disturbing the others. Nevertheless, a considerable cost saving was made. Breakeven point was 4700 pieces.

Layout drawings such as Figs. 7-8 and 7-9 are frequently made by the methods department, especially

Fig. 7-2 A saddle-type turret lathe with cross-sliding turret but no rear tool post. (*The Warner & Swasey Co.*)

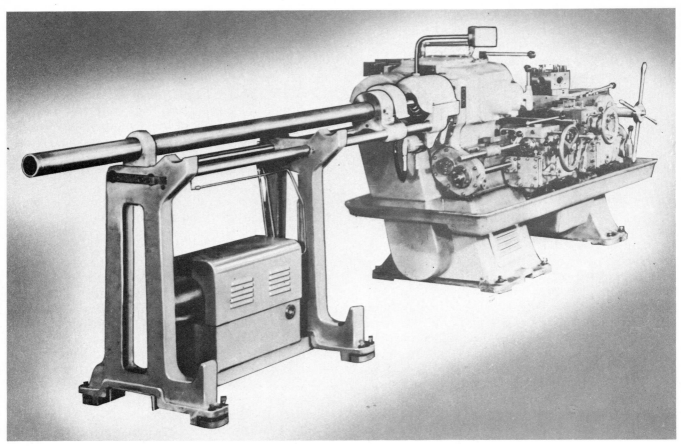

Fig. 7-3   A ram-type lathe equipped with hydraulic bar feed and collet chuck. [*Jones & Lamson Division of Waterbury Farrell (A Textron Co.).*]

(a)   *(b)*

Fig. 7-4   Turret lathe toolholders. (*a*) Four-sided front toolholder; (*b*) rear tool post.

on jobs which will be run repeatedly. They are not difficult to make, and these drawings assure that the proper setup will be used each time the job is run. Tool numbers and tool-bit sizes and compositions are added to the sketches shown. Often a complex assembly of cutters (such as on turret 1, Fig. 7-9) is never taken apart but stored as an assembled tool. This could save an hour or two on setup time.

Automatic Turret Lathes (ATL)

The ATL (automatic turret lathe) (Fig. 7-10) is built similar to a standard saddle-type turret lathe. It can be used for both bar-stock work and chucking. Bars up to about $4\frac{1}{2}$-in. (114-mm) diameter and chucks to 15 in. (381 mm) can be used. The motor is about 30 hp (22.4 kW), so that heavy cuts can be made. The bed, saddle, cross slide, etc., are extra solidly built in order to handle the high power and the continuous cycle and still maintain close tolerances of ±0.001 in. (0.025 mm). These machines cost from \$45,000 to \$85,000 plus \$700 to \$3000 per job for tooling.

 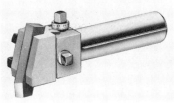

(a)   *(b)*

Fig. 7-5 *(a)* Plain long flanged toolholder. *(b)* Adjustable cutter holder, also made nonadjustable. (*The Warner & Swasey Co.*)

The ATL, once it has been set up, operates completely automatically. The operator merely loads and unloads the work and inspects it so that tools can be changed or sharpened as needed.

Indexing (which is *counterclockwise* so that the operator's hand cannot be pulled into the work) is automatic, and very fast. Speeds, feeds, and lengths of cut are all automatically controlled and changed automatically, for every station if necessary.

The panelboard (Fig. 7-11) shows one type. Some easily set dials on the turret and cross slides can be preset for any job. The methods engineer marks a setup chart (a line drawing of the panel), and the person preparing the setup easily follows this to "program" the machine.

The toolholders and the cutting tools used are basically the same as on the hand-operated turret lathe. Thus, setup time is about the same. However, many time-saving attachments such as a hexagon turret-tracing attachment, power-operated slide tools, hydraulic cutoff slides, and drill relieving unit (for "wood-peckering" in holes) are available.

The combination of quick, accurate setups, fast indexing, and versatile tooling makes the automatic turret lathe an economical choice for moderate-sized lots (about 100 to 500 pieces) of fairly complex, close tolerance lathe work.

Numerical Control Turret Lathe (NCTL)

The numerical control turret lathe is, like the N/C engine lathe, made in a wide variety of sizes and styles. The one shown in Fig. 7-12 is typical of the popular slant-bed type. The NCTL costs from $50,000 to $150,000. Tooling costs are low.

The NCTL and the ATL are in direct competition, as both can do much the same work. The ATL is faster to program because it does not require a tape. It is also easier to make changes on the ATL: simply change a panelboard setting instead of remaking a tape.

However, the N/C turret lathe can, with the proper

Fig. 7-6 Shank-mounting slide tool. The dial can have either English or metric scales. [*Jones & Lamson Division of Waterbury Farrell (A Textron Co.).*]

Fig. 7-7 *(a)* Multiple turning head. Note cross-slide tool also in use. Pilot bar at top. *(b)* Bar or roller turner. Rollers prevent deflection of unsupported bar stock. (*The Warner & Swasey Co.*)

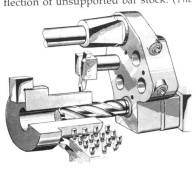

(a)

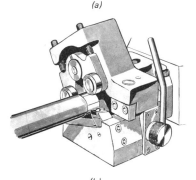

(b)

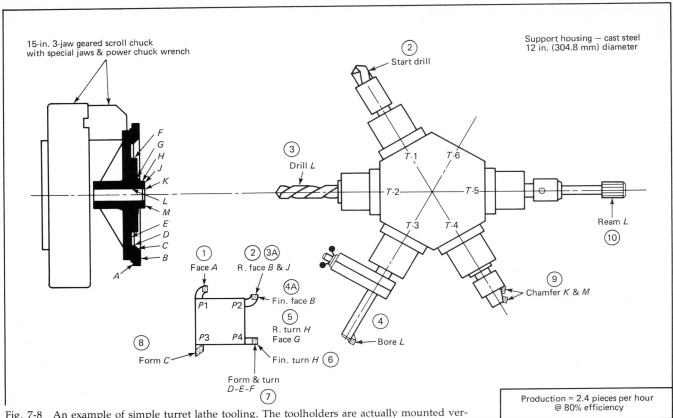

Fig. 7-8 An example of simple turret lathe tooling. The toolholders are actually mounted vertically, but are shown "flat" for easier identification. Circled numbers show sequence of operations. (*The Warner & Swasey Co.*)

Fig. 7-9 A fairly complex turret lathe setup which produced 50 percent more production. Note tools laid flat for easier understanding of the drawing. (*The Warner & Swasey Co.*)

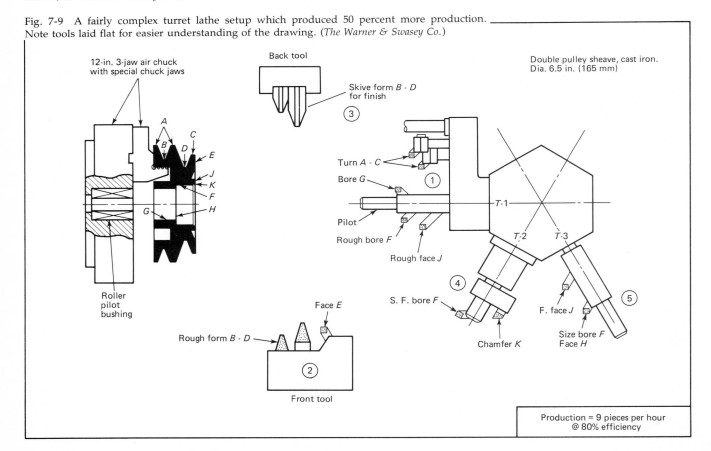

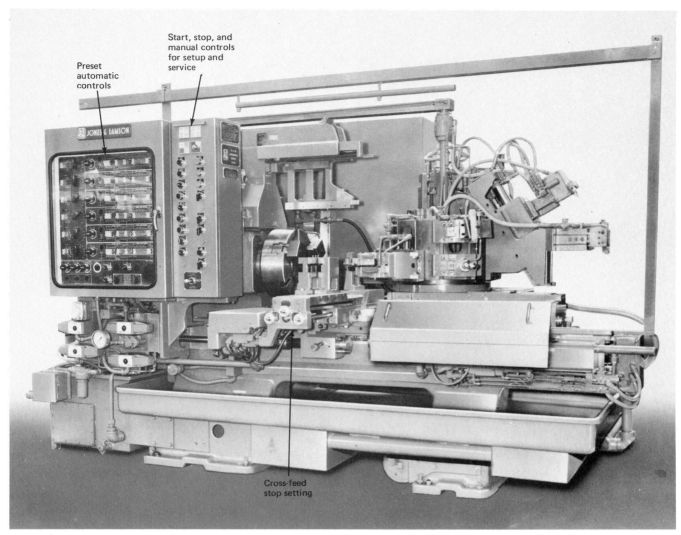

Fig. 7-10 An automatic turret lathe, setup speeds, feeds, and dimensions are set on the panel at the left. [*Jones & Lamson Division of Waterbury Farrell (A Textron Co.).*]

tape, cut curves, slopes, and profiles very accurately without any template. Also, once a tape is made and checked out, it is very simple to run the job again: no buttons, feed dials, or stops to set.

Using the NCTL does require some maintenance of electronic equipment and tape readers, which is an added expense and can cause "down time," although after the initial debugging, this is usually not more than 5 percent.

Possibly for work where either relatively long runs of complex parts or combined tooling at each turret can save a lot of time, the ATL would be the better buy.

For shorter runs (maybe 10- to 50-piece lots) of complex parts, where one tool can make several cuts when controlled by the tape, the NCTL would be preferred. Of course, once one or the other has been purchased, it can be used for a wide variety of work.

Small Single-Spindle Automatic Turret Lathes

There is a class of small automatic and semiautomatic turret lathes (Fig. 7-13) which is widely used for bar-stock work from under ¼-in. (6.35-mm) to 1-in. (25.4-mm) diameter. These are often called **automatic screw machines**, as making screws was one of their earliest uses.

A typical "automatic" is shown in Fig. 7-13a. These machines are especially economical for long runs of small parts (maybe 1000- to 5000-piece lots). "Small" parts might be from 0.020-in. (0.50-mm) diameter × ¼ in. (6.35 mm) long up to 1.00-in. (25.4-mm) diameter × 2 in. (50.8 mm) long.

The six-sided turret is most often mounted with a horizontal axis (as shown), although some have a vertical axis like standard turret lathes. The cross slide may have both front and rear tools similar to those used on conventional turret lathes but smaller in

111

Fig. 7-11   The setup sheet which is made out by a manufacturing or methods engineer. [*Jones & Lamson Division of Waterbury Farrell (A Textron Co.).*]

proportion to the work. Turning, drilling, boring, threading, facing, and cutoff may be done, just as on standard turret lathes.

All the motions of these automatic screw machines are controlled by cams (Fig. 7-13*b*) and trip dogs. The cams are designed and cut especially for each job. These and trip dogs control stock feeding, spindle speeds, infeeds, turret rotation, length of feed stroke, etc. The indexing is very fast, and the machines may run relatively unattended. The operator merely has to feed new bars of stock as needed, and check occasionally to see that tolerances are maintained and that dull cutters are replaced when necessary. Round, square, or hexagonal collets are available for standard sizes of bar stock, and specials can be supplied readily.

Figure 7-14 shows a few of the types of parts which are made on these machines. Some of these use one or more of the available special attachments. These include slot and flat milling and cross drilling.

swiss-type automatic screw machines
Figure 7-15 shows a rather unusual appearing but quite widely used type of small automatic bar-stock "lathe." It is called a "Swiss" automatic because it was first made in Switzerland especially to make small parts for watches.

These machines are used for long runs of relatively small work which can be made from bar stock. They do very accurate work, holding ±0.0005 in. (0.0125 mm) quite easily and ±0.0001 in. (0.0025 mm) without too much difficulty on both diameter and length. Once set up, they work completely automatically. The operator can often tend three to five machines since he or she merely puts in a new bar of stock as it

Fig. 7-12   A N/C turret lathe for chucking work only. Other models can do bar stock (shaft) work. [*Jones & Lamson Division of Waterbury Farrell (A Textron Co.).*]

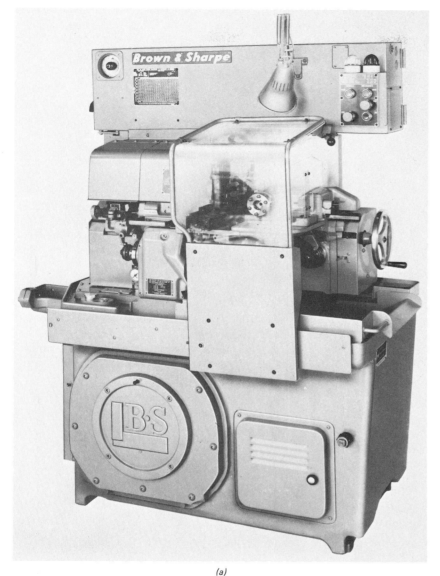

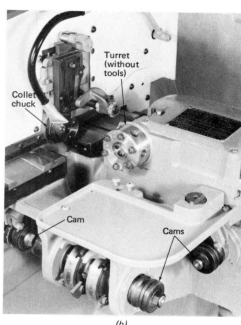

(a)

(b)

Fig. 7-13 (a) A ½- or ¾- in.-diameter (13 to 19 mm) bar capacity automatic screw machine.
(b) Some of the cams which control the cycle. (*Brown & Sharp Mfg. Co.*)

is needed and replaces cutting tools which get dull. Today an automatic bar feed is available. Cutters are simple ¼- or 5/16-in. (7- or 8-mm), HSS or carbide, square tool bits, ground with quite small rake angles.

Swiss automatics are made in metric sizes with a *maximum* capacity of 4 mm (0.157-in.) up to 32-mm-diameter (1.260-in.) bars. The smaller sizes can turn diameters as small as 0.020-in. (0.5-mm) diameter. The smaller machines can run up to 12,000 rpm and the larger ones to 3000 rpm.

The stock is held and rotated by an accurately made collet located at the back of the machine and additionally supported by a close-fitting guide bushing which is located only a few thousandths of an inch back of the cutting edge of the tools.

All motions of the Swiss automatics are controlled by cams which are especially designed for each job. Four of these cams feed the cutting tools, in and out, perpendicular to the work. This controls the diameter (see Fig. 7-16a). Another cam moves the work along endwise under the cutting tools. Thus, all turning is done at a point where the work is fully supported even though the part may be 2 to 8 in. (50 to 200 mm) long and with ¼-in. (6.35-mm) diameter. This long, thin part would be very difficult to turn on any other machine.

Setup time, if everything has to be changed, can take

Fig 7-14 Some parts which have been made on automatic screw machines. The holes were drilled with a special attachment. (*Brown & Sharpe Mfg. Co.*)

from 4 to 8 h. Cams will last for about 50,000 pieces if tolerance is very close, up to maybe 500,000 pieces on some jobs.

CAUTION: To get the accuracy which the Swiss automatic is capable of, only centerless ground stock should be used. You should not expect ½-in. (12-mm) cold-rolled steel with a tolerance of +0.000, −0.002 in. (0.05 mm) to center in the guide bushing to hold ±0.0005 in. (0.0125 mm).

Slopes and curves can easily be cut on the Swiss automatic since all that is required is for the cutting tool cam to move the cutter in or out at the desired ratio to the horizontal movement of the work. Examples of the kinds of work are shown in Fig. 7-16*b*.

A wide variety of attachments can be added to these machines for cross drilling, milling, tapping, threading, etc. Their cost without attachments is from $15,000 to $35,000.

LARGER AUTOMATIC LATHES

For larger bar-stock work and for machining parts which must be held in a chuck, there is a widely used group of heavy automatic and semiautomatic turning machines. These are called four-, six-, or sometimes eight-*spindle automatics*. They are used when lot size is 1000 pieces or more, though on the larger machines a lot size of 500 might be economical. The cost of these machines is from $40,000 to $71,000 plus tooling.

Multispindle Automatic Bar Machines

These machines can handle four, six, or eight bars of 12- or 20-ft (3600- or 6000-mm) lengths of round, hexagonal, or occasionally, square stock at the same time (Fig. 7-17). The stock is supported and is rotating in

Fig. 7-15 A Swiss automatic screw machine. (*American Bechler Corporation.*)

long tubes, with centering bushings, and gripped in quick-opening collets in the spindles which also rotate the stock.

The *size of the machine* is designated by the maximum diameter of round stock which it can handle. This may be 1 to $2\frac{5}{8}$ in. (25 to 67 mm). Some larger machines, up to $5\frac{1}{4}$-in.-diameter (133-mm) stock, are also made.

Production rate of the multispindle automatics is high because during every cycle, machining is being done at *every* spindle. Thus, each time the cutting tools back away from the work, another part has been completed and the bar stock at that station moves out to start the next piece. Thus, production of 100 to over 500 pieces per hour is normal except on the larger machines.

Rotation of the spindles and indexing of the spindle carrier is shown in Fig. 7-18. Of course, all the stock tubes are indexed at the same time.

Tooling of these machines can be quite elaborate. As shown in Fig. 7-17, and schematically in Fig. 7-18, each station has an *end slide* and a *cross slide* which can hold tools for work at that station.

The *end slide* carries six sets of tools (for a six-spindle machine). All these tools (drills, reamers, roller turners, etc.) are fed in the same distance at the same feedrate, actuated by a single feed cam.

The *cross slides* are fed in by separate, adjustable cams. They may be tooled up with grooving, facing, shaving, cutoff tools, etc., and can usually be cutting at the same time the end slide tools are being used. Thus, on a six-spindle machine, up to 12 separate operations may be done at the same time. Actually some stations may have more than one cutter similar to the setup on the turret lathe.

Setup time will vary according to the complexity of the job and how many tools must be used. If the stock size and all the tooling must be changed, it could take 6 to 12 h before production is started. An average of 4 to 8 h might be normal.

A sample job is shown in Fig. 7-19 along with the explanation of the work being done. This is a very simple job with short cycle time. Much more complex work is frequently done.

Multispindle Automatic Chucking Machines

If several operations must be done on castings, forgings, or the second end of bar-stock work, the parts are usually held in two-, three-, or four-jaw quick-acting chucks. If the quantity to be machined during one run (the lot size) is 500 pieces or over, the automatic chucking machine (Fig. 7-20) may be economical. This machine operates in exactly the same way as the bar-stock machine except that each piece must be

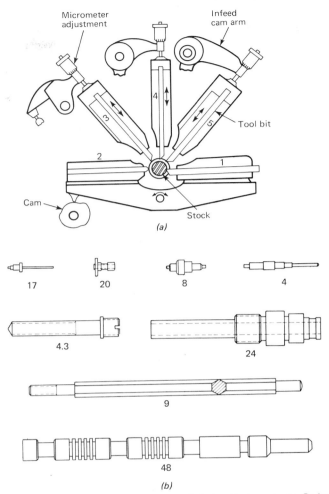

Fig. 7-16 (a) Detail of the cutters and their adjustment on a Swiss automatic lathe. (b) Typical parts made on this machine; shown actual size. Time to machine, in seconds, shown underneath. Much larger parts are also made.

loaded and unloaded from the chuck by hand, or sometimes by automatic feeding equipment.

The *chucks* are about 12-in. (300-mm) diameter for the four-spindle machines and 8-in. (200-mm) diameter for the eight-spindle machines since the spindle carriers are about the same diameter.

Speeds will be lower (26 to 900 rpm) because the workpieces are usually larger. As the parts become larger, from 20 to 40 hp (15 to 30 kW) is needed. *Setup time* is somewhat shorter than for the automatic bar machines since only the chucks need to be changed as compared with the bushings, fingers, and collets in the stock tubes.

Other Turning Machines

The same principles discussed above are used in other machines such as large single-spindle automat-

Fig. 7-17   A 1-in. (25-mm) six-spindle automatic bar machine. No tooling shown. (*Greenlee Bros. & Co., a subsidiary of Ex-Cell-O Corp.*)

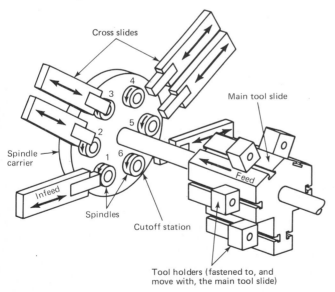

Fig. 7-18   A schematic drawing of a six-spindle automatic bar (or chucking) machine. The numbering system of the spindles is not standard.

ics. However, they are not used as often as the machines discussed in this chapter, and the student can easily understand their operation from their similarity to one or more of the machines studied here.

Vertical turret lathes and *vertical chucking machines*, while they are somewhat like the horizontal turret

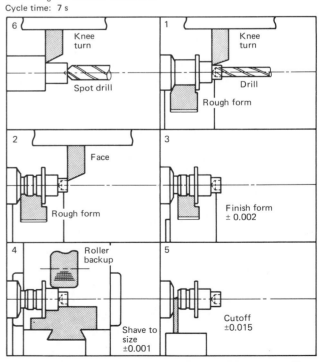

Material: $\frac{7}{8}$-in.-diameter leaded steel
Cycle time: 7 s

Fig. 7-19   A simple job which can be done on a six-spindle automatic bar machine. (*Greenlee Bros. & Co., a subsidiary of Ex-Cell-O Corp.*)

lathes, are here considered part of the "turning and boring machine" family, and are discussed in Chap. 18.

Review Questions

7-1. What are the advantages and disadvantages of a turret lathe?

7-2. Explain why the "cutting time" for a single cut will not be any shorter than with an engine lathe.

7-3. What factors would influence your choice of a turret lathe versus an ATL if both cost about $60,000?

7-4. a. What additional costs would you have to consider in choosing a numerical control lathe instead of an ATL?
b. What savings in cost would there be?

7-5. Why are the Brown & Sharp–type and the Swiss automatic–type screw machines recommended only for rather long runs of parts?

7-6. What factors would influence your choice of one of the two automatic screw machines described?

7-7. What two types of cutting tool motions are available on the multispindle automatic lathes?

7-8. What rpm would be needed, on the machines of Prob. 7-7:
a. If turning a 1¼-in.-diameter (31.8-mm) bar stock at 110 fpm (33.5 m/min)?
b. If turning an 8-in.-diameter (203-mm) iron casting at the same cutting speed?

7-9. Do you think that the machines discussed in this chapter have a headstock, tailstock, and cross slide like the turret lathes? Explain your answer.

Fig. 7-20   An eight-spindle automatic chucking machine. No tooling is shown. (*The New Britain Machine, a division of Litton Industries, Inc.*)

References

1. Berry, Albert E.: "Thinking Swiss," *Automatic Machining*, April–August 1963.
2. *Get the Most Out of Your Turret Lathe*, The Warner & Swasey Co.
   Note: W & S has several excellent brochures and tooling catalogs.
3. ASTME: *Tool Engineers Handbook*, 3rd ed., McGraw-Hill, New York, 1976.
4. The best sources of information are the manufacturers, some of whom are named in the illustrations. They will supply machine and tooling information and often excellent instructional brochures as well.

cutting tools for lathes

The cutting tools (such as drills, reamers, counter-bores, taps, etc.) which are described in the chapter on drill press cutting tools are also used on lathes of all types, although sometimes modifications in shank or length are made to adapt these tools for lathes. This chapter describes several of the specialized tools and toolholders which are used on the various models and styles of lathes.

TOOL BITS FOR LATHES

The *single-point tool bit*, in its several variations, is used for about three-quarters of the cutting done on all types of lathes.

Figure 8-1 shows a *positive rake* tool bit and the angles normally ground on it, with their names and abbreviations. The angles shown are satisfactory for cutting free-machining steel. Other materials require different angles. This nomenclature should be learned, as it applies to all single-point tools and is used in catalogs, references, and research work. The insert shows the relationship of the tool bit to the work when it is cutting.

Figure 8-2 shows a tool bit with similar angles, except that two of them are *negative rake* angles. As shown in Chap. 4, these two types of cutters make quite different chips.

Definitions of the Seven Features of the Tool Bit

Rake angles, side and back, form the cutting face and influence the direction of chip flow and the type of chip.

The front of the tool is the cutting end. Therefore, the **back rake angle** is the slope from the point (front) toward the back. This is perpendicular to the axis of the work.

The **side rake angle** (sometimes called the inclina-tion angle) is considered as starting at the cutting edge, the side which is in contact with the work.

If the rake starts from the above points and is paral-lel to the bottom of the tool bit, it is a *zero* rake.

One way to identify the rake angles is:

Positive rake decreases the lip angles, both side and end (as shown in Fig. 8-1), so that the lip angles will be less than 90°.

Negative rake increases the lip angles so that they are over 90° as shown in Fig. 8-2.

Zero rake (Fig. 8-3), which can be used for side rake or back rake or both, leaves the lip angle close to 90°.

Relief angles, often referred to in the shop as *clearance* angles, are usually fairly small, from 3 to 10°. Their only purpose is to prevent the end (*end relief, angle, ERf*), which is parallel to the work, and the side (*side relief angle, SRf*), which is at the cutting edge, from rubbing on the work. If these are too small, they wear down, rubbing starts, the tool and the work can get too hot, and chatter marks or a smeared surface may show up on the work. They should never be zero degrees.

Cutting edge angles refer to the angles at which the top face is shaped as shown in the top views of Figs. 8-1 and 8-2.

The *end cutting edge angle (ECEA)*, sometimes called the *auxiliary edge*, should never be zero. This is usu-ally small, from 5 to 12°, and it keeps the front edge of the cutter from rubbing on the work.

Note: There is an infrequently used technique called *broadnose* turning which does use practically a zero degree ECEA. Other than this, too small an ECEA can create difficulties.

The *side cutting edge angle (SCEA)*, sometimes called the *lead angle* or *principal edge angle*, must be 0° when a

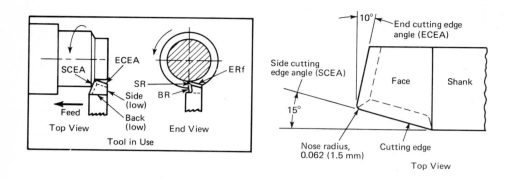

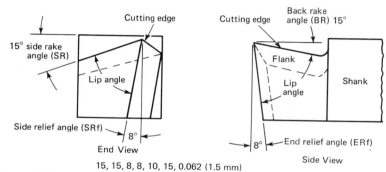

Fig. 8-1   A positive rake high-speed-steel tool bit showing the six angles, their abbreviations, and the nose radius.

Fig. 8-2   A negative rake tool bit showing typical angles.

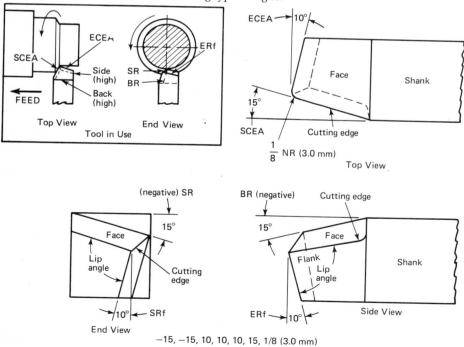

90° shoulder is required as in Fig. 8-4. However, whenever possible the SCEA should be more than 0°. Angles up to 45° are usually safe. Occasionally angles up to 60° have been used, but this is not recommended.

A large SCEA accomplishes two especially advantageous objectives:

1. It decreases the chip thickness, measured normal (perpendicular) to the cutting edge, as shown in Fig. 8-5a and c. If the SCEA is 45°, the chip thickness is only 0.707 times the feed per revolution. Thus, the load and wear on the tool bit is decreased, or a heavier feed, and thus faster production, can be used.

2. As shown in Fig. 8-5b and d, the tool bit with the side cutting edge angle first contacts the work at a position back of the point where the tool is quite strong. It then gradually picks up the full load. This is especially valuable if the surface is hard, such as in castings.

The zero SCEA picks up the full load immediately, almost as a shock or impact loading. Of course, this is being done every day in production. However, it may require heavier tool bits, and the tool life may be decreased.

CAUTION: Increasing the SCEA also increases the radial force. This can bend the work and cause

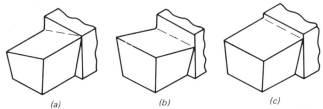

Fig. 8-3 Neutral (0°) rake angles as they may be used. (a) Zero back rake; (b) zero side rake; (c) zero side and back rake (neutral rake).

chattering unless the work is stiff or well supported.

The **nose radius** can be from 0 to $\frac{1}{2}$ in. (13 mm) and, in carbides, a circular piece of carbide may be used. As shown in Chap. 4, the nose radius, working with the feedrate, determines the finish on the work. Radii of 0.010, 0.032, and 0.060 in. (0.25, 0.8, and 1.5 mm) are quite commonly used.

Carbide-tipped tool bits, either the brazed-on type shown in Fig. 8-6 or the replaceable-tip types shown later, use the same angles described for HSS tools.

However, carbide is brittle, and the tips are relatively thin—$\frac{1}{8}$-in. (3-mm) to $\frac{1}{4}$-in. (6-mm) thick—so that too large an end relief angle would weaken the tip so that it would break more easily. Figure 8-7 shows this for positive and negative rake angles.

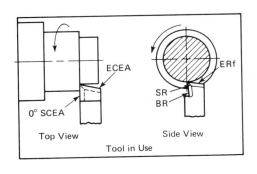

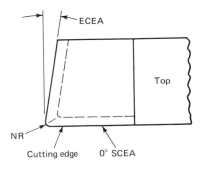

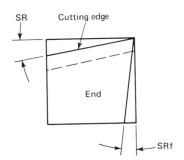

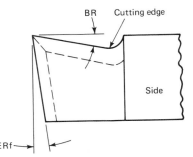

10, 10, 6, 6, 8, 0, 0.032 (0.8 mm)

Fig. 8-4 Positive rake tool bit with 0° SCEA.

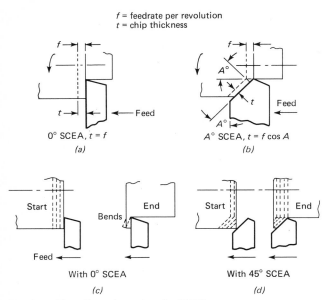

f = feedrate per revolution
t = chip thickness

0° SCEA, $t = f$
(a)

A° SCEA, $t = f \cos A$
(b)

With 0° SCEA
(c)

With 45° SCEA
(d)

Fig. 8-5   The effect of varying the SCEA.

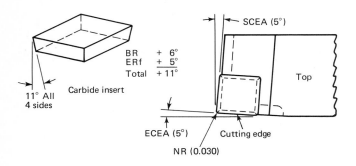

BR	+	6°
ERf	+	5°
Total	+	11°

11° All 4 sides   Carbide insert

SCEA (5°)

Top

ECEA (5°)   Cutting edge

NR (0.030)

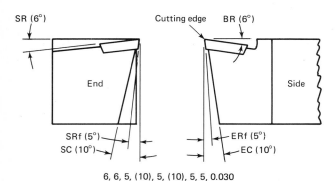

SR (6°)   Cutting edge   BR (6°)

End   Side

SRf (5°)
SC (10°)

ERf (5°)
EC (10°)

6, 6, 5, (10), 5, (10), 5, 5, 0.030

Fig. 8-6   Positive rake brazed-carbide-tipped tool bit. Notice that two ''clearance'' angles are added.

For this reason, the relief angles on the carbide tip are kept small, from 3 to 5°. However, to decrease the chance of rubbing and wearing out a $20 to $50 toolholder and to make it easier to sharpen the brazed tip, the tool steel holders are made with larger angles, referred to as the **_clearance angles_**. These are shown

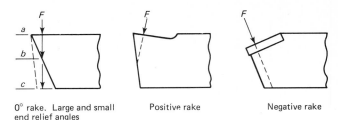

0° rake. Large and small end relief angles

Positive rake

Negative rake

Fig. 8-7   How various tool-bit angles effect the strength of the point of the cutter.

in Fig. 8-6. They are the *end clearance angle (EC)* and the *side clearance angle (SC)*.

Notice that the negative rake carbide tool bit can use the full 90° edge of the carbide which is much stronger than the shape for the positive rake.

Tool Signature

To avoid long written descriptions of tool bits, there is a universally agreed upon system of naming the angles on these single-point tools. This is shown in Fig. 8-8. This sytem is so widely used in the manufacturers' literature, technical articles in trade magazines, and research papers that it is worth memorizing.

Notice that there are three groups of numbers in this order:

1. *R*ake angles
2. *R*elief angles (with clearance angles in parentheses)
3. *C*utting edge angles

Notice also that the *side* angles are always the *second* of the two angles called out. Nose radius is the last number and may be a fraction, a decimal inch, or millimetres.

The rake and relief angles suggested for cutting various metals are shown in Table 8-1. These angles are not standard, and other recommendations may be different; however, they will establish a good starting point. The angles shown for carbide tools will apply to either the carbide-tipped or the inserted-blade type, which is mentioned later in this chapter.

Each of the seven items in the tool signature has its effect on the finish, the forces, and the total metal-removal rate. Table 8-2 shows the results of changing these angles and the nose radius. Of course, the figures in Table 8-1 often limit the amount of change possible.

Sizes and Shapes of Tool Bits

High-speed-steel tool bits are made most frequently with a square cross section. Because most toolholders are made with oversize slots and plenty of length on the fastening screw, metric size tool bits can often be used in the inch holders.

High-speed-steel tool bits are ground to whatever shape is needed for a particular job. Carbide-tipped tool bits cannot be easily ground this way. The small pieces of the proper grade, size, and shape of carbide are brazed or silver-soldered onto specially shaped tool-steel shanks.

Each shape of carbide-tipped tool has a standard letter designation of its "style." The ISO is suggesting the use of numbers instead of letters for each style, but these are not, at this writing, in general use. This letter or ISO number is followed by "R" for a right-hand tool bit and "L" for a lefthand tool bit. A right-hand tool bit is one used to *start* a cut at the right and cut toward the headstock or to start facing at the center and cut outward.

Some of the most used styles are shown in Fig. 8-9.

Toolholders for HSS and brazed-carbide tool bits are shown in Fig. 8-10. They are made in several sizes to fit the tool posts of various size lathes. Each size is slotted to hold a certain size tool bit. These are used

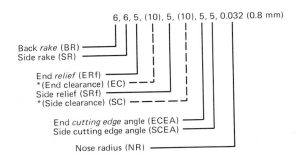

*Used with carbides and ceramics only.

Fig. 8-8   The "tool signature" used for describing tool bits. Typical values are also shown.

with engine lathes and sometimes in the cross slides of turret lathes.

The slots are broached (see Chap. 19) either parallel to the bottom or at a 15° angle as shown in the sectional views in Fig. 8-10. HSS tool bits may be used in either type, but carbide-tipped tool bits have only a small relief angle, so they must always be used with horizontally slotted holders.

TABLE 8-1   SUGGESTED ANGLES FOR TOOL BITS

High-Speed Steel

Material	Back Rake	Side Rake	End Relief	Side Relief	Comments
Mild steels	12	12	8	8	Use 15° SCEA when possible
Alloy steels	8	8	8	8	Use 15° SCEA when possible
High-strength steels	0	10	5	5	Use negative rake if over BHN 300
Free-machining stainless steel	5	8	5	5	Use up to 45° SCEA
Other stainless steel	0	10	5	5	Use up to 45° SCEA
Cast iron	5	10	7	7	Use smaller angles for harder C.I.
Aluminum alloys	20	15	10	12	Keep sharp edge
Copper alloys	0	0	8	10	Keep sharp edge
Nickel alloys	8	10	10	10	
Titanium	0	5	5	5	Use largest feed which is practical

Throwaway Carbide Inserts*

Material	Back Rake	Side Rake	End Relief	Side Relief	Comments
Mild steels	0	6	6	6	Neutral rake may also be tried
Alloy steels	−5	−5	5	5	
Free-machining stainless steel	0	5	8	8	Use 30° SCEA
Other stainless steels	−5	−5	5	5	Use 5° SCEA
Cast iron	0	6	5	5	Can also use circular inserts
Aluminum alloys	0	12	8	8	Difficult to break chips of 6061 and 7075
Copper alloys	6	10	6	6	
Nickel alloys	0	6	5	5	
Titanium	0 or −5	+5 or −5	6	6	

*Throwaway toolholders are made with fixed angles which may not be as shown here. In this situation, use the closest commercial angles, or have special clamp-on or brazed tools made for the job.

TABLE 8-2   SUMMARY OF THE EFFECTS OF CHANGING THE SEVEN CUTTING TOOL FEATURES

| Tool Feature | Can It Use Angles of | | | Effect of an increase in *Positive* Angle on | | |
	Positive	Negative	Zero	Forces	Finish	Strength of Tool Edge
Back rake angle	Yes	Yes	Yes	Longitudinal—Slight increase Tangential—No change Radial—Decrease, thus avoids chatter	Improves a little	Decreases
				Negative back rake is used especially with carbide tools to strengthen the edge.		
Side rake angle	Yes	Yes	Yes	Longitudinal—Decrease Tangential—Decrease Radial—Small decrease	Improves a little	Decreases
				Negative rakes used for heavy loads, harder metals, and carbide tools. Side rake controls chip flow, can create tightly curled or long, stringy chips.		
End relief angle	Yes	No	No	Slight decrease in all forces	Little change	Decreases, so use small angles for tough, hard, materials: 5 to 8°
				Ductile materials will plastically deform, so larger angles must be used. Slight wear on small angles will cause rubbing, heat, and tool wear.		
Side relief angle	Yes	No	No	Longitudinal—Decrease Tangential—No effect Radial—No effect	Slight improvement	Decreases
				Remarks on end relief angle also apply.		
End cutting edge angle	Yes	No	No	Little effect	Gets rougher because point angle decreases	Decreases
Side cutting edge angle‡	Yes	No†	Yes	Longitudinal—None* Tangential—Slight increase Radial—Large increase	Generally better	Increases due to larger point angle. Too much causes chatter
Nose radius	Does not apply			Longitudinal—Little effect Tangential—Increases Radial—Increases	Much better with larger radius	Large radius is stronger. Too large can cause chatter

*Unit pressure per inch *decreases* as SCEA increases due to longer edge in the cut, for a given depth of cut.
†Occasionally slight negative SCEA used on tracing tools.
‡Some SCEA helps considerably with interrupted cuts.

Throwaway Carbide Inserts

If a brazed-carbide tip is cracked or has been sharpened as much as possible, there are two choices: throw away the entire tool or unsolder the tip and put on another one. Either choice is expensive.

Thus, today, *throwaway inserts* (also called *replaceable or indexable*) have become the most widely used type of tooling for lathe work, milling, and many other operations. They are made in many shapes as shown in Fig. 8-11. Each shape insert is made in thicknesses from $\frac{1}{8}$ to $\frac{1}{2}$ in. (3 to 12 mm) and in sizes from $\frac{3}{8}$ to 1 in. (10 to 25 mm) square. Some styles have standard or special holes in them since different manufacturers have different methods of securing the inserts to the holders. A radius from $\frac{1}{64}$ to $\frac{1}{16}$ (0.4 to 1.5 mm), as desired, is ground on the corners.

Positive rake inserts have a 5 to 8° relief angle ground on all sides. Thus, they have three or four edges which can be used before a new insert is required. As these inserts cost only from $1.50 to $4.00 each, this makes quite a low cost per edge.

Negative rake inserts create the relief angle because of their position on the holder. Thus, they have all right-angle edges and possibly may have six to eight usable edges.

CAUTION: The second side can be used only if there is no roughness or burr due to the wear on

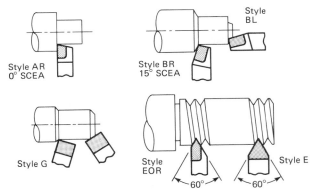

Fig. 8-9 Some of the styles of brazed-carbide cutting tools and their suggested uses.

the first side. Any such raised surface would make it impossible to fasten the insert flat against the toolholder's locating surface.

Regrinding these inserts with silicon carbide or diamond wheels is sometimes done if the wear is not too large. This requires that shims (spacers) be put under the insert to make up the original thickness.

Toolholders for carbide throwaway inserts (Fig. 8-12) come in a wide variety of shapes and sizes. There are three basic requirements for all of these:

1. A *pocket* for the insert to take the side longitudinal and end radial thrust from the cutting forces. This, in most styles, also positively locates the insert so that replacements will cut to the same size.

2. A *clamp* to hold the insert firmly against the bottom of the pocket and prevent it from being pulled out. Sometimes a special pin is used to hold the insert.

3. A solid *seat* for the bottom of the insert to take the tangential force and not allow the carbide to be subject to a bending force. This is either a carbide or hardened steel plate.

The *styles* of toolholders for inserts are lettered or numbered the same as those with brazed tips. The shapes and uses of some of these styles are shown in Fig. 8-13. Note that each style can be used for several different cuts.

The *accuracy* of the pockets is usually within ±0.001 or 0.002 in. (0.025 or 0.050 mm). However, the body of the toolholder is made to fractional dimensions. Sometimes, especially on numerical control lathes, it is advantageous to hold some lengths or widths much closer. Thus, *qualified* (accurately sized) toolholders are available. These hold specified dimensions to ±0.001 (0.025 mm).

Chipbreakers

The long curly or tangled chip is not desirable. The condition shown in Fig. 8-14 certainly should be avoided. Thus a chipbreaking groove is often cut in the edge of HSS tool bits. Some of the chipbreaker shapes used are shown in Fig. 8-15. The dimensions of these grooves vary with the feedrate and depth of cut. (These figures are given in many handbooks.)

For carbide-brazed-tip tools, these same shapes must be ground into the carbide. However, for throwaway inserts, the chipbreaker is formed as shown in Fig. 8-12*b* and *c*.

A square tool post (Fig. 8-16) similar to that shown on the turret lathes can be used on an engine lathe. The quick indexing and availability of four different cutting tools can greatly shorten the machining time. Either HSS or carbide tools may be used.

Boring Bars

The boring bar is made in a wide variety of styles. Previous illustrations have shown small, easily ground, single-point boring bars. It is difficult to adjust these when they are used in turret and automatic lathes.

Fig. 8-10 (*a*) Toolholder, 15° rake, lefthand offset. (*b*), (*c*) Cross-sectional views of toolholders. (*J. H. Williams, a division of TRW, Inc.*)

(a)

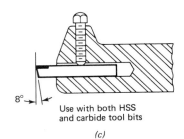

23°

*15°   8°

Use only with HSS tool bits

*Gives 15° back rake
Note: Need 23° end clearance on tool bit to give 8° ERf at workpiece

(b)

8°

Use with both HSS and carbide tool bits

(c)

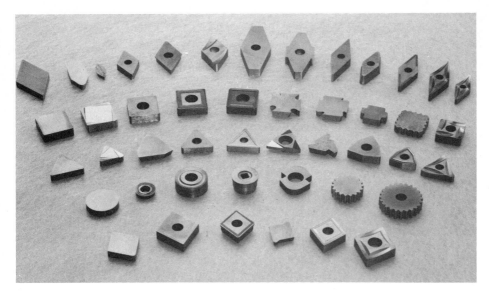

Fig. 8-11 Some of the shapes in which renewable "throwaway" carbide inserts are made. Not all styles are available in all grades. (*The Valeron Corporation.*)

(a)

(b)

(c)

Fig. 8-12 (*a*) Some of the styles of toolholders for use with replaceable carbide tips. (*b*) Positive rake with replaceable adjustable chip breaker. (*c*) Negative rake, with new double-positive molded-in chip breaker. (*Valenite Division of The Valeron Corporation.*)

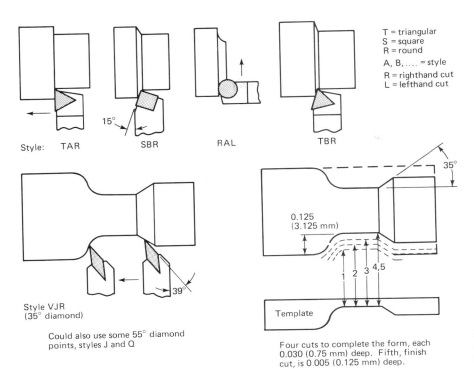

Style: TAR SBR RAL TBR

15°

T = triangular
S = square
R = round
A, B, ... = style
R = righthand cut
L = lefthand cut

Style VJR
(35° diamond)

Could also use some 55° diamond
points, styles J and Q

39°

35°

0.125
(3.125 mm)

1 2 3 4,5

Template

Four cuts to complete the form, each
0.030 (0.75 mm) deep. Fifth, finish
cut, is 0.005 (0.125 mm) deep.

Fig. 8-13   Letter identification and uses
of a few styles of throwaway carbide-
tipped tools.

Boring bars such as those shown in Fig. 8-17a have easily adjustable inserts built into them. These bars are made in standard sizes, each with a range of $\frac{1}{4}$ to $\frac{1}{2}$ in. (6 to 12 mm) on the diameter. A fine adjustment is included in increments of 0.001 in. (0.025 mm) or, in some cases, 0.0001 in. (0.0025 mm). These are standard up to about 6-in. (150-mm) diameter.

Double-edge boring tools are excellent for heavy cuts. The cutters are adjustable and give a balanced cutting action which prevents tool deflection and hole tapering.

Many times it is economical to order special bars with two or more preset diameters, set at the proper distance apart, as shown in Fig. 8-17b. These special bars may cost several hundred dollars, so they are used only when large quantities make it economical. Sometimes this is the only way to hold the required tolerances and concentricity. In this case, the expense simply raises the cost of the article.

Cutoff or **parting tools** use special long HSS blades in special holders as shown in Fig. 8-18a. The blade is extended only enough to complete the cut, as it has a thin cross section and breaks easily. If it does break, the blade is moved forward, locked into position, and resharpened.

Carbide cutoff tools are quite different. One style (Fig. 8-18b) uses a carbide tip brazed to a tool-steel shank. The extension of the narrow part limits the depth of cut. This is usually not over $1\frac{1}{4}$ in. (29 mm). Similar tools are used for cutting grooves on the OD (necking) or the ID of a part.

Knurling, as described in Chap. 6, is done on an engine lathe with the tool shown in Fig. 8-19a. The patterns produced are shown in Fig. 8-19b. Turret and automatic lathes use similar knurling wheels held in somewhat different holders.

Reaming may be done on lathes with the standard reamers described in Chap. 10. However, for deep holes and for diameters over 1 in. (25 mm), adjustable reamers (Fig. 8-20) are usually used. These cutters "float," that is, align themselves with the hole. Finishes as low as 50 μin. (1.25 μm) and accuracies to ±0.001 in. (0.025 mm) are considered practical in reaming.

Fig. 8-14   Result of an actual cut on CRS with positive rake tool bit, no chipbreaker. (*Photograph by Samuel Lapidge.*)

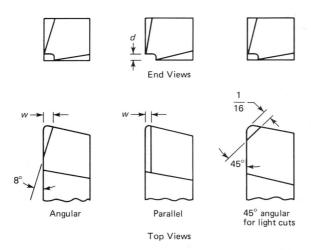

End Views

w = 1/16 to 1/8 in. (1.6 to 3.2 mm) for 0.008 to 0.032 in./rev
or 0.20 to 1.5 mm/rev. feedrates
d = 0.015 to 0.030 in. (0.38 to 0.75 mm)

Fig. 8-15 Some types of chipbreakers used on HSS and brazed-carbide tools.

signs should be able to acquire quickly any specialized knowledge needed.

N/C engine lathes use the same tooling, but HSS tools and the standard tool post are usually replaced by carbide-tipped tools and five-, six-, or eight-position automatically indexed toolholders.

N/C turret lathes use the same styles of cutting tools, but the turrets are often very different in design. You are accustomed to seeing a turning tool in a horizontal position at the front of the machine; however, the tool will cut equally well at 90°, 180°, 270°, or any angle in between as long as it is well supported. Many of the new N/C lathes do machine from the back or top of the work. This allows much greater freedom in designing the machine. The N/C lathe in Fig. 7-12 shows one type of variation.

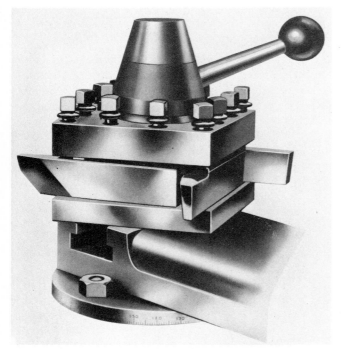

Fig. 8-16 A square tool post as used on an engine lathe.

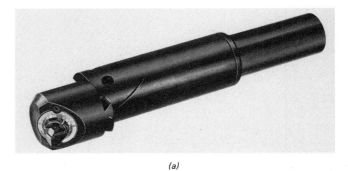

(a)

(b)

Fig. 8-17 Boring bars. (a) Single-point micrometer adjustment. (b) Multiple boring and reaming tool, made for a single job. (*Madison Industries, Inc.*)

Threading is described in Chap. 11 since lathes use the same or similar tooling for this work.

Special tools are often used to save time on certain jobs. They are usually designed by the company using them. There are also a great many variations on the "standard" tooling described in this book. However, a student who is familiar with these basic de-

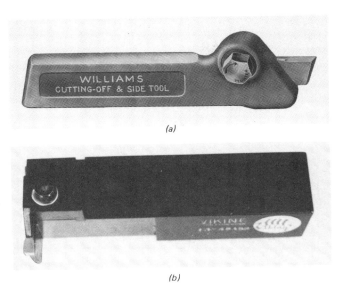

(a)

(b)

Fig. 8-18 (a) Straight shank cutoff or parting tool. (*J. H. Williams, a division of TRW Inc.*) (b) Carbide-tipped side cutoff or grooving tool. (*The Viking Tool Co.*)

Review Questions

8-1. Sketch a three-view drawing of a 1 in. (25 mm) square tool bit with angles -10, $+10$, -8, $+8$, -15, $+10$, 0.

8-2. Why can relief or clearance angles never be zero or negative?

8-3. Explain, in terms of rake angles, the type and flow direction of the chips shown in Fig. 8-12b and c.

8-4. a. Sketch a force diagram showing why a large SCEA increases the radial force.
   b. Why is the longitudinal force not increased?

Fig. 8-19 (a) Typical knurling tool used with engine lathe can make coarse, medium, and fine patterns. (b) Patterns which can be produced. (*J. H. Williams, and Co. a division of TRW Inc.*)

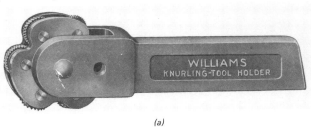

(a)

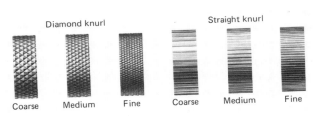

(b)

Fig. 8-20 An adjustable-blade floating reamer for holes of 1- to 6-in. (25- to 150-mm) diameter. Insert shows detail of the method of adjustment. (*Madison Industries, Inc.*)

8-5. Why are some SCEA recommended especially for carbide tools?

8-6. a. Why are relief angles on carbide tools kept as small as possible?
   b. Could you increase these angles if you were cutting brass? Why?

8-7. List several advantages in using throwaway (replaceable) carbide tips on lathe tools.

8-8. a. From Fig. 8-12b and c, would you judge one of the three styles of chipbreakers to be better than the others? Explain your answers.
   b. What disadvantages would this style have?

8-9. What would limit the depth of cut that can be taken with the boring bar shown in Fig. 8-17a?

8-10. Cutting speeds are often lower for cutoff and grooving tools than for turning tools. Explain why this is so.

8-11. How can you tell if a cutting tool is L.H. (left-hand) or R.H. (righthand) by just looking at it?

8-12. The type R (circular carbide disc) tool is most often used on iron castings. Why would you hesitate to use it on mild steel? Hint: What kinds of chips do you expect from each material?

8-13. What is the difference between the Madison Company's adjustable double-blade boring bar and adjustable reamer?

8-14. If the cutting speed of an adjustable double-blade boring bar is 25 percent less than that of the single-point tool but the feedrate is doubled, how many minutes per hour can be saved?

References

1. Catalogs of cutting tool manufacturers. See Thomas Register and the illustrations in this chapter for names.
2. Bhattocharyya, A., and J. Ham: *Design of Cutting Tools*, Society of Manufacturing Engineers, Detroit, Mich., 1969.
3. *Machining Data Handbook*, 2d ed., Machinability Data Center, Metcut Research Associates, Inc., 1972.
4. Publications of the British Hard Metals Association, Sheffield 10, England.
5. *Selection and Application of Single Point Metal Cutting Tools*, Metallurgical Products Dept., General Electric Co.. GT9-270, 1969.
6. ASTME: *Tool Engineers Handbook,* 3d ed., McGraw-Hill, New York, 1976.
7. *Why Tools Fail*, Metallurgical Products Dept., General Electric Co., Pamphlet No. TR-77.
8. *Why Do Tools Wear Out?*, Metallurgical Products Dept., General Electric Co., Pamphlet #TR-90.

the drill press

Making round holes in many types of materials is big business. About $130,000,000 is spent each year in the United States alone just on the various types of twist drills. This is the largest amount spent on any one class of cutting tools, according to the U.S. Department of Commerce.

Of course, in thin metals, a punch press may be preferred or, for certain types of work, electrical discharge machining (EDM), electrochemical machining (ECM), lasers, or electron beam machining (EBM) may be used. These are discussed in later chapters.

The twist drill is also used on lathe work and on large boring machines, both vertical and horizontal, as described later. However, the relatively simple, inexpensive drill press in its several forms is still one of the most frequently used machine tools in industry today.

TYPES OF CUTS

Drilled holes of diameters from under $\frac{1}{64}$ in. (0.4 mm) to over 2.0 in. (50 mm) may be made as "through" holes (Fig. 9-1a) or "blind" holes (Fig. 9-1b). If the depth of the hole is more than three or four times the diameter of the drill, it is a "deep" hole and may require special consideration.

Counterbored holes (Fig. 9-1c) are made with a larger opening at the top (and sometimes the bottom) of the hole to allow a bolt head or a nut to be set in flush with the top surface. These are usually made to fit socket-head cap screws.

Countersinking (Fig. 9-1d) is used either to provide a recess for a flat-head bolt or to burr or chamfer the edges of a hole. The included angle is usually 82° because that is the shape of a flat-head screw. Some countersinks, however, have 90° or, occasionally, 60° included angles.

Spot facing (Fig. 9-1e) is necessary when a bolt head or nut might rest on the rough or curved portion of a

casting, I beam, etc. The smooth "spot" must be at least $\frac{1}{32}$ in. (0.8 mm) larger than the bolt part resting on it. The depth is usually as shallow as possible. Spot facing is usually circular, but it may be a straight milling cut over a larger surface.

Spot drilling or *center drilling* (Fig. 9-1f and g) is often done to provide a stable starting point for a drill which otherwise would "skid" slightly off center or to locate centers on a lathe.

Reaming (to get close tolerance and good finish) and *tapping* (cutting an inside thread) are also frequently

Fig. 9-1 Several types of cuts which can be made with all types of drill presses.

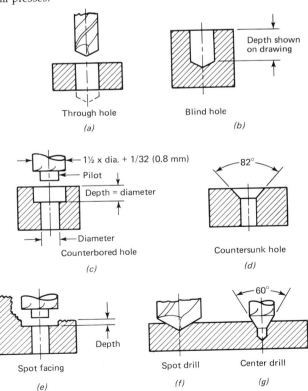

done on a drill press. These will be discussed in the next chapters.

Combination cuts, combining some of the above (such as drill and countersink), are sometimes done with special drills, illustrated in Chap. 10.

TYPES OF DRILL PRESSES

Simple Drill Press

A simple drill press (Fig. 9-2) may be floor-mounted, as shown, or have a shorter main post and be mounted on a bench as shown in Fig. 9-3.

The motions of this machine are very simple. The table, on a floor model, can be raised or lowered and rotated around the machine column. The spindle rotates and can be raised and lowered, with a stroke of 4 to 8 in. (100 to 200 mm). Stops can be set to limit and regulate the depth.

The *size* of a drill press is the distance from the center of the spindle to the column, that is, the radius of the largest circular part which can be drilled at its center. Maximum speed may be 2000 to 4000 rpm. Feed is by hand. The cost is from $350 to $1000.

Gang Drilling

An economical way to perform several different oper-

Fig. 9-2   (*a*) A "standard" drill press. (*b*) Eight-speed pulley drive. Note special narrow belt. (*c*) Infinitely variable speed drive. Note spline drive to spindle. Front pulley is cut off to show construction. (*Power Tool Division, Rockwell International.*)

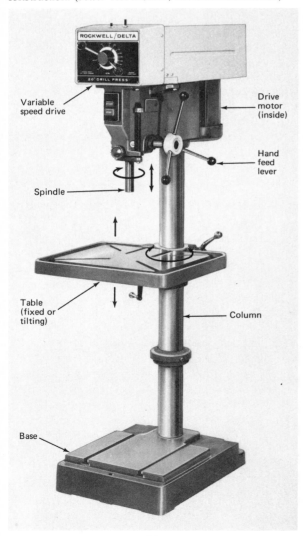

(a)

(b)

(c)

Fig. 9-3 A gang drilling setup. Each drill press will have a different cutting tool. (*Power Tool Division, Rockwell International.*)

ations on one piece is by gang drilling (Fig. 9-3). This might include drilling two or more sizes of holes, reaming, tapping, and countersinking. The work is held in a vise or special fixture and is easily slid along the steel table from one spindle to the next.

The drill presses usually run continuously, so the operator merely lowers each spindle to its preset stop to perform the required machining.

Sensitive Drill Press

Figure 9-4 shows a sensitive drill press. The name "sensitive" is used to indicate that the feed is hand-operated and that the spindle and drilling head are counterbalanced so that the operator can "feel" the pressure needed for efficient cutting.

This drill press has the same motions as the previous one plus a telescoping screw for raising and lowering the table and a sliding "drill head." These two features allow easier handling of parts of varying heights.

These machines can drill up to 1½-in. (38-mm) diameter in cast iron. Motors used are from 1 to 3 hp (0.75 to 2.2 kW).

Multispindle Drilling

This type of drilling can be done on drill presses by using attachments like those shown in Fig. 9-5. The spindle locations are adjustable, and the number of spindles may be from two to eight. Drills, reamers, countersinks, etc., can be used in the spindles.

Another system of multispindle drilling (a gearless system) is shown in Fig. 9-6. This system often requires a specially designed drilling machine and power feeding. The rpm and feedrate of all spindles in one drill head are the same, and the horsepower needed is the sum of the power for all cutting tools used. In this type of machine, a large number of holes may be drilled at one time, though the 3739 holes shown is rather unusual. Several different diameters of drills may be used at the same time. Similar machines are made for multispindle tapping such as in making nuts for bolts.

Turret Drills

Turret drills (Fig. 9-7) with either six or eight spindles enable the operator to use a wide variety of cutters

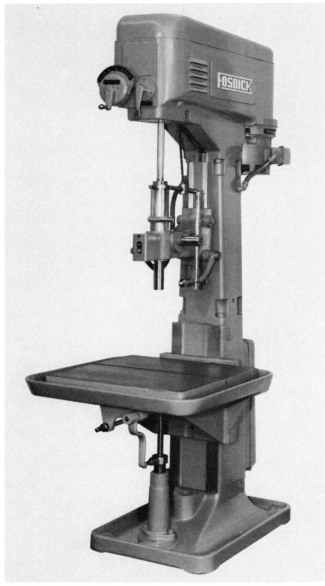

Fig. 9-4   A sensitive drill press, usually used for drilling holes of ¾ to 1½-in. (19- to 38-mm) diameter. Variable speed drive. (*Fosdick by LeBlond.*)

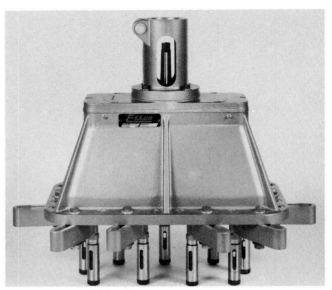

Fig. 9-5   A multispindle drill head with nine spindles. (*Ettco Tool & Machine Co.*)

drill, so that the operator merely has to load and unload the parts. The N/C turret drill can also make a variety of milling cuts.

Radial Drills

For handling medium- to very large-size castings, weldments, or forgings, radial drills are ideal (Fig. 9-8). Their size is specified by the length of the arm along which the spindle housing rides. This arm can be from 3 to 12 ft (900 to 3600 mm) long. The column which holds the arm may be from 10 to 30 in. (250 to 750 mm) in diameter. A 4-ft radial drill costs about $15,000 and an 8-ft one costs about $47,000.

The motions, power operated, are shown on the figure. A skilled operator will move the drilling head along the radial arm, at the same time swinging the arm around the column, and bring the spindle to its exact location in 15 or 20 s (seconds).

For very large work, the arm may be rotated 180°, and the work placed on the shop floor. Speeds and feeds are dialed in by the machine operator and are the same as for other drill presses. Drilling is either hand or power feed.

High-Speed Drilling

High-speed drilling is needed when using the smaller drills from 2 mm (0.079 in) down to 0.30 mm (0.012 in.), approximately No. 80 size and smaller. Theoretically, a No. 60 drill (1.0 mm, 0.040 in.) should be run at about 6000 rpm in mild steel and at 20,000 rpm in aluminum. Most drill presses have top speeds of 4000

and yet move the workpiece only a few inches, according to the hole spacing. These come in capacities from ½ to 1¾ in. (13 to 44 mm) drills in steel, and the cost is from $2000 to $35,000. The turret can be rotated (indexed) in either direction, then lowered by hand or automatically to make the cut.

Some turret drills have automatic, hydraulically controlled spindles. Speeds, feeds, and depths can be preset for fast production. Figure 9-7b shows an automatic machine. These machines are also made with the entire operation tape controlled, an N/C *turret*

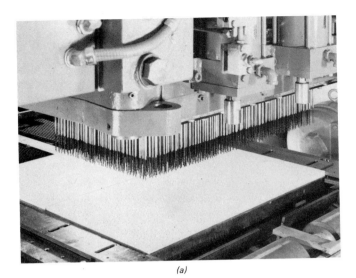

(a)

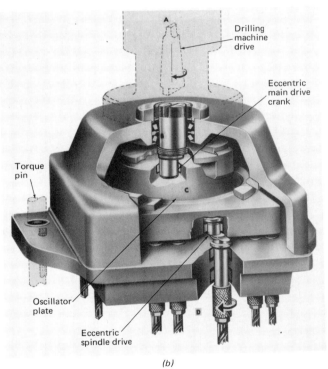

(b)

Fig. 9-6   (*a*) Gearless (eccentric) drive multispindle drilling head with 3739 spindles drilling 12 × 24 in. (300 × 600 mm) acoustical tile. (*b*) A smaller drill head showing drive mechanism. (*Zagar, Inc.*)

rpm or less. Thus, special machines such as that shown in Fig. 9-9 have been built which will safely run at over 20,000 rpm. In fact, special air-driven spindles with air bearings are made which will run, in production, at 175,000 rpm.

Gun Drilling

Originally only used for very long holes such as in gun barrels, gun drilling is now used for high-quality drilling of holes from $\frac{5}{64}$- to $1\frac{1}{2}$-in. (2- to 38-mm) diameter and from 1 to 30 in. (25 to 760 mm) long.

Excellent accuracy of location and diameter (±0.0005 to 0.001 in.) is possible because the machine is precisely made (Fig. 9-10). Accurately made, though often quite simple, work-holding devices (fixturing or tooling) are used. The fact that the gun drill (see Chap. 10) guides itself in a straight line also adds to the accuracy of the size and alignment of the hole.

Coolant is forced through a hole in the drill at from 400 to 1500 psi as described in Chap. 10. This cools the drill and forces the chips back through the long, straight flutes.

Speeds for the smaller drills may be over 10,000 rpm and feeds may be up to 20 ipm (500 mm/min). The gun drill usually has a solid-carbide or brazed-carbide tip so that it will hold to size during long drilling cycles.

ACCURATE HOLE LOCATION—METHODS AND MACHINES

Locating the holes by using any of the machines discussed in this chapter, cannot be done accurately, except for the N/C turret drill, because the tables have no X- or Y-axis movement. Accurate production work is done four ways.

1. *Lay out* and center-punch all locations before bringing the part to the drill press. This can be done quite accurately, and the holes should be within $\pm\frac{1}{64}$ in. (0.015 in. or 0.38 mm) or sometimes closer to the desired location.

This, however, is a rather slow process, so if there are more than maybe 10 or 20 parts to be drilled another method should be used.

2. A *drill jig* (Fig. 9-11) is an assembly constructed so that it can securely hold the workpiece. It also has hardened steel guide bushings accurately located wherever a hole is to be drilled. Using this drill jig on a drill press or a turret drill, an operator can accurately locate and quickly drill, ream, tap, etc., up to 12 or more holes in part after part.

The drill jig may cost from $200 to $2000, so they are usually ordered only if 50 to 500 parts are to be drilled at one time, and if annual production is high enough to amortize the cost.

3. *Multispindle drills*, after having been accurately located, will, of course, give reliable duplication of hole locations.

4. *N/C machines* will locate all holes to ±0.001 in. (0.025 mm) in any pattern.

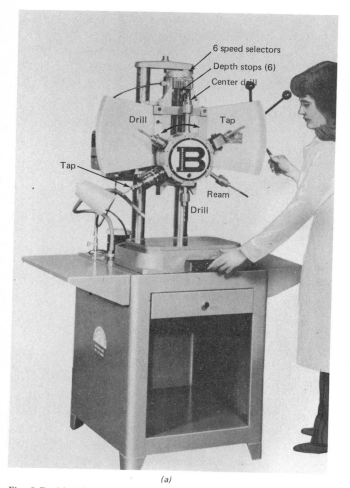

6 speed selectors

Depth stops (6)

Center drill

Drill

Tap

Tap

Ream

Drill

(a)

(b)

Fig. 9-7 (*a*) A ½-in. capacity turret drilling and tapping machine. (*Burgmaster Houdaille*) (*b*) A 20 in. (500mm) heavy duty drill press equipped with a multiple spindle head and automatic air feed. Operator only loads and unloads the part in the fixture. (*Rockwell International, Power Tool Division*)

The **jig borer** (Fig. 9-12) is sometimes listed as a boring machine but is included here because its main job is to locate very accurately a number of holes in the manufacture of a workpiece, a drill jig, or a punch and die set. Once the location is assured, the hole is drilled, reamed, or bored as needed. These machines can also do light milling and tapping, but this is often done more economically on other equipment.

The jig borer does not look very impressive, but it is built very ruggedly so that it will not deflect under the loads imposed by cutting. It is also very carefully built for accuracy in the "tenths" (0.0001 in. or 0.0025 mm) in all directions. Cost is from $10,000 to $20,000. The accuracy of the hole location is obtained in different ways on different makes of jig borers.

1. Some machines use *end measures* or rods which are like gage blocks of different lengths. These are used to determine the distance the table or carriage is moved. The final accuracy is sometimes achieved with a micrometer movement.

2. *Electronic "digital" read-out*, the same as used on the inspection machines shown in Chap. 2, gives the operator a fast, accurate way to locate either the table (*X* axis) or the carriage (*Y* axis).

3. *Precision lead screws* are the basis for the accuracy of the jig borer shown in Fig. 9-12. These are guaranteed accurate to a maximum error of 90 μin. (2.3 μm) over the full 18-in. (450-mm) table travel.

The jig boring machine shown has accurately graduated dials and verniers so that the operator can quickly read to 0.0001 in. or, on the metric model, to 0.001 mm (1 μm) on both the table and carriage. It is also made with either visual electronic read-out or numerical control.

The *spindle* on all types of jig borers must run true (no wobble) and must be exactly perpendicular to the machine table. These dimensions are held to between 50 and 100 μin. (0.001 25 to 0.0025 mm) on most machines.

Temperature can affect the accuracy of any machine,

but it is especially noticeable when trying to work so closely. Steel has a coefficient of expansion of about 0.000 006 5 in./in./°F. Thus, suppose that a 10-in. center-to-center hole location was drilled and reamed at the 68°F temperature of the air-conditioned room where the jig borer is kept. It is then used in the 78°F summer shop temperature. The 10-in. distance would quietly grow to 10.000 65 in. When drilled, it was made to ±0.0002 in. When extreme accuracy is necessary, manufacturing and inspection should be done under the same environmental conditions.

Final accuracy of work done on a jig borer (or any other machine) is the sum or difference of all the tolerances. This involves the tolerance allowed on the quill and spindle, the movements of the table and carriage, the taper in the toolholder, the taper on the tool shank, cleanliness, and temperature. Thus, a *practical* minimum requirement for jig boring is probably ±0.0002 in. (0.005 mm) in location.

Hole size, of course, depends on the cutting tools and on any spindle runout. This can be held to ±0.0001 in. (0.0025 mm), but it is much less expensive to allow two to five times this if possible in the design of the work.

Today *numerical control* jig borers are available. They have a built-in positioning tolerance of ±0.0001 or 0.0002 in. (0.0025 or 0.005 mm). While these N/C jig borers are more expensive, they are also much faster. Once the tape has been correctly made, these N/C machines can accurately do the drilling, reaming, boring, and tapping in one-tenth the time it takes to do it by hand.

Work holding on a drill press is often done by a drill jig (Fig. 9-11) or a large or small vise (Fig. 9-13) with or without simple attachments on the vise. If the piece is large or the drill is ⅜-in. diameter (9.5-mm) or over, the work or the vise should be bolted or clamped to the table or to the drill press.

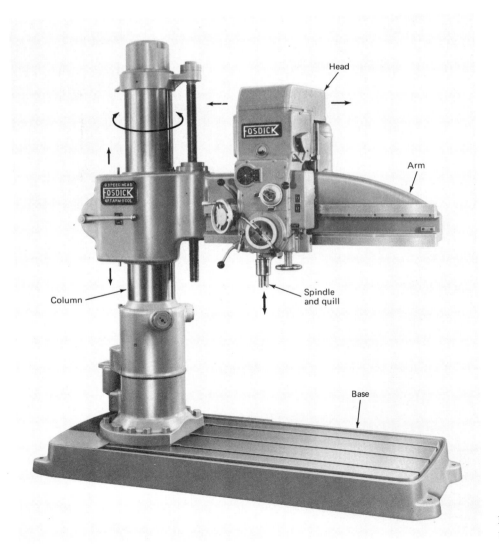

Fig. 9-8   A 4-ft (1200-mm) radial drill with base. (*Fosdick by LeBlond.*)

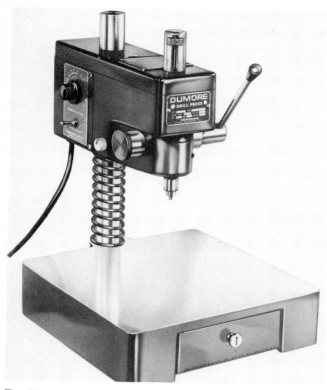

Fig. 9-9   A high-speed (21,500 rpm) drill press, $\frac{5}{32}$-in. (4.0-mm) drill maximum capacity. Table 12 in. (300 mm) square. Solid-state speed control. (*Dumore Co.*)

CAUTION: The "twist" (torque) of a twist drill is much greater than you think, and it is also trying to pull the workpiece *upward*. Thus, for safety's sake, do not try to hand-hold work which should be bolted down or held in a large milling vise. Operators have had some bad cuts from *parts* rotating at 1000 rpm because they broke free of their hand grip and went around with the drill.

Tool holding in a drill press may be done by several methods.

1. The simplest is with a drill chuck (Fig. 9-14). These are made in several sizes to hold up to $\frac{3}{4}$-in.-diameter (19-mm) straight shank drills. They are tightened with a single "key." Quick-change drill chucks are made so that drills may be changed without stopping the spindle rotation.

2. Most drill spindles have a tapered hole (see Morse taper) into which the tapered shank of the drill chuck is fastened. The drill, reamer, etc., can be bought with a taper shank which fits directly into this taper, eliminating the need for the drill chuck. This is a stronger and more accurate way to secure tools in the drill press.

3. There are several makes of fast-acting chucks and spindle attachments which can be advantageously used if production warrants it.

Operator's control panel

Gun drill

Guide bushing

Table for workpiece and holding fixture

High-pressure pump and tank for coolant

Fig. 9-10 A small, $\frac{1}{2}$-in.-capacity (12.7-mm) gun drilling machine. Much larger machines are available. Prices start about $12,000 fully equipped. (*Eldorado Tool & Mfg. Corp., a subsidiary of Litton Industries, Inc.*)

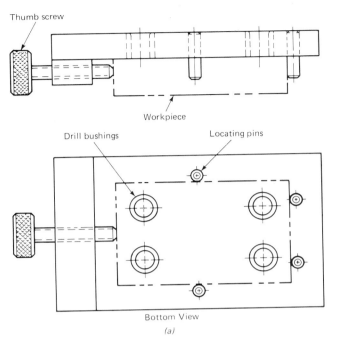

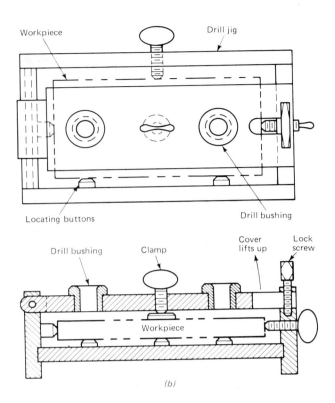

Fig. 9-11 (a) A simple "plate" type drill jig. (b) A "box" type drill jig.

POWER NEEDED FOR DRILLING

To be used for drilling, Eq. (4-8) and (4-9) must be changed slightly because the chip from a drilling operation can be assumed to be a circular disk with an area of $\pi d^2/4$ and thickness equal to the feed per revolution. Mathematically, the chip per revolution taken by the diagonal point of the drill is equal to that of a flat disk with the same side length. Thus, the volume per minute is

$$V = \frac{\pi D^2}{4} \times \text{feedrate} \times \text{rpm}$$

where D = in. or cm
feedrate = in./rev or cm/rev

$$\text{Horsepower} = hp = V \times hp_{cim} \times \text{efficiency}$$
or
$$kW = V \times kW/cm^3 \times \text{efficiency}$$

EXAMPLE   9-1

A $\frac{1}{2}$-in. (12.7-mm) drill is cutting aluminum at 200 fpm, with a feed of 0.005 in./rev (0.125 mm/rev) and with 80 percent efficiency. The hp constants in Table 4-4 are 0.35 hp or 0.016 kW.

$$\text{rpm} = \frac{4 \times 200}{0.5} = 1600 \quad \text{(in either English or metric)}$$

$$hp = \pi \left(\frac{0.5^2}{4}\right)(0.005)(1600)(0.35)\left(\frac{1}{0.8}\right) = 0.687 \text{ hp}$$

In metric,
$$kW = \left(\frac{\pi}{4}\right)\left(\frac{12.7^2}{10^2}\right)\left(\frac{0.125}{10}\right)(1600)(0.016)\left(\frac{1}{0.8}\right)$$
$$= 0.507 \text{ kW}$$

Check: (0.687 hp)(0.746 kW/hp) = 0.512 kW

The surprisingly wide variety of cutting tools which can be used with drill presses is described in Chap. 10.

Review Questions and Problems

9-1. Why make holes with a drill press instead of with a lathe?

9-2. What different cuts can be made with a drill press?

9-3. Sketch and name the parts on a simple drill press.

9-4. A part has 12 holes of various sizes to be drilled and 4 of them to be tapped. They are on three sides of the part. Would you use a gang drill setup or multispindle drilling?

9-5. Would a turret drill be suitable for use on the above job?

9-6. In what ways are a radial drill and a simple drill press alike and in what ways are they different?

9-7. Describe three ways of locating holes which

Fig. 9-12   An 11 × 24 in. (280 × 610 mm) table-size jig borer which uses precision lead screws and direct-reading °.0001-in. or °.001-mm dials. (*Moore Special Tool Co.*)

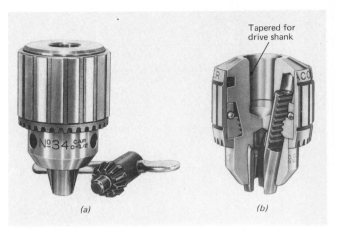

Fig. 9-14   A 0- to ½-in. (0- to 12.7-mm) drill chuck. (*Jacobs Manufacturing Company, a subsidiary of Chicago Pneumatic Tool Co.*)

are to be drilled on a drill press of any type. What is the accuracy of each?

9-8.   What factors influence the final accuracy of work done on a jig borer (or on any drill press)?

9-9.   What are the advantages and disadvantages of using a numerical control jig borer?

9-10. Twenty No. 60 (0.04-in.-diameter) holes are to be drilled in a 6 × 3 in. aluminum sheet at 300 fpm.
   a.   What rpm is needed?
   b.   What kind of drill press should be used?
   c.   What horsepower will be needed?

9-11. A 30-mm hole is to be drilled in stainless steel at 15 m/min. What rpm and which machine should be used?

9-12. If, in Prob. 9-11, feed is 0.10 mm/rev, what kW capacity motor is needed?

9-13. Change Probs. 9-11 and 9-12 to English units and solve them.

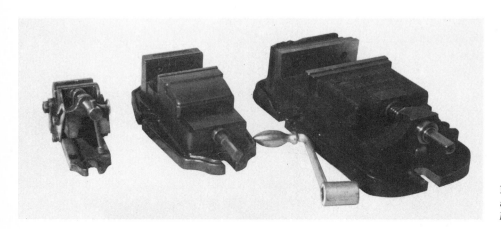

Fig. 9-13   Vises which may be used on a drill press. (*Photograph by Samuel Lapidge.*)

References

1. *Drilled Holes for Tapping*, rev. ed., Metal Cutting Tool Institute, 1967.
2. *Metal Cutting Tool Handbook*, Metal Cutting Tool Institute, 1965.
3. *The Metal-Cutting Tool Industry*, 1965–1969, U.S. Dept. of Commerce, Washington, D.C., 431-996/65, August 1970.
4. *Metal Cutting Tool Nomenclature*, Metal Cutting Tool Institute, New York, 1957.
5. ASTME: *Tool Engineers Handbook*, 2d ed., McGraw-Hill, New York, 1965.

cutting tools for drill presses

Even though it is convenient to describe these drills and related cutting tools as being used principally with drill presses, they are also used frequently with all types of lathes and with numerically controlled (N/C) machines, and (in the larger sizes especially) some are used with the various types of boring mills.

It is estimated that up to 80 percent of the metal-removing dollar is spent in making holes. Thus any methods, materials, or special geometries which can lower the cost of drilling holes can be very profitable.

DRILL MATERIALS

Plain high-carbon tool steel, can be made hard enough but will not stand the heat generated in drilling most metals.

High-speed steel is used for about 90 percent of all twist drills. Most of these are of M2 or M1 high-speed steel, although for drilling the more difficult metals, HSS alloys such as M7, M10, T15, and the M40 (high cobalt) series are used.

Tungsten carbide drills account for about 10 percent of the drill production. In the small sizes (0.012 to 0.125 in., or 0.3 to 3.0 mm) and occasionally to $\frac{1}{2}$ in. (12.7 mm) solid-carbide drills are often used as they are stiffer and will stand the heat of the high speeds. Larger drills are more often made with brazed-on carbide inserts at the point.

DRILL SIZES

In the United States there are three series of twist drill sizes. Since they "interlock," one must refer to a table such as the one shown in Appendix F to determine the actual diameter needed.

The line graph of Fig. 10-1 shows the relative sizes of the number, letter, fraction, and metric sizes.

Number-size drills are stocked in No. 1 (0.228) to No. 60 (0.040) in. They are commonly made up to No. 80 (0.0135), and some manufacturers make up to No. 97 (0.0059 in.).

Letter-size drills are from A (0.234) to Z (0.413) in., and are made and stocked by most manufacturers.

Fractional sizes are made in steps of $\frac{1}{64}$ in. (0.4 mm) up to 1 in., and in $\frac{1}{16}$-in. (1.59-mm) steps to 2.0 in. Twist drills over 2-in. (50.8-mm) diameter are seldom used as they are heavy and very expensive (from \$100 to \$450 each), and there are less expensive ways to do the work.

The ***metric-size*** drills are a single series, from 0.20 mm (0.0079 in.) up. The most used sizes are shown in Appendix F with their inch equivalents.

Note: Not all styles of drills are made in all sizes. Check the manufacturers' catalogs before ordering or specifying special drills.

Tolerances on drill diameters are held quite close, averaging +0, −0.001 in. (+0, −0.02 mm). The included angle of the point must be within ±3° for sizes $\frac{1}{2}$ to $1\frac{1}{2}$ in. (12.5 to 38 mm). Lip height, centrality of web, and flute spacing must be within 0.003 to 0.007 in. TIR (total indicator reading), depending on the drill size.

Many manufacturers make premium quality, closer tolerance drills which are used for accurate work, especially numerical control. However, the common twist drill is *not* a precision tool, as indicated by the above tolerances, and this fact can affect actual hole sizes.

DRILL NOMENCLATURE

Figure 10-2 indicates the major terms and angles used in discussing drills. Notice that the basic principles used in lathe tool design also apply to drills. The *lip*

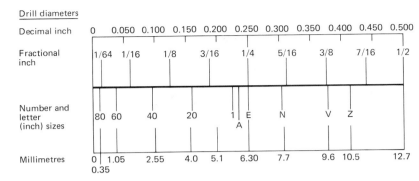

Fig. 10-1 Comparison chart of drill sizes, inch and metric. All horizontal scales are in proportion.

relief angle provides clearance so that only the cutting *lip* touches the cut. If this were 0°, the entire point would rub, and the drill would not cut.

The *margin* is quite narrow ($\frac{1}{64}$ to $\frac{1}{16}$ in. or 0.4 to 1.6 mm) and as the body clearance is a smaller diameter, only the margin of the drill touches the walls of the hole. The drill diameter is measured over the margins at the drill point.

The combination of the *helix angle* and the *point angle* creates the rake angle, and the point angle acts the same as the side cutting edge angle on a lathe-turning tool.

The *standard drill*, with 118° point angle, about 12° lip relief angle, and 29° helix angle, standard margin and flute sizes, will drill 80 percent of all the materials used with high to fair efficiency. However, variations of several features can make some work more productive.

The *point angle* can be ground from 90° up to 150° as shown in Fig. 10-3. In general, the 90° point angle is used for Bakelite, hard rubber, and some fibrous plastics. The 135 to 150° point angles are used for hard steels and nickel alloys and are a necessity for drilling through thin sheet metal.

The *spiral or helical point* (Fig. 10-3c) was developed to solve an expensive problem. Because of the flat chisel edge on regular drills, they "walk" off center when they first touch the work surface. This causes the holes to be out of line sometimes as much as 0.005 to 0.010 in. (0.125 to 0.250 mm). Solutions are:

1.   *Center punch* the hole location so that the chisel edge is held in place. This is impractical in production work.

2.   *Spot drill* (or center drill) with a short, stiff drill which cannot wander. This works well, but it is time-consuming.

3.   *The spiral-point drill* will not "walk" because the flat chisel edge has been eliminated. Three variations of this point are shown in Fig. 10-3c.

These points must be sharpened on special cam-controlled drill-pointing machines, which involves some added expense. However, where these can be used, they save considerable time, which is money.

The *helix angle* can vary as shown in Fig. 10-2. The slow or low helix drills are most often used for brass, bronze, Bakelite, and hard rubber. The fast or high helix angles are used on low-strength materials such

Fig. 10-2 Drill nomenclature, including variations in helix angle.

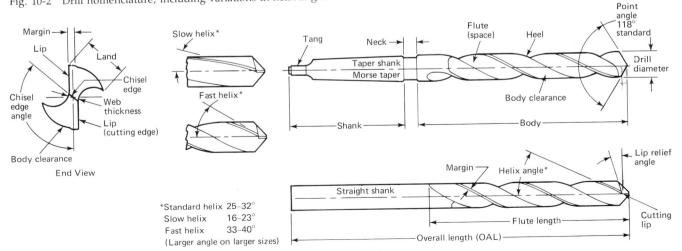

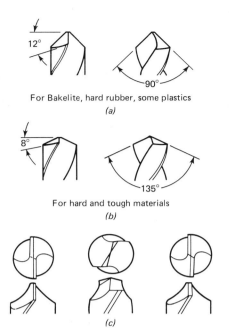

Fig. 10-3   Some possible variations in twist drill point angles. (*a*) Steep 90° angle. (*b*) Flat 135° angle. (*c*) Three types of self-centering drill points. (*Mohawk Tools, Inc.*)

as zinc alloys, magnesium, and soft copper. They can also be used in deep holes (four to six times the drill diameter) to bring the chips up faster.

The *web* of a drill is the center column of metal; to give the drills added strength, it is made thicker toward the shank. As shown in Fig. 10-4*a*, after much sharpening, this makes a long chisel edge making

Fig. 10-4   (*a*) The reason drills must have their webs thinned. (*b*) Three methods of web thinning. (*Mohawk Tools, Inc.*)

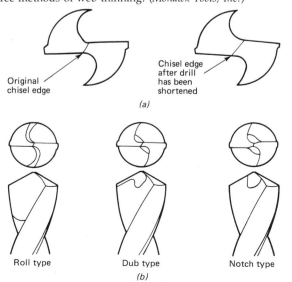

cutting difficult and generating additional heat. Thus, after sharpening a few times, the web must be *thinned* back a short distance from the point as shown in Fig. 10-4*b*. Other thinning methods are used to create shapes for special jobs even on new drills.

The *flutes* are sometimes made wider, making a smaller land. Sometimes they also are polished. These two changes are especially desirable when drilling soft aluminum, copper, and some plastics.

TYPES OF DRILLS

Combinations of the variables just described make up the many drill types. There are nearly 50 different types listed in the manufacturers' catalogs. However, many of these are for special industries and special jobs, such as the railroad and structural steel industries and special aircraft work.

Fig. 10-5   (*a*) General-purpose jobbers length drill. (*b*) High helix drill for aluminum. (*c*) Low (slow) helix drill for brass. (*d*) General-purpose drill with taper shank. (*e*) Combined center drill and countersink, 60°. (*Brown & Sharpe Mfg. Co.*) (*f*) Stub pattern screw machine drill. (*g*) Four-flute core drill. (*h*) Oil hole drill with side feed. (*Union/Butterfield Div., Litton Industries.*)

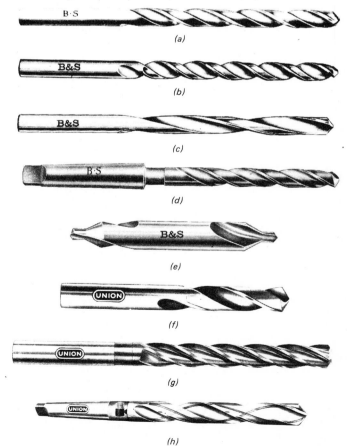

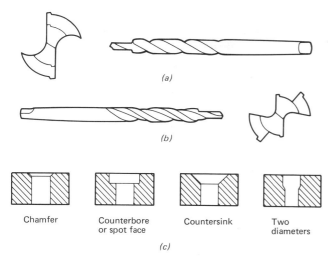

Fig. 10-6 (a) Step drill. (b) Subland drill. (*Greenfield Ampco Drill Division.*) (c) Some combined cuts which can be made with these drills.

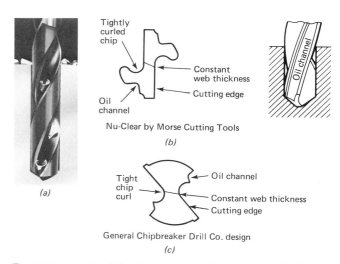

Fig. 10-7 (a) The "ideal" action of a chipbreaker drill. (*General Chipbreaker Drill Company.*) (b), (c) Cross-sectional views of two types of chipbreaker drills.

The most frequently used drills are shown in Fig. 10-5. The short *screw machine* drill (Fig. 10-5f) is used in automatic lathes. Because of its short length, it is stiffer than the standard "jobbers length" drill; thus, it is often used in regular and N/C machines because it will not "wander" or bend under pressure. It also is less expensive.

The *core drill* (Fig. 10-5g) is sharpened to the 118° angle but cannot start a hole because it has no point. It is used for enlarging previously drilled holes or for drilling out cored holes in castings. Because of its three or four flutes and heavier web, it can cut fast, give good finish, and make a straight hole.

The *oil hole* drill (Fig. 10-5h) is used for deep holes or for metals difficult to machine. Forcing the oil down to the actual cutting point cools the drill and forces the chips out. A special connection is needed to allow the drill to rotate while the oil line stays fixed. The oil can be fed from the end or the side of the shank.

Multidiameter Drills

These drills (Fig. 10-6) can often save considerable amounts of time. They have two or more diameters, and the top (larger) diameter may be shaped to countersink, counterbore, tap drill, or clearance drill. The saving in setup and handling may frequently save much more than the extra drill cost.

The step drill (Fig. 10-6a) is often ground down from a stock drill right in the shop's own toolroom. Both diameters are on the same land. This style is most economical when the small diameter is long since this is always nicked when resharpening the larger diameter.

The *subland drill* (Fig. 10-6b) has two separate lands, each continuous along the flute; thus, the end view is quite odd looking. However, it can be resharpened the full length of the flutes with no damage to the smaller diameter. This is a more expensive drill, but it might be cheaper in the long run.

CAUTION: The rpm of these multidiameter drills must be based on the *largest* diameter, and often the feed must be based on the *smallest* diameter. This would seem obvious, but some sorry looking drills have resulted from failure to observe this basic rule.

Five Special Drills

"chipbreaker drills" (Fig. 10-7)

These drills were created to solve the problem of the long stringy chips shown in Fig. 10-8. These can wind around the machine spindle and become dangerous to the operator.

The patented extra rib shown in Fig. 10-7c reinforces the drill's strength and helps to break up chips by forcing material into the smaller fluted area. In Fig. 10-7b a specially shaped flute space curls the chips to help them break up. Other advantages are: a constant-thickness web (no thinning needed), a stronger drill, and an improved flow of coolant to the working area.

Because of these factors, the *feed* can be more than doubled (which also helps to break up the chips), and deeper holes can be drilled, with longer life between sharpenings. Cutting speeds remain about the same

Fig. 10-8   Typical stringy drill chips on the right, the preferred broken up chips on the left. (*General Chipbreaker Drill Company.*)

or somewhat higher. Sizes up to 4-in. diameter are listed.

Other chipbreaker drills are made by different companies using similar basic principles.

Radial Lip[1] drills (Fig. 10-9a)

Radial Lip drills have a very different point geometry. If a drill is run until it is excessively worn, the sharp top outer edges of the cutting lips will be rounded over because of the high forces and thin cross section at this outer edge.

The patented point form redistributes the stress over the cutting lips so that the heaviest stress is closer to the center. This is described as a "radial lip" grind. Any make or size drill can be ground with this specially shaped point.

The usual included point angles are used since these are involved in the drill's ability to penetrate the work. Thus, the *thrust* forces are not materially changed; however, the *torque* is decreased by the radical grind.

Extensive testing indicates that these drills can give four to ten times the drill life; smoother, closer-to-size, and practically burr-free holes at the same speeds; and up to double the feeds of regular drills. Industry reports as high as 7000 holes per grind in automobile engine blocks, and some reaming was eliminated.

Sharpening of these drills must be done with a special machine. A patented sharpening machine is on the market (see Fig. 10-9b). The rather high cost of this machine and a somewhat longer sharpening time do add to the cost appreciably. However, only one-

quarter to one-tenth as many drills will be purchased; the longer time between sharpening can create very worthwhile savings; and the lowered burring cost can be an important economy if a large number of holes are being drilled.

spade drills

These drills are *not* twist drills. Their geometry resembles that of a lathe-turning tool. They vary in size from $\frac{3}{32}$ in. (2.4 mm) "flat" drills to over 10-in.-diameter (250-mm) spade drills.

The *flat drills* or *stub drills* are short, sturdy, spade-type drills used for drilling short holes or for center drilling in tough hardened materials. They are made in sizes up to $\frac{1}{2}$ in. (12.7 mm) and are often made of solid carbide. They use the standard 118° angle.

The *spade drills* and their special holders (Fig. 10-10) are made in diameters from $1\frac{1}{32}$ to 6 in. (26.2 to 150 mm) as stock items and up to 15 in. (381 mm) and over as specials.

The *advantages* of spade drills are:

1. Large-sized holes can be made in a single pass.
2. Rigid holders allow very heavy feeds (two or three times the feeds for twist drills), which saves considerable time.
3. They can drill deep holes with low-pressure coolant. Holes up to 30 in. (750 mm) deep have been drilled under proper conditions. They cut into solid metal or enlarge cored holes.
4. Low cost. The drill bits (blades) cost from $9 to $50, and one holder will serve for several sizes. The blades can be resharpened.
5. Accuracy of the hole diameter is as good or better than twist drills. A 4-in. (100-mm) hole will be only 0.006 to 0.012 in. (0.150 to 0.300 mm) oversize.

[1]Trademark of Radial Lip Machine, Inc.

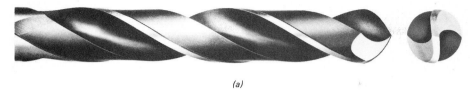

(a)

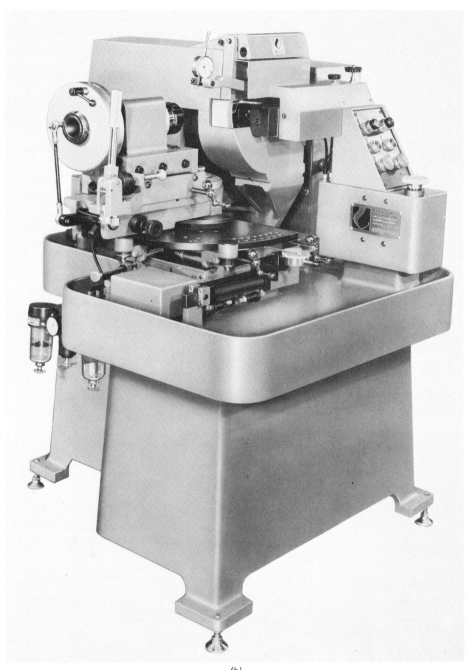

(b)

Fig. 10-9   (*a*) A Radial Lip point on a drill. (*b*) The machine used to grind these points. (*Radial Lip Machine, Inc.*)

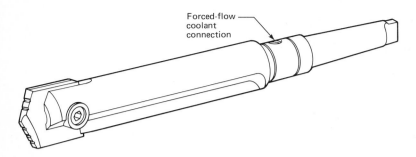

Fig. 10-10 Spade drill with forced-flow coolant for deep holes. (*Erickson Tool Company.*)

Cutting speeds should be about the same or slightly lower than those used for regular drilling.

The *power needed* can be very high, so these are seldom used on an ordinary drill press. Using 4-in. (100-mm) spade drills, with 0.030 to 0.060 in./rev (0.75 to 1.5 mm/rev) feed may require a 10- to 30-hp (7.5- to 22.5-kW) motor, depending on the material being drilled. Therefore, a milling or boring machine is more often used for these large drills.

Material for the blades is usually M2 or other HSS alloys as needed. Carbide blades are made also for the occasional difficult job or longer tool life.

gun drills

These single-lip drills (Fig. 10-11) are made especially for drilling holes 3 to 30 in. (75 to 750 mm) deep. However, their potential for straight, true, round holes with good finish and closer tolerances than twist drills is making their use in "shallow" (4 diameters or less in depth) holes more frequent.

Sizes are from $\frac{5}{64}$- to 2-in. (2- to 50-mm) diameter. The cutting tips may be high-speed steel, but most frequently they are solid carbide or carbide tipped. Gun drills have a single cutting edge and are straight flute drills. A hole goes the full length of the drill so that cutting fluid may be forced through. This fluid, most importantly, forces the chips up the flutes and out of the hole at pressures of 200 to 1500 psi (1.38 to 10.35 mPa).

Cutting speeds are often double those used for solid-carbide twist drills. Feedrates, however, are less than 0.001 in./rev (0.025 mm/rev). This is necessary in order to produce fine chips which will easily flush out the narrow flute openings. Finish may be 32 to 64 μin. (0.8 to 1.6 μm) in steel and as low as 4 μin. (0.1 μm) in aluminum and copper.

Properly designed fixtures and carefully sharpened drill points enable a gun drill to cut within 0.001 in. (0.025 mm) of specified diameter and maintain parallelism with excellent finish and small burrs at breakthrough.

Convoflute[1] drills (Fig. 10-12)

Convoflute drills with very open, auger-like flutes, special point geometry, and extra length are made for drilling holes 10 to 12 diameters deep. The flute shape moves chips (and thus heat) out of the bottom of the hole much more efficiently than standard drills.

Fig. 10-11 (*a*) A gun drill. (*b*) Some of the cuts made with gun drills. (*Eldorado Tool and Manufacturing Corp., a subsidiary of Litton Industries, Inc.*)

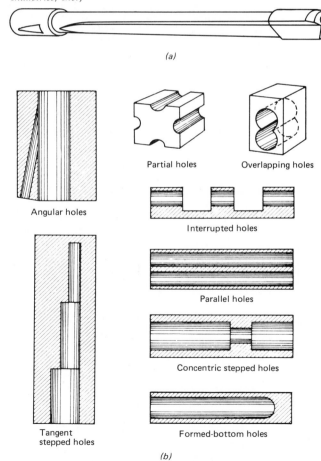

(a)

Angular holes

Partial holes   Overlapping holes

Interrupted holes

Parallel holes

Concentric stepped holes

Tangent stepped holes

Formed-bottom holes

(b)

[1]Trademark of Chicago-Latrobe.

Enlarged view of point

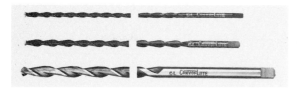

Fig. 10-12   Patented Convoflute deep-hole drill. (*Chicago-Latrobe, a division of TRW Inc.*)

The Convoflute drill is made from M10 or cobalt HSS in sizes $\frac{3}{16}$- to 1½-in. (4.8- to 38-mm) diameter. Tests indicate many more holes between sharpenings, at usual or higher speeds, and no need to withdraw the drill to "clear the chips." Cost is higher than for conventional drills, but the time saved in the actual drilling of deep holes plus the fact that fewer drills are needed could make these a good investment. They are used on any drilling machine.

ACCURACY OF DRILLED HOLE SIZES

Many thousands of twist drills are sharpened "off-hand" by an operator applying the drill point to the grinding wheel. The resulting point may be close to perfect or merely adequate for drilling a few holes. For production work, more predictable drill life, and control of hole diameter, twist drills must be machine sharpened. Figure 10-13 illustrates what happens to hole size when drills are incorrectly sharpened.

A study by a number of drill manufacturers using *new*, HSS drills, in average condition drill presses, found that the actual hole diameter made by a twist drill in iron or steel is always *oversize* (Fig. 10-14). In many jobs these seemingly small amounts may not make any difference. They do, however, indicate that a twist drill should not be relied on for accurately sized holes.

Fig. 10-13   Two types of improper twist drill sharpening. Both will cause oversize and out-of-round holes. (*a*) Lips at unequal angles; (*b*) lips of unequal length.

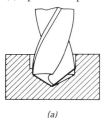
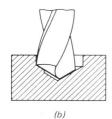

(a)   (b)

REAMING

As twist drills do not make accurately sized or good finish holes, a *reamer* of some type is often used to cut the final size and finish. *A reamer will not make the original hole*, it will only enlarge a previously drilled or bored hole. It will cut to within +0.0005 in. (0.013 mm) of tool size and give finishes to 32 μin. (0.8 μm).

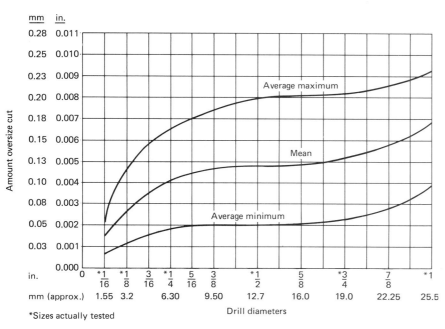

Fig. 10-14   Results of tests, showing amount of oversize in holes made by standard HSS twist drills in cast iron and steel. (*Metal Cutting Tool Institute.*)

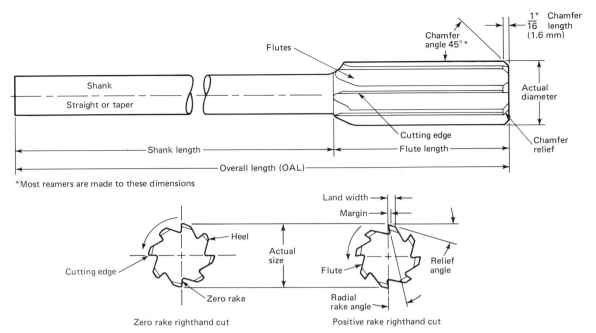

Fig. 10-15   Nomenclature of a standard chucking, or jobbers, reamer and the usual rake angles.

Reamers are usually made of HSS although solid-carbide and carbide-tipped reamers are made in many sizes and styles.

The regular *chucking or jobbers reamers* (Figs. 10-15 and 10-16) are made in number and letter sizes, in fractional inch sizes, and in millimetre sizes. They can be purchased (for additional cost) ground to any desired diameter.

The basic construction and nomenclature of reamers is shown in Fig. 10-15. This shows the most frequently used style for holes up to 1 in. (25 mm), called a *chucking* reamer. It has 1 to 2½ in. (25 to 68 mm) of flute length.

These, and some other reamers, are made with three shapes of flutes, and all are for standard right-hand cutting.

1. *Straight flute* is satisfactory for most work and the least expensive. Do not use if a keyway or other interruption is in the hole.

2. *Righthand spiral* fluted reamers (Fig. 10-17b) give freer cutting action and tend to lift the chips out of the hole. *Do not* use on copper or soft aluminum because these reamers tend to pull down into the hole.

3. *Lefthand spiral* fluted reamers (Fig. 10-17a) require slightly more pressure to feed but give a smooth cut and *can* be used on soft, gummy materials, as they tend to be pushed *out* of the hole as they advance. It is

not wise to use these in blind holes, as they push the chips down into the hole.

As shown in Fig. 10-17c, solid reamers do almost all their cutting with the $\frac{1}{16}$-in. (1.6-mm), 45° chamfered front end. The flutes guide the reamer and slightly improve the finish. Thus, reamers should *not* be used for heavy stock removal.

The drilled hole should be $\frac{1}{64}$ (0.0156) in. or 0.5 mm (0.020 in.) *smaller* than the desired final size for holes up to 1-in. (25-mm) diameter. Over this, about $\frac{1}{32}$ (0.031 25) in. or 1.0 mm (0.0394 in.) *undersized* drill or bored holes may be used before reaming.

Fig. 10-16   Some styles of (a) chucking reamers straight shank and (b) taper-shank jobbers reamers. (*Glenbard Tool Mfrs., Inc., a subsidiary of Anixter Bros. Inc.*)

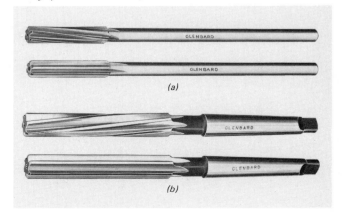

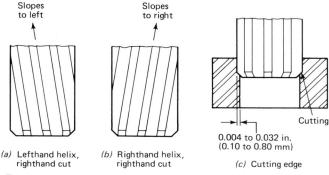

(a) Lefthand helix, righthand cut

(b) Righthand helix, righthand cut

Slopes to left

Slopes to right

Cutting

0.004 to 0.032 in. (0.10 to 0.80 mm)

(c) Cutting edge

Fig. 10-17   Method of identifying lefthand and righthand reamers and a diagram of the cutting action.

Types of Reamers

The Metal Cutting Tool Institute lists about 45 different types of reamers; however, those shown in Fig. 10-16 are by far the most frequently used.

Shell reamers (Fig. 10-18a) are frequently used for hole sizes over 1 in. (25 mm). They are less expensive than large solid reamers, and the arbors which hold them can be made with shanks to fit any machine spindle. They are made in fractional sizes up to 3-in. diameter with HSS or carbide-tipped blades.

Inserted blade reamers come in several styles including Fig. 10-18b and can be used instead of the shell-type reamers for accurately sizing large holes. Large sizes of the style shown in Fig. 8-20 can be used only in radial drills because of the power required.

Counterbores (Fig. 10-19a and b) are used especially for setting the heads of socket-head cap screws flush or below the surface (Fig. 9-1c). The hole diameter should be about $\frac{1}{32}$ in. (0.80 mm) larger than the diameter of the bolt head. They are also made for hexhead and flathead bolts.

As counterbores are not self-centering, a fixed or a renewable pilot, the diameter of the drilled hole, is always included with the cutter. Counterbores are usually made of HSS. Feeds and speeds are about 75 percent of those used in drilling.

Solid counterbores are often used up to 1-in. (25-mm) diameter and are available in some styles up to 2-in. (50-mm) diameter.

Spotfacing is the same as counterboring but is quite shallow. Over $\frac{1}{8}$ in. (3.18 mm) deep, it could be called a counterbore.

Countersinks (Fig. 10-19c) are used to make a small chamfer or bevel at the edges of a hole, to remove burrs, and also to allow bolts, dowel pins, taps, or reamers to enter the hole more easily. These chamfers may be 82° for flathead bolts or 90°. A 60° angle is also made.

Chatter is often a problem when cutting with countersinks, and several manufacturers make models which help avoid this.

The material used for countersinks is usually HSS, though sometimes carbide is used. The *size* is the maximum diameter of the cutter, which must, of course, be larger than either the diameter of the bolt head being used or the outside of the chamfer.

Countersinks are made in sizes from $\frac{1}{2}$ to 1 in. (12 to 25 mm) graduated in $\frac{1}{8}$-in. (3-mm) steps. Over this size, a 45° boring tool would be used.

Fig. 10-18   (a) An adjustable-blade shell reamer. (b) An adjustable-blade tapered-shank chucking reamer. (*McCrosky Tool Corp.*)

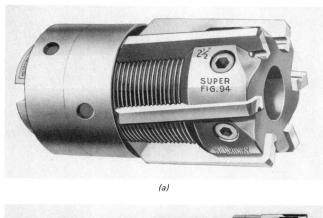

(a)

(b)

Fig. 10-19   (a) Solid pilot counterbore. (b) Taper-shank counterbore with interchangeable pilot. (c) Countersink, chatterless, six flute. (*DoAll Company.*)

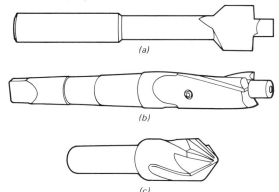

(a)

(b)

(c)

COMPUTING DRILLING TIME

As shown in Fig. 10-20, the *depth* of a drilled hole means depth of the full drill diameter. Thus, the length of the point is an additional distance which the drill must travel. In "through" holes a distance must also be allowed to be certain that the drill has completely driven through and, especially in N/C machines, a clearance is needed at the top to prevent a collision between the drill and the work.

EXAMPLE   10-1

Several $\frac{1}{2}$-in. (12.7-mm) holes are to be drilled through a $\frac{3}{4}$-in.-thick (19-mm) piece of stainless steel. Cutting speed is 35 fpm (10.7 m/min), and feedrate is 0.005 in./rev (0.125 mm/rev). How long will it take to drill each hole?

$$\text{rpm} = \frac{4(35)}{\frac{1}{2}} = 280 \text{ (for inch and metric)}$$

$$\text{Feed} = (280)(0.005) = 1.4 \text{ ipm}$$

or $= (280)(0.125) = 35.0 \text{ mm/min}$

The total drill distance (use 0.020 in. or 0.50 mm top and bottom clearances), referring to Fig. 10-20*a*, is

in.:
$$0.020 + 0.750 + (0.3 \times \tfrac{1}{2}) + 0.020 = 0.94 \text{ in.}$$

mm:
$$0.50 + 19 + (0.3 \times 12.7) + 0.050 = 23.81 \text{ mm}$$

$$\text{Time} = \text{distance} \div \text{rate}$$

$$\text{(in.)} \; t = \frac{0.94}{1.4} = 0.67 \text{ min}$$

$$\text{(metric)} \; t = \frac{23.81}{35.0} = 0.68 \text{ min}$$

More information (including dos and don'ts, thrust, horsepower, torque, etc.) can be found especially in the *Metal Cutting Tool Handbook* available from the Metal Cutting Tool Institute.

For complete information on the almost limitless variations and combinations of the tools discussed in this chapter, refer to the references at the end of this chapter or, better still, browse through the catalogs of the manufacturers of this kind of tooling. Some of their names are given in the illustrations.

Review Questions and Problems

10-1.   Which is the larger of each of these pairs of drills: No. 7 or No. 25, $\frac{1}{16}$ or A, F or No. 52, 4.5 mm or K?

10.2.   If a drawing specified some hole diameters of
   a.   0.3750 to 0.3755
   b.   1.500 to 1.502
   what tools and machines would you use to machine them?

10-3.   Make a rough sketch of a drill and label the principal parts.

10-4.   Describe three ways of locating holes on a workpiece. What is the accuracy of each method?

10-5.   Would a numerical control drill press locate holes within 0.002 in. (0.05 mm)?

10-6.   When would you specify the use of a core drill or oil hole drill?

10-7.   Why is a subland drill more expensive than a step drill? Why might it be worth the extra cost?

10-8.   How does a chipbreaker drill cause the chips to break up?

10-9.   Why will a radial lip grind on a drill make it last longer between sharpenings?

10-10.   What kind of a drill can make a hole 2.720-in. (69.1-mm) diameter in solid metal? Describe this drill.

10-11.   Describe a gun drill and its uses. How deep is a "deep" drilled hole?

10-12.   If you were drilling mild steel with a new standard $\frac{1}{2}$-in. (12.7-mm) drill, which would be the most likely size of the hole: 0.498, 0.501, 0.504 in. or 12.75, 12.65, 12.83 mm?

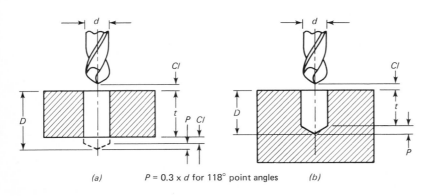

(a) $P = 0.3 \times d$ for 118° point angles   (b)

Fig. 10-20   Sketch showing elements needed in calculating the total depth of a drilled hole. (*a*) Through hole; (*b*) blind hole.

10-13. What two reasons are there for using reamers?

10-14. If you were reaming a blind hole, would you use a straight, righthand, or lefthand spiral-fluted reamer? Why?

10-15. Would you use a countersink or a counterbore to

    a. Make room for the head of a flathead bolt?

    b. Make a chamfer on the edge of a hole?

    c. Make space for the head of a socket-head bolt?

10-16. You are to drill some ⅜-in. (9.50-mm) holes through a ⅝-in.-thick (16-mm) carbon-steel plate. Cutting speed is 60 fpm (18 m/min), and feed is 0.004 in./rev (0.10 mm/rev). Standard drill is used. How long (in minutes) will it take to drill one hole? Use both inch and metric figures and explain the difference.

10-17. Using a spade drill, you are drilling some 3.450-in. (87.6-mm) holes in medium cast iron. Cutting speed is 50 fpm (15 m/min), feedrate is 0.020 in./rev (0.50 mm/rev), and the hole is 3.10 in. (79 mm) deep. How long (in minutes) will it take?

References

1. Blach, Frederick S.: *Gundrilling*, Eldorado Tool & Manufacturing Corp., Milford, Conn., 1968.

2. Butrick, Frank M.: "Drilling Big Holes," *Modern Machine Shop*, January 1970.

3. Greuner, Bernd: *Needed: More Information on Drilling and Boring*, Society of Manufacturing Engineers, Detroit, Mich., Ms72-156, 1972.

4. *Gundrilling, Trepanning, and Deep Hole Machining*, Society of Manufacturing Engineers, Detroit, Mich., 1969.

5. *Machining Data Handbook*, Machinability Data Center, Metcut Research Associates Inc., 1972.

6. *Metal Cutting Tool Handbook*, Metal Cutting Tool Institute, 1965.

7. *Use and Care of Twist Drills*, The Cleveland Twist Drill Co., Cleveland, Ohio, 1965.

cutting threads

Threading tools are big business. Over $106,000,000 worth were sold in 1968, according to the U.S. Department of Commerce. These tools are used in drill presses, engine lathes, turret and automatic lathes, and special equipment. They include taps for internal (female) threads, dies for external (male) threads, and special rolling equipment for bolts.

The terminology used and the types of tools used are, therefore, an important part of the knowledge of manufacturing which any engineer should have.

NOMENCLATURE

Both U.S. and metric threads use the same basic thread form, and nomenclature, as shown in Fig. 11-1. The tolerances and class numbers specified by the ISO are somewhat different than those in the U.S. systems.

Thread angle (included angle) in all systems is now 60° for the vee-type thread. This is used for over 80 percent of all threaded fasteners.

Fig. 11-1   Standard U.S. and metric thread form and terminology.

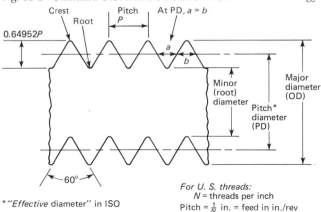

*"Effective diameter" in ISO

Note: There is no metric equivalent of the U. S. "threads per inch"; only pitch is specified.

Major diameter, or outside diameter (OD), is the "size" of the bolt, nut, or tapped hole at its crest. This is specified as ½ in. or 12.5 mm, etc. It is sometimes referred to as the *nominal* diameter because the actual diameter of a bolt is slightly *under* the nominal.

Minor diameter is the diameter at the *root* or bottom of the threads. Originally, this and the crest were sharp vees. However, a sharp vee is impossible to maintain due to the wear of the cutting tools, and it is not a good shape for strength because sharp vees are "stress raisers" and cause premature failure. Therefore, the international standard (ISO) calls for rounded or flat roots and crests as shown in Fig. 11.1.

Pitch diameter (PD), or effective diameter, is somewhat more than halfway down between crest and root. It is the diameter at which the nut and bolt threads are both the same thickness. This is carefully controlled because threads mating accurately at this point give the greatest strength.

Helix angle is the angle at which the thread is inclined from a perpendicular to the centerline. It is built into the thread and seldom has to be computed.

Lead is the amount a screw thread advances in a 360° turn.

Pitch is the distance from center to center of two adjacent crests (or roots). On the usual single-thread screw, the pitch and the lead are the same. On double- or triple-thread screws, the lead is two or three times the pitch. The reciprocal of the pitch is the threads per inch (sometimes abbreviated tpi) in the U.S. system.

Other thread shapes are shown in Fig. 11-2. The *acme thread* (Fig. 11-2a) or a variation of it is often used for jack screws, lead screws, and other heavy work. The *square thread* (not shown) is seldom used today, as it is difficult to machine. A 10° modified square thread, with the sides sloped 5°, is sometimes used (Fig. 11-2b).

The *buttress thread* (Fig. 11-2c) is used to hold heavy loads applied axially on the screw with practically no radial component of thrust.

Pipe threads (National Taper Pipe, NPT) have a 60° thread angle, the same as machine threads. However, they are cut on a taper of 1:16 or 0.75 in./ft. The taper provides a way of making pipe joints seal watertight. A variation is the *dryseal* (NPTF) taper pipe thread, which has a close fit between crest and root of external and internal threads.

CAUTION: Pipe sizes are very different from bolt sizes. They are based on the inside diameter of the standard weight pipe and not very close to that in many cases. Steel pipe is made in three weights, Standard (std), Extra Strong (XS), and Double Extra Strong (XXS). The outside diameter of all weights is the same. The extra thickness is taken from the inside dimension.

Right- and lefthand threads are cut in all the above styles. A righthand thread is one in which the bolt enters the tapped hole when the bolt is rotated clockwise.

Classes of Fit

These classes are specified for inch thread sizes, followed by the letter A for external (male) threads and B for internal (female) threads.

Class 1A and B are loose fit, used where fast, easy assembly is necessary, even if the threads are slightly damaged or dirty. Farm plowing and harrowing equipment and military equipment such as tanks could use this fit.

Class 2A and B are medium fit. This is the fit which is supplied as "standard" unless otherwise specified. The bolts are usually slightly undersize, and there

will be some "shake" or "play" between the bolt and the nut or tapped hole.

Class 3A and B are made closer to basic sizes with closer tolerances; thus, they will fit together with almost no play. These may be required where the screw thread is used for adjustment of timing, spacing, etc.

There is no class 4 in use now.

Class 5A and B are made for *interference* fit. The bolt is actually a few thousandths larger than the hole. Thus, they must be assembled with a wrench. This class is used especially for studs (headless bolts threaded on both ends). One end would have the class 5 fit so it would stay on the machine and the other end a class 2 fit for easy assembly and disassembly of an attached part.

metric (ISO) classes of fit

The ISO uses a different numbering system and, within this numbering system, the tolerances (at this time) are somewhat greater than those in use in the United States.

The ISO metric fits are:

Free fit: Bolt, Class 8g   Nut, Class 7 H

Medium fit: Bolt, Class 6g   Nut, Class 6 H

Close fit: Bolt, Class 4h   Nut, Class 5 H

The uses of these fits are the same as classes 1, 2, and 3 in the U.S. Standard.

SPECIFYING THREADS

Since the 60° Unified National Standard and Metric ISO threads are by far the most used, the following will refer only to these and only to single pitch threads.

These 60° thread forms are made in *coarse*, *fine*, and *extra fine* threads. The inch series are referred to as *Unified* because of an agreement between Great Britain and the United States on the specifications. Thus, we have UNC (Unified National Coarse), UNF (fine

Fig. 11-2   Four commonly used styles of threads.

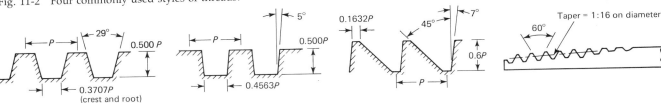

American Standard *acme* thread. (There is a similar metric thread called "trapezoidal" which has a 30° included angle.)

(a)

American Modified *square* thread.

(b)

American Standard *buttress* thread (Metric is similar, but different proportions or angles.)

(c)

American National *taper pipe thread.*

(d)

thread), and UNEF (extra fine thread). The UNEF is seldom used, and the "standard" thread is the *coarse* series in most sizes.

The ISO also has a coarse and fine thread series, but the coarse series is used for most work.

Complete specification of a threaded bolt involves:

1. Nominal diameter or size
2. Pitch of the threads or threads per inch
3. Group (fine, coarse)
4. Class of fit required
5. Length of the bolt

Thus, a ¼-in. "standard" hexagonal head bolt or screw would be specified as shown in Fig. 11-3a, and its metric equivalent (6-mm diameter) is completely specified in Fig. 11-3b.

In industry it is usually assumed that standard bolts and nuts and thread-cutting equipment will be supplied. The shortened specifications in this case would be:

¼—20 × 1½—in inch dimensions

M6 × 38   —in metric dimensions

TAPPING (CUTTING INTERNAL THREADS)

Some internal threads can be cut with a single-point tool as shown previously. However, most frequently a *tap* of some type is used as it is faster and generally more accurate.

Taps are made in many styles; however, a few styles do 90 percent of the work. Figure 11-4a shows the general terms used. As shown in Figs. 11-4b and 11-5, the cutting end of the tap is made in three different tapers.

The *taper tap* is not often used today. Occasionally, it is used first as a starter if the metal is difficult to tap. The end is tapered about 5° per side, which makes eight partial threads.

The *plug tap* is the style used probably 90 percent of the time. With the proper geometry of the cutting edge and a good lubricant, a plug tap will do most of the work needed. The end is tapered 8° per side, which makes four or five incomplete threads.

The *bottoming tap* is used only for blind holes where the thread must go close to the bottom of the hole. It has only 1½ to 3 incomplete threads. If the hole can be drilled deeper, a bottoming tap may not be needed. The plug tap must be used first, followed by the bottoming tap.

All three types of ends are made from identical taps. Size, length, and all measurements except the end taper are the same.

Styles of Taps

Today, the **hand tap** (Fig. 11-5) is used both by hand and in machines of all types. This is the basic tap design: four straight flutes, in taper, plug, or bottoming types. The small, numbered machine screw sizes are standard in two and three flutes depending on the size.

If soft and stringy metals are being tapped, or if horizontal holes (where chips can bother) are being made, either two- or three-flute taps can be used in the larger sizes. The flute spaces are larger, but the taps are weaker. The two-flute especially has a very small cross section.

CAUTION: The chips formed by these taps cannot get out; thus, they accumulate in the flute spaces. This causes added friction and is a major cause of broken taps.

The **spiral point**, or **"gun," tap** (Fig. 11-6a) is made the same as the standard hand tap except at the point. A slash is ground in each flute at the point of the tap. This accomplishes several things:

Fig. 11-3   Complete specifications for (a) inch and (b) metric bolts.

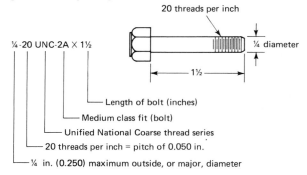

¼-20 UNC-2A × 1½

20 threads per inch

¼ diameter

1½

— Length of bolt (inches)
— Medium class fit (bolt)
— Unified National Coarse thread series
— 20 threads per inch = pitch of 0.050 in.
— ¼ in. (0.250) maximum outside, or major, diameter

(a)

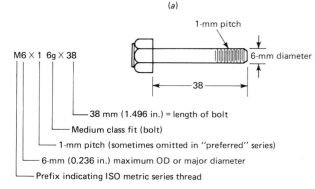

M6 × 1 6g × 38

1-mm pitch

6-mm diameter

38

— 38 mm (1.496 in.) = length of bolt
— Medium class fit (bolt)
— 1-mm pitch (sometimes omitted in "preferred" series)
— 6-mm (0.236 in.) maximum OD or major diameter
— Prefix indicating ISO metric series thread

Note: The "class" is omitted from both specifications when "standard" commercial bolts and nuts are used.

(b)

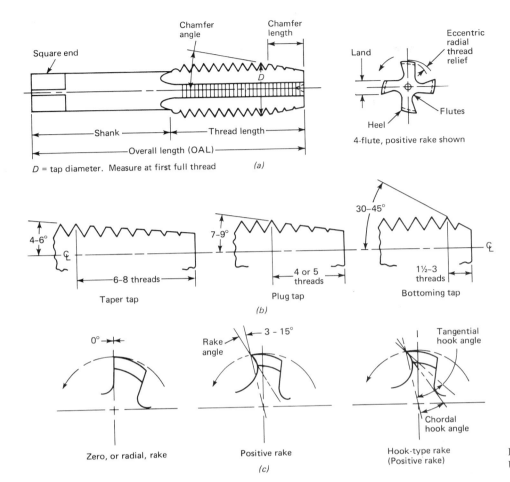

Fig. 11-4 Tap nomenclature for both U.S. unified and metric taps.

1. The gun tap has fewer flutes (usually three), and they are *shallower*. This means a stronger tap.

2. The chips are forced out *ahead* of the tap instead of accumulating in the flutes as they will with a plug tap.

3. Because of these two factors, the spiral point tap can often be run faster than the hand tap, and tap breakage is greatly reduced.

The gun tap has to quite an extent replaced the "standard" style in industry, especially for open-ended through holes in mild steel and aluminum. Both regular and spiral-point taps are made in all sizes including metric.

The **spiral-flute tap** (Fig. 11-6b) is made in regular and fast spirals, that is, with a small or large helix angle. They are sometimes called *helical-fluted taps*. The use of these taps has been increasing since they pull the chips up out of the hole and produce good threads in soft metals (such as aluminum, zinc, and copper) yet also work well in Mónel metal, stainless

steel, and cast steel. They are made in all sizes up to 1½ in. and in metric sizes up to 12 mm.

CAUTION: While the "standard" taps will efficiently do most work, if a great deal of aluminum, brass, cast iron, or stainless steel is being tapped, the manufacturer can supply "standard" specials which will do a better job.

Fluteless taps do not look like taps, except for the spiral "threads." Figure 11-7 shows one brand of these. As shown in the cross section, these taps are *not* round. They are shaped so that they "cold form" the metal out of the wall of the hole into the thread form with no chips. The fluteless tap was originally designed for use in aluminum, brass, and zinc alloys. However, it is being successfully used in mild steel and some stainless steels. Thus, it is worth checking for use where BHN is under 180. They are available in most sizes, including metric threads.

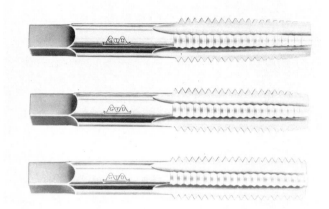

Fig. 11-5   A "set" of taps. *Top*, taper; *center*, a plug tap; *bottom*, a bottoming tap. (*Greenfield Tap & Die.*)

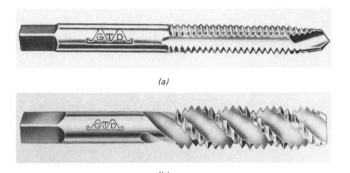

(a)

(b)

Fig. 11-6   (*a*) Spiral-point (gun) tap. (*b*) Fast spiral-flute tap. (*Greenfield Tap & Die.*)

These taps are very strong and can often be run up to twice as fast as other styles. However, the size of the hole drilled before tapping must be no larger than the pitch diameter of the thread. The cold-formed thread (Fig. 11-7*b*) often has a better finish and is stronger than a cut thread. A cutting oil *must* be used, and the two ends of the hole should be countersunk because the tap raises the metal at the ends.

Collapsing taps (Fig. 11-8) collapse to a smaller diameter at the end of the cut. Thus, when used on lathes of any kind, they can be pulled back rapidly. They are made in sizes from about 1 in. up in both machine and pipe threads. They use three to six separate "chasers" which must be ground as a set. The tap holder and special dies make this assembly moderately expensive, but it is economical for medium and high production work.

Pipe taps (not shown), both straight and taper, are made like hand taps. They come in standard pipe sizes from $\frac{1}{8}$ in. diameter, 27 threads per inch to 2 in. diameter, $11\frac{1}{2}$ threads per inch and sometimes to 3 in., 8 threads per inch.

Material used for taps is usually high-speed steel in the M1, M2, M7, and sometimes the M40 series cobalt high-speed steels. A few taps are made of solid tungsten carbide.

Most taps today have *ground* threads. The grinding (see Chap. 15) is done after hardening and makes much more accurate cutting tools. "Cut thread" taps are available at a somewhat lower cost in some styles and sizes.

Toolholders for taps used by hand are called *tap wrenches*. There are two types, as shown in Fig. 11-9. Each *wrench* will hold the square shank of several sizes of taps.

When taps are used in drill presses, a special head with a reversing, slip-type clutch is used. These *tapping heads* (Fig. 11-10) can be set so that if a hard spot is met in the metal, the clutch slips and the tap will not break. They are constructed so that when the hand-feed lever or the automatic numerical control machine cycle starts upward, the rotation reverses (and often goes faster) (Fig. 11-10*b*) to bring the tap safely out of the hole. Tapping heads cost from $75 to $350 each.

Work holding for tapping is the same as for any drill press or lathe work: clamps, vises, fixtures, etc., as needed. It is necessary to locate the tap centrally and straight in the hole. This is difficult in hand tapping but relatively easy in machine tapping.

Numerical control is especially efficient, as it will locate over a hole, regardless of when it was drilled, if it was drilled from the same tape and on the same setup.

Fig. 11-7   (*a*) A standard and a bottoming fluteless tap with oil grooves. (*b*) Showing cold-forming action of fluteless tap. (*Bendix Industrial Tools Division.*)

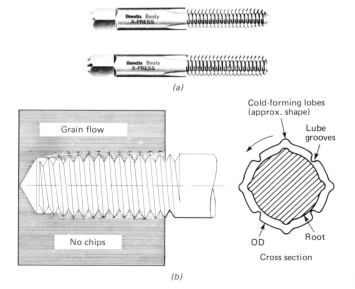

(a)

(b)

Sizes of Threads

In the inch system (Table 11-1), a ¼-in. bolt is the smallest fractional size which is standard. Below this size a numbering system is used, and the sizes are referred to as *machine screw sizes*. The smallest is No. 0 (0.060-in. diameter) and they go to No. 12 (0.216-in. diameter). A No. 14 is listed but seldom used. There is a 0.013-in. increase in diameter for each larger number. Notice in Table 11-1 that No. 5 is exactly ⅛-in. (0.125-in.) diameter and that No. 10 (0.185 in.) is very close to 3/16-in. (0.1875 in.) diameter.

In the metric system, the ISO lists a first choice, a second choice, and a few third choice diameters. The hope is that most equipment can be designed using first choice. These, the *coarse series*, are listed in Table 11-2 with comparisons with our "threads per inch."

The *accuracy of taps* (or tap limits), as manufactured, is specified for ground thread taps by H (meaning high) followed by a number. These numbers specify the maximum amount that the tap is *over* the basic minimum pitch diameter. Each number is 0.0005 in. maximum.

H1 = +0 to +0.0005 in. over basic pitch diameter
H2 = +0.0005 to +0.0010 in. over basic pitch diameter
H3 = +0.0010 to +0.0015 in. over basic pitch diameter
H4 = +0.0015 to +0.002 in. over basic pitch diameter

Fig. 11-8   A collapsing tap with a plate trip and hand reset. (*Geometric Tool.*)

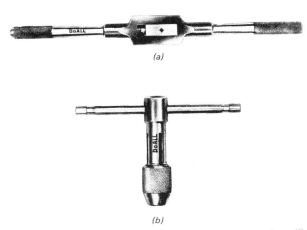

Fig. 11-9   (*a*) Straight and (*b*) T-handle style tap wrenches. (*DoAll Company.*)

TABLE 11-1   INCH SIZES OF MACHINE THREADS

Size	Threads per Inch		OD, in.	OD, mm	Pitch, UNC only		Tap Drill* in. UNC only—75%
	NC and UNC	NF and UNF			in.	mm	
0	—	80	0.060	1.524	0.0125	0.32	3/64 (0.0468)
2	56	64	0.086	2.184	0.0179	0.46	No. 50 (0.070)
4	40	48	0.112	2.845	0.025	0.64	No. 43 (0.089)
5	40	44	0.125	3.175†	0.025	0.64	No. 38 (0.102)
6	32	40	0.138	3.505	0.0313	0.80	No. 36 (0.107)
8	32	36	0.164	4.166	0.0313	0.80	No. 29 (0.136)
10	24	32	0.190	4.826†	0.0417	1.06	No. 25 (0.150)
12	24	28	0.216	5.486	0.0417	1.06	No. 16 (0.177)
¼	20	28	0.250	6.350†	0.0500	1.270†	No. 7 (0.201)
5/16	18	24	0.3125	7.938	0.0556	1.41	F (0.257)
⅜	16	24	0.375	9.525†	0.0625	1.59	(0.313)
7/16	14	20	0.4375	11.113	0.0714	1.81	U (0.368)
½	13	20	0.500	12.700†	0.0769	1.95	(0.422)
9/16	12	18	0.5625	14.288	0.0833	2.12	(0.484)
⅝	11	18	0.625	15.875†	0.0909	2.30	(0.531)
¾	10	16	0.750	19.050†	0.1000	2.540†	(0.656)
⅞	9	14	0.875	22.225†	0.1111	2.82	(0.766)
1	8	14	1.000	25.400†	0.125	3.175†	(0.875)

Over 1 in.   Sizes increase every 1/16 in. to 2 in.

*For complete table of tap drill sizes see Appendix G.
†Exactly equivalent.

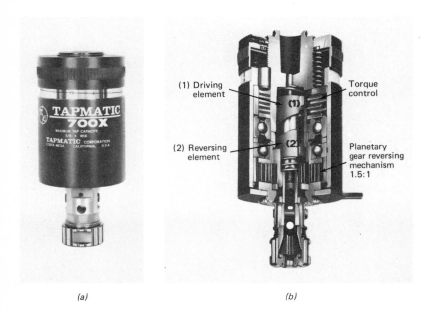

(a)   (b)

Fig. 11-10 (*a*) A reversible tapping attachment for drill press or numerical control use. (*b*) A reversing mechanism. (*Tapmatic Corporation.*)

Labels in figure (b): (1) Driving element; Torque control; (2) Reversing element; Planetary gear reversing mechanism 1.5:1

up to H6 which is 0.0030 in. maximum over basic pitch diameter. This may seem large, but actually considerable tolerance is allowed in the various classes of threads. The H3 and H4 sizes are the most used.

Taps are bought oversize to allow for wear. Thus, a GH4 (the G means ground thread) tap can wear up to 0.002 in. on the diameter before it starts to cut undersize.

In the *metric* system, five "zones" are available which specify somewhat similar oversize taps up to 0.10 mm (0.004 in.) oversize.

When buying taps, the manufacturer usually ships the H3 or H4. The "limit" or, in metric, the "fundamental deviations" are specified only if a special situation exists.

Tap Drills

It is quite obvious that the taps shown here cannot cut their own opening. Thus, a hole of the proper size must be made before the tap can be used. Usually this hole is drilled. A tap drill is *not* a special kind of drill. A tap drill is merely a convenient way to refer to the proper *size* drill to be used before using a tap.

If the hole was drilled the same as the minor diameter of the thread, then the tap would have to cut out a full, approximately triangular section of metal as shown in Fig. 11-11*a*. The tap simply is not strong enough to do this, so it is impractical.

For years industry has used a drill which cuts a hole which leaves only about 75 percent of the depth of the triangle to be taken out by the tap. As shown in Fig. 11-11*b*, this leaves a flat-topped internal thread.

Since the strongest position of nut and bolt is when both meet at the *pitch diameter*, it would seem possible that drilling a larger hole to leave only a *50* percent thread (as in Fig. 11-11*c*) would provide sufficient strength.

The Metal Cutting Tool Institute conducted tests of the effects of changing the percent of thread of the internally threaded member (i.e., the nut on a bolt or a tapped hole). The two curves in Fig. 11-12 are from a published report on these tests. Notice that the load required to either break the bolt or strip the threads in the nut with only 50 percent of the thread left in the nut is very close to the load achieved with 75 percent thread. Incidentally, the "fit" is the same for any percent thread. The "shake" or play in the fit will be close to the same.

The conclusion in the report was "for a length of thread engagement equal to a standard nut height, the static strength of a threaded fastener is not improved by a thread percentage greater than 60% in the internally threaded member."

The *torque* on the tap is what breaks it. Another part of the same report is shown in Fig. 11-13. A $\frac{3}{8}$-16 NC HSS plug-type hand tap was used, and results on the 1020 steel might not have been as dramatic if a spiral-point (gun) tap had been used. However, it illustrates another reason for using larger tap drills to leave smaller percentages of metal in the hole to be tapped.

Tap drill *sizes* shown in Tables 11-1 and 11-2 are for about 75 percent threads. The tap drills for other percentages are shown in Appendix G. The trend today in many factories is to save taps, time, rejects, and dollars by following up on the results of tests such as

TABLE 11-2 ISO METRIC THREADS—COARSE THREAD SERIES

	Diameter, mm	Pitch, mm	Diameter, in.	Equivalent Threads per inch (Approx.)	Tap Drill* mm
	M1	0.25	0.039	101	0.75
	M1.2	0.25	0.047	101	0.95
Equivalent to numbered machine screw sizes	M1.6	0.35	0.063	73	1.25
	M2	0.40	0.079	64	1.6
	M2.5	0.45	0.098	56	2.05
	M3	0.50	0.118	51	2.5
	M4	0.70	0.157	36	3.3
	M5	0.80	0.197	32	4.2
Equivalent to fractional sizes to 1 in.	M6	1.00	0.236	25	5.0
	M8	1.25	0.315	20	6.8
	M10	1.50	0.394	17	8.5
	M12	1.75	0.472	14.5	10.3
	M16	2.00	0.630	13	14.0
	M20	2.50	0.787	10	17.5
	M24	3.00	0.945	8.5	21.0
	M30	3.50	1.181	7.25	26.5

NOTE: As these become standard, they will be listed simply as M2.5, or M8, and the pitch will be assumed to be as above.
*To give 70 to 80 percent thread.
See Appendix G for more complete listing.

those just mentioned and using 60 to 65 percent threads.

Depth of Tapped Holes

The deeper a hole is threaded, the longer it takes to drill and tap and the more likely it is that the tap will break. Yet if there are too few threads holding the bolt, the threads will strip. Somewhere in between there is a depth of thread engagement which is the minimum which will hold enough so that the bolt will break before the threads let go. This is called the *optimum depth*.

There are many rules of thumb for depths of thread. In general the following rules apply:

1. If bolt material and tapped material are the same, then depth of thread equals diameter of bolt or thickness of a standard nut.

2. For steel bolts in aluminum or brass, the depth of thread is $1\frac{1}{4}$ to $1\frac{1}{2}$ times the diameter of the bolt.

3. If the full strength of high-strength bolts is needed, depth up to $1\frac{1}{2}$ times diameter can be used. *Do not* use a tighter (higher percent) thread as it just will not improve conditions.

Tap drilling must be deep enough in blind holes to allow for the two to five tapered threads on the tap plus chip clearance, plus the drill point (see Fig. 11-14, which is self-explanatory).

If the thread is, for example, $\frac{1}{2}$-13 UNC, and the workpiece is $\frac{7}{8}$ in. thick, it is often cheaper to drill all the way through (as in Fig. 11-14b) if conditions allow it. There is then no danger of the tap hitting bottom, and chips from drilling and tapping can fall freely through the hole.

Computing the time for tapping requires consideration of a number of factors. Assume a $\frac{3}{8}$-16 UNC thread to be tapped in aluminum through a $\frac{1}{2}$-in.-thick plate. The proper $\frac{5}{16}$-in.-diameter tap-drilled hole has been made. The tap is a HSS spiral-point

Fig. 11-11 The meaning of "percent thread" as applied to tapped holes.

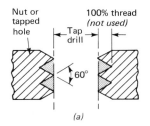
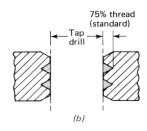
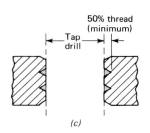
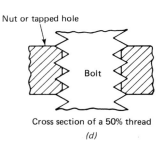

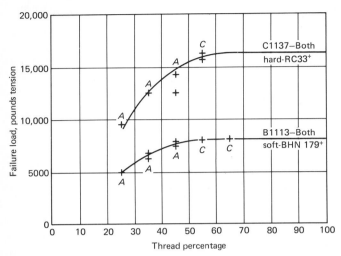

$\frac{3}{8}$ - 24 UNF-2 Threads, 0.312 in. depth of engagement

A — External thread (bolt) *stripped*
C — External thread member (bolt) broke

Fig. 11-12 The results of one of several tests made of "percent thread" vs. thread strength. (*Metal Cutting Tool Institute.*)

plug tap with a maximum of five tapered threads. Referring to Fig. 11-15a, the distance which the tap must travel is 0.875 in. using the clearances shown. Cutting speed chosen is 80 fpm.

$$rpm = \frac{4(80)}{0.375} = 860 \text{ rpm}$$

$$Feed = pitch = \frac{1}{threads\ per\ inch} = \frac{1}{16} = (0.0625) \text{ in./rev}$$

$$Time = \frac{distance, in.}{feed\ (in./rev) \times rpm} = \frac{0.875}{(0.0625)(860)} = 0.0164 \text{ min (1 s)}$$

This is obviously too fast. It is generally agreed that most tapping (especially with N/C machines) is done at 400 rpm or less. Thus, a revised computation would be

$$Time = \frac{0.875}{(0.065)(400)} = 0.035 \text{ min} = 2.1 \text{ s}$$

This is still quite fast. The return stroke may take 60 percent of this time.

Stroke   2.1 s
Return at 60%   1.3 s
Total time   3.4 s

The *metric equivalent* of the above specifications would be an M10 × 1.5 thread through 12 mm of stock. Referring to Fig. 11-15b, the length the tap must travel is 21 mm. Cutting speed will be 24 m/min.

$$rpm = \frac{300(24)}{10} = 720$$

This is too high, so 400 rpm will be used.

$$Feed = pitch = 1.5 \text{ mm/rev}$$

$$Time = \frac{distance, mm}{feed\ (mm/rev) \times rpm}$$

$$= \frac{24}{1.5 \times 400} = 0.04 \text{ min} = 2.4 \text{ s}$$

Return at 60% = 1.4 s
Total time = 3.8 s

THREAD CHASING (OUTSIDE THREADS)

Hand threading of outside (male) threads is done with dies held in a *diestock*. One type is shown in Fig. 11-16. The diestock will hold dies of a certain outside diameter such as 2 in. The dies are slightly adjustable for oversize or undersize threads and are made in all sizes of threads. Often a smaller set of 1-in.-diameter dies in a small diestock is used for threads from No. 0 (M1.6) to No. 12 (M6).

Instead of round dies, square, single segments, double cutter, etc., dies are sometimes used, but the basic idea is the same. Usually four segments or cutter sections are used for threads up to 1-in. (24-mm) diameter.

Hand threading is slow, and it is difficult to align

Fig. 11-13 Effect of hole size on tapping torque. (*Metal Cutting Tool Institute.*)

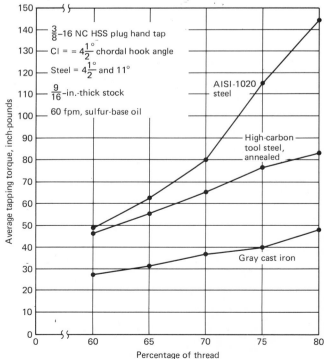

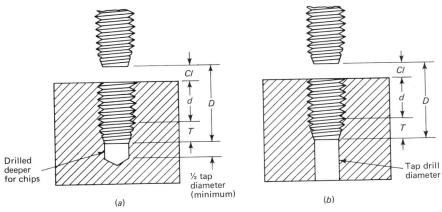

(a)   (b)

Drilled deeper for chips

½ tap diameter (minimum)

Tap drill diameter

d = depth of thread specified on the drawing
T = number of incomplete (tapered) threads on the tap
Cl = clearance needed for "slowdown" on N/C and automatic tapping equipment which has a rapid traverse approach
D = total distance tap must travel at feedrate

Fig. 11-14   Dimensions to be included in total depth of tapping when computing the time for tapping. (*a*) Blindhole. (*b*) Drill through tap to specified depth.

the die exactly with the part, so threading is done on a machine if at all possible.

Machine thread chasing can be done with a single-point tool on a lathe. However, this is slower than using a hand diestock, so it is used only for special threads or for threads not on the end of a part such as a shaft.

For even moderate quantities of threads done on an engine lathe, turret lathe, or automatic lathe, **quick-opening dies** (sometimes called *self-opening dies*) are used.

Figure 11-17*a* shows a self-opening die with *radial* chasers, or cutting members. This is a simple construction, easily operated type. Chasers and dies are made in thread sizes (from $\frac{3}{16}$- to 10-in. diameter) and can be changed in less than 5 min. The die head costs from $300 up, and a set of four dies costs $35. The type shown has a pull-off trip. When the turret stops advancing, the die continues to cut, which pulls the

front end forward and tips a latch allowing the dies to snap open enough to clear the threads. A push on the knobbed handle resets the chasers to size.

Dies can also be snapped open by "pin trips" fastened to the die or to the lathe. Contact points are made on the lathe or the die so that thread length is controlled.

Figure 11-17*b* is a similar die with tangential chasers. These are sharpened on the cutting face, which means a longer chaser life, though the chasers with holders cost about twice as much as radial blades. The die head costs about $400.

Figure 11-17*c* shows a die with *circular* chasers. Due to their shape, no relief needs to be ground into the cutters, and friction is very low. These are made with either four or five chasers. The five-chaser models, while more expensive, do give longer cutter life and

Fig. 11-16   (*a*) Adjustable round die. (*b*) Solid square bolt die. (*c*) Diestock for holding round dies.

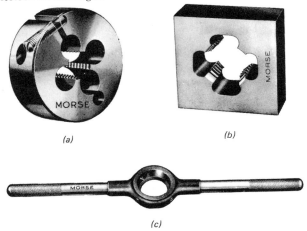

(a)

(b)

(c)

Fig. 11-15   Drawing showing total tap travel for computing time. (*a*) Inch threads. (*b*) Metric threads.

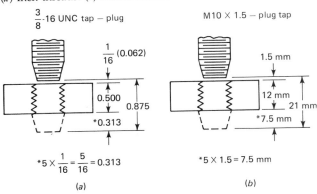

$\frac{3}{8}$-16 UNC tap – plug

$\frac{1}{16}$ (0.062)

0.500

0.875

*0.313

*5 × $\frac{1}{16}$ = $\frac{5}{16}$ = 0.313

(a)

M10 × 1.5 – plug tap

1.5 mm

12 mm

21 mm

*7.5 mm

*5 × 1.5 = 7.5 mm

(b)

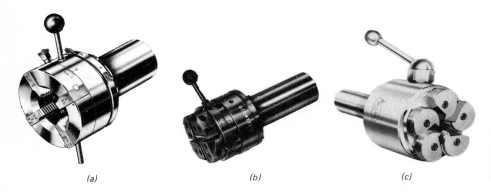

(a)   (b)   (c)

Fig. 11-17 Self-opening diestock. (*a*) With radial chasers. (*b*) With tangential chasers. (*Geometric Tool.*) (*c*) With circular chasers. (*Cleveland Twist Drill Co., a subsidiary of Acme Cleveland Corp.*)

will cut very accurate threads even on slightly out-of-round stock. Costs are about the same as for the tangential die.

Some small automatic and semiautomatic lathes have a quick-reversing motor so that the dies (or taps) can be backed off very fast. In this case, fixed dies, which are less expensive to buy and maintain, can be used.

Rolled threads can also be made on the lathe with quick-opening die heads. No chips are made, but the OD of the part must be turned to a smaller diameter (as was the case with fluteless taps.) These are available from 28 to $4\frac{1}{2}$ threads per inch and diameters from 0.062 to 2.0 in. (1.5 to 50 mm).

These, of course, can only be used on the softer metals. These can also be used for *burnishing* (smoothing and polishing) a previously cut thread.

Bolt threads are made commercially with special *flat* rolling dies shown in Fig. 11-18. The head and body of the bolt are usually cold-formed (see Chap. 25), and the body diameter of the section to be threaded is formed to just under the pitch diameter.

The preformed bolts are automatically fed into the reciprocating die thread plates which roll the threads. This is done at rates from 60 to 200 per minute as shown in Fig. 11-18*b*.

Other systems involving rotary feeders and rotary dies are also used, but the principle is the same.

Metric threads can be made with any of the above equipment simply by cutting or grinding chasers or dies to the correct pitch and setting them to the correct diameter.

Lubrication

The cutting edges on both taps and dies are buried in the material, so lubrication is quite necessary. For aluminum, kerosene or light lard oil is used; other metals require a sulfur-based oil, sometimes chlorinated also.

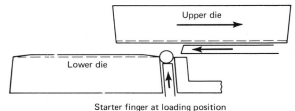

Upper die

Lower die

Starter finger at loading position

Starter finger returning for next blank after follow-through

Fig. 11-18 (*a*) High-production flat thread-rolling dies. (*b*) One method of automating bolt threading. (*Reed Rolled Thread Die Company.*)

Copper alloys are stained by sulfur, so mineral oils or soluble oil must be used. Cast iron is often threaded without any lubricant.

There are several synthetic tapping fluids on the market today. They are somewhat more expensive but may save their cost in better threads and fewer broken taps.

CAUTION: Realize that the first few teeth—the tapered ones and the first full tooth—do all the cutting. Thus, the lubricant must get down into the hole, through the flutes, at all times.

Threads are also frequently ground into a workpiece as described in Chap. 15. Sometimes threads are milled, by a method similar to that described in Chap. 28 on gear cutting. However, the most widely used threading methods are those just described in this chapter.

Review Questions and Problems

11-1. What is a "tap drill"?

11-2. Is the "nominal diameter" larger or smaller than the actual diameter of a bolt or screw?

11-3. What is the difference between pitch and lead of a thread?

11-4. What example, in addition to those given here, can you suggest for uses of the four "classes" of fits?

11-5. Using a bottoming tap requires an extra operation. How can this be avoided
   a. In a blind hole
   b. When tapping a $\frac{3}{8}$-16 thread $\frac{1}{2}$ in. deep in a $\frac{3}{4}$-in.-thick piece of steel?

11-6. Explain the advantages and limitations of a spiral-point (gun) tap as compared with a conventional three- or four-flute tap.

11-7. Should you use larger or smaller tap drills for fluteless taps as compared with spiral-flute taps?

11-8. Are new taps usually oversize or undersize?

11-9. What are the advantages and disadvantages of using a 60 percent instead of a 75 percent thread.

11-10.
   a. Sketch the drill and thread depth for a $\frac{1}{2}$-13 UNC bolt in a blind hole using a steel bolt in a steel casting. Maximum fpm = 60, for drilling.
   b. Do the same for a M8 × 1.25 steel bolt in an aluminum casting. Maximum m/min = 90, for drilling.

11-11. Compute the time for tapping (including return) the two threads in Prob. 11-10.

11-12.
   a. Compute the time for chasing a $\frac{3}{4}$-10 thread $1\frac{1}{2}$ in. long with a quick-opening die. Fpm = 45.
   b. Compute the time to cut an M30 × 3.5 thread 50 mm long with the same die head. Use 12 m/min cutting speed.

11-13. Is proper lubrication in tapping and threading of greater importance than in turning? Give reasons for your answer.

References

1. *Bendix Cutting Tool Handbook*, Industrial Tools Division, The Bendix Corporation.
2. *Drilled Holes for Tapping*, rev. ed., Metal Cutting Tool Institute, 1967.
3. *Facts about Taps and Tapping*, Greenfield Tap and Die Corp., Greenfield, Mass.
4. *Machinery's Handbook*, 19th or later ed., Industrial Press, Inc., New York.
5. *Machinery's Screw Thread Book*, 20th ed., The Machinery Publishing Co. Ltd., 1969.
6. *Metal Cutting Tool Handbook*, Metal Cutting Tool Institute, 1965.
7. *The Use and Care of Taps*, Threading Tools Div., The Cleveland Twist Drill Co., Cleveland, Ohio.

milling machines

Modern milling machines look much the same as they did 25 years ago. However, they now must cut superalloys, titanium, and high-tensile steels to closer tolerances and at faster rates than previously. To handle these requirements, the new milling machines provide higher horsepower, greater stiffness, and wider speed and feed ranges than before. In addition, more accurate lead screws, closer alignment, and numerical control all result in faster work with better finish and greater accuracy than ever before attained.

TYPES OF MILLING MACHINES

There are hundreds of styles, sizes, and specially designed milling machines. However, they are all derived from three basic types. These are:

1. Knee-and-column
   a Vertical spindle
   b Horizontal spindle
2. Bed type
3. Planer type
   a Vertical spindle
   b Horizontal spindle

Knee-and-Column-type Milling Machines— Common Factors

These milling machines are used for small- and medium-sized work in toolrooms and prototype shops and for production work for lots of about 5 to 50 pieces.

The knee-and-column style is made with both *horizontal* and *vertical* spindles as shown schematically in Figs. 12-1 and 12-2.

The *column* in these machines is the "backbone" or large vertical section, which is often cast integral with the *base*. The column supports the *overarm*. The *knee* is

a heavy section which supports the *saddle* and *table*. The knee can be raised and lowered by the *elevating screw*. This moves the work up or down, brings the workpiece to the cutters, and sets the depth of cut.

The *table*, in both styles of machines, moves left and right and supports the workpiece. The *saddle* supports the table and moves in and out. The *overarm* can be moved in and out to help position the cutters when wide or narrow workpieces require more adjustment than can be made with the saddle.

Feed on horizontal-spindle machines may be by the handwheels or by power feed in any direction or

Fig. 12-1 Schematic drawing of a vertical-spindle knee-and-column-type milling machine.

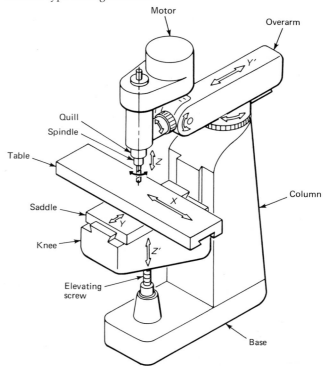

combination of directions. The feedrate is usually in inches per minute (ipm) or millimetres per minute (mm/min) on the feed selector dial. Feedrates from $\frac{3}{4}$ to 30 ipm (20 to 760 mm/min) are usually available. Power feed on vertical-spindle machines is usually used only on the table and spindle. This may be in inches per minute or inches per revolution (millimetres per minute or millimetres per revolution). *Speed* on both vertical and horizontal machines is usually in the range of 25 to 2000 rpm and sometimes up to 3000 rpm.

Measuring and *locating holes* or milling cuts on both vertical- and horizontal-spindle machines is done by using the scales on dials which are located just behind the handwheels. These dials are graduated in 0.001-in. (0.025-mm) or 0.02-mm (0.0008-in.) divisions.

Fig. 12-2 Schematic drawing of a horizontal-spindle knee-and-column-type milling machine.

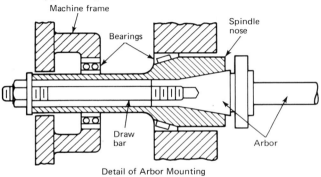

Detail of Arbor Mounting

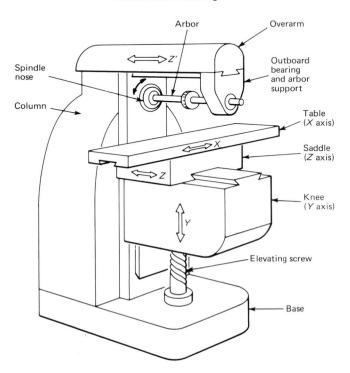

One complete turn of the handwheel moves the table, saddle, or knee some even amount such as 0.200 or 0.250 in. (5.0 or 6.0 mm).

work which can be done

These milling machines are very versatile. As shown in Fig. 12-3, they can mill slots and keyways, cut shelves, and do face milling. The vertical-spindle miller can also drill, ream, bore, tap, and counterbore. With a special short "stub" arbor, the horizontal-spindle machine can also drill, ream, and bore.

Notice that several types of cuts can be made by either the vertical- or the horizontal-spindle machine. The choice of machine would depend on the depth and width of the cut, the power needed, the preferred style of cutter, and the setup time. In some cases it makes very little difference which machine is used.

vertical-spindle knee-and-column milling machines

The milling machine shown in Fig. 12-4 is typical of this type of milling machine. They may have from $\frac{1}{2}$- to 5-hp (0.37- to 3.7-kW) motors, and tables up to 72 in. (1800 mm) long. If the tables are made much longer, they will "droop" because of elasticity of steel and the overhanging load. Sometimes the knee and saddle are made extra wide to hold the long table more rigidly. These machines cost from $2000 to $8500.

As can be seen in both Fig. 12-1 and Fig. 12-4, the heads on these milling machines can be rotated either left and right or forward and back. This makes it possible to make single- or compound-angle cuts.

The *numerical control* versions of these machines often do not have any handles for moving the work. One style of N/C milling and drilling machine is shown in Fig. 12-5. These milling machines must not have any *backlash*, so special ball screws like the one shown in Fig. 12-6 are used.

The machine control unit (MCU) reads the tape and moves the table and saddle to the positions called for on the tape. Some N/C knee-and-column millers also control the vertical movement (Z axis) of the spindle. The knee must be moved into position by hand operation of the elevating screw during setup. An N/C milling machine costs from $16,000 to $40,000.

horizontal-spindle knee-and-column milling machines

These milling machines will take heavier cuts than the vertical-spindle machines. As shown in Fig. 12-7, the knee and the column are very heavy and rigid. The *horizontal arbor* is supported by an outboard bear-

Fig. 12-3 Some of the types of work which can be done on knee-and-column milling machines.

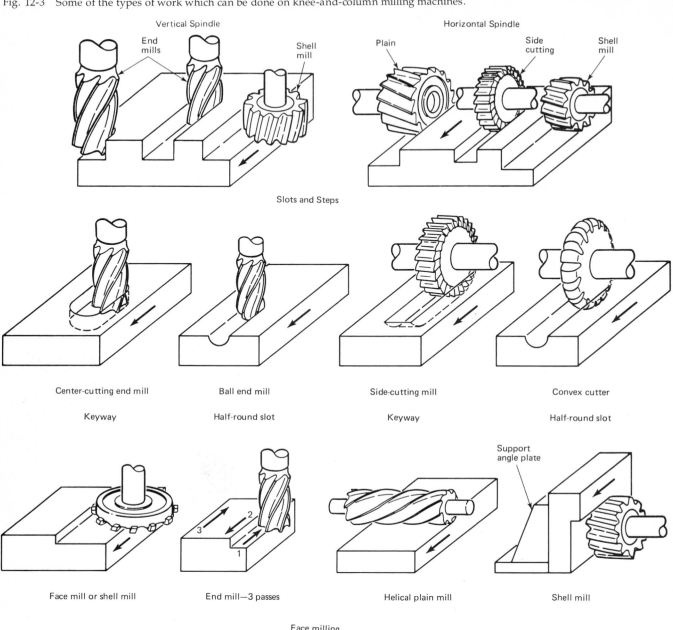

Vertical Spindle

End mills

Shell mill

Horizontal Spindle

Plain

Side cutting

Shell mill

Slots and Steps

Center-cutting end mill

Keyway

Ball end mill

Half-round slot

Side-cutting mill

Keyway

Convex cutter

Half-round slot

Support angle plate

Face mill or shell mill

End mill—3 passes

Helical plain mill

Shell mill

Face milling

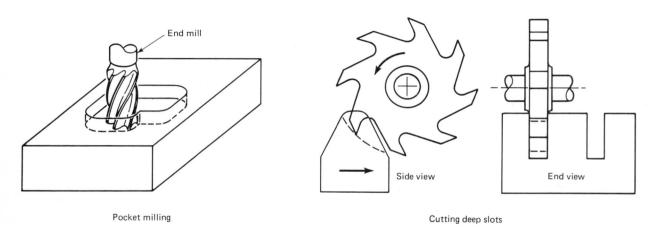

End mill

Pocket milling

Side view

End view

Cutting deep slots

ing and, for heavy cuts or long extensions of the overarm, a center bearing can be added.

The horizontal-spindle milling machine comes in two styles:

1. *Plain*, with the table fixed in position at right angles to saddle. This is the miller shown in Fig. 12-7.

2. *Universal* (Fig. 12-8), where the entire table can be rotated to about 25°. This makes it possible to cut helical gears and threads. A dividing head is shown attached to this machine. This is an attachment which is used for evenly spacing gear teeth or splines around a circle.

The universal knee-and-column machine can be used for the same cuts done on the plain machine. However, the addition of the rotary plate does decrease its rigidity and adds to the cost. Thus, for production work, the plain type is more often used. These machines cost from $6000 to $20,000.

The *size* of horizontal-spindle knee-and-column machines is basically the *table size*. These sizes are sometimes numbered, but these numbers are *not* universally used. Some manufacturers make tables of quite different sizes. The range of sizes is shown in Table 12-1. Selection of these machines may also be based on motor horsepower, 5 to 10 hp (3.7 to 7.5 kW), and the arbor diameter, $\frac{3}{4}$ to $1\frac{1}{2}$ in. (20 to 40 mm).

This same heavy knee-and-column machine is also made with a vertical spindle for work which requires more power and rigidity than is available on the regular vertical-spindle milling machines.

Bed-type Milling Machines

Knee-and-column milling machines have two limiting factors: (1) their tables are subject to deflection, so very heavy workpieces cannot be machined; and (2) their rigidity is limited so that high feedrates and deep cuts can cause chatter.

Fig. 12-4 Vertical-spindle milling machine with $1\frac{1}{2}$-motor, table and quill power feeds only. (*Bridgeport Machines, a division of Textron, Inc.*)

The *fixed-bed-type* or *production-type* milling machine is made to fill the need for machining fairly heavy work, in the 200 to 2000 lb [90 to 900 kg (mass)] range, with heavier feeds.

TABLE 12-1 APPROXIMATE SIZES OF SOME KNEE-AND-COLUMN HORIZONTAL-SPINDLE MILLING MACHINES*

Machine Number	Table Size Length and Width	Travel, Table and Saddle (X and Y)	Vertical (Knee) Travel (Z)	Hp (kW)
1	46 × 12 (1160 × 300)	22 × 10 (560 × 250)	15 (380)	5 (3.7)
2	56 × 12 (1420 × 300)	28 × 10 (710 × 250)	17 (430)	5 (3.7)
4	74 × 15 (1880 × 980)	42 × 14 (1070 × 350)	18 (460)	10 (7.5)
6	108 × 18 (2700 × 460)	60 × 16 (1500 × 400)	21 (530)	10 (7.5)

* Dimensions in inches with millimetre equivalents in parentheses.
NOTE: These are not standard dimensions, or a standard numbering system. The figures do show the range of sizes available, and are approximately the figures used by some manufacturers.

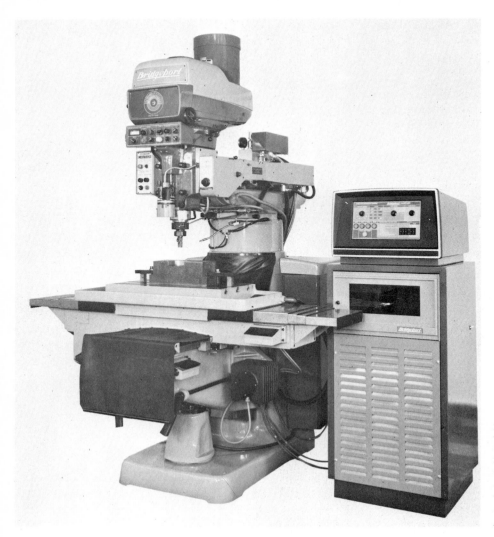

Fig. 12-5 The numerical control version of a vertical-spindle milling machine with 4-hp motor, variable speed drive, and power feeds. Note covered ways and absence of handwheels. Workpiece is clamped to table. (*Bridgeport Machines, a division of Textron, Inc.*)

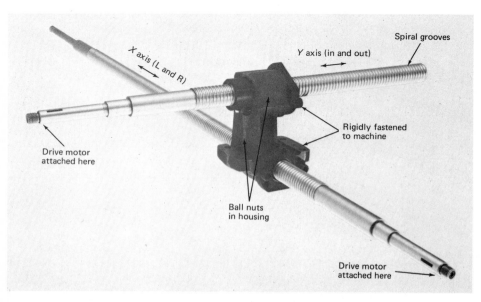

Fig. 12-6 Lead-screw assembly for numerical control, using zero backlash ball screws. (*Saginaw Steering Gear Division, General Motors Corp.*)

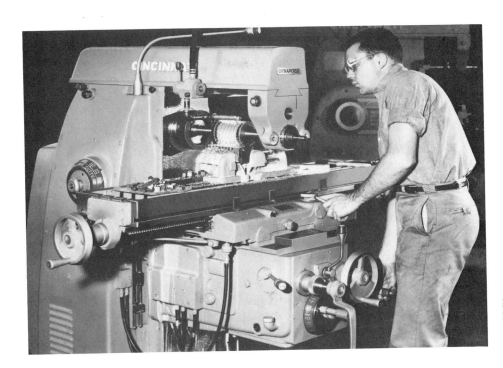

Fig. 12-7 A plain horizontal-spindle knee-and-column milling machine. Shown cutting nine slots in one pass. (*Cincinnati Milacron Inc.*)

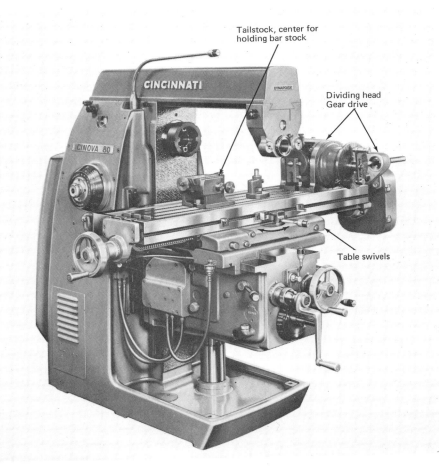

Fig. 12-8 A universal-spindle knee-and-column milling machine with dividing head. (*Cincinnati Milacron Inc.*)

Figure 12-9 shows a medium-sized, plain bed-type, horizontal-spindle milling machine. Its principal features are:

1. The table travels on a solid bed. There is no saddle and no in and out movement of the table.

2. The *spindle carrier* can be moved up or down on vee or flat ways to adjust the depth of cut. The outboard bearing carrier is tightened into position after the depth is set.

3. The location of the cutters, in and out, is set by spacers on the arbor.

A *rise* and *fall* bed-type milling machine is the same as the plain bed type except that the spindle carrier and the outboard bearing carrier can be moved up and down together *during* the cutting cycle.

The sizes of these milling machines cover a wide range. They may have tables from 15 × 36 in. (380 × 915 mm) up to 26 × 168 in. (660 × 4300 mm). They are driven by motors up to 50 hp (37 kW) at *feeds* from ½ to 150 ipm (12 to 3800 mm/min). *Speeds* are as low as 10 rpm on large models and up to 3000 rpm on the smaller models. Cost is from about $20,000 up.

Since these are production machines and may be used for lots of 500 to 5000 or more parts, they are usually equipped with *automatic table feed cycles*. A cycle for the rise and fall machine might be:

Spindle rapid down to position—table rapid traverse to work—feed across, making the cut—table rapid forward to clear the work—spindle rapid up to clear the work—table rapid traverse back to start.

The plain machine might have a similar cycle, without the vertical motions.

With this ability, the operator merely loads and unloads the work. In fact, sometimes two pieces can be put on a long table and one can be cut while the other is being secured in position.

The vertical travel of the rise-and-fall milling

Fig. 12-9   Plain bed-type milling machine for medium-to-high production lots. (*Cincinnati Milacron Inc.*)

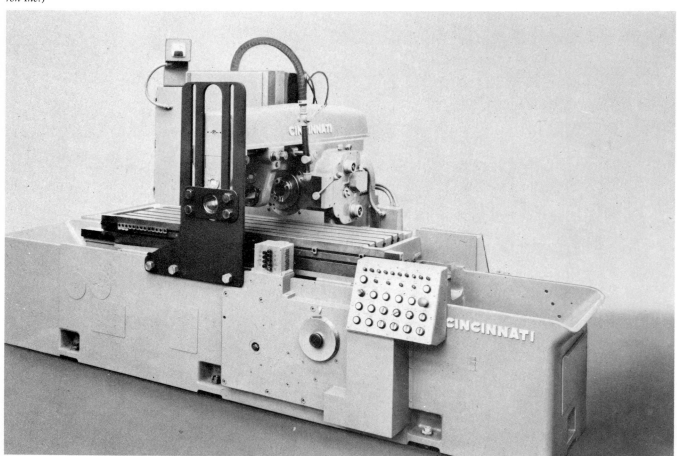

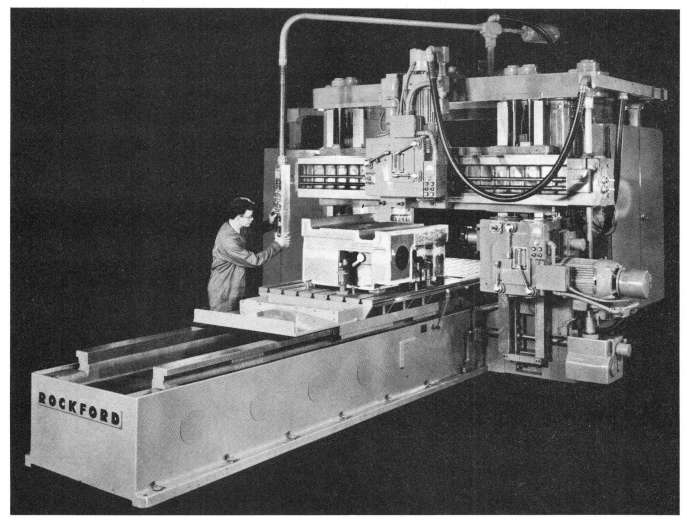

Fig. 12-10   A planer miller with one railhead and one sidehead, both 30 hp (22 kW), milling a cast-iron part. (*Rockford Machine Tool Co., Subsidiary of Greenlee Bros. & Co.*)

machines can also be tracer controlled, using a 1:1 template cut to the desired shape. Both types of machines can mount face-milling or end-milling cutters directly on the spindle nose so that they can cut on the side of the workpiece.

The *bed-type* milling machine is less versatile than the knee-and-column machine and costs considerably more. However, for production work, they have several advantages:

1.   They are more rigid, so that a long table is fully supported and thus can take heavier cuts and larger work.

2.   The table is always at the same level, which means easier handling of heavy workpieces.

3.   With automatic cycling, much faster production is obtained.

Planer-type Milling Machines

For very heavy or very large work, the *planer-miller* or "double-housing adjustable-rail miller" is made. One of these machines is shown in Fig. 12-10.

These machines are made with tables up to 20 ft long × 96 in. wide (6000 × 2500 mm) and use motors of 50 hp (37 kW) and over per spindle. They are sometimes called *skin mills* or *spar mills* because of the work they originally did on large airplane parts.

These milling machines have two heavy columns which are held at the top with a heavy cross beam and at the bottom by the bed. The large *crossrail* moves up and down on ways on the column.

Supported on the crossrail are two (occasionally only one) *milling heads*. These heads contain the drive motors and spindles for milling cutters and can each be independently moved to the right and the left. The

8- to 20-in. (200- to 500-mm) spindles or quills can also be moved vertically. Often one or two milling heads are mounted to cut on the sides of the work; these are called *sideheads*. Thus, cutting can be done on three sides of the work at the same time. This same type of construction also is used for large planers (Chap. 17) and vertical boring mills (Chap. 18).

All spindles can be used for cutting at rates from ½ to 50 ipm (13.0 to 1270 mm/min). Speeds from 25 to 900 rpm are available for use with large milling cutters from 10- to 30-in. (250- to 750-mm) diameter. These machines cost from $70,000 to $350,000.

Today, these double-housing rail millers are often numerically controlled. The three or four heads can be programmed to cut independently or in unison.

With numerical control and cuts up to ½ in. (13 mm) deep and 12 in. (300 mm) wide, these machines can remove a lot of metal in a short time on medium- to very large-sized workpieces.

Horizontal-spindle milling is possible on the planer millers with the sideheads and by using 90° adapters on the vertical heads. A great deal of milling is also done with the bed-type and floor-type horizontal boring machines described in Chap. 18.

Other Styles of Milling Machines

A wide variety of milling machines, based on the styles shown, is available today. Smaller knee-type machines are made for very light work. Vertical and

Fig. 12-11 A hydraulically operated, three-dimensional duplicating milling machine. Two cutting spindles are shown, but three or four may be used. (*Bridgeport Machines, a division of Textron, Inc.*)

universal attachments are available on several styles of knee-and-column machines.

Multispindle small tracer millers and milling machines with rotary tables to bring work sequentially under different cutters are special variations of the vertical-spindle milling machine.

Duplicating or Tracer Milling Machines

There are many shapes, especially in aircraft parts and making dies for forging and plastics, which are too complicated to program economically for numerical control, so tracer milling machines of many styles are used.

One type, for parts up to about 12 × 30 in. (300 × 760 mm), is shown in Fig. 12-11. The stylus at the right is run over the "model" either automatically or by hand. Every movement of the stylus is duplicated by the milling cutter in one or several spindles. This motion is transmitted hydraulically or, sometimes, electronically.

These duplicating millers are made both two dimensional (*X* and *Y* axis only) and three dimensional (including the *Z* axis, vertical travel). To cover a large surface, many passes are made, each overlapping the previous path. An example of the very wide variety of work this type of milling machine can do is shown in Fig. 12-12.

Another well-known vertical-spindle duplicating miller is the *Hydrotel* made by Cincinnati Milacron. This does work the same as, but slightly larger than, the two machines shown.

For very large work, the horizontal-spindle *Keller Machine* (made by Colt Industries, Pratt and Whitney Machine Tool Division) is very well known. Most work is mounted on a large (maybe 72 × 96 in., 1800 × 2400 mm) angle plate. *Kellering* is used especially for making large dies, molds, and aircraft parts.

While the above are especially well-known makes of tracer mills, several other high-quality machines are on the market.

WORK HOLDING ON MILLING MACHINES

Workpieces machined on milling machines vary in size from smaller than a walnut to larger than an automobile. However, the great majority of the workpieces are in the 5 to 500-lb [2.3- to 225-kg (mass)] range, such as automobile engine blocks, levers of all kinds, and other large items mentioned.

The *milling machine vise*, like the one holding the work in Fig. 12-7, is frequently used for holding work up to 6 in. (150 mm) wide. The vise is bolted securely

(a)

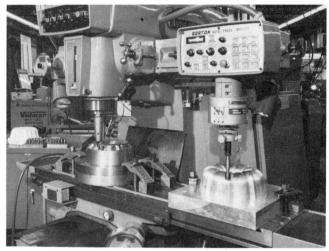

(b)

Fig. 12-12 (*a*) "Auto-Trace-Master" tracer miller cutting an 8-point Geneva wheel from a square blank. (*b*) The same machine cutting a mold for a die casting, in three dimensions. [*Kearney & Trecker Machine Tools. (Gorton Machine Tools.)*]

to the machine table, using the tee slots in the table. The vise can be rotated 360° so that cuts can be made at any angle.

Often special jaws, shaped to hold irregular workpieces, are fastened into the vise. Locating bars and pins are sometimes added so that each piece of a lot of pieces (either castings or cut bar stock, etc.) can be positioned accurately each time.

Parts too large for the vise or of irregular shape are *bolted down*, either directly to the table or secured to a fixture which is bolted to the milling machine table as shown in Figs. 12-4 and 12-12*b*.

Sometimes the work is fastened to a *rotary table*, like the one shown in Fig. 12-13. This table can be rotated *during cutting*, either by hand or by power. It can also be indexed any desired number of degrees between

Fig. 12-13 Rotary table. By using an adapter base, this can also be mounted vertically. (*Bridgeport Machines, a division of Textron, Inc.*)

cuts so that several identical cuts (such as slots) can be made with the specified angular spacing between them.

The rotary table can also be mounted on an angle plate so that it can be used with horizontal-spindle N/C machines.

Holding the Cutters on Milling Machines

Vertical-spindle milling machines may use collets similar to those used on lathes. One type is shown in Fig. 12-14a. A separate collet must be used for each diameter of cutter shank. Figure 12-14b is a setscrew-type holder, and Fig. 12-14c is a shell mill arbor.

There are several "quick change" toolholders made

Fig. 12-14 Milling cutter holders. (*a*) Collet. (*b*) Setscrew type. (*c*) Shell mill arbor. (*d*) Quick change holder with adapter. (*Bridgeport Machines, a division of Textron, Inc.*)

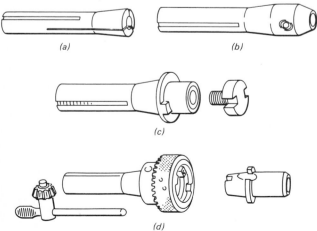

by different manufacturers. These, as shown in Fig. 12-14d, require only a quick turn with a key or spanner wrench to release or secure a cutter holder. Each holder is sized to fit one cutter diameter.

Horizontal-spindle milling machines basically use an arbor such as the one shown in Fig. 12-15. These arbors have diameters in the range of $\frac{3}{4}$ to $1\frac{1}{2}$ in. (20 to 40 mm). Milling cutters must be ordered with the proper hole size to fit the specific machine's arbor.

These arbors have a long, standard-size keyway so that milling cutters will have a positive lock to prevent them from slipping around under loads. The cutters, of course, have a correspondingly sized keyway.

Spacers (which are cylinders which fit over the arbor and which are ground flat on both ends) are placed between the milling cutters and the ends of the arbor and may be ground to a specific width between cutters used for straddle milling.

Instead of the long horizontal arbor, a short shell mill adapter or collet-type toolholder may be fastened directly into the milling machine's nose plate. This may be held by its taper and the drawbar, or it may be bolted directly to the nose plate.

Most of the work done on milling machines involves facing or straight-line cutting. However, many numerically controlled milling machines can cut arcs of circles and sloped lines quite accurately. The arcs are actually a series of very short straight lines. However, they are at varying angles, and the result is often a 100 to 200 μin. (2.5 to 5.0 μm) finish. Thus, numerical control has considerably increased the versatility of the milling machine.

Chapter 13 discusses some of the wide variety of cutting tools which can be used with both conventional and N/C milling machines.

Review Questions and Problems

12-1. What are the three basic types of milling machines?

12-2. Describe the vertical-spindle knee-and-column type of milling machine.

12-3. What type of work can be done on both the vertical- and horizontal-spindle-type milling machines?

12-4. What is the advantage of the horizontal-spindle knee-and-column milling machine?

12-5. What is the difference between the plain and the universal milling machines? What is the disadvantage of a universal-type milling machine?

12-6. What is the advantage of the bed-type milling machine over the knee-and-column type?

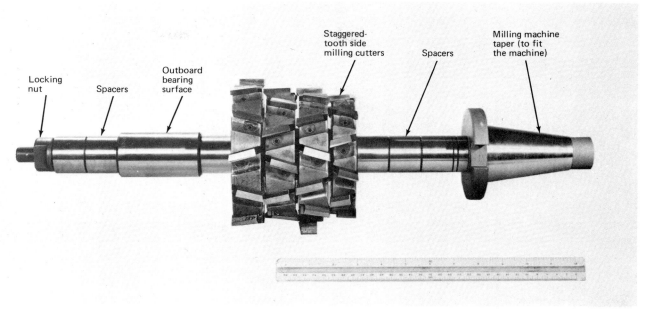

Fig. 12-15 An arbor for horizontal-spindle knee-and-column type milling machine. Gang milling machine. Gang milling cutters (Chap. 13) shown mounted in place. (*Viking Tool Co.*)

12-7. How do the planer-type milling machines differ from the bed-type milling machines:
   a. in construction.
   b. in their uses.
12-8. How is the workpiece held on milling machines?
12-9. How are milling cutters mounted on milling machines?

References

1. *A Treatise on Milling and Milling Machines*, 3d ed., The Cincinnati Milling Machine Co. (now Cincinnati Milacron), 1951.
2. Hall, Frank, Sr. (ed.): "Choosing a Milling Machine—Special Report," *Machinery*, September 1971.
3. Manufacturers' catalogs.

milling cutters

One manufacturer catalogs 75 different varieties of milling cutters, which would seem to indicate that this is a complicated subject. However, all these cutters are actually variations of the four basic styles:

1. Plain milling cutters, including side mills and special shapes.
2. Shell mills
3. Face mills, inserted tooth
4. End mills

This chapter will illustrate and describe each style and some of the variations of each.

Keep in mind that all milling is "interrupted" cutting, as the teeth enter and leave the cut on every revolution. Thus, milling cutters are subject to more stress and more rapid wear than the lathe tools, which are usually cutting on smooth work. A milling cutter is only as strong as each individual tooth. Thus, the feedrate depends on the number of teeth.

All milling cutters cut on their periphery (the outside of the cylindrical shape). Some also cut on their end, side, or face surfaces.

TYPES OF MILLING CUTTERS

Plain and Side Milling Cutters

Plain milling cutters (Fig. 13-1) have teeth *only* on their periphery. The teeth may be straight across or may be sloped at helical angles of approximately 25°, 45°, or 52° (low, medium, and high helix). The helix can be either left- or righthand; there is no advantage either way.

As you would judge by their shape, these plain milling cutters are used principally for facing wide flat surfaces using a horizontal-arbor milling machine.

These milling cutters, sometimes called *slab mills*, are made in diameters from 2¼ to 4 in. (60 to 100 mm)

and widths of face from ½ to 8 in. (12 to 200 mm), and sometimes larger and longer for special jobs. They are most often made of high-speed steel, though some have brazed-carbide inserts. Cost is from $20 to $250 each.

Side milling cutters (Fig. 13-2) have cutting teeth on the periphery and on one or both sides. If the cutting teeth are only on one side, they are called *"half-side" cutters*, and are made left- and righthand.

These very versatile cutters are used on horizontal-arbor milling machines to cut notches, slots, and some keyways. As shown in Fig. 13-3, they can be accurately spaced for straddle milling, or several widths and diameters can be combined for gang milling.

Staggered-tooth side mills are excellent for any deep cuts, and are especially good on aluminum. Widely spaced, high-positive-rake staggered-tooth side mills will make deep slots in aluminum at high speeds and feeds. A 6-in.-diameter (150-mm) cutter of this style can cost $85, but it will pay its way.

Most *side mills* are made of HSS, though some are carbide-tipped (Fig. 13-4). They are made in diameters of 2 to 8 in. (50 to 200 mm) and from ¼ to 1 in. (6 to 25 mm) face width.

Fig. 13-1 Three styles of "plain" milling cutters. (*Colt Industries, Inc., Pratt and Whitney Machine Tool Division.*)

Light duty — straight tooth   Light duty R.H., 20° helix   Heavy duty L.H., 45° helix

The nomenclature used to describe these and other similar milling cutters is shown in Fig. 13-5.

There are two major sets of angles:

1. *Radial*, in the same plane as that in which the radius or diameter is measured, which means the flat surface in many cutters.

2. *Axial*, that is, in relation to the axis of the arbor through the cutter. When the peripheral surface is wide, as in some plain milling cutters, or long, as will be seen in end mills, the axial rake is referred to as the *helix angle*. This term also was used with drills and taps.

Rake angles are similar to those used in turning. Small or negative rake angles are required for tough, hard materials. Quite large positive rakes are required for aluminum and magnesium.

Tooth spacing is not considered in Fig. 13-5, but it is very important. 4-in.-diameter (100-mm) plain or side milling cutters are made with 3, 6, 8, 16, 18, 20, and 24 teeth, though not all by the same company. General Electric Company, Metallurgical Products Department recommends:

Milling steel: maximum of 3 teeth per inch of diameter (1 per 8 mm diameter)

Milling cast iron: maximum of 5 teeth per inch of diameter (1 per 5 mm diameter)

Milling aluminum: use $\frac{1}{2}$ to 2 teeth per inch of diameter (1 per 16 or 13 mm diameter)

If more than eight teeth are used on a 4-in.-diameter (100-mm) cutter when cutting aluminum, the teeth soon clog up due to the high speeds and heavy chips. Conversely, such wide spacing used on steel or cast iron would destroy the cutter fairly soon.

Relief and clearance angles are needed for the same reasons they were used in lathe work, that is, to prevent cutter surfaces from rubbing on the workpiece. All rubbing creates heat, and heat shortens cutter life.

Fig. 13-2 Two styles of "side" milling cutters. (*Colt Industries, Inc., Pratt and Whitney Machine Tool Division.*)

Regular

Staggered tooth

(a)

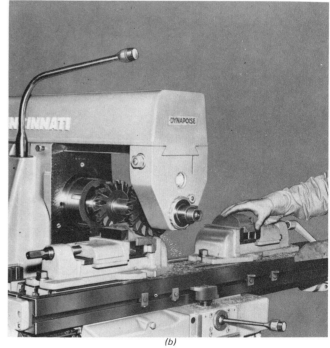
(b)

Fig. 13-3 (*a*) Gang milling a recessed cut. (*b*) Straddle milling, on a horizontal-arbor milling machine. (*Cincinnati Milacron Inc.*)

Slitting saws, or slotting cutters (Fig. 13-6), are really thin plain or side milling cutters. Their use is sometimes to cut off an end of a casting or forging for good finish and correct angularity. They also are used singly or in gangs of 2 to 10 or more for cutting slits in thin metal or for slotting castings.

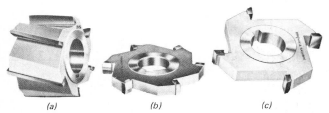

Fig. 13-4   Some styles of carbide-tipped milling cutters (*a*) Plain, or slab mill. (*b*) Side milling cutter for tough alloys. (*c*) Side milling cutter for aluminum. (*Brown & Sharpe Mfg. Co.*)

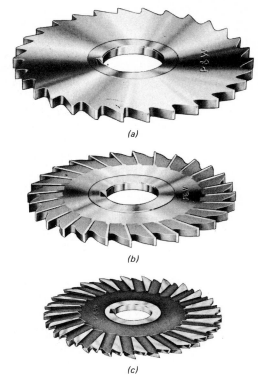

Fig. 13-6   Metal slitting saws. (*a*) Plain, hollow ground. (*b*) Side chip clearance. (*c*) Staggered side chip clearance for deep cuts. (*Colt Industries, Inc., Pratt and Whitney Machine Tool Division.*)

The 2-in.-diameter (50-mm) slitting saws are made as thin as 0.020 in. (0.50 mm). Diameters up to 8 in. (200 mm) and thicknesses up to $\frac{1}{4}$ in. (6 mm) are standard.

These slitting saws must be handled carefully. They are brittle and the slightest side pressure will break off a section of a $20 to $50 investment.

Form cutters in standard styles are shown in Fig. 13-7. These are often made as *"form-relieved"* cutters, that is, they are ground back in an eccentric relief but keep the same form throughout the length of each tooth. Thus, to sharpen a form-relieved cutter, it is necessary to grind only the *front* of each tooth. The diameter will decrease slightly with each grind, so the cutter will have to be reset in the machine before using it again.

Fig. 13-5   Nomenclature of plain and side milling cutters.

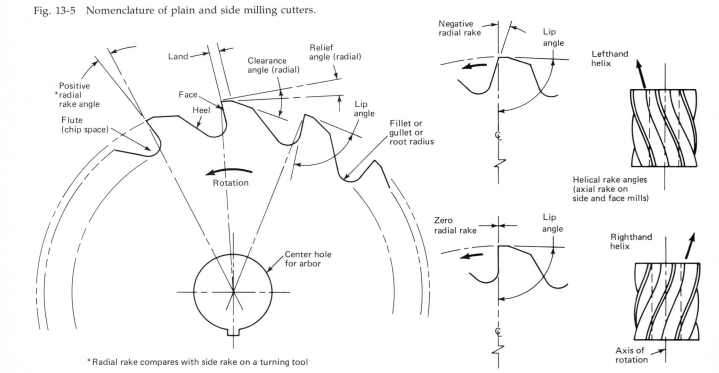

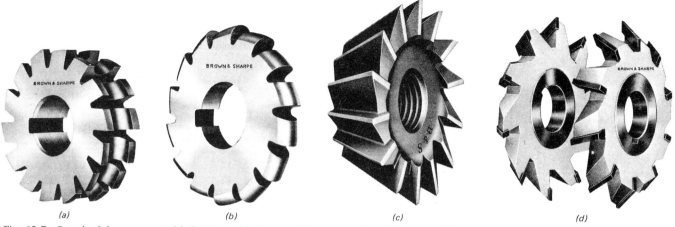

Fig. 13-7   Standard form cutters. (*a*) Concave. (*b*) Convex. (*c*) Angle or dovetail. (*d*) Carbide-tipped angle cutter. (*Brown & Sharpe Mfg. Co.*)

The common semicircular and quarter-circle forms and the 60° and 45° single-, double-vee, and dovetail cutters can be purchased from stock. An unlimited variety of shapes can be made to order. Figure 13-8 shows one method of cutting a dovetail like those used for lathes, milling machines, etc. The "relief" may be on either part of the dovetail slide. It reduces surface contact and thus reduces sliding friction. Look at some of the machines in these illustrations or in your shop.

Shell Mills

Sometimes called *shell-end mills*, shell mills were originally the type shown in Fig. 13-9*a*. This is a very versatile tool, as it can be used for simple facing work or for cutting "steps" which are too wide for a side mill and too deep for a face mill. This type of shell mill can also be used as a peripheral cutter for smoothing an edge to a depth not greater than the cutter's height.

These cutters can be mounted in either horizontal or vertical milling machines by using an arbor with a suitable tapered end.

These high-speed steel or carbide-tipped cutters are made in diameters from $1\frac{1}{4}$ to 6 in. (30 to 150 mm) with corresponding side heights of 1 to $1\frac{1}{4}$ in. (25 to 32 mm).

Face Mills

As the name *face mill* implies, these are used only for facing flat surfaces. They are used on both horizontal- and vertical-spindle milling machines.

All face mills today are made with inserted blades, either a long HSS or cast-alloy blade which can be repositioned for sharpening (Fig. 13-10*a*) or a carbide "throwaway" insert (Fig. 13-10*b* and *c*). The depth of cut, especially with the throwaway inserts, is limited by the $\frac{1}{2}$- to $\frac{3}{4}$-in. (13- to 19-mm) size of the insert.

The nomenclature and geometry of both types are similar and are shown in Fig. 13-11. The rake angles are similar to those used in other milling cutters. The *axial* rake is comparable to the *back* rake in lathe tools, and the *radial* rake is comparable to the lathe tool's *side* rake. The double-positive and double-negative rakes are illustrated. However, combinations of one posi-

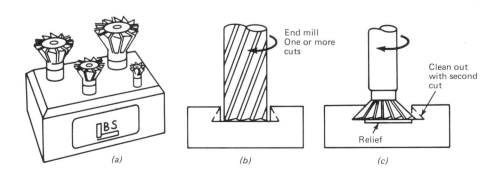

End mill
One or more cuts

Clean out with second cut

Relief

Fig. 13-8   Small (up to 1-in., 25-mm) dovetail cutters and one way in which they are used. (*a*) Cutters; (*b*) Cut out center with end mill or side mill; (*c*) Cut dovetail, one side at a time.

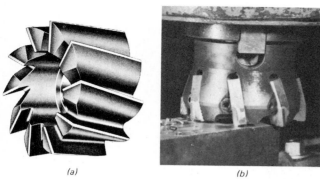

(a)   (b)

Fig. 13-9   Shell mills. (a) Standard HSS. (b) With replaceable HSS teeth.

Fig. 13-10   (a) Serrated regrindable blade face milling cutter. (b) Positive rake throwaway-insert face mill. (c) Negative rake throwaway-insert face mill. (*Viking Tool Co.*)

tive and one negative rake angle are often used. For example, the combination of negative radial and positive axial rake helps throw the chips away from the cutter as shown in Fig. 13-12.

The *relief angles* can be small, 2 to 8°. Notice that, theoretically, eight edges of a negative rake insert may be used but only four with a positive rake. The lead angle, or corner angle, is formed by tilting the square carbide insert. This "tilt" can be up to 45° and

helps the cut in the same way that the SCEA does in turning and facing.

It is impossible to cut to a 90° shoulder with a square insert, since some tilt must be used so that the entire insert does not scrape the surface. Figure 13-13 shows two ways of getting the "square" cut. Both the triangular and the 86° diamond inserts are standard from most manufacturers.

A different way to mount throwaway indexable

Fig. 13-11   Nomenclature of face mills. (a) Positive rake. (b) Negative rake.

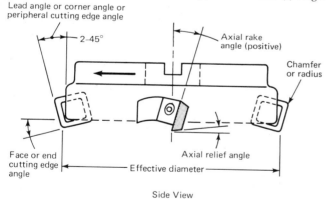

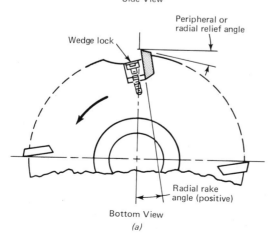

Bottom View
(a)

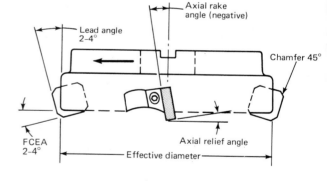

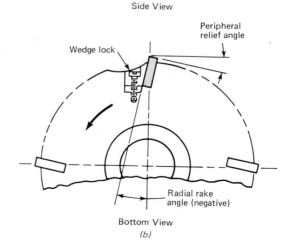

Bottom View
(b)

carbide inserts is shown in Fig. 13-14. The inserts are set flat and do their cutting with the narrow edge. This creates an unusually strong cutter, which is claimed to take up to twice the feedrate of the usual style at equal or higher cutting speeds in cast iron, steel, or stainless steel. Cuts up to 0.200 in. (5.0 mm) deep and over have been made.

Figure 13-14 shows a cutter with flat finishing blades in the center set 0.005 in. (0.125 mm) below the roughing blades. On medium to light cuts, this cutter will rough and finish to 63 μin. (1.6 μm) in one pass. Tool life, in test runs, has been more than double that of standard cutters.

Replaceable cutter-type face mills are made in diameters from 3 to 12 in. (75 to 300 mm) as standard. Many 18 to 24-in. (450- to 600-mm) cutters are in use, and much larger ones have been made for special jobs.

Besides various combinations of rake angles, face mills are also available with close-, medium-, and wide-spaced teeth for efficient cutting of fine finishes on cast iron, steel, and aluminum.

Fig. 13-12   A "Shear-Clear" negative-positive face milling cutter. Note chips leaving the cutter. (*The Ingersoll Milling Machine Co.*)

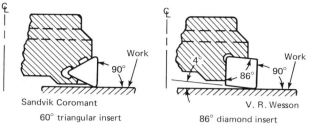

Fig. 13-13   Two ways of achieving a 90° shoulder with a face mill.

End Mills

These mills cut on the end as well as on the periphery, and are today the largest selling type of milling cutter. The growing use of N/C machines has almost doubled the use of many styles of end mills.

Numerical control machines also use milling cutters for cutting two- and three-dimensional shapes. End mills do a great variety of work, and as one manufacturer said, "Few tools are called upon to do the variety of jobs or to stand the punishment expected of end mills."

The *geometry* and *nomenclature* of end mills is similar to that of the plain mills, as shown in Fig. 13-15. The most used *sizes* of end mills are from $\frac{1}{4}$ to 1 in. (6 to 25 mm), though fairly frequent use is also made of larger cutters up to 2-in. (50-mm) diameter in some styles. End mills as small as $\frac{1}{32}$ in. (0.8 mm) are used in tool- and diemaking and some production work.

The usual helix (axial rake) angle is 30°, and the axial relief angle is about 5°. Radial relief angle varies with the diameter of the cutter. For $\frac{3}{8}$- to 1-in.-diameter (9- to 25-mm) end mills, the radial relief is about 8 to 10°.

The cost of end mills is reasonable. Two-flute single-end end mills cost about $5 for $\frac{1}{2}$ in. (12 mm) and $12 for 1 in. (25 mm). Four-flute end mills cost

Fig. 13-14   Cutter with flat positioned inserts and center finishing teeth. (*The Ingersoll Milling Machine Co., Cutting Tool Division.*)

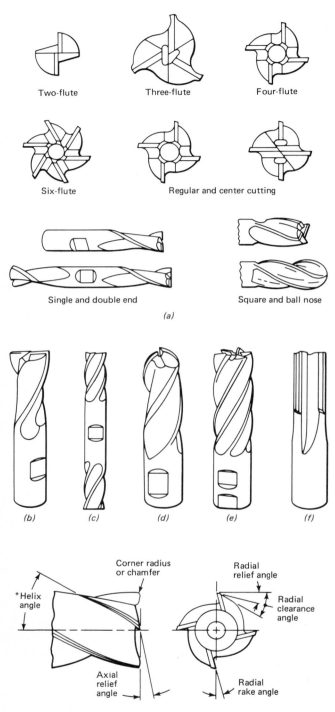

Two-flute   Three-flute   Four-flute

Six-flute   Regular and center cutting

Single and double end   Square and ball nose

(a)

(b)   *(c)*   *(d)*   *(e)*   *(f)*

Corner radius or chamfer

*Helix angle

Radial relief angle

Radial clearance angle

Axial relief angle

Radial rake angle

*Same as axial rake angle in plain and face mills. R.H. helix shown. This four-flute end mill is not center-cutting and cannot be used for plunge cuts

(g)

Fig. 13-15   (*a*) Outlines of the major classifications of end mills. (*b*) Two-flute end mill. (*c*) Four-flute, double end. (*d*) Three-flute ball end mill. (*e*) Six-flute end mill. (*f*) Straight-flute end mill. (*g*) Nomenclature for end mills. (*Colt Industries, Inc., Pratt & Whitney Machine Tool Division.*)

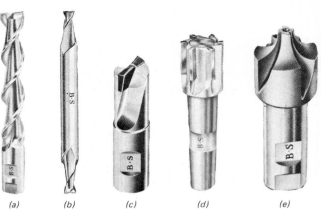

(a)   *(b)*   *(c)*   *(d)*   *(e)*

Fig. 13-16   (*a*) Extra long two-flute end mill with high helix for aluminum. (*b*) Tiny mills $\frac{1}{32}$- to $\frac{3}{16}$-in. (0.8- to 4.8 mm) diameter. (*c*) Carbide-tipped, positive rake, for aluminum. (*d*) Carbide-tipped, for cast iron, brass, and bronze. (*e*) Corner-rounding end mill. (*Brown & Sharpe Mfg. Co.*)

about \$6 to \$18 each. Carbide-tipped 2-in. (50-mm) diameter can cost up to \$85, and a $\frac{1}{2}$-in. (12-mm) solid-carbide two-flute end mill will cost about \$35.

Material for making end mills is usually high-speed steel, M1, M2, or M7. Some are made from T15 or M42 HSS. The *style* of end mills is based on a few basic shapes. The most used styles are shown in Figs. 13-15 and 13-16. The **two-flute** end mill can plunge cut like a drill or counterbore. The **three-flute** end mill has one tooth cutting all the way to the center, so it also can plunge cut. The regular **four-flute** end mill has no teeth going in to the center, so that it *cannot* plunge cut. Four-flute center-cutting end mills are made; however, they cost about 50 percent extra and are more difficult to sharpen, so they are infrequently used.

The six- to eight-flute end mills are standard in sizes from 1 to 2 in. (25 to 50 mm). The larger number of flutes when used with steel, cast iron, etc., gives much smoother, faster cutting.

Milling aluminum efficiently requires special end mills. The speeds are so high and the chips so thick that the chips from many aluminum alloys will plug up the flutes of regular milling cutters. In end mills, the fast (40°), spiral open-fluted end mill in Fig. 13-16*a* is especially designed for aluminum. Most have polished flutes to prevent chips from sticking.

Ball-end end mills are frequently used in three-dimensional contouring with tracers or numerical control as well as for round bottom cuts.

Long lengths are available in most styles and sizes, for jobs requiring the milling cutter to reach into a cavity or below an obstruction.

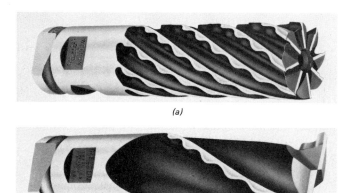

Fig. 13-17 (a) Crest-Kut end mill for roughing and finishing, 2 in. eight flute. (b) Two-flute Crest-Kut end mill for aluminum, etc. (c) Crest-Kut end mills used in N/C contouring work. Note the finish on the top piece. (*The Weldon Tool Co.*)

CAUTION: The long and extra long end mills will bend as much as 0.030 in. (0.75 mm) out of line if feeds are too heavy, so go slowly. Solid-carbide, long end mills will deflect only one-third as much as HSS. For accurate work or for long runs, they can be a good investment.

Carbide end mills may save money even when cutting aluminum. One job on a tracer mill, using ½-in. diameter by 6-in. long solid-carbide end mills, not only held size at a fast feed but stayed sharp for two shifts, cutting 50 to 60 pieces per hour.

Straight-flute end mills are sometimes used for cutting slots and keyways because they cut straighter sides and leave ends with no "dog-bone." Their use requires lower feeds and speeds.

Roughing end mills (Fig. 13-17), also called *chipbreaker end mills* or *hog mills,* look quite different from the conventional type. The grooves or scallops around the body are either semicircular or sinusoidal shaped. These produce well broken up chips.

The side thrust on the roughing end mills is less than on conventional end mills, and they have a heavier cross section. This combination gives the effect of considerably greater stiffness as compared with the conventional end mill. Deeper cuts and heavier feeds may be used. Speed is higher only if cobalt HSS is the cutter material.

Because of the combination of these factors, stock removal by the chipbreaker end mill, is at up to three times the rate of the usual end mill on steel, aluminum, stainless steel, etc.

These cutters are made with regular helix angles or with a 40° helix for aluminum. Diameters are to 2 in. (50 mm), and up to 6 in. (150 mm) in the shell mill design.

The *finish* obtained is not as good as you would get with a standard end mill, but finishes from 80 to 160 μin. (2.0 to 4.0 μm) are possible by using modest feedrates.

Sharpening is somewhat more difficult, but these are form-relieved cutters so only the front of the tooth needs to be ground. Incidentally, the gap widens each time the cutter is sharpened, which can actually improve the cutter's performance in aluminum.

The *cost* is high, from about $18 for a ½-in. diameter to $90 for the 2 in. However, the long life of the cutters and the large volume of metal they can remove makes them economical when large amounts of metal must be removed.

CLIMB AND CONVENTIONAL MILLING

The *direction of travel* of the work in relation to the rotation of the cutter can be very important. This relationship is shown in Fig. 13-18. There are two ways the cutter can work:

1. .*Conventional* or *up milling*
   a. On a flat surface, with a horizontal plain

Fig. 13-18   Typical examples of conventional and climb milling.

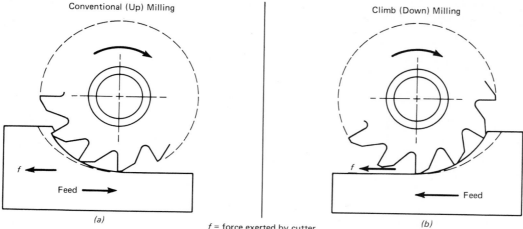

Conventional (Up) Milling

Climb (Down) Milling

f = force exerted by cutter

(a)   (b)

Slab milling with plain mill. Also *slotting* with a side milling cutter

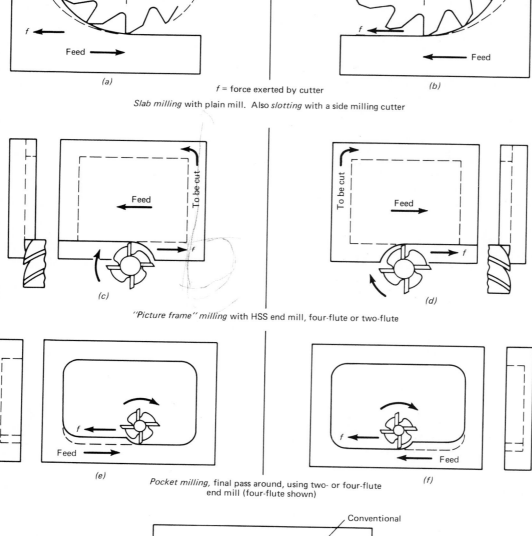

"Picture frame" milling with HSS end mill, four-flute or two-flute

Pocket milling, final pass around, using two- or four-flute
end mill (four-flute shown)

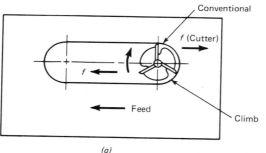

Keyway or *slot milling*, with two-, three-, or four-flute center-cutting
end mill (three-flute shown)

mill, the cutter tooth starts at the bottom and breaks out at the top. Thus, it moves *up* through the material as in Fig. 13-18a.

   b. With vertical-axis cutters, it is easier to recognize conventional or up milling by noticing that the force exerted by the cutter due to its rotation is *against* the force due to the feed of the work. This is illustrated in Fig. 13-18c and e.

2. *Climb or down milling*

   a. On a flat surface, the cutter moves *downward* through the material. It takes a heavy bite at the start and ends with zero chip thickness, as shown in Fig. 13-18b.

   b. With vertical cutters, the rotational force pulls *with* the force due to the feed of the work, as shown in Fig. 13-18d and f.

CAUTION: Always use climb milling when contouring or cutting pockets in aluminum as the finish will be much better. When cutting slots, one side will be up milling and one side down milling (Fig. 13-18g). The up-milling side may have a much poorer finish. Use of the proper cutting fluid will help considerably. In some cases it is necessary to use a smaller diameter cutter and climb mill both sides. Table 13-1 lists the advantages and disadvantages of the two types of cuts.

MILLING COMPUTATIONS

A round cutter does not make a full cut when it first contacts the edge of a workpiece. This condition is shown in Fig. 13-19. It shows the condition which exists when using plain or side milling cutters to face a surface or cut a slot on a horizontal-spindle milling machine.

The distance traveled between first contact and cutting to full depth is called the *approach* distance. The same condition exists as the cutter leaves the work-

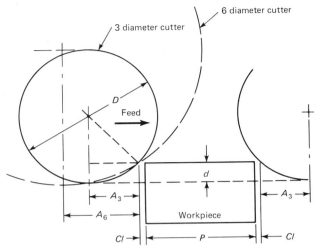

$$A = \sqrt{d(D - d)}$$

A = approach distance, in. or mm
D = cutter diameter, in. or mm
d = depth of cut, in. or mm
Cl = clearance, to avoid collisions at rapid advance
Length of cut = $L = P + 2A + 2Cl$

Fig. 13-19 Method of computing total distance traveled when milling with plain or side milling cutters.

piece at the end of the cut. This is sometimes called the *overtravel*. Actually both distances are the same, and the two terms are used interchangeably.

The construction lines show how the computations can be made to arrive at the equation:

$$\text{Approach} = A = \sqrt{d(D-d)} \tag{13-1}$$

Notice that A increases as the cutter diameter gets larger. Thus, it saves time to use the smallest cutter possible, but care must be taken that no part of the work can bump the arbor or the outboard arbor support.

Figure 13-20 shows the condition existing when face milling or some end milling cuts are made. The cutter is cutting the full width before it is halfway into the work. The distance needed to cut full width is

TABLE 13-1 COMPARISON BETWEEN CONVENTIONAL AND CLIMB MILLING

CONVENTIONAL (UP) MILLING	CLIMB (DOWN) MILLING
Advantages	
Must be used on older machines which have looseness (backlash) in their leadscrews.	Downward force helps keep work flat. Especially helpful when machining thin parts.
On sand castings, cutter is not damaged as much by the rough, sandy surface.	Chips are thrown away from the direction of the cutter's travel.
Usually gives a somewhat better finish on steel, though not on aluminum.	There is less normal pressure on the material, which is especially advantageous for milling stainless steels and other work hardening materials.
Disadvantages	
Cutter force tends to lift the work off the table.	The milling machine must have zero backlash or there will be chatter as the cutter tries to pull the table faster than the feedrate.
Chips sometimes get picked up and carried around by the cutter. This injures the finish.	On steel, the finish may be slightly rougher.

$$A \text{ (face milling)} = \tfrac{1}{2}(D - \sqrt{D^2 - W^2}) \qquad (13\text{-}2)$$

However, in the real world of manufacturing, this equation is seldom used. It is standard practice to run the face mill or end mill completely off the work as shown by the cutter positions in Fig. 13-20. This extra distance takes time, but it produces a better finish, flatter work, and leaves the milling cutter clear of the work so that the piece can be removed. The equation shown for L in Fig. 13-20 is the one most often used.

Clearance is needed between the edge of the milling cutter and the work. This allowance is a matter of judgment. If the work has smooth, flat edges, then 0.010 to 0.020 in. (0.25 to 0.50 mm) is plenty of clearance. If the workpiece is a rough casting, $\frac{1}{8}$ to $\frac{1}{2}$ in. (3 to 13 mm) may be needed to allow for irregularities in the size of the work.

Speed, Feed, and Time Computations

After calculating the total length of cut, it is necessary to compute the feedrate. **Feedrate** as given in the handbooks is in inches per tooth (sometimes abbreviated as ipt) or millimetres per tooth (sometimes abbreviated as mmpt). This is logical because it is the strength of *one* tooth which limits the strength of the

cutter. Moreover, the finish is related to the distance between each tooth's contact with the work.

However, many conventional and N/C machines are calibrated in inches per *minute* (ipm) or millimetres per minute (mm/min). Moreover, the advance per minute is the basic figure used to compute the *time* for the cut.

To compute the feedrate for milling:

1. Decide, from published tables or from experience, the *feed* per tooth = f (in. or mm)

2. Distance traveled in *one revolution* = D

D = feed per tooth × number of teeth on the cutter n

or, $D = f \times n$ (in. or mm) = in./rev or mm/rev

3. Feed per minute = F (in./min or mm/min)

$$F = f \times n \times \text{rpm} \qquad (13\text{-}3)$$

Speed in rpm is calculated the same as in lathe or drill press work. The diameter used is, of course, the diameter of the milling cutter.

Time is basically distance divided by rate. Thus in milling, the time for cutting is

$$\text{Time} = T = \frac{\text{total length at feedrate (in. or mm)}}{\text{feed per minute (in. or mm)}}$$
$$(13\text{-}4)$$

These formulas can be combined into one long

Fig. 13-20   Theoretical and practical approach distance and total travel for (*a*) single and (*b*) multiple face or end milling cuts.

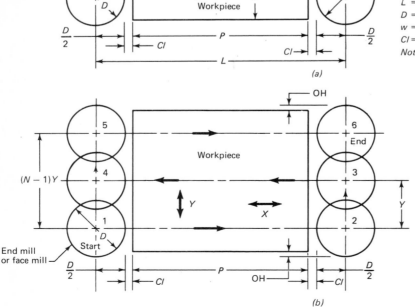

A = theoretical approach distance
L = P + D + 2Cl
L = length of travel at *feedrate*, in. or mm
D = diameter of cutter, in. or mm
w = width of work, in. or mm
Cl = clearance
Note: This also applies when cutting slots with end mills

L   = N(P + D + 2Cl) + (N − 1)Y
N   = number of passes across the work
Y   = may be at rapid traverse, and must be less than the cutter diameter. Approximately 80–90% of D is often used
OH = overhang, to ensure a full cut

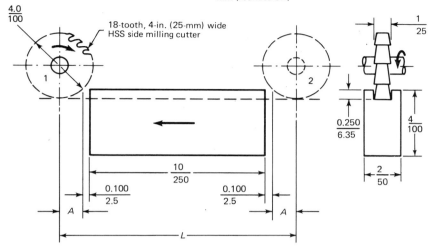

Dual dimensions; $\frac{inch}{mm}$ (rounded off)

18-tooth, 4-in. (25-mm) wide
HSS side milling cutter

Material: Soft C.I.

Cutting speed: $\frac{100\ fpm}{30\ m/min}$

Feed per tooth: $\frac{0.005\ in.}{0.125\ mm}$

Fig. 13-21 An example of slot milling computations. Same figures would be used for face milling with plain mill.

"plug in" formula. However, you will find that it takes no longer to do the work as shown in the example, and you will more easily catch errors in rpm, feedrate, etc., because you can check each figure against past experience or common sense.

EXAMPLE   13-1

Note: The metric dimensions are rounded off as they might be if the *original* dimensions were metric. They are *not* precise metric equivalents.

Refer to Fig. 13-21. Find cutting time.
Inch system:

$$rpm = \frac{4(100)}{4} = 100\ rpm$$

Feed $F = (0.005)(18)(100)$
$\qquad = 9\ ipm$

Approach $A = \sqrt{0.25(4 - 0.25)} = \sqrt{0.9375}$
$\qquad\qquad = 0.97\ in.$

Length of cut $L = 2(0.97) + 2(0.100) + 10.0$
$\qquad\qquad\qquad = 12.14\ in.$

Time $T\ = \frac{L}{F} = \frac{12.14}{9.00} = 1.35\ min$

Metric system (Fig. 13-21):

$$rpm = \frac{300(30)}{100} = 90\ rpm$$

Feed $F = (0.125)(18)(90) = 202\ mm/min$

Approach $A = \sqrt{6.35(100 - 6.35)} = \sqrt{595}$
$\qquad\qquad = 24.4\ mm$

Length of cut $L = 2(24.4) + 2(2.5) + 250$
$\qquad\qquad\qquad = 303.8\ mm$

Time $T\ = \frac{L}{F} = \frac{303.8}{202} = 1.5\ min$

The same computations would be made if a 4-in. (100-mm) plain spiral mill were used to face 0.250/ 6.35 off of the top of the casting.

EXAMPLE   13-2
Refer to Fig. 13-22. Find cutting time.
Inch system:

$$rpm = \frac{4(300)}{0.750} = 1600\ rpm$$

$F = (0.004)(4)(1600) = 25.6\ ipm$

$A = \frac{D}{2} = \frac{0.750}{2} = 0.375\ in.$

$L = 0.375 + 0.03 + 4.0 = 4.405\ in.$

$T = \frac{4.405}{25.6} = 0.172\ min = 10.3\ s$

Note that depth of cut does not affect the time.
Metric system:

$$rpm = \frac{300(90)}{20} = 1350\ rpm$$

$F = (0.100)(4)(1350) = 540\ mm/min$

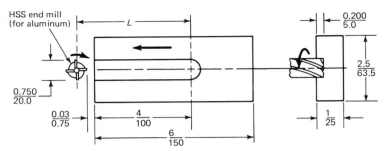

Fig. 13-22 An example of keyway cutting computation.

$$A = \frac{D}{2} = \frac{20}{2} = 10 \text{ mm}$$

$$L = 10 + 0.75 + 100 = 110.75 \text{ mm}$$

$$T = \frac{110.75}{540} = 0.205 \text{ min} = 12.3 \text{ s}$$

EXAMPLE 13-3

Refer to Fig. 13-23. Find cutting time.

Inch system:

$$\text{rpm} = \frac{4(250)}{6} = 167 \text{ rpm}$$

$$F = (0.003)(10)(167) = 5.01 \text{ ipm}$$

$$A = \text{diameter} = 6 \text{ in.} \quad (3 \text{ in. per side})$$

$$L = 6 + (2 \times 0.02) + 20 = 26.04 \text{ in.}$$

$$T = \frac{26.04}{5.01} = 5.2 \text{ min}$$

Metric system:

$$\text{rpm} = \frac{300(76)}{150} = 152 \text{ rpm}$$

$$F = (0.075)(10)(152) = 114 \text{ mm/min}$$

$$A = \text{diameter} = 150 \text{ mm} \quad (\text{both sides})$$

$$L = 150 + (2 \times 0.50) + 500 = 651 \text{ mm}$$

$$T = \frac{651}{114} = 5.7 \text{ min}$$

Horsepower Calculations

Most milling machines are used at far less than their maximum horsepower. However, when wide cuts or gang milling cuts are used, it is occasionally wiser to check the needed horsepower before starting the job.

Fig. 13-23 An example of face milling computations.

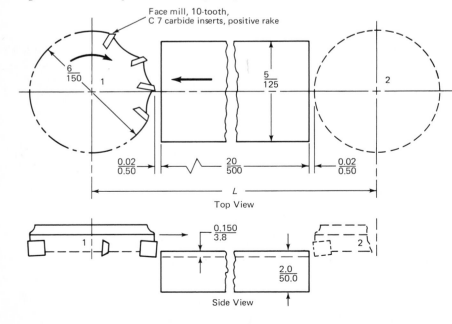

hp = Horsepower needed

w = Width of cut, in. or mm

d = Depth of cut, in. or mm

f = Feedrate, in./min or mm/min

K = Horsepower necessary to remove one cubic inch, or one cubic centimetre per minute, at 100 percent efficiency

See Table 4-4 (p. 72) for values of K.

E = Efficiency of the machine, the percent of motor horsepower which is available at the spindle. Expressed as percent in decimal form

Volume of metal removed per minute = wdf (in.³ or mm³)

$$HP = wdfK \div E$$

EXAMPLE   13-4

See Fig. 13-23, use 70 percent efficiency and $K = 0.8$.

Inch system:

$$hp = (5)(0.150)(5.01)(0.8) \div 0.70$$
$$= 3.00 \div 0.7 = 4.3 \text{ hp needed (3.2 kW)}$$

Metric system:

$$hp = (125)(3.8)(114)\left(\frac{0.8}{16,400^*}\right) \div 0.70$$
$$(^*1 \text{ in.}^3 = 16,387 \text{ mm}^3)$$
$$= 2.6 \div 0.7 = 3.7 \text{ hp needed (2.77 kW)}$$

Gang milling, using two or more cutters, requires that the volume be computed for *each* cutter. The *total* volume is then multiplied by K and divided by E.

CAUTION: The rpm is computed for the *largest* diameter milling cutter in gang milling, and the feed per revolution is figured for the cutter with the *fewest* teeth. These figures are then used for all cutters on the same arbor regardless of their size when computing the cubic inches of metal removed.

Review Questions and Problems

13-1. What are the four basic types of milling cutters?

13-2. Describe plain milling cutters and their principal uses.

13-3. What types of milling cuts can be made with side milling cutters? When would you use staggered-tooth side milling cutters?

13-4. What are slitting saws? What precaution must be taken when using a slitting saw?

13-5. How are form cutters made and how are they sharpened?

13-6. Compare face mills and shell mills.

13-7. What are the basic styles of end mills? What type of cut cannot be made with a four-flute end mill that some other end mills can make? Why?

13-8. Why can more teeth per inch of diameter be used for cutting cast iron than for cutting steel?

13-9. What changes are made to a standard end mill to improve its ability to mill aluminum?

13-10. How do "roughing" end mills differ from the conventional type of end mill? What are their advantages and disadvantages?

13-11. By the use of a sketch, show the difference between conventional (or up milling) and climb (or down milling). When should climb milling always be used?

13-12. Two keyways are to be machined as shown in the sketch below. Both horizontal- and vertical-milling machines are available. If the following cutters are available, which machine and which tool would you use for each keyway?

a. $4 \times \frac{3}{8}$ side milling cutter

b. $\frac{3}{8}$ two-flute end mill

c. $\frac{3}{8}$ four flute end mill

d. $4 \times \frac{3}{8}$ plain milling cutter

e. $\frac{3}{8}$ drill

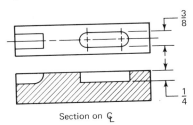

Section on ₵

13-13. A manufacturer wishes to mill a $\frac{5}{8}$-in.-wide (16-mm) slot $\frac{3}{4}$ in. (19 mm) deep in a piece of mild steel 20 in. (508 mm) long at 125 fpm (38 m/min), using a 6-in.-diameter (150-mm) side milling cutter having 20 teeth. If the feed is 0.003 inches per tooth (0.08 millimetres per tooth) and the clearance is 0.100 in. (2.5 mm) calculate:

a. rpm of cutter

b. Table feed

c. Length of cut

d. Time of cut

e. Horsepower required at 75 percent efficiency.

13-14. A piece of aluminum 2 in. (50 mm) wide by 18 in. (450 mm) long is to be face milled using a 4-in.-diameter (100-mm) face mill having 6 teeth at 250 fpm (76 m/min). The feed is 0.006 inches per tooth (0.15 millimetres per tooth), and the clearance is 0.050 in. (1.25 mm). Calculate:

a. rpm of cutter
b. Table feedrate
c. Length of cut
d. Time of cut
e. If the amount of stock removed is 0.125 in. (3.0 mm), what horsepower will be required?

13-15. A slab of steel 12 in. (300 mm) wide by 24 in. (600 mm) long is to be face milled using a 5-in.-diameter (125-mm) cutter having 8 teeth at 100 fpm (30 m/min) using a feed of 0.002 inches per tooth (0.05 millimetres per tooth). If the clearance and cutter overrun on each side is 0.125 in. (3.0 mm), calculate:

a. rpm of cutter
b. Table feedrate
c. Total length of cut
d. Time of cut.

13-16. Slots are to be gang milled, as shown in the sketch, in a piece of mild steel 30 in. (760 mm) long. The two cutters used are side milling cutters, a 6-in. (150-mm) diameter by 1 in. (25 mm) wide having 20 teeth, and an 8-in.-diameter (200-mm) cutter ¾ in. (19 mm) wide having 22 teeth. The machining is to be done at 60 fpm (18 m/min) and a feed of 0.004 inches per tooth (0.10 millimetres per tooth).

If 0.100 in. (2.5 mm) is allowed for clearance, calculate:

a. rpm used
b. Table feedrate
c. Length of cut
d. Time of cut
e. Horsepower required.

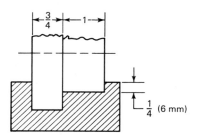

References

1. Bhattacharyya, A., and J. Ham: *Design of Cutting Tools*, Society of Manufacturing Engineers, Detroit, Mich., 1969.
2. "How to Figure Milling Times, Special Report No. 581," *American Machinist*, Jan. 3, 1966.
3. Kang, T. H.: *Investigation of Numerically Controlled End Milling*, Society of Manufacturing Engineers, Detroit, Mich., N572-165, 1972.
4. *Metal Cutting Tool Handbook*, Metal Cutting Tool Institute, 1965.
5. *Milling Cutters and End Mills*, ANSI B94-1968, American National Standards Institute, New York.
6. *Milling Cutters and End Mills, Application and Sharpening*, Metal Cutting Tool Institute, New York, 1962.

abrasives and grinding wheels

Grinding is a cutting process, just like the cutting process done by lathes or milling machines. The "cutters" get dull and require "sharpening" as do single-point tool bits. The difference is that the "cutters" in grinding and other abrasive processes are irregularly shaped grains which in most grinding wheels vary in size from 0.0787 in. (2.0 mm) coarse grain to as small as 0.00079 in. (0.02 mm), which is used for very fine finishing grinding wheels. The abrasive grit usually cuts with a zero to negative rake angle (Fig. 14-1) and produces a large number of short, small, curly or wavy chips.

WEAR OF A GRINDING WHEEL

The grinding wheel "wears out" due to three factors:

1. *Dulling* (rounding) of the abrasive grains because of rubbing over the work material. This does not cause much actual decrease in wheel diameter, but it does decrease the amount of cutting, as is typical of any dull tool.

2. *Grain fracture* is the breaking off of *part* of one of the abrasive grains. This accounts for most of the "wear" on the wheel. Actually it is a very desirable condition. The grains get dull and the force on them increases until they break, thus exposing a new, sharp cutting surface.

3. *Grain pullout* refers to whole grains of abrasive breaking away from the wheel. This is unlikely in precision grinding, but more common in very rough grinding.

The G ratio is the ratio of the amount of stock removed versus the amount of wear on the wheel, measured in cubic inches per minute (or cubic centimetres per minute). This is called the *grinding ratio.*

$$G = \frac{V \text{ work removed}}{V \text{ wheel removed}}$$

where V = volume.

This G ratio will vary from 1.0 to 5.0 in very rough grinding up to 25.0 to 50.0 in finish grinding, and much higher with some of the newer grinding wheels.

Even though grinding wheels are fairly expensive, a high G ratio is not necessarily economical, as this may mean a slower rate of stock removal, and time is costly. It often takes some experimenting to find the wheel-metal combination which is the most economical for a long-run job. Some typical shapes and sizes of grinding wheels are shown in Fig. 14-2.

ABRASIVES USED IN GRINDING WHEELS

Aluminum Oxide (Al_2O_3)

The material most used for grinding abrasives is aluminum oxide, made by refining bauxite, mixed with coke and iron borings. This results in a very large mass which is crushed and screened into grit sizes.

This aluminum oxide material is refined to different degrees, and the resulting wheels may be black, gray, pink, or white. The white wheel is the most pure Al_2O_3. It is very friable (brittle) and free cutting.

Fig. 14-1 The way an abrasive grain "cuts" material.

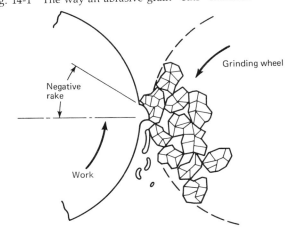

193

Aluminum oxide will grind a wide variety of materials such as hardened steel (to HRC 80), hard bronze, steel billets, stainless steel, ball bearings, etc., as well as softer materials. It is not as hard as silicon carbide, but it is tougher and more shock resistant.

Modifications of the basic aluminum oxide are constantly being investigated. Adding chromium created a ruby or purplish-red colored wheel which is longer lasting and freer cutting on tool steels and tough metals.

Alloying aluminum oxide with vanadium yields a green-colored grit which makes a strong, hard abrasive with high grinding ratios on alloys of HRC 55 or higher. This abrasive cuts faster and lasts longer on steels such as M10, T15, and 52100 ball-bearing steel.

Zirconia and other alloys are being combined with aluminum oxide in a constant effort to improve production rates and grinding wheel life. All these alloyed wheels are more expensive, so their use must be justified by cost and production studies.

Silicon Carbide (SiC)

Silicon carbide is harder than tungsten carbide and is made by combining pure quartz sand, coke, and sawdust in an electric furnace. The resulting crystalline mass is crushed and graded by particle size. It is made in two grades: the relatively pure, green silicon carbide and a blue-black grade.

The well-known "green wheel" is used for grinding many carbide tools. Carbide wheels are also recommended for grinding gray cast iron, some aluminum, brass, chilled iron, some glass, and materials like ceramics and brick. It is not often used on

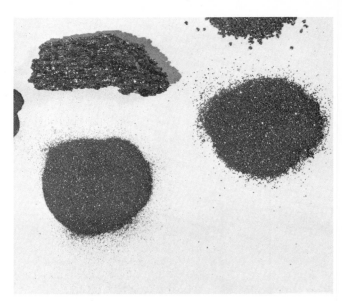

Fig. 14-3   A chunk of fused silicon carbide (upper left) and three sizes of abrasive grain made from it. *(Norton Co.)*

steel. The fused SiC and some abrasive grains are shown in Fig. 14-3.

Borazon

General Electric's name for their patented *cubic boron nitride* is Borazon, and it was first put on the market about 1969. The material is much harder than any other substance known, except the diamond. Many grinding wheel manufacturers are making Borazon wheels, mostly in the sizes and shapes used for tool and cutter grinding. Although it is somewhat more difficult to dress, it is so hard that it will stay sharp for unusually long runs. Its grinding ratio is claimed to be over 100 times greater than the ratio for presently used abrasives, yet it cuts stock faster, cooler, either wet or dry, uses less power, and needs very little dressing. Thus, accuracy is much easier to maintain especially on long jobs.

Principal uses for Borazon wheels are:

1.   Grinding HSS cutters during their manufacture
2.   Grinding tool-steel punch-press dies
3.   For some hardenable stainless steels
4.   Internal grinding in all ferrous metals

Borazon is *not* recommended for:

1.   Snagging operations
2.   Use on nonferrous materials, carbides, or steels with high chrome content
3.   Making "corner" or interrupted cuts

This material is not made into an entire grinding

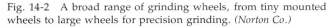

Fig. 14-2   A broad range of grinding wheels, from tiny mounted wheels to large wheels for precision grinding. *(Norton Co.)*

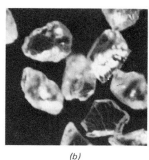

(a)   *(b)*   *(c)*

Fig. 14-4 (*a*) Synthetic, 200-mesh diamond for use with metal-bonded wheel. (*b*) A blocky-shaped natural diamond, 100 mesh, for grinding carbides. (*c*) A coated (nickel cladding) 100-mesh synthetic diamond for resin bond, wet grinding of steel and carbides. All at 70X magnification. (*Industrial Diamond Information Bureau, London, England.*)

wheel. It is resin bonded about $\frac{1}{8}$ in. (3.2 mm) thick around the rim of a metal wheel. (See illustrations of diamond wheels.) It is efficient even at lower grinding speeds, about 3500 to 4000 fpm (1100 to 1200 m/min). The cost per wheel is comparable to that of diamond grinding wheels.

Diamonds

Both natural and manufactured diamonds are the hardest material known today. The natural diamonds which are used for grinding are made from *Bort*, i.e., diamonds not suitable for ornamental uses. These diamonds are sorted for shapes: slivers and flats for best finish and blocky shapes for roughing and cutting off uses. Synthetic diamonds come in similar shapes and in various grades of friability. Figure 14-4 shows some of the diamonds used.

Diamond grinding wheels are used for sharpening carbide and ceramic cutting tools as well as for dressing flat and shaped wheels (see *crush dressing*). They are also used for working on glass, germanium and silicon slicing, boron carbide, slate, etc.

Even though the Bort costs only about 1 to 5 percent of the value of gem diamonds, these wheels are expensive. Thus, the diamonds are used only in the outer surface of a metal grinding wheel, about $\frac{1}{8}$ to $\frac{1}{4}$ in. (3.2 to 6.4 mm) thick, embedded in resinoid or metal binders. Diamond grinding wheels cost upwards of $200 each.

SHAPES OF GRINDING WHEELS

The familiar disk-shaped grinding wheel is only one of many available shapes. A few of the most frequently used wheel shapes are shown in Fig. 14-5*a*.

Type 1—By far the most used, for surface, OD and ID cylindrical, free-hand cutter grinding, and some tool grinding. Type 5 is used similarly but for larger wheels. It is recessed for flange mounting.

Type 2—Used for surface grinding on vertical-spindle grinding machines, often made in several sections.

Type 4—Used for some tool sharpening and in "off hand" snagging.

Types 6, 7, 11, 12—Used mostly for cutter grinding and sharpening in the toolroom.

Diamond grinding wheels are made in shapes 1, 6, 11, and 15 but, as shown in Fig. 14-5*b*, the grit is contained only in a band on the actual grinding surface.

Fig. 14-5 (*a*) A few of the standard grinding wheel shapes. (*b*) Some diamond or Borazon wheel shapes. Note the location of the abrasive sections.

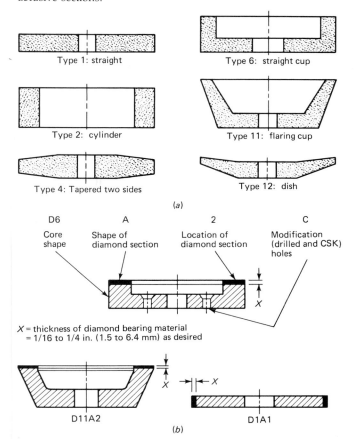

THE GRINDING WHEEL NUMBERING SYSTEM

A complete grinding wheel specification, in both the U.S. and ISO systems, consists of outside diameter, thickness, hole size, and the items shown in Fig. 14-6. The next few pages describe the significance of the numbers and letters shown in Fig. 14-6.

Abrasive Used

The first letter signifies the abrasive used and is frequently preceded by a number. The first *number* is not standard, but refers to the manufacturer's modification of the basic material. The *letters* mean:

A—for *aluminum oxide*

C—for *silicon carbide*

D or ND—for *natural diamonds*

SD—Sometimes used for *synthetic diamonds*

CB or CBN—for cubic boron nitride, *Borazon*

Grit Size

Grit size is signified by a number specifying the sieve screen size which the grain will pass *through*. Thus, a 60 grit will pass through a screen having 60 holes per inch but will be held on the 70-mesh screen. A 60-mesh sieve has 60 × 60 = 3600 openings per square inch. The grit sizes available are shown in Table 14-1*a*. Suggested metric sizes and the U.S. equivalents are shown in Table 14-1*b*. Metric sizes specify the *opening* in the sieve (between the wires) in millimetres.

CAUTION: A 30 grit is *not* $\frac{1}{30}$-in. diameter. There are 30 spaces per inch in the sieve, but the wires take up some of the space. Thus, a 30 grit will pass through a square hole 0.0234 in. (0.595 mm) on a side.

Fig. 14-6   A complete wheel specification and the range of values for the five standard specifications.

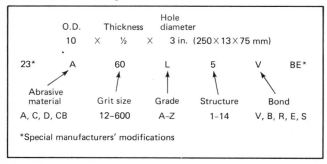

TABLE 14-1a   GRIT SIZES FOR GRINDING WHEELS

Coarse	Medium	Fine	Very Fine
12	**30**	70	220
14	**36**	80	240
16	**46**	90	280
20	**54**	100	320
24	**60**	**120**	400
		150	500
		180	600

NOTE: The divisions between coarse, medium, etc., are according to ANSI B74.13-1970. Only the grit sizes shown are available, though occasionally *mixtures* of two or three grits are used. Boldface numbers indicate the most used sizes.

TABLE 14-1b   PROPOSED ISO GRIT SIZE NUMBERING SYSTEM (PARTIAL LIST)

Proposed ISO Standard*	Approx. U.S. Equivalent Sieve Number	Sieve Opening, in.
2.00 mm	10	0.0787
1.41	14	0.0555
1.00	18	0.0394
707 μm†	24	0.0278
500	36	0.0197
354	(45)	0.0139
250	60	0.0098
177	80	0.0070
125	120	0.0049
88	(170)	0.0035
63	(230)	0.0025
44	(325)	0.0017

NOTE: Sizes in parentheses are close to but not exact equivalent U.S. Standard grain sizes.
*Also sieve opening in millimetres.
†A micrometre or micron = 0.001 mm.

Grits 14 through 24 are used for snagging (grinding rough castings, etc.).

Grits 24 through 80 are by far the most used for all purposes. An 8- to 16-μin. (0.2- to 0.4-μm) finish can be obtained quite easily with an 80 grit.

Grits 90 through 240 are used mostly for very fine finishes on some hardened steel work, glass, and thread grinding.

Grits 320 through 600 are "flour" sizes and are used for 1- to 4-μin. (0.025- to 0.1-μm) finishes, lapping, and some honing.

Grade

The grade of a wheel is specified by a letter, from A (soft) to Z (hard). Soft and hard refer to the holding power or strength of the *bond* which is used. If a wheel is "soft," the grains will break free under a relatively small force. If a heavier or tougher bond

TABLE 14-2   GRADE CLASSIFICATION FOR GRINDING WHEELS

Soft		Medium		Hard	
A	F	J*	M	Q*	U
B	G	K	N*	R*	V
C	H*	L*	O	S	W
D	I		P*	T	(X, Y, Z)
E					

*Used also for diamond wheels.
NOTE: The division between soft, medium, and hard is not standard. Boldface numbers indicate the most used grades.

coating is used, the grains are "glued" in more tightly and are "hard" (difficult) to break loose.

The difference in the construction of soft and hard grades is shown in Fig. 14-7. The grades are listed in Table 14-2, with the most used grades in boldface type.

Grades A through F (very soft), *X, Y, Z* (very hard) are seldom used, in fact, not used at all with carbide wheels.

Grades G through R are used by 90 percent of the grinding wheels.

Grades S through W are used for cutoff and snagging wheels.

The *grade* is determined by the percent by weight of the bonding agent to the abrasive grains when the mix is made, and the pressure used to compact the wheel when it is molded. The lower percent of bonding agent and lower pressures will make the "soft" grinding wheels.

Structure

The structure, or grain spacing, of a grinding wheel is specified by a number from 1 to 14. The lower the number is, the denser the structure or the more closely spaced the grains are, as shown in Fig. 14-8.

Structures 1 and 2, very dense, are used mostly for cutting and snagging wheels.

Structures 3, 4, 7, 9, 10, 11, and 13 are infrequently used.

Structures 5, 6, 8, medium density, are used for about 90 percent of all grinding wheels. No. 8 is the most frequently specified for silicon carbide wheels.

Structures 12, 14, very open, are used in wheels for grinding ball-bearing races, grinding brass and bronze, and some surface grinding of high-speed steels.

No structure specified. This is the custom on several

aluminum oxide wheels and on about two-thirds of the silicon carbide wheels. This is because the manufacturer has found that there is one particular structure which does best the work for which it is specified, with a particular grit and bond.

The structure is partly obtained from the bond/grit mix and pressure, plus the shrinkage from being heated to 1700 to 2300°F (900 to 1200°C) for up to five days when using "vitrified" bonds.

For the more open structures, "induced porosity" is created by mixing an easily burned substance into the mix of bond and grits. This substance may be many products, including sawdust, ground peanut shells, or even a substance resembling moth ball flakes. These are inexpensive, burn off at the temperatures used, and leave little, if any, residue.

Type of Bond

The *first* (or only) letter at the end of the grinding wheel specification indicates the type of bond used. If two or three letters are used, the last one or two are special modifications which the manufacturer has made.

There are *five* basic types of bonds:

V = vitrified bond—This is used in over 75 percent of the grinding wheels manufactured. "Vitrify" means to change into glass by heat and fusion. Thus, when clay, feldspar, or flint are mixed with the abrasive grains and heated to 2300°F (1200°C), the ceramic material melts and forms a glasslike coating and bonding agent for the grains.

This material is medium strength and is not affected by water, acid, or oils at normal working tempera-

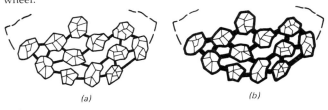

Fig. 14-7 (*a*) Illustrating the small amount of binder in a "soft" grade wheel. (*b*) The larger amount of binder in a "hard" grade of wheel.

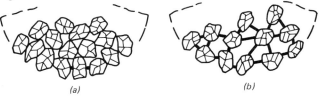

Fig. 14-8 The "structure" of a grinding wheel.

tures. Except for special wheels, vitrified bond should not be run over 6500 fpm (2000 m/min).

B = resinoid bond—This is the next most frequently used bond. It is made from a powdered synthetic resin such as phenol formaldehyde. This is mixed, pressed, and heated to about 350°F (177°C). After cooling, this makes a wheel which is less brittle, tougher, and more flexible than the vitrified bond and which can be run up to 9500 fpm (2900 m/min). Resinoid-bonded wheels are extensively used for weld grinding, cutoff wheels, and dry drill grinding (during manufacturing), and for diamond and Borazon wheels.

R = rubber bond—This bond is used for very fine finishing and polishing of metals (such as ball-bearing races) and for cutoff wheels where burr and burn must be avoided. Rubber-bonded wheels can be run at up to 9500 fpm (2900 m/min) and over.

E = shellac bond—This bond is used for only a few special applications such as very high finishes on paper mill rolls and some knife and scissor grinding. The wheel is made by mixing shellac and the grit in a heated mixer, molding, and baking at about 300°F (150°C).

S = silicate bond—This is the least used bond today. It is made by mixing sodium silicate (Na_2SiO_3) with the abrasive, pressing in molds, drying, then baking at 500°F (260°C) for a day or more. Grinding wheels with this bond are used where the generated heat must be kept at a minimum, such as for keen-edged tools. Silicate bond is also used for some large (over 24 in., 600 mm) wheels since they do not warp or crack in the baking process.

Now, listing three typical grinding wheel specifications, they would line up this way:

Special Symbol	Abrasive	Grit Size	Grade	Structure	Bond	Special Symbol
23	A	60	L	5	V	BE
37	C	36	K	*	V	K
37	C	24	J	12	V	P

*"Standard" structure by the manufacturer.

The first wheel shown above is recommended for cylindrical grinding of hardened steel. It could be a 10-in.-diameter (250-mm) by ½-in.-wide (13-mm) wheel, costing about $12.

The second grinding wheel is suggested for cylindrical grinding of brass or aluminum. It might be used on a larger grinder and be 14 in. (355 mm)

*Man-Made (capitalized) is General Electric Company's trademark for their manufactured diamonds.

diameter by 2 in. (50 mm) wide. It would cost about $55.

The third wheel listed is for use on a vertical-spindle grinder for flat surfaces on cast iron. A common size would be 18-in. (460-mm) diameter with a 1½-in. (38-mm) rim. It would cost about $95.

Identifying Diamond Grinding Wheels

The abrasive symbol is D with prefixes specifying natural or man-made,* as described earlier in this chapter.

The *grit size* is specified the same as other grinding wheels.

The *grade* is the same, except that most diamond grinding wheels use only grades H, J, L, N, P, Q, R.

The structure is not used; instead the *concentration of diamond*, which is on an arbitrary scale of the quantity of diamonds used in relation to the bond is specified. The scale is:

25—50—75—100 concentration (not percent)

One hundred "concentration" means that 72 carats, ½ oz or 14.4 g, of diamond grit is included in every cubic inch of bonding material. Other concentrations are in proportion.

The *bonds* used for diamond wheels are:

B = resinoid–similar to those used in conventional grinding wheels. These diamond wheels are very cool cutting, though they wear more rapidly than other bonds.

V = vitrified—the same as previously described. This gives a fast cut and slow wear when properly selected, though this bond is little used now.

M = metal bond—The metal may be soft steel or copper and is quite frequently used. Metal-bonded wheels are not free cutting, but they are very efficient. They are used for severe applications such as ceramics and finishing carbide tools.

Cubic Boron Nitride (Borazon) wheel specifications are somewhat similar to those for diamond wheels. However, there are fewer grades and different modifications, and no standard system has, at this time, been established.

A diamond grinding wheel for grinding chipbreakers in carbide tools could be:

Diamond	Grit	Grade	Concentration	Bond	Special Symbol	Section Depth
D	150	R	100	B	11	⅛
For Finishing Carbide Milling Cutters, Reamers, etc.						
SD	180	R	75	B	56	1/16

Today, many diamond and Borazon wheels use coated grits, referred to as "clad" or "armoured" grit. The coating of nickel, copper, or composites on each grain helps absorb the heat generated during grinding, improves the grit-bond adhesion, and holds the crystal in place so that it will fracture instead of pulling out of the wheel.

DRESSING AND TRUING GRINDING WHEELS

Dressing a grinding wheel means:

1. Removing metal or foreign matter which has lodged in and loaded (filled up) the pores of the wheel. Fine-grit wheels may load up quite quickly.
2. Removing dull grains which did not break off. The dull grains cause a wheel to become *glazed*. Such a wheel will *burn* the work, that is, raise the work temperature enough to cause discoloration and, sometimes, fine heat cracks.

Truing a grinding wheel may simply mean dressing it so that the face of a flat surface wheel is truly in line with the table traverse. It may, as discussed later, mean re-creating a special form or shape on the wheel.

A theoretically perfect match between the grinding wheel and the work material would not require dressing. The grains would become just dull enough to give a good finish and would then break off, creating fresh sharp points. Thus, it would only require adjustment for size as the wheel "breaks down."

The *frequency of dressing* will vary considerably. A few jobs work best with continuous dressing. Often a wheel is dressed after grinding one to four pieces. Some of the new wheels may last several hours between dressings.

Dressing tools are most frequently single-point

Fig. 14-9 (a) A single-point diamond tool for straight or form dressing. (b) Cluster-type diamond dressing tool for roughing and commercial finishes. (*Parsons Diamonds Products, Inc.*)

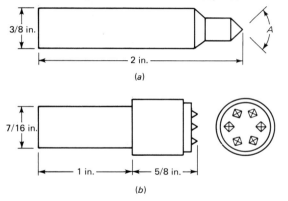

diamonds (Fig. 14-9) which are mounted in metal holders held on the grinding machine. These are traversed across the face of a straight wheel to dress and true it.

Crush forming and the use of templates to guide the diamond are discussed in Chap. 15.

SELECTING THE RIGHT GRINDING WHEEL

The Norton Company lists six factors to be considered when selecting a grinding wheel:

1. The material to be ground and its hardness
2. The amount of stock to be removed and the finish required
3. The operation, wet or dry
4. The wheel speed
5. The areas of grinding contact
6. The severity of the grinding operation

Briefly, the general recommendations for selecting grinding wheels for these six factors are as follows:

1. The material:
   Steel and alloy steel: use aluminum oxide abrasives.
   Cast iron, brass, aluminum, etc.: use silicon carbide abrasives.
   HSS tools: use aluminum oxide or Borazon.
   Carbide tools: use silicon carbide or diamond wheels.
   For hard materials: use finer grits so more grains are cutting at once and a softer grade so that the wheel will break down and stay sharp.
   For soft materials: use coarser grit so the wheel will not load up as quickly and the harder grades since they will stay sharp longer because of the lower force needed.
2. Stock removal rate and finish required:
   For rapid stock removal where finish is not important: use coarse grits.
   Fine grits give high finishes, though too fine a grit will slow the job. Excellent finishes can be achieved with 60 to 80 grit.
   For highest finishes: use resinoid, rubber, or shellac bonds.
3. Wet or dry operation:
   Most grinding works better if done wet. This also may permit the use of one grade harder wheel, giving longer wheel life. Soluble oils, sulfurized oil, and synthetic compounds are used to cool and clean the work and the wheel.
4. Wheel speed:
   This is limited by the strength of the bond. The

maximum safe rpm is marked on each grinding wheel, and this is often the best running speed.

Slower wheel speeds will cause the wheel to act as if it were harder.

5. Area of contact:

Large contact areas, such as those occurring in internal grinding, require the use of coarser, softer grades of grinding wheels.

Small contact areas, such as those occurring in OD grinding, can generally use finer and harder wheels.

6. Severity of the grinding operation:

A "severe" grinding operation is one where deep cuts at high feedrates are used (see "snagging" as one example).

At this point the friability of the various compositions of aluminum oxide, silicon carbide, and Borazon must be considered.

Severe conditions require tough (not friable) abrasives.

Fine finishing cuts can often be best done with more friable abrasive compositions.

All grinding wheel companies publish lists of recommended wheel types for dozens of jobs. One company, at the beginning of its list states, "These specifications are subject to slight or large modification with any change in the conditions."

Thus, trial and error, around a preliminary well-thought-out-first recommendation is still the only way to get optimum results in the choice of the grinding wheel.

SAFETY IN GRINDING

1. Never exceed the rpm marked on the grinding wheel or the wheel may "explode" (fly apart) from the centrifugal force.

2. Keep heavy wheel guards securely fastened in place in case the wheel breaks from any cause.

3. Be certain that the work is very secure before starting the wheel. Flying pieces can be dangerous.

4. Vitrified wheels can be "rung" to be certain there are no cracks.

5. Be sure to use the "blotters" on the sides of the wheel so the fastening flanges will tighten securely without cracking the wheel.

Chapter 15 will describe some of the machines which use the extremely wide variety of shapes, sizes, and compositions of grinding wheels which are available today.

Review Questions and Problems

14-1. Which three factors cause wear on a grinding wheel?

14-2. What are the principal abrasive materials used in the manufacture of grinding wheels, and what are the properties of each?

14-3. A grinding wheel is marked A54H8V. Give the full meaning of each letter and number in the marking.

14-4. How is the size of an abrasive grain measured, and what are the most used sizes?

14-5. Describe the principal bonds used in grinding wheels. Which one is used most often with aluminum oxide grit?

14-6. What is meant by
a. The "grade" of a grinding wheel?
b. The "structure" of a wheel?

14-7. What number or letter system is used to specify grade and structure?

14-8. What factors influence the selection of the abrasive used in a grinding wheel?

14-9. When are coarse grains used? When are fine grains used?

14-10. What determines the grain spacing a grinding wheel should have?

14-11. What factors determine the type of bond a grinding wheel should have?

14-12. What is meant by concentration as applied to diamond wheels? How is it shown in the numbering system? Give an example.

14-13. Four grinding wheels are available having the following specifications: (1) A46R12V, (2) C20Q9V, (3) C16G2V, (4) A54F2V.
a. Which of the four wheels listed above would you use to finish grind the OD of a piece of hard tool steel of 3-in. diameter?
b. Which wheel would you use to rough grind a similar piece of soft brass?
Note: Refer to the six factors listed.

14-14. What are truing and dressing of grinding wheels? How and why are they done?

14-15. List the safety precautions which should be followed when grinding.

14-16. What is the maximum rpm which can be used for a vitrified-bonded grinding wheel 12 in. (300 mm) in diameter if the limit for safe operation is 6500 fpm (1980 m/min)? Solve with both English and metric units.

References

1. *Grinding Ratios for Aerospace Alloys*, Air Force Machinability Data Center.

2. Hahn and Lindsay: "Principles of Grinding," *Machinery*, July–November 1971.

3. Krar, S. F., and Oswald: *Grinding Technology,* Delmar Publishers (Litton Educational Publishing, Inc.), New York, 1974.

4. *Lectures on Grinding,* 9th ed., Norton Grinding Wheel Division, 1961.

5. *Use, Care and Protection of Abrasive Wheels: Safety Code for*: B7.1-1970, American National Standards Institute, New York.

6. See also the B74 Series of Standards from American National Standards Institute.

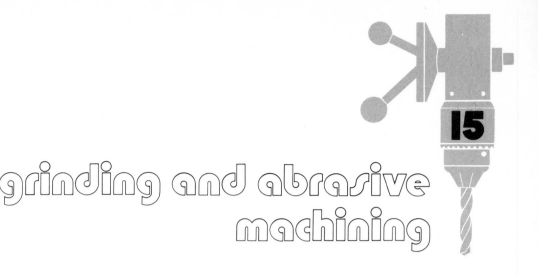

grinding and abrasive machining

The use of grinding is usually considered when one of the following four conditions is needed.

1. A surface finish of from 32 to 8 μin. (microinches) (1.0 to 0.2 μm) AA is required.

2. Size tolerances on flat or round work from 0.0005 to 0.000050 in. (0.013 to 0.0013 mm) are required.

3. "Flatness" must be held very closely, about 0.001 in. (0.025 mm) or less, especially over fairly large surfaces such as the ways of machine tools.

4. Work has been hardened so that any other method of machining is difficult or impossible.

Grinding is a rather slow way to remove stock; thus other methods are used to bring the work quite close to its final size and finish. An exception is *abrasive machining,* which is discussed later in this chapter.

GRINDING MACHINES

There are many different varieties of grinding machines, but those most used in industry are:

1. *Surface grinders*, reciprocating for flat workpieces

2. *Cylindrical (OD) grinder*, for finishing the outside of round workpieces

3. *Internal grinder*, for finishing the inside diameters of round holes

4. *Centerless grinders*, for high production grinding of round work on the OD

5. *Vertical-spindle surface grinders*, usually with rotary tables, occasionally with reciprocating tables, for flat surfaces in large size or large quantities

Other grinding processes which will be discussed in this chapter are:

6. *Abrasive cutoff wheels*, for cutting hardened metals or large-size stock

7. *Snagging*, very rough grinding of defects in slabs and stubs left on castings

8. *Abrasive machining*, the newest high production use of the grinding process

Surface Grinders (Reciprocating Table)

A *surface grinder* (Figs. 15-1 and 15-2) is basically a movable table with a horizontal-spindle grinding wheel mounted above it. These machines are made fully hand-operated, partly automatic, and fully automatic.

The size of a surface grinder is given as the width and length of the table on which the work is placed. The grinding wheels used are somewhat in proportion to the table sizes, as shown in Table 15-1.

The motions of the surface grinder are shown in Fig. 15-1.

The *grinding wheel* is mounted on a column which allows it to be raised and lowered at least 12 in. (300 mm) and more on the larger machines. The positioning handwheel is usually marked off in 0.0005-in. or 0.01-mm divisions so that accurate light cuts may be taken when close size control is required.

This vertical motion sets the *depth of cut*, which for average work is about 0.001 in. (0.025 mm), although it may be 0.003 in. (0.075 mm) for heavy cuts or 0.0002 in. (0.005 mm) for finish cuts.

TABLE 15-1 SURFACE GRINDING MACHINE DATA

	Table Size		Wheel Size*		Cost
	Width	Length	Diameter	Thickness	
Small	6 ×	18 in.	7 ×	½ in.	$1500 to
	(150 ×	450 mm)	(180 ×	12 mm)	$3500
Medium	24 ×	120 in.	20 ×	3 in.	$40,000
	(610 ×	3100 mm)	(500 ×	75 mm)	
Large	48 ×	240 in.	36 ×	12 in.	$95,000
	(1220 ×	6100 mm)	(900 ×	300 mm)	or more

*Diameter, thickness, and center hole size vary with the manufacture.

The *table* may be moved on the X axis, left and right. This is called the *longitudinal motion,* or *table traverse,* and is stated in feet per minute (fpm) or metres per minute (m/min). A typical 6 × 18 surface grinder has hydraulic traverse speeds of 5 to 100 fpm (1.5 to 30 m/min).

So many variables of material and wheel types enter into the grinding wheel's cutting action that trial and error (experience) is still the final answer in deciding on speed of traverse. Keep in mind that the work is more likely to "burn" at too slow a table traverse speed.

The amount of *infeed,* cross feed, or transverse feed per pass of the carriage is the width of cut, the amount the work is moved in or out between successive passes along the work. Surface grinders come equipped with automatic transverse feeds of 0.010 to 0.500 in. per pass (0.25 to 12.7 mm per pass). When hand feed is used, the cross feed is often $\frac{1}{16}$ to $\frac{1}{8}$ in. per pass (1.5 to 3 mm per pass).

Work holding is usually by means of a magnetic "chuck" or table (Fig. 15-3). These may be permanent or electromagnetic. If the work is small or nonmagnetic, properly sized steel plates are placed around the work to hold it securely. Sometimes the work can be held in a steel vise which is held by the magnetic

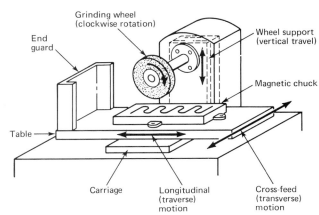

Fig. 15-1 The principal parts and motions of a plain surface grinder.

chuck. Magnetic chucks are made from 5 × 10 in. (125 × 250 mm) up to 42 × 96 in. (1070 × 2400 mm).

It is important that care be used in securing the work. The large vertical baffle at the left end of the table has stopped many a piece from flying into the shop at 60 mi/h (96 km/h).

Safety precautions include keeping the wheel almost completely enclosed, checking the wheel (as mentioned in Chap. 14), securing the work carefully, and wearing safety glasses.

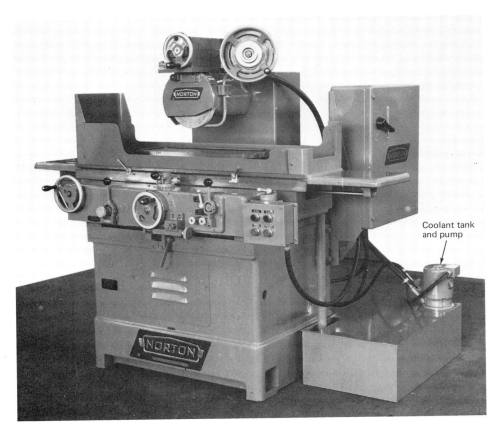

Fig. 15-2 A small (8 × 24 in., 200 × 600 mm) table surface grinder with automatic traverse and cross feed. [*The Warner & Swasey Co., Grinding Machine Division. (Formerly Norton Co.)*]

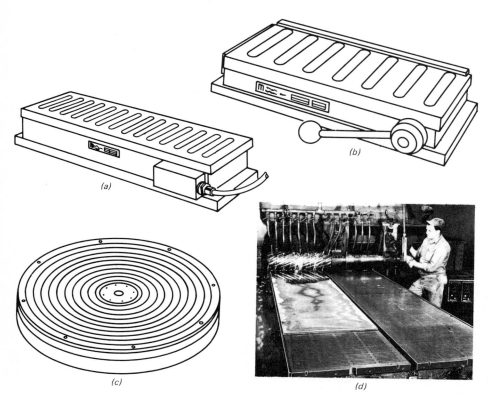

Fig. 15-3 Magnetic chucks. (*a*) Electromagnetic. (*b*) Permanent magnet. (*c*) Rotary electromagnet chucks available to 96-in. (2400-mm) diameter. (*d*) A very large electromagnet chuck used for a face milling job. Similar chucks are used on large grinding machines. (*Magna-Lock Corporation.*)

Cylindrical Grinding (OD)

A *cylindrical grinder* (Figs. 15-4 and 15-5) is used for grinding the outside diameters (OD) of a workpiece. The cuts it can make include straight, stepped, and forms of all kinds (see Fig. 15-6). Thread grinding can also be done on some cylindrical grinders. Tapers can also be ground.

The cylindrical grinder resembles a simplified lathe, with the grinding wheel at the back instead of a tool bit in front. These machines can be fully automated, fully hand-operated, or a combination of the two.

The *size* of a cylindrical grinder is given in the same way as the size of a lathe—by the diameter of the work which will clear the table (the "swing") and the length of the table. For example a cylindrical grinder might be 8 × 24 in., that is, 8 in. swing × 24 in. long table (200 × 600 mm). Sizes up to 14 × 72 in. (350 × 1800 mm) are fairly common. Cost is from $5000 to $40,000.

The motions of a cylindrical grinder are shown in Fig. 15-4. The *grinding wheel* can be raised and lowered on its supporting column. It should rotate downward tangent to the work. On some cylindrical grinders, the wheel head can be rotated so that plunge cuts can be made at an angle.

The *table* moves in both left and right (longitudinal) and in and out (transverse) directions just as the surface grinder does. Traverse and infeed dials and rates are also the same.

Work holding is done most frequently on centers. A dog fastened to the left end of the work is used to rotate the work. Figure 15-5 shows a typical automatic traverse, cylindrical grinding machine.

Plunge and traverse grinding can both be done on the cylindrical grinder. In *plunge* cutting, the work is fed into the wheel at a steady rate with no transverse (left and right) movement of the table. Thus, the wheel must be either wider than the work (Fig. 15-6*b*) or the exact width and shape of the cut as in Fig. 15-6*a*.

In *traverse* grinding, the wheel is narrower than the work, so, as in Fig. 15-6*d*, the table must be moved left and right (longitudinally) the same as in surface grinding. The depth of cut (infeed) for each pass should be about 0.0002 in. (0.005 mm) for finish and 0.001 to 0.003 in (0.25 to 0.75 mm) for roughing.

Wheel speeds are the same as for surface grinding and are dependent on the bonding agent used in making the wheel.

Work speed, in rpm, can make a difference in the speed of cutting and the finish. Work speeds vary from 25 to 70 fpm (7.6 to 21.3 m/min) depending on the type of material, whether it is a rough or finish

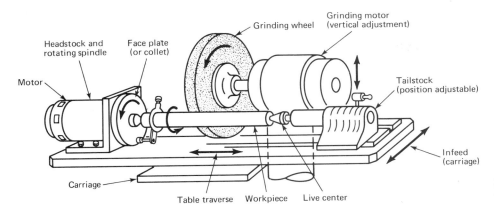

Fig. 15-4 The principal parts and motions of a cylindrical grinder.

cut, and the type of grinding wheel being used. A table of cutting speeds and suggested variations are given in some books and manufacturers' literature.

Traverse speed (longitudinal) is about the same as in surface grinding—5 to 100 ipm.

Universal cylindrical grinders are the same as the plain grinder except that there is an added section to the table which can be rotated about 10° so that tapers can be ground.

Form grinding is a method of plunge cutting an infinite variety of workpiece shapes by using a special cylindrical grinder and formed grinding wheels. Some of these shapes are shown in Fig. 15-7.

For form grinding, the grinding wheel must be shaped to the exact form needed. One way of doing this is to make a hardened steel or carbide **crushing roll** which is the reverse of the profile needed (Fig. 15-8).

This hardened wheel, when pressed by hydraulic force against the rotating grinding wheel, breaks off grains of the wheel until the wheel is the correct shape. The wheel is then plunged into the workpiece to grind the desired shape.

This same method is used for grinding threads of all sizes and diameters, using special crush forming rolls. The threads can be plunge ground or "pass-

Fig. 15-5 A 10 × 36 in. (250 × 900 mm) automatic cylindrical grinder. (*The Warner & Swasey Co., Grinding Machine Division.*)

Top Views

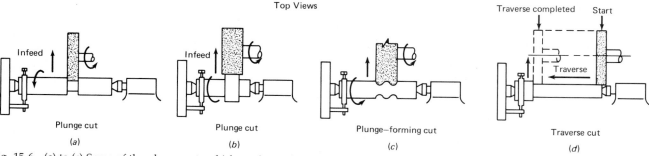

Fig. 15-6   (a) to (c) Some of the plunge cuts which can be made with a cylindrical grinder. (d) A traverse cut.

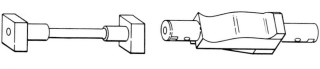

Test Specimen
Made on a cylindrical grinder
Material: 300 series stainless steel
Production: 9 to 10 per h, removing
about 8 in.^{3} of material
9 to 10 pieces per crushing
20 crushings per crushing roll regrind
30 regrinds per roll life

Cam -- from Preform
Made on a reciprocating surface
grinder
Material: steel, HRC 60
Production: 24 per hour at 12
parts per load
216 parts (18 loads) per crushing
6 crushings per roll regrind

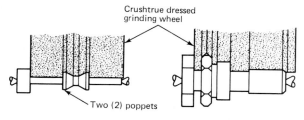

Other parts which are form ground.

Fig. 15-7   Some parts which have been ground to shape by crush-formed wheels. *(Bendix Corp., Automation and Measurement Division.)*

over" ground. The latter requires that the table be traversed past the wheel.

The crush rolls cost $100 to $500, can be reground several times, and will last from 10 to 50 crushings (average 25 crushings) per grind. Each crushing creates very sharp grained wheels which will grind from 20 pieces per dressing, if the cut is quite deep (0.180 in., 4.5 mm), up to 3500 parts per dressing on shallow (0.006 in., 0.150 mm) cuts.

The same process can be used to form flat parts using special surface grinders.

Accuracy is excellent. Tolerances on diameters within 0.0002 in. (0.005 mm) and on radii within 0.002 in. (0.05 mm) can be crush ground. *Surface finish* down to 8 μin. (0.2 μm) can be ground under ideal conditions.

The crushing rolls may also be made from tool steel

and then coated with a thin layer of diamonds or Borazon.

Another way of getting the desired shape in a grinding wheel is shown in Fig. 15-9. This equipment uses a template, usually ten times actual size, with a follower and linkage which causes a single-point diamond to cut the shape into the grinding wheel. This can be used for very small wheels and up to 20-in.-diameter (500-mm) and 2-in.-thick (50-mm) wheels.

Fig. 15-8   A typical Crushtrue form grinding setup. Part shown was ground over 0.200 in. (5 mm) deep in one plunge. *(Bendix Corp., Automation and Measurement Division.)*

Internal Grinding (ID)

Accurate sizing and finishing of the inside diameter of hardened workpieces or work requiring extreme accuracy of inside diameters is done on the *internal grinding machine*. Figure 15-10 shows the relative motions of the work and wheel in the most used method, called a *chucking-type* internal grinder, and Fig. 15-11 shows a typical machine. The work done may vary from small [¼ in. (6.3 mm)] holes to holes 12 in. (300 mm) or larger, though most work is in the 1- to 4-in. (25- to 100-mm) range. The length of cut is preferably not over three or four times the diameter, though longer holes can be ground.

The *wheels* are usually softer than those used in surface or cylindrical grinding since they contact a large surface area, as shown in Fig. 15-12. As the grinding wheels are often quite small, they must run at very high rpm in order to cut efficiently. For example, a ¼-in. (6.35-mm) wheel to cut at 5000 fpm (1520 m/min) should rotate at 80,000 rpm.

The *size* of an internal grinder is the diameter of work which it can hold and the length of the stroke it can produce. They cost from $12,000 to $40,000, with large machines to $125,000.

The *motions* of the parts of the chucking internal grinder are as follows. The grinding wheel is mounted on a metal shaft called a quill or mandrel

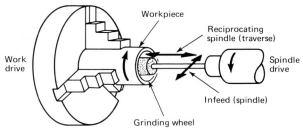

Fig. 15-10   The principal components and movements of an internal grinding machine.

which *rotates* at high speed against the rotation of the work. The wheel also reciprocates in and out over the depth of the hole and is *fed outward* 0.0002 to 0.0015 inch per stroke (0.005 to 0.038 millimetres per stroke) to achieve the required diameter. The workpiece is held in a collet, chuck, or special fixture which rotates at 100 to 200 fpm.

CAUTION: The quill or mandrel can be as small as ⅛ in. (3.2 mm) in diameter and the pressure of grinding will deflect the wheel, due to bending of the quill. This makes a tapered hole.

The cure may be to have the grinding wheel traverse a few times without infeeding. This is called *sparking out or sizing.* It may also be helpful to use a tungsten carbide mandrel which will deflect

Fig. 15-9   Single-point diamond and template method of dressing grinding wheels for form grinding. *(Engis Corp.)*

only one-third as much as a steel shaft under the same load.

Some of the latest machines have built-in sensors to control infeed by the pressure on the quill. This *adaptive control* is used on other machines also.

Centerless Grinding (OD)

When cylindrical work is to be ground but no centers may be drilled in the ends or when long pieces of round stock must be ground to size, then centerless grinding is used. Accuracy to ±0.001 in. (0.025 mm) is frequently held. These machines, once they are set up, will grind many long lengths of stock or hundreds of parts, with very little attention except for feeding the parts, dressing the wheel, and adjusting the size settings. In some jobs, feeding can be done automatically.

Centerless grinding machines are made to handle diameters from ¼ in. (6.4 mm) to 8 in. (200 mm) and sometimes larger. The grinding wheel is often 20 in.

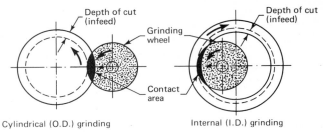

Fig. 15-12 Comparison of the area of contact between wheel and work during cylindrical and internal grinding.

(500 mm) in diameter and may be from 2 to 8 in. (50 to 200 mm) wide. The regulating wheel may have a 12-in. (300-mm) diameter and be from 3 to 8 in. (75 to 200 mm) wide. The grinding wheel is often standard vitrified bonded, and the regulating wheel is usually rubber bonded.

A typical centerless grinder is shown in Fig. 15-13. The schematic diagram in Fig. 15-14 shows the arrangement and motions of the parts for grinding any length of straight bars of stock.

Fig. 15-11   A universal internal grinder for straight or tapered ID and some OD grinding. *(Bryant Grinder Corporation, a subsidiary of Ex-Cell-O Corp.)*

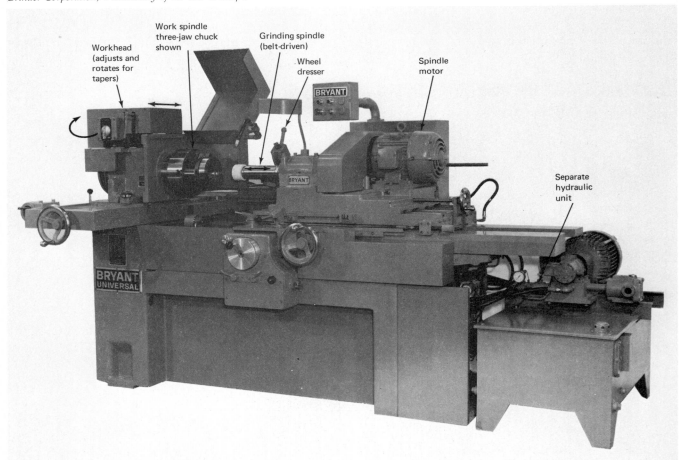

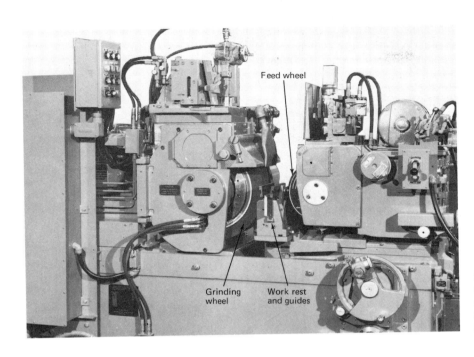

Fig. 15-13 A typical centerless grinder. Supports for holding finished work must be added at the front of the machine. *(The Warner & Swasey Co., Grinding Machine Division.)*

The **grinding wheel** simply rotates so that it is moving downward as it touches the work. It rotates at about 5500 fpm (1700 m/min).

The **regulating wheel** rotates at 50 to 200 fpm (15 to 60 m/min) and is set at an angle as shown in Fig. 15-14. This angle is usually adjusted to from 0° (for infeed work) to 8° in order to feed the work past the grinding wheel at the desired rate. The rate of feed can be computed from the equation

$$f = CN \sin a$$

where f = feedrate of work, in./min
 C = circumference of regulating wheel, in.
 N = rpm of regulating wheel
 a = angle of inclination of regulating wheel

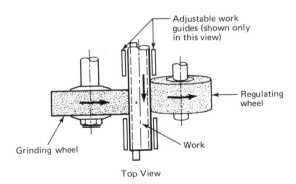

Adjustable work guides (shown only in this view)

Regulating wheel

Grinding wheel

Work

Top View

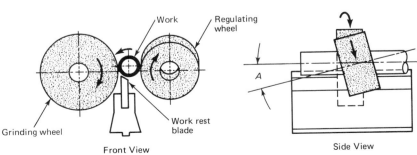

Work Regulating wheel

Grinding wheel

Work rest blade

Front View

A

Side View

A = regulating wheel angle of inclination

Fig. 15-14 Three views of the arrangement and motions of the principal parts of a centerless grinder.

TABLE 15-2 VERTICAL-SPINDLE ROTARY-TABLE GRINDER DATA

	Size (Table Diameter)	Grinding Wheel Diameter	Table Motor hp	Wheel Motor hp	Cost*
Small	16 in. (400 mm)	11 in. (280 mm)	1 hp (0.75 kW)	15 hp (11 kW)	$18,000 $38,000
Average	30–48 in. (760–122 mm)	18–32 in. (450–800 mm)	5–10 hp (3.8–7.5 kW)	30–75 hp (22–56 kW)	$26,000 $54,000
Large	96–120 in. (2400–3000 mm)	48–60 in. (1200–1500 mm)	20–30 hp (15–22 kW)	150–300 hp (112–225 kW)	$125,000 $200,000

*Cost depends on options purchased with the machine, such as automatic gaging or automatic cycle controls.

The regulating wheel is also moved toward and away from the grinding wheel to regulate the diameter of the finished work.

The *work support* is made of hardened tool steel or carbide and must be as long as the grinding wheel. This support is set so that the centerline of the work is slightly above center and is sloped at an angle of 30° for most work. It must be exactly parallel to the face of the grinding wheel and is adjusted to accommodate varying diameters.

Infeed centerless grinding is shown in Fig. 15-15. This is used whenever a taper or shoulder on the workpiece makes through grinding impractical.

Vertical-Spindle Rotary-Table Surface Grinders

A vertical-spindle grinding machine cuts with the *side* of the wheel. However, the wheels (Fig. 15-16) are hollow or in segments, so the side is only from 1.5 to 4 in. (38 to 100 mm) thick.

The vertical-spindle grinder is sometimes used with a reciprocating table similar to that used on the surface grinder. However, the most widely used style of machine is the one shown schematically in Fig. 15-17 and illustrated in Fig. 15-18. It uses a rotating table for holding the work.

The size of these grinders is specified by the diame-

Fig. 15-15   A typical example of infeed centerless grinding.

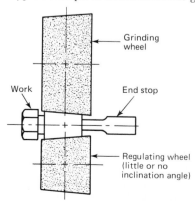

ter of the rotating magnetic table. Basic range of data is as shown in Table 15-2.

Two of the many types of work which can be done on these vertical-spindle grinders with rotating tables are shown in Fig. 15-19. Figure 15-19a illustrates how a large number of small parts may be ground at one time. They are shown in the loading position, before the chuck is moved under the wheel.

Figure 15-19b shows a few larger parts being ground. The magnetic chuck is in the grinding position. The machine can also handle small, thin parts or large single castings up to 60 in. (1500 mm) high and up to 10,000 to 60,000 lb [4500 to 27,200 kg (mass)]. Finishes of 32 μin. (1.0 μm) and better can be achieved.

The *motions* of the machine are simple as shown in Fig. 15-18. The grinding wheel spindle rotates at from 300 to 1200 rpm and is mounted to the motor on the *grinding head*, which can be raised and lowered, either at rapid traverse or in down-feed from 0.004 to 0.080 ipm (0.1 to 2.0 mm/min).

Work holding is by the magnetic chuck (work table). This is similar to the magnetic chuck used on the surface grinder except that it is circular. It rotates at from 3 to 40 rpm. It is moved into the clear for loading and under the wheel for grinding. The "pull" (amperage) of the chuck can be varied so that a heavy pull can be used to hold parts for roughing and a lighter pull can be used during finishing to avoid distortion of odd-shaped parts. Nonmagnetic materials are held by blocking with steel pieces or in a steel holding fixture.

Accuracy of these machines is excellent. Tolerances to ±0.005 in. (0.125 mm) are easily held. With some extra care ±0.001 in. (0.025 mm) and even, on some machines, ±0.0005 in. (0.0125 mm) can be held. Coolant is usually required. This cools the work and also washes away the chips, which is important, as these grinders can remove stock very rapidly.

Abrasive Cutoff Saws

Quite often hardened or very tough metals, tile, concrete, or brick must be cut. Steel saws either cannot

Fig. 15-16 Some solid and segmented shapes of wheels for vertical-spindle grinding. *(Cone-Blanchard Machine Company.)*

cut these at all or cut very slowly. This is where cutoff saws using aluminum oxide or silicon carbide wheels become a necessity.

One type of cutoff saw is shown in Fig. 15-20. These saws may be small laboratory models, using 7-in.-diameter (180-mm) wheels, or large saws for cutting big pipe, using wheels of 20-in. (500-mm) diameter or larger.

The grinding "wheels" are usually resinoid bonded, rubber bonded, or a combination of the two.

Fig. 15-17 Outline of the principal parts and motions of a vertical-spindle, rotary-table grinding machine.

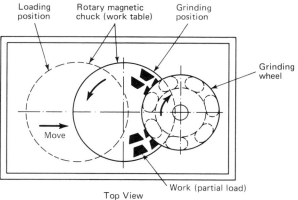

Top View

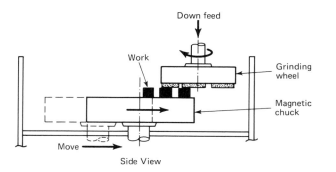

Side View

As they are only from $\frac{1}{16}$ in. (1.6 mm) to $\frac{1}{8}$ in. (3.2 mm) thick, these "saw blades" break fairly easily. To minimize breaking, these wheels may be purchased with glass fiber and other reinforcements molded into the wheel.

These "sawing" wheels are run at quite high speeds of 9000 to 16,000 fpm (2700 to 4900 m/min), though the higher speeds can be used only if bearings and protective devices are adequate.

Either *wet* or *dry* cutting can be done. The rubber-bonded wheel, used wet, can give an excellent finish.

The *wheels* are usually quite hard grade, P through T, of medium density (5 to 8), and 36 to 70 grain size. The selection available from the manufacturers is limited, as the conditions of use are quite standardized.

Snagging

Castings of all kinds usually need to have the stubs of the sprues ground smooth, and often the parting line is rough. Steel slabs must have slag pockets removed; welds may require grinding smooth. This type of rough "off-hand" grinding is referred to as *snagging*.

Most of this type of grinding is done with portable, hand-held machines. One type is shown in Fig. 15-21. These are either air-powered or use high-speed electric motors.

The wheels are usually straight or flaring cup type and are from 4 to 9 in. (100 to 230 mm) in diameter. They may be made of either aluminum oxide, silicon carbide, or one of the newer variations of either, running at 6500 to 9500 fpm (2000 to 2900 m/min). The grain sizes are coarse, 14 to 20 grit, and resinoid or rubber bonded. They are hard, Q to S grade.

Similar work is done in a *swing grinder*, which uses 14- to 24-in.-diameter (350- to 600-mm) wheels mounted in a frame which can be "swung" by the operator so that it snags out bad spots in metal ingots.

Abrasive Machining

Also called *high efficiency grinding*, abrasive machining refers to the use of grinding equipment to get fast, high volume removal of material in competition with milling, planing, and turning.

Several things have happened since 1963 which make this idea very practical.

1. Grinding machines are now made more rigid, stronger, with higher horsepower and higher speeds.

2. Grinding wheels have been developed so that some wheels can now safely be run at up to 16,000 fpm (4900 m/min). Harder and tougher abrasives have also been developed.

3. The "discovery" that if speed and feed were increased *drastically* beyond what was being used for rough and finish cuts, the condition shown in Fig.

15-22 made high volume metal removal possible. Sometimes over 30 in.³/min (490 cm³/min) can be removed.

Note in Fig. 15-22 that as the infeed was increased, the part "burned," that is, it turned brown or purple due to the high temperature produced.

However, *increasing* the infeed from 0.015 in. (0.38 mm) per pass to 0.020 in. (0.5 mm) caused the burn to disappear. Further increasing, to 0.025 in. (0.63 mm), infeed actually *decreased* the horsepower needed and the work was not burned.

What happened, in brief, is that the higher pressures, and greater power used, caused the grinding wheel to be *self-sharpening*, that is, the grains were breaking off before they became dull enough to burn

the work. This leaves new, sharp grinding edges which continue cutting.

Advocates of abrasive machining say that if the part is *finished* with abrasives, it can be economically *machined* with abrasives. There are many instances where abrasive machining has proven more economical than milling or planing. This is especially true on flat or cylindrical surfaces on parts made of cast iron, steel, or some of the hard-to-machine alloys.

Some *advantages* of abrasive machining are:

1. Scale and hard spots do not bother a grinding wheel.

2. Interrupted cuts often help a grinding wheel to be self-dressing.

3. Handling time is reduced. Often a rough cut

Fig. 15-18   A medium-sized, vertical-spindle surface grinder with rotating magnetic table. Table shown in loading position. *(Cone-Blanchard Machine Company.)*

No. 22

Fig. 15-19   (a) Over 100 small steel parts being surface ground in one loading. (b) Long, heavy bars being surface ground on a vertical-spindle grinder. (*Cone-Blanchard Machine Company.*)

Fig. 15-20   An abrasive cutoff saw with chop (straight down) stroke and drive rolls to rotate the work. Handles work 5- to 12-in. (125- to 300-mm) diameter. (*Allison-Campbell Division, ACCO.*)

can be made, then the wheel "fine dressed," and an excellent finish obtained in a few additional passes—all without removing the work from its original chucking.

4. Distortion and residual stresses are usually less than those created by milling, turning, or planing.

Some *disadvantages* of abrasive machining are:

1. The grinding wheel wears much faster. In one test, at 26-ipm feed, wheel wear equaled depth of metal removed. However, harder wheels can be used, which decreases wear.

2. About four times as much power per cubic inch of stock removed is needed as compared with milling.

3. The grinding machine must be stiff enough to handle the heavy loads without vibration. Thus, many older grinders cannot take advantage of the heavy feedrates required.

4. A special coolant nozzle (which is inexpensive) must be used on high-speed wheels. Moreover,

handling the chips, or swarf, can be a problem because of the large amount of both coolant and chips.

Actually many jobs using crush forming and many jobs on vertical-spindle, rotating table grinders have been attaining high metal-removal rates for years. However, new grinding machines and grinding wheels, and the knowledge gained from research have made abrasive machining, or high efficiency machining, much more common in industry. The reason? Because on some flat and cylindrical work (even some centerless grinding) it can produce a better product less expensively.

PROCESS CONSIDERATIONS IN GRINDING

Because of the relatively low metal-removal rate of the grinding process, only 0.005 to 0.030 in. (0.125 to 0.75 mm) of material should be left to be ground off in either surface or cylindrical grinding.

If work is to be heat-treated, extra stock may be left on to allow for any distortion which may occur. If the work has been carburized or nitride hardened, care

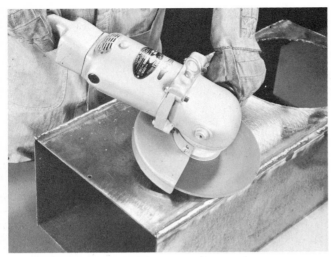

Fig. 15-21   A hand-held, 2-hp electric-motor-driven off-hand grinder being used to smooth a weldment. (*Sioux Tools, Inc.*)

must be taken not to grind off the rather shallow depth of heat treatment.

Grinding cannot make sharp inside corners because the edges of a grinding wheel wear quite rapidly. Thus grooves must be cut, usually in a lathe, before grinding.

Do not specify any smoother finish than is absolutely necessary. Finishes less than 32 μin. (1.0 μm) can add up to 50 percent to the cost of grinding.

AUTOMATION IN GRINDING

Most types of grinding machines can be purchased with hydraulic, electrical, or numerical controls to automate the rapid and feed motions and the table motion.

Automatic in-process gaging and visual in-process gaging of diameters in cylindrical grinding and thick-

Fig. 15-22   Graph of feed to power relationship in surface grinding and its effect on the workpiece.

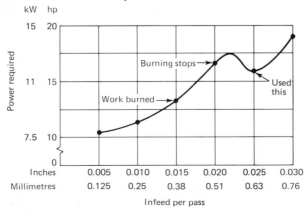

ness in surface grinding are being done on many machines today.

Especially on cylindrical and internal grinding, automatic feeding and unloading devices are being successfully used throughout industry today. These may have been designed by either the machine builder or the user. They are usually made for a specific high-volume workpiece or "family" of similar workpieces.

Much of the full or partial automation is not very expensive and should be considered when purchasing a grinding machine.

Review Questions and Problems

15-1.   Under what four conditions is the use of grinding usually considered?

15-2.   Name the five types of grinding machines most used in industry.

15-3.   Discuss the work which can be done with a surface grinder. Consider size of work, types of cuts, and stock removal.

15-4.   How is the workpiece held on the table of a surface grinder?

15-5.   What is the purpose of the large vertical baffle at the left end of the table?

15-6.   What safety precautions should be followed when using the surface grinder?

15-7.   Compare the cylindrical (OD) grinder with an engine lathe, considering machine motions available and types of work done on each.

15-8.   Differentiate between plunge and traverse grinding.

15-9.   How does a universal cylindrical grinder differ from a plain cylindrical grinder?

15-10.   How is form grinding done? How is the grinding wheel dressed? What dimensional accuracy can be held?

15-11.   Describe internal grinding. What precautions can be taken to reduce taper?

15-12.   a.   What work can be done on a centerless grinder which cannot be done on a lathe?
        b.   What work might possibly be done on either machine?

15-13.   How do vertical-spindle surface grinders differ from plain surface grinders? What types of work are they best suited for?

15-14.   How is the work held on the table? What accuracy are they capable of holding?

15-15.   What kinds of cuts can an abrasive cutoff saw make that cannot be done with a saw using a blade? What cuts might be done on either type of "saw"?

15-16.   What is snagging and where is it used?

15-17. What is abrasive machining? What three factors have made abrasive machining practical?

15-18. What are the advantages and disadvantages of abrasive machining?

15-19. What are some of the considerations to keep in mind when specifying grinding?

References

1. Fisher, R.C.: "High Wheel Speed Is Not Enough," *Manufacturing Engineering and Management*, February 1970.

2. Haggett, J. E., and R. L. Smith: "How to Find Profit in Abrasive Machining," *Steel*, Sept. 16, 1963.

3. Krar, S. F., and Oswald: *Grinding Technology*, Delmar Publishers (Litton Educational Publishing, Inc.), New York, 1974.

4. *Milling—Grinding Confrontation*, Society of Manufacturing Engineers, Detroit, Mich., Tech. Paper MR67-538.

5. Merritt, R. H.: "Cutting Force, Key to a Systems Approach," *Abrasive Engineering*, November 1968.

6. Navarro, N. P.: "How to Analyze Abrasive Wheel Costs," *Manufacturing Engineering and Management*, October 1971.

finishing methods

"Finishing" in this chapter refers to the final machining operations on certain types of work. These operations are to produce precise size, shape, and finish or to remove burrs and sharp edges on flat, round, or irregularly shaped parts.

These machining processes are quite specialized in their application, and not all work requires their use. However, some knowledge of them is needed as every step in producing usable products is important.

HONING

Machining a hole to within less than 0.001 in. (0.025 mm) in diameter and maintaining true roundness and straightness with finishes under 20 μin. (0.50 μm) is one of the more difficult jobs in manufacturing.

Finish boring or internal grinding *may* do the job, but spindle deflection, variations in hardness of the material, and difficulties in precise work holding make the work slow and the results uncertain, as shown in Fig. 16-1. *Honing,* because it uses rectangular grinding stones instead of circular grinding wheels (as shown in Fig. 16-2) can correct these irregularities.

Advantages of Honing

Honing can consistently produce finishes as fine as 4 μin. (microinches) (0.1 μm) and even finer finishes are possible. It can remove as little as 0.0001 in. (0.0025 mm) of stock or as much as 0.125 in. (3.18 mm) of stock. However, usually only 0.002 to 0.020 in. (0.05 to 0.50 mm) stock is left on the diameter for honing.

As the workpiece need not be rotated by power, there are no chucks, faceplates, or rotating tables needed; thus there are no chucking or locating errors. The hone is driven from a central shaft, so bending of the shaft cannot cause tapered holes as it does when boring. The result is a truly round hole, with no taper or high or low spots, provided that the previous operations left enough stock so that the hone can "clean up" all the irregularities.

Honing uses a large contact area at slow speed compared with grinding or fine boring, which use a small contact area at high speed. Because of the combined rotating and reciprocating motion used, a cross-hatched pattern is created which is excellent for holding lubrication. Diameters with 0.001 to 0.0001 in. (0.025 to 0.0025 mm) and closer accuracies can be repeatedly obtained in production work. Closer tolerances will, of course, take somewhat longer.

Honing can be done on most materials from aluminum or brass to hardened steel. Carbides, ceramics, and glass can be honed by using diamond stones similar to diamond wheels.

Disadvantages of Honing

Honing is thought of as a slow process. However, new machines and stones have shortened honing times considerably. Horizontal honing may create oval holes unless the work is rotated or supported. If the workpiece is thin, even hand pressure may cause a slightly oval hole.

Fig. 16-1 Conditions left by previous operations which can be corrected by honing. *(Sunnen Products Co.)*

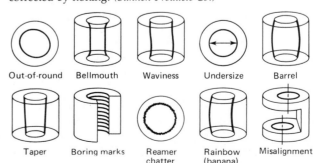

Out-of-round   Bellmouth   Waviness   Undersize   Barrel

Taper   Boring marks   Reamer chatter   Rainbow (banana)   Misalignment

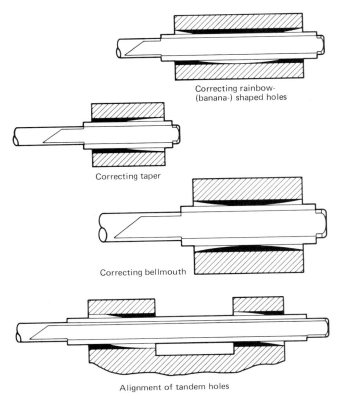

Fig. 16-2   How honing corrects some of the undesirable conditions. (*Sunnen Products Co.*)

(a)

CAUTION: Honing *cannot* change the location of the hole or correct a sloped (not perpendicular to a surface) condition of a hole.

Honing Machines

For most work, honing machines are quite simple (Figs. 16-3 and 16-4). They are relatively inexpensive, costing from about $1000 to $10,000 depending on size and complexity. The most used honing machines are made for machining internal diameters from 0.060 in. (1.50 mm) to 6 in. (150 mm). However, large honing machines are made for diameters up to 48 in. (1200 mm). Larger machines are sometimes made for special jobs.

The *length of the hole* which can be honed may be anything from $\frac{1}{2}$ in. (12 mm) to 6 or 8 in. (150 to 200 mm) on smaller machines and up to 24 in. (600 mm) on larger machines. Special honing machines are made which will handle hole lengths up to 144 in. (3600 mm) (Fig. 16-4).

Horizontal-spindle honing machines, for hand-held work with bores up to 6 in. (150 mm) (Fig. 16-3*b*), are among the most widely used. The machine rotates the hone at from 100 to 250 fpm (30 to 76 m/min).

The machine operator moves the work back and

(b)

Fig. 16-3   (*a*) Horizontal honing machine with power-stroking attachment. Foot-pedal pressure control. (*b*) The same machine equipped for hand honing. Note work support and built-in sizing gages. (*Sunnen Products Co.*)

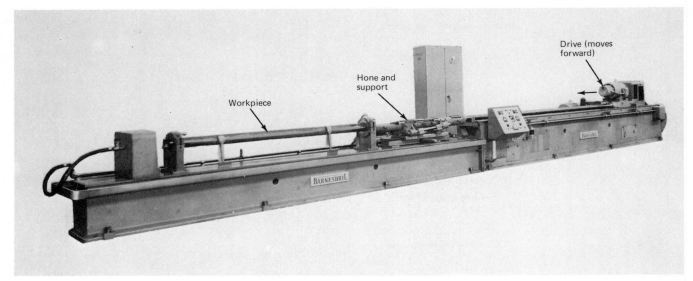

Fig. 16-4   An unusually large horizontal honing machine for 105-mm (4.14-in.) gun barrel. Stroke is 144 in. (3660 mm). *(Barnes Drill Co.)*

forth (strokes it) over the rotating hone. The operator must "float" the work, that is, not press it against the hone or the hole will be slightly oval. Sometimes the workpiece must be rotated.

Horizontal-spindle honing machines are also made with "power stroking" (Fig. 16-3*a*). In these, the work is held in a self-aligning fixture and the speed and length of the stroke are regulated by controls on the machine.

As a hone is being used, it is *expanded* by hydraulic or mechanical means until the desired hole diameter is achieved. Various mechanical and electrical devices can be attached to the honing machine to control the rate of expansion, and stop it when final size is reached.

On the simplest hand-held machines, the operator may check the bore size with an air gage, continue honing, recheck, etc., until the size is correct.

Vertical-spindle honing machines are used especially for larger, heavier work (Fig. 16-5). These all have power stroking at speeds from 20 to 120 fpm (6 to 36 m/min). The length of the stroke is also machine controlled by stops set up by the operator.

Vertical honing machines are also made with multiple spindles so that several holes may be machined at once, as in automobile cylinders, Fig. 16-5*b*.

The *hone body* is made in several styles (Fig. 16-6) using a single stone for small holes, and two to eight stones as sizes get larger. The *stones* come in a wide variety of sizes and shapes, a few of which are shown in Fig. 16-7. Frequently there are hardened metal guides between the stones to help start the hone cutting in a straight line.

The *material* for the honing stones is the same as

that used for grinding wheels: aluminum oxide, silicon carbide, or diamond. The grades and bonds are similar, and the most used grain sizes are from 150 to 600 grit.

A *fluid* must be used with honing. This has several purposes: to clean the small chips from the stones and the workpiece, to cool the work and the hone, and to lubricate the cutting action.

A fine mesh filtering system must be used, since recirculated metal can spoil the finish.

The *time* for honing is sometimes less than for grinding. Reports mention fairly heavy stock removal with fine finishes:

1.   Removing 0.030 in. (0.75 mm) of carbide on a $1\frac{1}{4}$-in.-long (31-mm) bore in 8 min to an 8-μin. (0.2-μm) finish and within 0.000050-in. (0.00025-mm) roundness tolerance

2.   Removing 0.020 in. (0.50 mm) from a 2-in. (50-mm) long stainless-steel part in 6 min

3.   Honing a 20-in.-long (500-mm) part in 45 min to a 1.5-μin. (0.038-μm) finish

Of course, the finer finishes take extra time, and sometimes both a rough and a finishing operation are required.

LAPPING

The principal use of the lapping process is to obtain surfaces which are truly flat and smooth. Lapping is also used to finish flat and round work (such as precision plug gages) to tolerances of 0.0005 to 0.000020 in. (0.0125 to 0.0005 mm).

Work which is to be lapped should be previously finished close to the final size. While rough lapping can remove considerable metal, it is customary to leave only 0.0005 to 0.005 in. (0.0125 to 0.125 mm) of stock to be removed.

Lapping, though it is an abrasive process, differs from grinding or honing because it uses a "loose" abrasive instead of bonded abrasives like grinding wheels.

The abrasive may be silicon carbide, aluminum oxide, diamonds, boron carbide, or crushed natural stones such as garnet, quartz, or rouge.

These abrasives are often purchased "ready mixed" in a "vehicle" often made with an oil-soap or grease base. These vehicles hold the abrasive in suspension before and during use. The paste abrasives are generally used in hand-lapping operations. For machine lapping, a light oil is mixed with dry abrasive so that it can be pumped onto the lapping surface during the lapping operation.

Lapping Machines

These machines (Fig. 16-8) are fairly simple pieces of equipment consisting of a rotating table, called a *lapping plate*, and three or four *conditioning rings*. Standard machines have lapping plates from 12- to 48-in. (300- to 1200-mm) diameter. Large machines up to 144 in. (3660 mm) are made. These tables are run by $\frac{1}{4}$- to 20-hp (0.2- to 15-kW) motors.

The lapping plate is most frequently made of high-quality soft cast iron, though some are made of copper or other soft metals. This plate must be kept "perfectly" flat. The work is held in the conditioning rings. These rings rotate as shown in Fig. 16-9. This rotation performs two jobs. First, it "conditions" the

Fig. 16-5   (a) Vertical honing machine with up to 6-in. (150-mm) capacity and automatic size control. *(Sunnen Products Co.)* (b) A multispindle vertical hone for cylinder blocks. *(Barnes Drill Co.)*

(a)

(b)

(a)

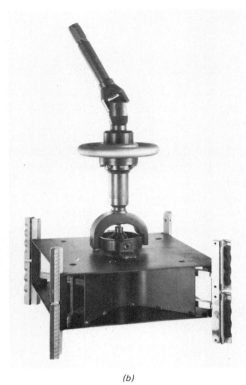
(b)

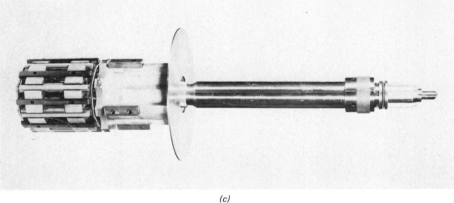
(c)

Fig. 16-6 (a) Several sizes of hone bodies, stones not shown. (b) Remote-feed giant hones, made 10- to 60-in. (250- to 1500-mm) diameter. (Sunnen Products Co.) (c) Multiple stone hone with bearing bar spacers. (Barnes Drill Co.)

plate, that is, it distributes the wear so that the lapping plate stays flat for a longer time. Secondly, it holds the workpiece in place. The speed at which the plate turns is determined by the job being done. In doing very critical parts, 10 to 15 rpm is used, and when polishing, up to 150 rpm is used.

A *pressure* of about 3 psi (20.7 kN/m²) *must be applied* to the workpieces. Sometimes their own weight is sufficient. If not, a round, heavy pressure plate is placed in the conditioning ring. The larger machines use pneumatic or hydraulic lifts to place and remove the pressure plates.

The workpieces must be at least as hard as the lapping plate, or the abrasive will be charged into the

work. It will take from 1 to 15 or 20 min to complete the machining cycle. Time depends on the amount of stock removed, the abrasive used, and the quality required. Some of the types of work done by lapping are shown in Fig. 16-10.

CAUTION: Flatness, surface finish, and a polished surface are not necessarily achieved at the same time or in equal quality. For example, silicon carbide compound will cut fast and give good surface finish, but will always leave a "frosty" or matte finish.

The *grits used* for lapping may occasionally be as

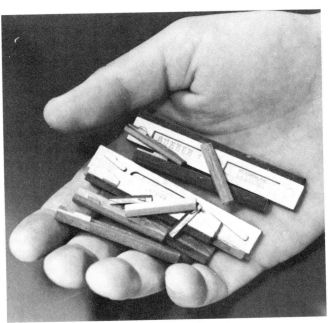

Fig. 16-7   A few of the styles of stones which are used in hones. *(Sunnen Products Co.)*

coarse as 100 to 280 mesh (170 to 40 μm). More often, the "flour" sizes of 320 to 800 mesh (30 to 6 μm) are used. The grits, mixed in a slurry, are flowed onto the plate to replace worn-out grit as the machining is in process.

Advantages and Disadvantages

Advantages of lapping are:

1.   Any material, hard or soft, can be lapped, and any shape, as long as the surface is flat.

2.   There is no warping, since the parts are not clamped and very little heat is generated.

3.   No burrs are created. In fact, the process removes light burrs.

4.   Any size, from $\frac{1}{2}$- to 36-in. (12- to 900-mm) diameter, and from a few thousandths thick up to any height the machines will handle can be lapped. Figure 16-10 shows some of the work which is finished by lapping.

Disadvantages are mainly that lapping is still somewhat of an art. There are so many variables that start-

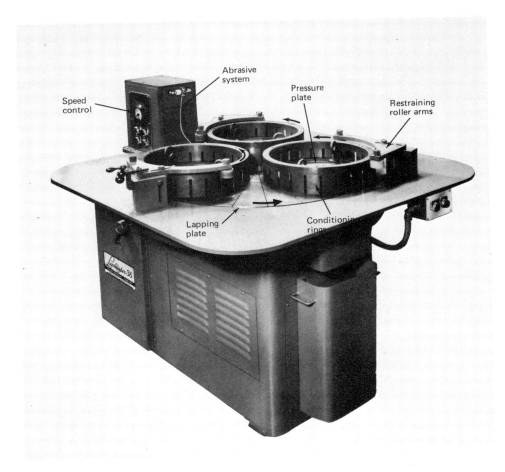

Fig. 16-8   A 36-in. (900-mm) plate lapping machine. Can lap 2240 $\frac{1}{2}$-in.-diameter (13-mm) parts or 52 3-in. (75-mm) parts per load. *(Lapmaster Division of Crane Packing Co.)*

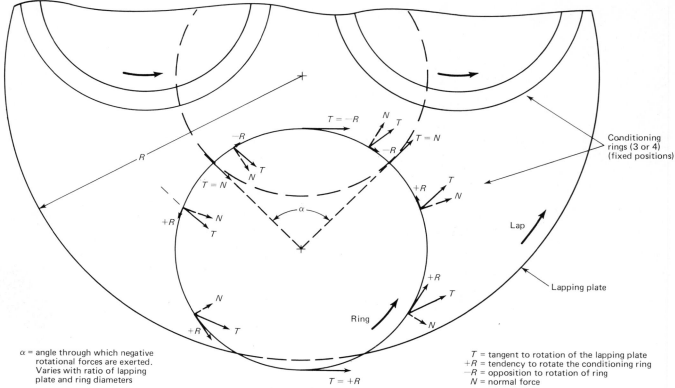

α = angle through which negative
rotational forces are exerted.
Varies with ratio of lapping
plate and ring diameters

T = tangent to rotation of the lapping plate
$+R$ = tendency to rotate the conditioning ring
$-R$ = opposition to rotation of ring
N = normal force

Fig. 16-9   Diagram of the forces which cause the conditioning rings to rotate on the lapping plate.

ing a new job requires experience and skill. There are excellent general recommendations and assistance from the manufacturers, and past experience is an excellent guide, but trial and error still may be needed to get the optimum results.

Free-Abrasive Machining (FAM)

Machines for free-abrasive machining use a hardened steel "lap" plate and are produced by one company. This hard plate forces the abrasive particles to roll freely over the wheel surface, thus presenting the maximum number of new, sharp cutting edges to the workpiece.

This process uses a pressure pad, pressure plates, and conditioning rings in much the same way as the lapping process. Flatness and accuracy are said to be the same as lapping, and somewhat faster cutting may be achieved.

BUFFING

Buffing is basically a polishing operation. Surprisingly it requires a machine with considerable power. Buffing machines (Fig. 16-11a) use 3 to 40 hp (2.2 to 30 kW).

The surface to be buffed may be a die casting or stamping which does not require any preparatory treatment except cleaning. A machined part (such as cams or shafts) may require roughing (smooth grinding) and/or *prepolishing* on a smooth abrasive belt backed by rubber-faced contact wheels.

The buffing machine (usually called a *buffing lathe*) has one or usually two arbors which are 1¼- to 2-in. (32- to 50-mm) diameter. Two flanges are mounted on the arbor to hold the 3- to 6-in.-thick (75- to 150-mm) buffing wheel.

Buffing wheels are made of cloth (cotton, wool, flannel, or muslin) or fiber firmly sewed together in ¼- to ½ in.-thick (6- to 13-mm) sections (Fig. 16-11b). These sections are closely pressed together by the flanges on the lathe arbor. These wheels or *buffs* are usually 12 to 30 in. (300 to 760 mm) in diameter.

The **abrasive** usually has almost no cutting ability. It may be red rouge (ferric oxide) or green rouge (chromium oxide) for soft metals, or alumina, and similar compounds. These abrasives are made up (in tubes) with a heavy grease or wax base.

The buffing wheel is "charged" by pressing the grease stick against the cloth wheel until the wheel is thoroughly coated. This also protects the wheel from tearing when the part is pressed against it.

Buffing is usually a *hand operation*, though it is occa-

Fig. 16-10 Some of the variety of parts which are lapped. They include aircraft, automobile, camera, and pump parts. (*Lapmaster Division of Crane Packing Co.*)

sionally automated. The operator presses the part against the charged buff with considerable pressure. This heavy pressure gives fast action (time is money) and also heats up the work. The temperature may, on some types of work, rise to 200 or 300°F (93 to 149°C), which will not usually warp the work but does require careful handling. The buffing wheel turns at 1000 to 3000 rpm.

The *size of the work* is changed very little, if any. Sometimes a 0.0001- to 0.0003-in. (0.0025- to 0.0075-mm) change will occur. Basically, buffing *flows* or smears the surface metal. It takes the peaks (asperities) and flows them into the valleys. This gives very low surface finish readings and a mirror-like polish.

CAUTION: Buffing does *not* maintain flatness or roundness. It is used *only* to obtain very smooth refective surfaces.

BARREL FINISHING (MASS FINISHING)

Industry has moved a long way ahead of the old-style tumbling barrel filled with parts, gravel, and water in an attempt to remove burrs by some method which was faster and more reliable than hand deburring.

This mass finishing process is used for low-cost, fast work on relatively large numbers of parts. It is used:

1. To remove the burrs left by the machining or stamping process.
2. To eliminate tool marks left by machining.
3. To remove "flash" left around die castings.

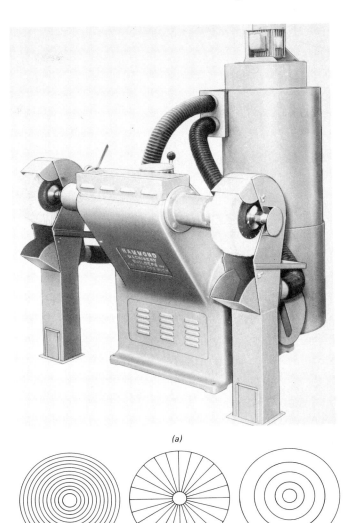

(a)

Concentric sewed Radial (spoke) sewed Leather center (concentric sewed)

(b)

Fig. 16-11 (*a*) A 7½-hp (5.6-kW) buffing lathe with attached dust collector. (*Hammond Machinery Builders, Inc.*) (*b*) Some of the types of buffing wheels. (*Divine Bros. Co. and Schaffner Mfg. Company.*)

4. To remove scale on castings (though more often this is sand-blasted off).
5. To remove sharp edges, that is, to radius these edges for safety in use and assembly.
6. To provide good finish, sometimes referred to as *burnishing*, and to give good "color," that is, to bring out the natural color of the material.
7. To prepare, by the above, parts for painting or plating.
8. Sometimes heavy steel balls are used to "peen" a steel part to increase its strength and to help eliminate surface cracks.

Many times several of the above objectives are achieved at the same time. Sometimes the deburring and burnishing are done as two separate operations in the same machine but with different "media" or stones.

Tumbling Barrel

The tumbling barrel (Fig. 16-12) is still in use today, either the tilting type or the horizontal rotating type.

The load in tumbling barrels, or the vibratory barrels which will be discussed next, consists of three elements:

1. Workpieces
2. Media
3. Compound

The **workpieces** may be of any size and shape, limited only by the size of the equipment. Small, thin parts, complex sand castings, stampings, and die castings can all be barrel finished. Any *material* (steel, aluminum, brass, ceramics, even molded rubber) can be handled.

The **media** or the "stones," which do most of the finishing job are today made in several materials and in numerous shapes. Natural stones are seldom used because the newer molded shapes and bonded stones can accomplish real precision finishing with excellent repeatability.

Figure 16-13 shows a few of the shapes which are

Fig. 16-13   Some of the "media" shapes which can be used in all types of barrel finishing. *(Norton Co.)*

available. These come in a variety of sizes from $\frac{1}{8}$ to $1\frac{7}{8}$ in. (3.2 to 47.6 mm) on a side and $\frac{3}{32}$ to $\frac{5}{8}$ in. (2.4 to 15.9 mm) thick. Hardened steel and abrasive balls are made in sizes from about $\frac{3}{16}$- to $\frac{3}{4}$-in. (4.8- to 19-mm) diameter for some work.

Materials used may be aluminum oxide, silicon carbide, ceramics, quartz, even carpet tacks and ground nutshells. The abrasives are bonded with various resins, with different grits and compositions, thus making available a very wide assortment of media.

The **compound** may be abrasive or nonabrasive. It is a liquid which may be acid, alkaline, or neutral. This compound "cushions" the action between the parts and the media. It may also wash the parts, add rust inhibitors, and flush away the "fines" (small particles of burrs or media).

The *speed of rotation* of the barrel is important. Figure 16-14 shows how the *turnover point* varies with

Fig. 16-12   An oblique tumbling barrel with drainage piping. Can be used to dry and degrease as well as tumble the parts. *(U.S. Baird Corp.)*

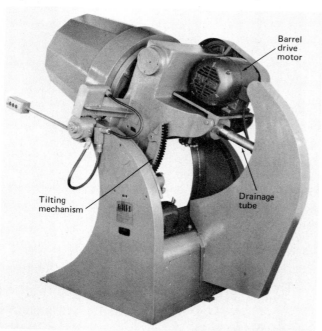

Fig. 16-14   The effect of rotational speeds on barrel finishing work. *(Norton Co.)*

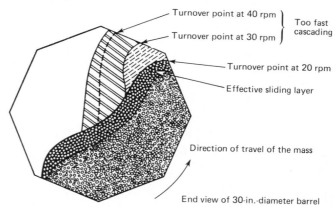

Fig. 16-15 Samples of work which has been deburred or finished by tumbling or vibratory finishing. *(Norton Co.)*

speed. This point is important, as the burring action will not take place unless the work *slides* down through the media. Moreover, if the work *drops* from the high point, it may be damaged by the impact. Cycle times to complete a load will vary from 1 to 18

h. Some of the work which has been tumbled is shown in Fig. 16-15.

Vibratory Finishing

A newer method of mass finishing is vibratory finishing. The vibrating "barrel" may be rectangular (Fig. 16-16) or doughnut shaped (Fig. 16-17). These machines are more expensive, but they will do the job many times faster than barrel finishing. They are made in sizes from 4 to 100 ft³ (85 to 2800 litres).

The frequency and amplitude of the vibrations can be changed to produce fast cut (high frequency, high amplitude), or with slower and smaller motions, a high luster can be obtained.

Because of the much shorter processing time, and the ability to "feed through" the work continuously on some machines, these vibratory machines are replacing the tumbling barrel in many shops. Cycle times will be from 10 min to 2 h.

The media and compounds used are the same as described for tumbling. They wear down in size more rapidly so that it is necessary to recharge the media more often, but this is a minor disadvantage.

SAND BLASTING

To remove the scale formed by casting or heat-treating, to blend in grinding marks, to remove corrosion from used parts, or to blast off burned-on carbon from automobile cylinders, sand blasting is often a

Fig. 16-16 (*a*) A dual shaft, batch-type vibratory finishing machine. (*b*) The deburring or finishing action of the machine. *(Materials Cleaning Systems Division, Wheelabrator-Frye Inc.)*

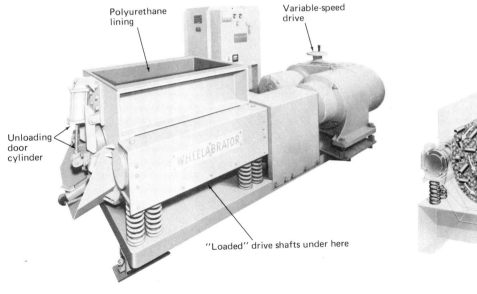

(a)

(b)

Fig. 16-17   Rotary-shaped vibratory finishing barrel with automatic part and media separation, and continuous feed. (*Roto-Finish Co.*)

quick, inexpensive method. It is also used for deburring, creating a satin finish, and preparing surfaces for painting or plating.

Sand blasting consists of bombarding one or several workpieces with sand, other abrasives, or steel shot. The abrasive is forced onto the work by air pressure or centrifugal force, as described later.

The sand-blast *cabinets* may be only 24 in. (600 mm) square, or large enough to hold 10-ft (3000-mm) castings. Typical cabinets are shown in Fig. 16-18.

The work may be steadied by the operator's hand

so that a stream of abrasive can be directed on the material through a hose. The heavy rubber arms and gloves shown in Fig. 16-18 protect the operator. Sometimes several workpieces are loaded on a rotating table inside the cabinet, and the operator directs the stream on one after the other. The nozzles have ceramic or carbide linings.

For automation, one or several blasting nozzles are fixed in position so that the abrasive strikes on parts moving through the cabinet on a conveyor.

The velocity may be achieved by suction (venturi tube) or by a higher velocity, *pressure blast system*. Very high velocities up to 300 ft/s (90 m/s) are obtained by centrifugal "slingers" as shown in Fig. 16-19. One or more of these centrifugal units can be positioned to handle a continuous run of parts as shown in Fig. 16-19b.

The abrasive media may be *micro-grit* or larger glass beads, clean sand of various sizes, aluminum oxide grains, or steel balls. The steel balls are most often used for peening. Most work is done dry, but wet blasting is sometimes used for finer work.

There are numerous sizes, and many variations on the mass finishing methods described here. However, understanding these basic ideas will enable the student to easily grasp the detailed knowledge needed to actually use or specify this type of equipment.

Review Questions and Problems

16-1.   Why are "finishing" operations necessary?
16-2.   What are the principal uses for honing?
16-3.   Describe vertical and horizontal honing and the advantages of each.

Fig. 16-18   (*a*) Inside view of hand cabinet for blast cleaning a number of parts. Note foot-pedal blast control, turntable to hold several parts, blast gun, and heavy-duty gloves. (*b*) A medium-sized, two-station, feed-through blasting cabinet. (*c*) Airblast car-type room for large, heavy workpieces. Operator wears protective clothing and directs blast hose by hand. (*Peenamatic Division, Metal Improvement Company.*)

(a)

(b)

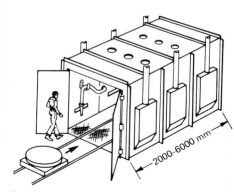

(c)

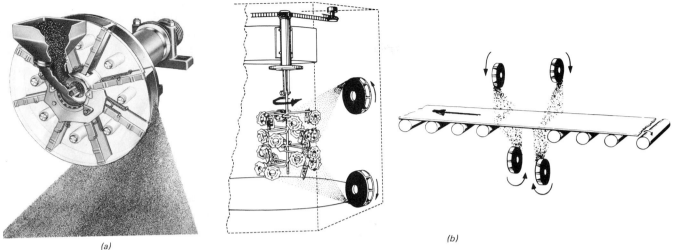

(a)

(b)

Fig. 16-19 *(a)* The centrifugal "Wheelabrator" or slinger unit used for grit blasting or peening workpieces. Used often for continuous processing of plate, rod, and parts. *(Materials Cleaning Systems Division, Wheelabrator-Frye Inc.)* *(b)* Positioning of centrifugal units to handle a continuous run.

16-4. How is the size of the hole controlled in the honing operation?

16-5. What materials are used to make the honing stones?

16-6. Why must a fluid be used in honing?

16-7. What are the advantages of honing as compared with grinding?

16-8. What is the principal use of and the advantages of lapping?

16-9. How much material is removed in lapping?

16-10. How does lapping differ from grinding?

16-11. Describe the principal parts of a lapping machine and their uses.

16-12. What size grits are used for lapping?

16-13. What are the disadvantages of lapping?

16-14. Describe a buffing machine and its principal uses.

16-15. What are buffing wheels made of? How and with what are they "charged"?

16-16. How much material is removed in a buffing operation?

16-17. Describe barrel finishing and some of its uses.

16-18. What does the load in the tumbling barrel consist of?

16-19. What are the limitations on the size and material of the workpiece which can be tumbled?

16-20. What materials are used for the media or stones in a tumbling barrel or vibratory finisher?

16-21. How much time does the barrel finishing operation require per load? How does this compare with vibratory finishing?

16-22. What is the advantage of vibratory finishing?

16-23. What is sand blasting used for?

16-24. In what ways is the force developed for sand blasting?

16-25. What abrasive materials are used for sand blasting?

16-26. You are producing an aluminum die casting which must have the burrs (called *flash*) removed and some other projections smoothed off. They are about $4 \times 8 \times 2$ in. ($100 \times 200 \times 50$ mm) in size. How would you do this if production is:

    a. 1000 pieces per year

    b. 25,000 pieces per year

    c. 200,000 pieces per year?

Explain your reasons for selecting each process.

References

1. Abrasive Grain Association, 2130 Keith Building, Cleveland, Ohio, 44115.

2. *Grinding and Lapping Compounds*, United States Products Co., Pittsburgh, Pa. (eight-page article).

3. *Handbook of Processing Techniques for More Effective Precision Vibratory Finishing*, Materials Cleaning Div., Wheelabrator-Frye Inc., Mishawake, Ind., Bulletin No. 624.

4. Kelso, Thomas D.: "Power Stroking—A New Image for Honing," *Manufacturing Engineering and Management*, January 1971.

5. *Tumblex Finishing*, Abrasive Materials Div., Norton Company, Worcester, Mass., 1970.

6. "What's Happening in Hand-Held Honing," *Grinding and Finishing*, April 1968.

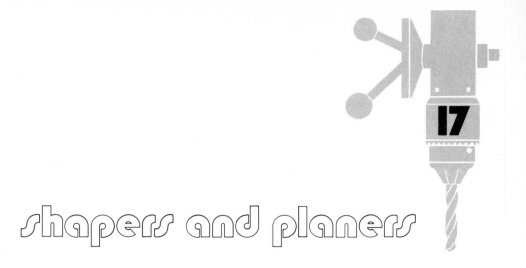

shapers and planers

Both the shaper and the planer cut only in straight lines, and they both make the same types of cuts. Their principal differences are:

1. The shaper handles relatively small work. The planer handles work weighing up to several tons.

2. The cutting stroke of the shaper is made by moving the tool bit attached to the ram. The cutting stroke of the planer is achieved by moving the work past a stationary tool bit.

The *types of cuts* which can be made with either machine are shown in Fig. 17-1. Contouring cuts can be roughed out on a shaper by a skillful machinist moving the table and crossrail before each stroke, or by using a tracer attachment to the shaper.

Both the shaper and the planer usually cut only in one direction; thus, the return stroke is lost time. However, the return stroke is made at up to twice the speed of the cutting stroke. The planer can, for simple work, be tooled for *double cutting*, that is, cutting on the return stroke, but this is not done very often.

THE SHAPER

The shaper is a relatively simple machine. It is used fairly often in the toolroom or for machining one or two pieces for prototype work. Setup time is often only 30 min. Tooling is simple, and shapers do not always require operator attention while cutting. The horizontal shaper is the commonest type, and its principal parts and motions are shown in Fig. 17-2.

The **ram** slides back and forth in dovetail or square ways to transmit power to the cutter. The starting point and the length of the stroke can be adjusted.

The **toolhead** is fastened to the ram on a circular plate so that it can be rotated for making angular cuts. The toolhead can also be moved up or down by its hand crank for precise depth adjustments.

Attached to the toolhead is the toolholding section. This has a *tool post* very similar to that used on the engine lathe. The block holding the tool post can be rotated a few degrees so that the cutter may be properly positioned in the cut.

The **clapper box** is needed because the cutter drags over the work on the return stroke. The clapper box is hinged so that the cutting tool will not dig in. Often this clapper box is automatically raised by mechanical, air, or hydraulic action.

The **table** is moved left and right, usually by hand, to position the work under the cutter when setting up. Then, either by hand or more often automatically, the table is moved sideways to feed the work under the cutter at the end or beginning of each stroke.

The **saddle** moves up and down (Y axis), usually manually, to set the rough position of the depth of cut. Final depth can be set by the hand crank on the tool head.

The **column** supports the ram and the rails for the saddle. The mechanism for moving the ram and table is housed inside the column.

The size of a shaper is the maximum length of stroke which it can take. Horizontal "push"-type shapers are most often made with strokes from 16 to 24 in. (400 to 600 mm) long, though some smaller and larger sizes are available. These shapers use from 2- to 5-hp (1.5- to 3.7-kW) motors to drive the head and the automatic feed.

The *maximum width* which can be cut depends on the available movement of the table. Most shapers have a width capacity equal to or greater than the length of stroke. The maximum *vertical height* available is about 12 to 15 in. (200 to 380 mm).

Work holding is frequently done in a vise. As shown in Fig. 17-3, the vise is specially made for use on shapers and has long ways which allow the jaws to open up to 14 in. (355 mm) or more, thus quite large work can be held. The vise may also have a swivel base so that cuts may be made at an angle.

Work which, due to size or shape, cannot be held in the vise is clamped directly to the shaper table in much the same way as parts are secured on milling machine tables.

Toolholders are the same as the ones used on an engine lathe, though often larger in size. The cutter is sharpened with rake and clearance angles similar to lathe tools, though the angles are smaller because the work surface is usually flat. These cutters are fastened into the toolholder, just as in the lathe, but in a vertical plane.

Two Types of Drives

mechanical drive

The less expensive shaper, thus the one most often purchased, uses a *mechanical drive* (Fig. 17-3). This drive uses a crank mechanism (Fig. 17-4). The *bull gear* is driven by a pinion which is connected to the

Fig. 17-1 The principal types of cuts which can be made with shapers and planers.

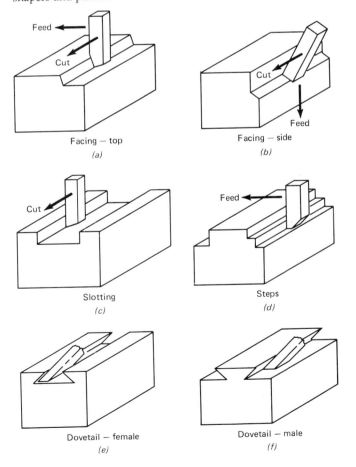

Facing — top
(a)

Facing — side
(b)

Slotting
(c)

Steps
(d)

Dovetail — female
(e)

Dovetail — male
(f)

motor shaft through a gear box with four, eight, or more speeds available. The rpm of the bull gear thus becomes the *strokes per minute* (sometimes abbreviated spm) of the shaper.

The *cutting speed* of the tool across the work will vary during the stroke as shown by the velocity diagram in Fig. 17-4. The maximum is at the center of the stroke. However, if the cutting speed chosen is somewhat on the slow side, the *average* speed may be used, and computations are greatly simplified.

Although the ratio varies somewhat, several shapers have a linkage which uses 220° of the cycle for the cutting stroke and 140° for the return stroke. This is close to a 3:2 ratio. Thus, about three-fifths (60 percent) of a complete cycle is used for cutting.

In setting up a mechanically operated shaper, the length of cut (in inches or millimetres) is known, and the cutting speed in feet per minute or metres per minute is selected according to the kind of metal being cut. It is then necessary to compute the strokes per minute since that is how the shaper speed is controlled. The *working* distance traveled by the cutter in 1 min is

$$\text{ipm or mm/min} = \text{length of cut (in. or mm)} \times \text{strokes per minute} \qquad (17\text{-}1)$$

However, the cutting speed is in *feet* per minute, so the length of cut L must be divided by 12. Moreover, as noted above, the shaper is cutting only three-fifths of the time, so the strokes per minute must be divided by $\frac{3}{5}$ or multiplied by $\frac{5}{3}$. Equation (17-1) now becomes

$$\text{fpm} = \left(\frac{L}{12}\right)(\text{spm}) \div \frac{3}{5}$$

$$= \frac{(L)(\text{spm})}{7} \qquad \text{approximately}$$

$$(17\text{-}2)$$

Solving for strokes per minute:

$$\text{spm} = \frac{(7)(\text{fpm})}{L} \qquad \text{for a 220° cutting cycle}$$

$$(17\text{-}3)$$

Note: L = actual length of cut plus an allowance at the end of the cut for chip clearance and at the beginning of the cut to allow the clapper box to drop down. Using $\frac{1}{2}$ in. (13 mm) for each is usually enough.

This approximate equation is sufficiently accurate especially considering the fact that most shapers have only six to eight different speeds available, in the range from 10 to 150 strokes per minute.

A general equation which can be used with any

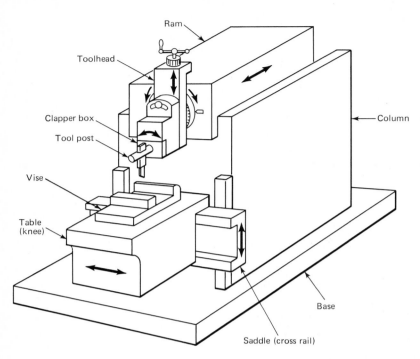

Fig. 17-2 Motions and principal parts of a shaper.

shaper or planer when the ratio of cutting to return speed is known is as follows.

With t = time (min) for cutting stroke only
  R = ratio of return stroke speed to cutting stroke speed
  A = fpm (average) for cutting stroke (or m/min)
  N = complete strokes per minute = spm = $1/T$
  T = time for: cut + return stroke = min/ stroke = $1/N$
  L = length of cut, in. (or mm)

$$T = t + \frac{t}{R} \quad \text{and} \quad t = \frac{L/12}{A}$$

or metric $t = \dfrac{L/1000}{A}$

So,

$$T = \frac{L/12}{A} + \frac{L/12}{AR} = \frac{L}{12A}\left(1 + \frac{1}{R}\right) = \text{min/stroke}$$

$$= \frac{1}{N} \tag{17-4}$$

Inverting

$$N = \frac{12A}{L + L/R} \quad \text{or metric } N = \frac{1000A}{L + L/R} \tag{17-5}$$

EXAMPLE 17-1
Find the number of strokes per minute to make a

15-in.-long cut in cast iron at 60 fpm. The exact $R =$ 220/140 = 1.57; use Eq. (17-5).

$$N = \frac{12(60)}{16 + 16/1.57} = \frac{720}{26.2} = 27.5 \text{ spm (use 27)}$$

In *metric units*:

$$N = \frac{1000(18)}{405 + 405/1.57} = \frac{18\ 000}{663} = 27.1 \text{ spm}$$

If the spm (N) is known, the *average* cutting speed is:

$$fpm = \frac{N}{12}\left(L + \frac{L}{R}\right) \tag{17-6}$$

or metric, $\text{m/min} = \dfrac{N}{1000}\left(L + \dfrac{L}{R}\right)$ (17-7)

The strokes per minute available on a shaper will vary according to the size of the shaper. The larger shapers will have lower speeds. A 16-in. (400-mm) shaper may have speeds of 27 to 150 strokes per minute, while a 24-in. (600-mm) shaper will have 10 to 90 strokes per minute speeds available.

Feed per stroke on a shaper is comparable to the feed per revolution on a lathe. Coarse feeds for roughing range up to 0.100 inch per stroke (sometimes abbreviated as ips) or 2.5 millimetres per stroke, and finish cuts to 0.005 to 0.015 inch per stroke (0.125 to 0.375 millimetre per stroke). Finish would also depend on the nose radius of the cutting tool.

Depth of cut can be quite heavy: roughing cuts up to 0.250 in. (6.35 mm) deep are not unusual.

Fig. 17-3 A horizontal push-type mechanical shaper, 22-in. (560-mm) stroke, with universal vise for work holding. [*Austin Industrial Corp. (J & N Shafer, Sweden).*]

hydraulic shaper

The hydraulic shaper (Fig. 17-5*a*) has the same major parts as the mechanical one. However, the ram is driven by a hydraulic cylinder as shown in the simplified sketch in Fig. 17-5*b*. These shapers use 5- to 10-hp (3.7- to 7.5-kW) motors.

The *cutting speed* of the hydraulic shaper is infinitely variable by means of hydraulic controls, as is the cross feed. The reverse stroke is made faster than the power stroke because of the smaller area in the return side of the cylinder (due to the presence of the piston rod) if a constant volume pump is used. Another method is to have the rate of fluid flow increased to speed up the return stroke. The return speed is often twice the cutting speed, though this is variable.

Speed and feed on a hydraulic shaper are controlled by simple dials. Speed is read directly in feet per minute or metres per minute. Feed is read directly in decimal inches or millimetres. The cutting speed remains nearly constant through the full stroke.

The **vertical shaper**, sometimes called a *slotter*, has a vertical ram, with table and saddle similar to the horizontal shaper. If a rotary table is mounted on the regular table, a number of slots can be made at quite accurately spaced intervals. This machine can work either outside or inside a part, provided that the interior opening is larger than the tool head.

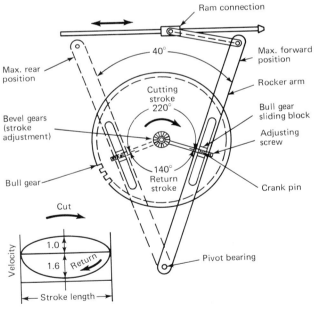

Fig. 17-4 Sliding block drive mechanism for mechanical shapers, typical angles shown, and proportional velocity diagram.

comparison of mechanical and hydraulic shapers

The mechanical shaper is less expensive, and the flywheel effect of the bull gear furnishes a sort of

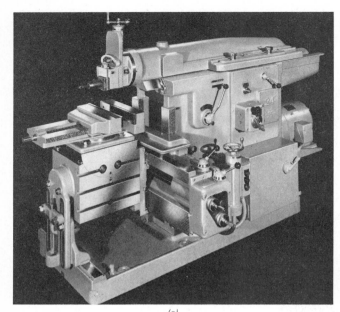

(a)

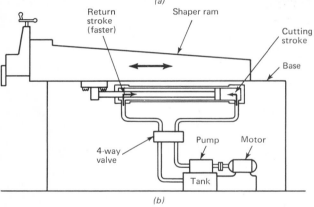

(b)

Fig. 17-5 (*a*) A 24-in.-stroke (610-mm) hydraulic, push-type shaper. (*b*) Simplified layout of the hydraulic circuit. (*Rockford Machine Tool Co., Subsidiary of Greenlee Bros. & Co.*)

reserve power for heavy cuts and hard spots. Thus, some shops prefer them for heavy work.

The hydraulic shaper has constant cutting speeds and fast returns, so that it can make more strokes per minute at the same surface speed. It is easier to adjust for ram stroke and position, and it has infinitely variable speeds and feeds.

The final choice would depend on the available funds, and the type of work to be done. The hydraulic shaper might possibly be chosen for somewhat faster speed and finer finish if these are important. Shapers cost from $7000 to $20,000 depending on size and equipment.

Time to Make Cuts

Calculating the time to face a flat surface, either horizontal or vertical, requires only simple calculations.

EXAMPLE 17-2

If the 10-in.-long cast-iron part previously mentioned is 8 in. wide, the cross feed for roughing is 0.050 inches per stroke, and the cut is 0.100 in. deep, then

Number of strokes needed

$$= \frac{8.000 \text{ (width of work)}}{0.050 \text{ (feed per stroke)}} = 160 \text{ strokes}$$

Previously calculated strokes per minute = 27

$$\text{Time} = \frac{\text{no. of strokes}}{\text{strokes per minute}} = \frac{160}{27} = 5.92 \text{ min}$$

In the *metric system*, the width would be 200 mm, and feed 1.25 millimetres per stroke.

$$\text{Number of strokes needed} = \frac{200}{1.25}$$

$$= 160 \text{ strokes}$$

$$\text{Time} = \frac{160}{27} = 5.92 \text{ min}$$

THE PLANER

A planer like the one shown in Fig. 17-6 makes the same types of cuts as a shaper. However, it is a production-type machine for certain types of work.

1. It can machine any flat or angular surface, including grooves and slots, in medium- and large-sized workpieces. Typical work would be machine beds and columns, marine diesel engine blocks, and bending plates for sheet metal work. These parts are usually large iron castings or steel weldments and may weigh a few hundred pounds or several tons.

2. Two to six identical smaller parts (small being about 3-ft or 900-mm cubes) can be machined at the same time by aligning them on the long planer bed.

3. As shown in Fig. 17-7, up to four different tools may be cutting at the same time. This can save hours of production time, even though it takes longer to set up and requires a lot of horsepower.

The *size of planers* is often spoken of as a 30-in. (760-mm) planer or a 60-in. (1500-mm), etc., planer. This specifies the approximate width of the table, which ranges from 30 to 72 in. (760 to 1800 mm). A more complete specification is:

Width of table × height under rail × length of table

Fig. 17-6   A medium-sized planer equipped with two railheads and two sideheads. *(The G. A. Gray Company.)*

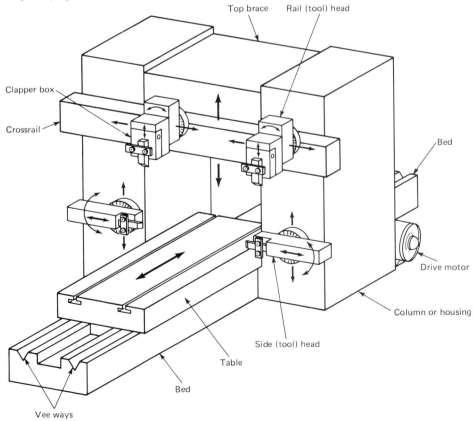

Fig. 17-7   The motions and principal parts of a double-housing planer.

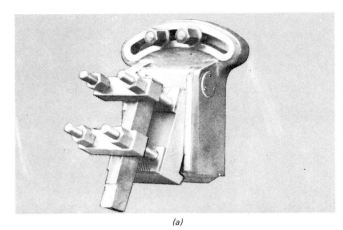

(a)

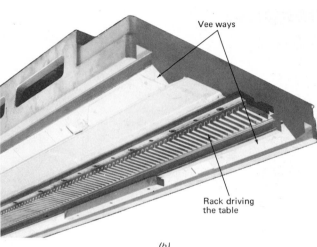

(b)

Fig. 17-8   (*a*) Clapper box and toolholder. (*b*) Bottom view of table, showing vee ways and driving rack for a mechanical planer. *(The G. A. Gray Company.)*

EXAMPLE

48 in. × 48 in. × 14 ft or in *metric*, 1200 × 1200 × 4200 mm

Notice that the width and height are usually, though not always, the same. Table length is often made to order and may be as short as 8 ft (2400 mm) or as long as 20 ft (6000 mm) or more. The drive may be 15 hp (11 kW) on the smaller planers, and 100 hp (75 kW) or more on the larger models.

In spite of their large size, planers can hold tolerances of ±0.005 in. (0.125 mm) and, with care, even closer. The larger planers may weigh 100,000 lb (45,000 kg) and cost over $100,000. Planers may be purchased with metric lead screws and dials at no extra cost.

Construction of the Planer

The most frequently used type of planer is the *double-housing planer* shown in Fig. 17-6 and schematically in Fig. 17-7.

The **frame** is the same hollow-box type used on large milling machines and also, as will be shown later, on large vertical boring machines.

This frame is basically two heavy columns fastened together at the top with a large bracing section and fastened at the bottom to the machine bed. This creates a very strong, rigid structure which will handle heavy loads without deflection.

The **crossrail** is also a heavy box, or similar construction. It slides up and down on vee or flat ways, controlled by hand or by power-operated screws. These crossrails are so heavy that they are counterweighted, with either cast-iron weights or hydraulic cylinders, in order that they may be moved easily and positioned accurately. After being positioned, they are clamped in place.

The two **railheads** can be moved left or right across the crossrail, each controlled by a separate lead screw, which can be turned by hand but usually by power feed.

The railheads can be rotated, and vertically adjusted for depth of cut the same as the shaper heads. They also have a clapper box (often with power lift) like the shaper's (Fig. 17-8*a*).

The **sideheads** (either one or two) are independently moved up or down by hand or by power feed and can also be rotated and moved in or out for depth of cut.

The **table** is a heavy casting which carries the work past the cutting heads. It runs on vee or flat ways. The table is driven either by a very long hydraulic cylinder (like the hydraulic shaper) or by a pinion gear driving a rack which is fastened under the center of the table as shown in Fig. 17-8*b*. The motor driving the pinion gear is the reversible type with variable speed.

The **bed** of the planer must be a weldment or casting twice as long as the table. Thus, a 12-ft (3660-mm) table requires a 24-ft (7300-mm) bed. The gearing or hydraulic cylinder for driving the table is housed under the bed.

A large planer is set on a concrete foundation, up to 6 ft (1800 mm) deep, containing tons of concrete. Leveling bolts are supplied with the machine and, using a laser or other accurate equipment, the entire bed is leveled before the table is set on it. This may take a few hours or a couple of days.

Another frequently used machine is the **open-side**

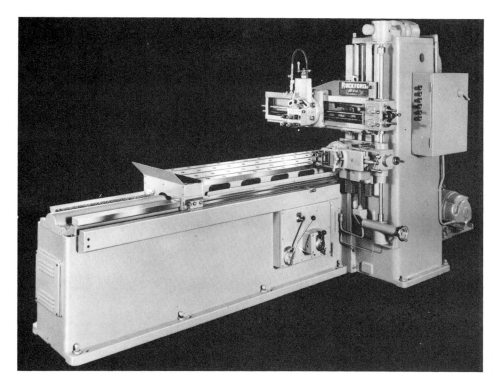

Fig. 17-9 A small openside, hydraulic planer with one railhead and one sidehead. (*Rockford Machine Tool Co., Subsidiary of Greenlee Bros. & Co.*)

planer (Fig. 17-9). This is especially designed to hold extra wide workpieces, but it is also used for general planing work.

Cutting Tools, Feeds, and Speeds

Planers use high-speed steel or carbide-tipped cutting tools similar to those used on shapers. However, since planers make heavy cuts, their tools are much larger as shown in Fig. 17-10.

Rake and relief angles are similar to those used on lathes for cutting cast iron or steel, though relief angles are often only 3 to 5° as all cuts are on flat surfaces.

Cutting speeds are about the same as used for turning. Feed per stroke can be quite large, up to ½ in. (13 mm) for roughing and 0.020 inch per stroke (0.5 millimetre per stroke) for finish cuts. Feeds may be applied at either the beginning or the end of the stroke.

CAUTION: If cross feed is engaged at the beginning of the cut, the stroke must start 6 to 12 in. (150 to 300 mm) before the work to allow sufficient time for the heads to move to the new position.

Depths of cut may be as much as 1 in. (25 mm) for roughing cuts or only a few thousandths for finish cuts. Most cuts will range from 0.010 to 0.500 in. (0.25

to 13 mm) deep. Some typical planer cuts are shown in Fig. 17-11.

Holding the Work

Making such heavy cuts at 60 to 100 fpm (18 to 30 m/min) requires considerable force; thus, the workpieces must be solidly fastened to the table. As the reversal of direction occurs quite rapidly, the work must be especially well braced at the ends. Several of the illustrations show the clamps and spacers used.

The table has tee slots, both lengthwise and across, in which heavy bolts and clamps may be used. Sometimes holes are drilled in the table so that large pins can be used to prevent the workpiece from going off the table when the machine reverses.

Setup can take from 2 to 10 h, as the workpiece must be lifted onto the table by a crane. Then the work must be squared to the table by measuring or by using a dial indicator. If several parts are placed end-to-end, each must be aligned with the others.

Calculating Time and Power

EXAMPLE 17-3

A planer is cutting an 8 ft long × 3 ft wide × 2 ft high iron casting, using one railhead and both sideheads. Depth of cut is 0.500 in. on the top and 0.300 in. on each side. Speed is 80 fpm cutting and

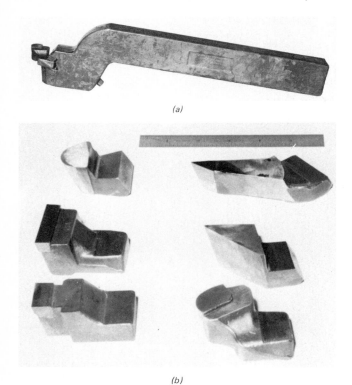

(a)

(b)

Fig. 17-10 (*a*) Toolholder and tool for a large planer. (*b*) Some of the types of tool bits used on planers. *(Photograph by Samuel Lapidge.)*

160 fpm on the return stroke. Feed is 0.250 inch per stroke on all cuts. Omit any consideration of over-travel at the ends. This block weighs approximately 22,000 lb [9990 kg (mass)] if it is solid.

Computations:

$$\frac{\text{Width}}{\text{Feed}} = \text{number of cuts}$$

As the top is the widest cut

$$\frac{36}{0.250} = 144 \text{ cuts required}$$

Each cut is 8 ft long, so $8 \times 144 = 1152$ ft of cutting is needed.

$$\text{Time} = \frac{\text{distance}}{\text{rate}} = \frac{1152}{80} = 144 \text{ min}$$

$$\text{Return stroke} = \frac{1152}{160} = \underline{\;72 \text{ min}}$$

$$\text{Total time} = 216 \text{ min}$$
$$(3.6 \text{ h})$$

Power

$$\text{in.}^3/\text{min} = (w)(d)(\text{fpm})(12)$$

Fig. 17-11 Two ram heads making different cuts at the same time. *(Rockford Machine Tool Co., Subsidiary of Greenlee Bros. & Co.)*

TABLE 17-1 COMPARISON OF SHAPER AND PLANER

Item	Shaper	Planer
Size of machine	Length of stroke = size 12–24 in. (300-600 mm)	Width × length of table size. Maximum "standard" is 72 × 240 in. (1800 × 6100 mm)
Size of work	Maximum about 24-in. (600-mm) cube	Maximum about 60 × 216 in. (1500 × 5500 mm); longer are machined
Number of toolholders	One	One to four
Cutting stroke	The ram moves the cutter through the work	The table moves the work past the cutters
Feed per stroke	The table moves	The toolheads (railheads and sideheads) move

$$\text{Top} = (0.250)(0.500)(80)(12) = 120 \text{ in.}^3/\text{min}$$
$$\text{Two sides} = 2(0.250)(0.300)(80)(12) = \underline{144 \text{ in.}^3/\text{min}}$$
$$\text{Total} = 364 \text{ in.}^3/\text{min}$$

Assume that planing cast iron takes 0.48 hp/(in.³/min)

Power needed = (364)(0.48) = 175 hp

If efficiency = 80 percent, required hp = $\dfrac{175}{0.8}$

= 220 hp (164 kW).

If the planer has only a 100-hp (75-kW) motor, either the depths of cut or the feed will have to be halved.

Table 17-1 shows a comparison of the basic features of the shaper and planer.

Review Questions and Problems

17-1. Compare shapers and planers in regard to:
  a. What moves to create the cutting force
  b. Size of work which can be done
  c. Cutting speeds and feeds
  d. Type of cuts which can be made
  e. Major uses in industry
  f. Size of the actual machine

17-2. What is the reason for having a clapper box? Why would a positive mechanical operation of it be desirable?

17-3. How is work held in each machine? Would a vise be useful on a planer?

17-4. What two types of drives can be used on shapers and planers? Give the advantages of each.

17-5. In what ways is the planer similar to the planer-miller discussed in Chap. 12? In what way is it different?

17-6. Why must a planer's bed be twice as long as its table?

17-7. A shaper is facing an aluminum block 8 × 15 in. (200 × 380 mm), with feed of 0.012 inch per stroke (0.30 millimetre per stroke), and a cutting speed of 220 fpm (67 m/min). Depth of cut is 0.125 in. (3.2 mm).
  a. At what strokes per minute should the machine run if the 220:140 ratio is used?
  b. How long will the machining take?
  c. If it takes 0.30 hp/(in.³/min) and efficiency is 70 percent, what horsepower will be needed?

17-8. A large hydraulic planer is facing an iron casting 1500 × 4500 mm top dimensions. The cut is 10.0 mm deep, feed is 2.5 millimetres per stroke, and the cutting speed is 25 m/min.
  a. How long will it take to face the casting, if return speed is 2:1?
  b. What horsepower or kilowatts will be needed? Use 0.228 kW/cm³/min.

References

The best references are the catalogs and other literature published by the manufacturers of the machines.

boring and turning machines

If you are working in an industry which makes small motors, typewriters, household appliances, or similar products, you may never see any of the metalworking machines described in this chapter. However, if your company makes large pumps, turbines, railroad car wheels, or many different styles of machinery, these *boring machines* will be an important part of the equipment you will use. They are some of the largest of the machine tools and are able to machine workpieces weighing up to 40,000 lb. (18,000 kg).

Boring is, of course, done with all types of lathes and with milling machines. The machines discussed here do similar work but on larger workpieces and to larger diameters.

The term *boring machines* is used to describe several quite different pieces of equipment. *Jig borers*, which are used for very accurate, relatively small work, were discussed in Chap. 9. The vertical boring machine (VBM) and the horizontal boring machine (HBM) are both used for large workpieces—from 24 in. (600 mm) to 120 in. (3000 mm) on a side. Parts as large as 40-ft (12,200-mm) diameter have been machined.

VERTICAL BORING MACHINES (VBM)

A general description of a vertical boring machine would be that it is a lathe turned on end with the headstock resting on the floor. This machine is needed because even the largest engine lathes cannot handle work much over 24-in. (600-mm) diameter.

Today's VBMs are often listed as *turning and boring machines*. If facing is added to that name, it pretty well describes the principal uses of this machine. As for any lathe, these machines can make only *round* cuts plus facing and contouring cuts.

Figure 18-1 shows the general construction and the motions available on the VBM. The construction is the same as that of the double-housing planer, except that a *round table* has been substituted for the long reciprocating table, and the toolholders are different since the VBM does not need clapper boxes. Figure 18-2 shows two typical vertical boring machines, though neither has the "sideheads" which are often used.

The *size* of a vertical boring machine is the *diameter* of the revolving worktable. The double-housing VBM is most often made with table diameters from 48 in. (1200 mm) to 144 in. (3600 mm). Larger machines have been made for special work.

Vertical Turret Lathe (VTL)

When a vertical-type turning and boring machine with a table smaller than about 48-in. (1200-mm) diameter is needed, a single-column machine called a *vertical turret lathe* (VTL) is used. As shown in Fig. 18-3, this has the same motions and many of the construction features of the VBM. However, as its table is smaller, from 24 in. (600 mm) to 48 in. (1200 mm) in diameter, a *heavy central column* is all that is needed to support the much shorter crossrail. Figure 18-4 shows a typical vertical turret lathe.

There is one other difference: the *turret* on the VTL. This turret can be either four-, five-, or six-sided (pentagonal or hexagonal), similar to that found on a standard turret lathe. However, in recent years, a large percentage of the large vertical boring machines have been made with a turret on one of the rams, usually the righthand vertical ram. Because of this, some shops and catalogs refer to *all* turning, facing, and boring machines as vertical boring machines. It is understood that the smaller machines will have the center column instead of the double column. Moreover, both the VBM and the VTL may have either one or two railheads and either one or two sideheads, whichever combination is needed by the customer.

Thus, the VBM and VTL today really differ only in the size of the work which can be done on each, and the names are, in some shops, used interchangeably. However, in this book, the two machines will be referred to separately.

Cuts which Can Be Made

Some of the types of cuts which can be made on both machines are shown in Fig. 18-5. Where "turret head" is noted, the VBMs might use a straight ram head instead of the turret.

Figure 18-5 illustrates the large savings which can be realized by combining cuts. Each head can be fitted with two cutting tools so that, theoretically, eight tools could be cutting at the same time. Actually, it is seldom that more than three or four cutters are working at one time.

Construction of the Machines

Either the two heavy columns of the VBM or the heavy center column of the VTL supports the **crossrail**, which is moved up or down by power-driven lead screws (Figs. 18-1 and 18-3). This very heavy part is counterbalanced, often with cast-iron weights,

so that it can be moved by handwheels for close adjustment.

The **toolheads** are mounted on the crossrail and can be moved left or right and up or down. Most of the large vertical boring machines are equipped with two railheads. The second head is frequently omitted on the VTLs, especially the smaller ones.

These tool heads may be straight or swivel type. They may be **ram heads**, with a single toolholder, or four-, five-, or six-sided **turret heads** for multiple tooling.

The vertical travel of either the ram heads or turret heads may be from 12 to 48 in. on standard machines (300 to 1200 mm), or up to 60 in. (1500 mm) if specially ordered.

The movement of each head is independently controlled, both left and right (X axis) and up and down (Z or W axis). Thus boring, turning, taper turning, or facing cuts can be made with one head while the other is making a totally different cut. The righthand head can be brought to the center of the table so that a twist or spade drill can drill a hole at the *center only*.

The **sideheads** (usually one, occasionally two) can be moved up or down on ways, and the tool slide can move in or out for facing, turning, and grooving. The

Fig. 18-1   General construction and part names of a double-column vertical boring machine.

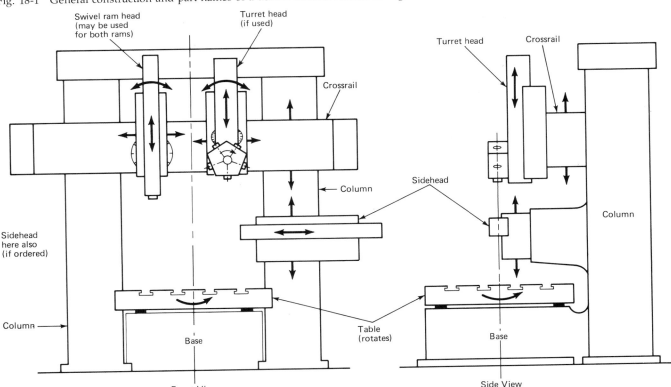

(a)

(b)

Fig. 18-2   (a) A VBM with numerical control. Table has 16-ft (4800-mm) diameter. For machining gas turbines. *(The G. A. Gray Company.)* (b) A Bullard 60-in. (1500-mm) VBM machining centrifugally cast sections. *(Wisconsin Centrifugal Inc.)*

sideheads also are moved independently of each other and independently of the heads on the crossrail. The sidehead cannot usually be tilted. There is often a four-sided turret installed on this head.

The **round worktable** is supported on bearings both at the center and near the outside. It is of heavy, well-braced construction to support heavy loads. A radial pattern of tee slots is cut into the table for hold-

ing clamps, bolts, special fixtures, or jaws. Motors of 20 to 100 hp (15 to 75 kW) and sometimes larger are used to turn the table through a gear drive.

Work holding is often accomplished with heavy chuck jaws furnished with the worktable or with separate adjustable jaws bolted to the table. Three or four jaws may be used as shown in Fig. 18-6.

A casting may be centered, for the first roughing

Fig. 18-3   General construction and part names of a vertical turret lathe.

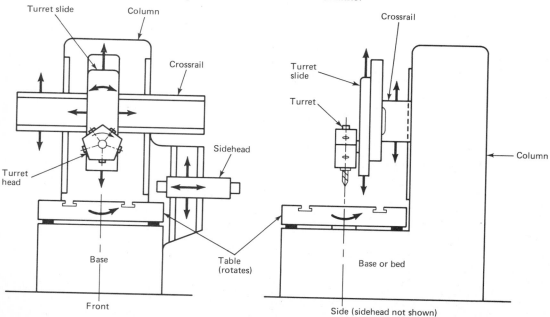

cuts, by merely measuring to the edge of the table with a steel rule. For second operations (completing cuts which could not be made with the first setup), a large centering plug in the table may be used. Frequently large fixtures are bolted to the round table. These have locating surfaces onto which the workpiece is positioned.

Practically all workpieces are so heavy that they are handled with overhead cranes with capacities from 500 to 20,000 lb (230 to 9200 kg).

Cutting tools are basically the same as those used on turret lathes. However, they are much larger in size. Boring bars are often 2 to 3 in. (50 to 75 mm) in diameter. Cutting tools may be made with steel bodies of 4 × 2 in. (100 × 50 mm) cross section and 6 to 12 in. (150 to 300 mm) long. Today most VBM and VTL cutting is done with carbide-tipped tools. Some tools have brazed-on tips, although the "throwaway" carbide tip is usually much less expensive to use and is most often used today.

A great deal of thought is needed to "tool up" a job most efficiently, using combined cuts whenever possible. Figure 18-5 shows two possible combination cuts. Figure 18-7 illustrates a complete job using only three sides of a four-sided turret and two stations on a sidehead turret. This was done on a 30-in. (760-mm) vertical turret lathe with semiautomatic controls.

Available *speeds* and *feeds* will vary according to the size of the VBM or VTL. For large diameters, very low speeds are needed. For example, a stainless-steel casting of 80-in. (2000-mm) diameter with an allowable cutting speed of 35 fpm (10.7 m/min) should run at only 1.75 rpm. Table 18-1 shows the general ranges available. The higher speeds are to accommodate parts which are quite a bit smaller than the table diameter. Metric feedrates are readily available.

Depths of cut can be quite heavy. It is not unusual to make $\frac{3}{16}$- to $\frac{3}{8}$-in.-deep (4.76- to 9.53-mm) cuts when rough turning or facing an iron casting. Of course, only 0.010- to 0.030-in.-deep (0.25- to 0.75-mm) cuts will be used for finish cuts. Metric dials and lead screws are, when specified, furnished as "standard" by all manufacturers.

Finishes attained on these machines can be excellent. A 125-μin. (3.2-μm) finish is normally expected on cast iron and often on steel. With cutters with large nose radius and with the use of smaller feeds, a 64-μin. (1.6-μm) or better finish is not difficult to achieve.

Tolerances of 0.005 in. (0.125 mm) are about normal, though 0.001- to 0.002-in. (0.025- to 0.050-mm) tolerances can be held if the tooling is rigid and the machine is in good condition.

Automation

Both the VTL and the VBM are frequently automated. One method is by the use of drums or boards into which plugs can be inserted to control the movement

Fig. 18-4 A 46-in. (1170-mm) manually operated vertical turret lathe. Note digital read-out boxes for fast, accurate setting of dimensions. (*Bullard Co.*)

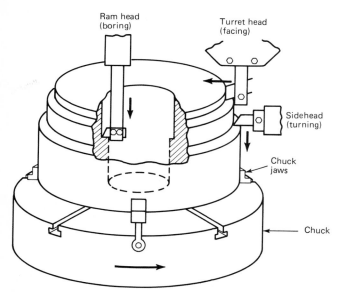

The above three cuts may be made at the same time.

Note: Either ram or turret head can face, turn, bore, or cut inside or outside grooves. The sidehead can face, turn, or cut outside grooves.

(a)

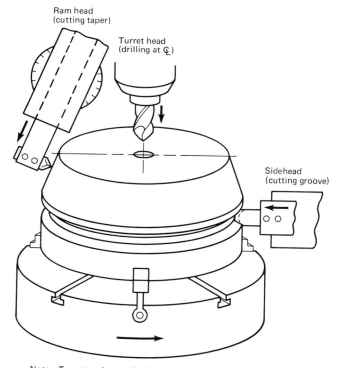

Note: Taper turning can be done by swiveling the cross head (as shown), N/C programming, or controlling X and Z movements by digital feed attachment or pickoff gear attachment.

(b)

Fig. 18-5 Some of the cuts which can be made on either the VBM or the VTL.

of the various heads. Sometimes the operator must set feeds and speeds.

Numerical control is applied to many vertical boring machines today. With numerical control, one cutter may make a series of cuts (horizontal, vertical, or at an arc or angle) before it leaves the work.

Movement in the Z and X axes, and often speeds and feeds too, are controlled by the coded holes in the tape. Tape-controlled motion is more consistently accurate than any other method of control. The tapes may take 20 min or several hours to prepare, but they can be used many times, so they are economical.

The N/C machine control unit plus the special lead screws may add $15,000 to $30,000 per machine to the cost. However, if there is much work for your VTLs or VBMs, the numerical control will usually pay for itself in savings within 2 years or less.

Vertical Chuckers

There are many round parts weighing from 10 to 40 lb (4.5 to 18 kg) which could be machined on a turret lathe. However, it is often easier and faster to use a one- or two-spindle, *vertical chucker* like the ones shown in Fig. 18-8. These are always automatic, which makes them comparable to the automatic turret lathe (ATL). The control may be with a prewired plug board, a plug-in punched card reader, or numerical control or selector switch programming very similar to that shown with the ATLs. The chucks are usually of 10- to 24-in. (250- to 600-mm) diameter with some, for heavy work, up to 30-in. (760-mm) diameter.

The *vertical automatic chuckers* (VAC) may have sideheads as well as five- or six-sided turrets. Special attachments, such as tracers and extra cross slides,

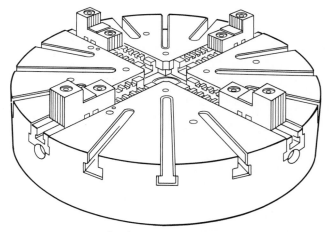

Four-jaw independent chuck

Fig. 18-6 Typical VBM or VTL rotating table chuck. (*The G. A. Gray Company.*)

TABLE 18-1 APPROXIMATE SPECIFICATIONS FOR TYPICAL VERTICAL BORING AND TURNING MACHINES

Table Diameter		rpm		Feedrate	Weight (Mass)	Cost*
in.	mm	Min.	Max.			
30	760	10	300		25,000 lb 11,300 kg	$55,000
48	1200	7	250	In general, 0.003–0.500 in./rev or, metric 0.05–13 mm/rev	35,000 lb 15,900 kg	$70,000
60	1500	1.0	75		74,000 lb 33,600 kg	$90,000
144	3700	0.6	25		230,000 lb 104,000 kg	$250,000

*Cost will vary widely depending on equipment ordered. N/C will add $15,000 to $30,000.

can be added to the VAC. Thus, combined cuts can be made.

The *vertical numerical control chuckers* (VNC) are made in the same sizes as the VAC. On this machine all X and Z motions, feeds, speeds, and dwell are programmed on the tape. Most of the newer machines also have linear and circular interpolation controls. With these controls, angular and circular cuts may be programmed easily and made accurately. Both English and metric machines are available in all styles of controls.

"Two table" models permit machining two identical parts at once, controlled by the same tape. All kinds of materials (such as cast iron, alloy steel, steel forgings, aluminum, copper, etc.) may be machined accurately on the VAC and the VNC.

With carefully planned tooling, these chuckers can do very fast, accurate work on medium- and large-lot sizes.

HORIZONTAL BORING MACHINES (HBM)

In spite of the similarity in the names, the horizontal boring machine (HBM) is totally different in appearance and operation from the vertical boring machine (VBM).

The HBM is made to handle medium- to very large-sized parts, but these parts are usually some-

Fig. 18-7 Three operations performed on a railroad car wheel in one setup on a 30-in. (760-mm) VTL. *(Bullard Co.)*

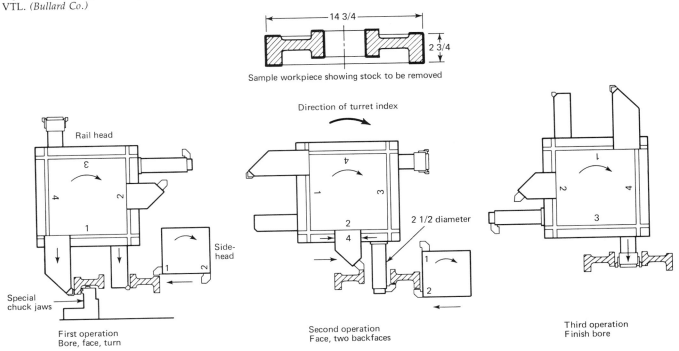

Fig. 18-8   A two-spindle vertical chucker. Control is either by a template (cam) or by numerical control. *(Bullard Co.)*

what rectangular in shape, though they may be asymmetrical or irregular.

Work Done on Horizontal Boring Machines

The HBM with its *rotating horizontal spindle* can:

Bore (enlarge a previously drilled or cored hole) up to about a 12-in. (300-mm) diameter

Drill, with twist drills or spade drills

Ream, with solid or inserted tooth reamers

Counterbore, with large or small cutters

Tap holes from ¼- to 3-in. (6.35- to 75-mm) diameter

Cut slots, with end mills of all sizes

Face large or small areas, using shell or face mills

The size of cut is limited only by the available cutting tools, the stiffness of the spindle, and the available horsepower.

CAUTION: The HBM, with its rotating spindle, can only bore out to the largest diameter of boring tool available. Any boring bars over about 12-in.-diameter (300-mm) capacity are usually too heavy to use. Thus, work requiring larger holes must usually be laid flat and centered on a *vertical* boring machine. As the *table* rotates on the VBM, the boring tool can be moved out to any diameter needed.

Types of Horizontal Boring Machines

There are two principal types of HBM. The **table-type** HBM (Figs. 18-9 and 18-10) is built on the same principles as the horizontal-spindle milling machines. The base and column are fastened together, and the column does not move. The tables are heavy, ribbed castings which may hold loads up to 20,000 lb (9100 kg).

The basic size of an HBM is the *diameter of the spindle*. Table-type machines usually have spindles from 3- to 6-in. (75- to 150-mm) diameter. The larger sizes will transmit more power and, equally important, the spindle will not sag or deflect as much when using a heavy cutting tool while extended.

The *size* is further specified by the *size of the table*. Although each machine has a "standard" size table, special sizes may be ordered. These machines are available in inch or metric sizes.

The principal parts of the horizontal boring machine are clearly shown in Fig. 18-9. This diagram illustrates a N/C machine, but the same style is also made with standard hand controls as shown in Fig. 18-10.

Notice that either the Z-axis spindle or the W-axis saddle may be moved to obtain the depth of cut. (The W axis is a standard N/C term designating a motion parallel to the Z axis of the spindle.)

Speeds and feeds cover a wide range because of the wide variety of cutters which may be used on the HBM. Speeds from 15 to 1500 rpm and feedrates from 0.1 to 40 ipm (2.5 to 1000 mm/min) are commonly used.

Work holding is with clamps, bolts, or fixtures, the same as with other machines. *Rotary tables* (as shown in Fig. 18-10) allow machining of all four faces of a rectangular part or various angle cuts on any shape of part. Rotary tables up to 72 in. (1800 mm) square or round are used for large work.

If large, rather flat work is to be machined, an angle plate such as shown in Fig. 18-11 is used. The workpiece is bolted or clamped onto the angle plate so that the "flat" face is toward the spindle.

Cutting tools are held in the rotating spindle by a tapered hole and a drawbar, just as illustrated for the horizontal-spindle milling machine. To speed up the process of tool changing, either or both of two things are done:

1.   The drawbar (which pulls the tapered tool-holder tightly into the spindle hole) can be power operated. Thus, the holder is pulled tight or ejected very quickly.

2.   Quick-change tooling is used. A basic holder is

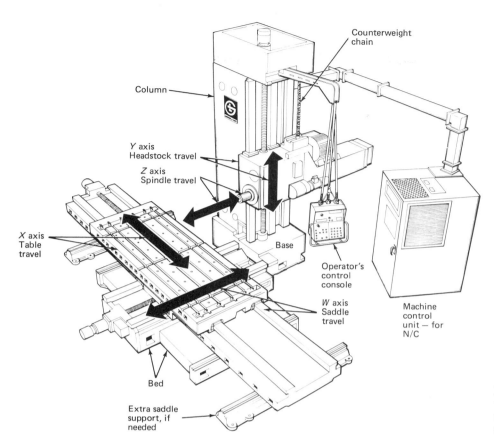

Column

Counterweight chain

Y axis
Headstock travel

Z axis
Spindle travel

X axis
Table travel

Base

Operator's control console

W axis
Saddle travel

Machine control unit — for N/C

Bed

Extra saddle support, if needed

Fig. 18-9 Diagram and motions of a table-type N/C horizontal boring machine. Very large table shown. *(Giddings & Lewis, Inc.)*

Fig. 18-10 A small hand-control horizontal boring machine in use. Note air-lift rotary table mounted on standard table, so holes can be bored in all four sides with only one setup. *(Lucas Machine Division, Litton Industries.)*

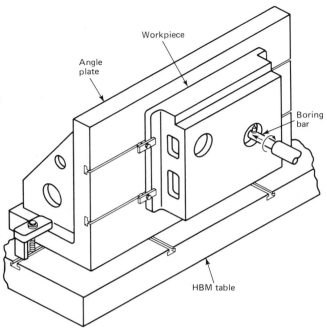

Fig. 18-11   A small angle plate in use, mounted on a milling machine or HBM table.

secured in the spindle. It has a taper into which tools may be secured by a quarter to half turn of the locking collar. Thus, the operator can change preset tools in 10 to 30 s. Some typical quick-change tools are shown in Fig. 18-12.

The **floor type** HBM (Fig. 18-13) is used for especially tall or long workpieces. The "standard" 72-in. (1800-mm) runway can be made almost any length required for special jobs. Lengths of 20 ft (6000 mm) are in use today.

The height of the column, which is usually 60 to 72 in. (1500 to 1800 mm), can be made to order up to twice this height if the work requires it.

The *table* is separate from the boring machine though it is, of course, fastened to the floor. It may be bolted to the runway.

The entire *column* and *column base* move left and right (*X* axis) along special ways on the *runway* (Fig. 18-14). The runway must be carefully aligned and leveled when it is first installed, and then checked at intervals as the machine is used. If ordered (as an extra cost option), the column may also move in and out (*W* axis). These motions can be controlled to about ±0.001 to 0.005 in. (0.025 to 0.125 mm).

Note: The foundation for a machine of this size will be reinforced concrete and will cost several thousand dollars.

The *headstock* can be moved accurately up and down the column (the *Y* axis). The 6- to 10-in.-diameter (150- to 250-mm) spindle rotates to do the machining. It is moved in and out (*Z* axis) up to 48 in. (1200 mm) for boring cuts, drilling, setting the depth of milling cuts, etc. As in the *table*-type HBM, the spindle diameter and table size specify the machine size.

Cutting tools are the same as those used on the table-type machine. Work holding is also the same, and angle plates are frequently used.

Numerical control is, today, widely used with these floor-type HBMs because of its speed, accuracy, and repeatability.

The large N/C HBMs may cost from $100,000 to over $500,000, but the considerable amount of time saved and the much decreased chance of spoiling a part make the cost worthwhile if a company is machining very much large work.

A HIGH-PRODUCTION BORING AND TURNING MACHINE

The turning and boring machine shown in Fig. 18-15*a* is ordered only if over 100,000 parts per year, of one kind, are to be made. It can produce 300 to 650 machined parts per hour.

The heart of this vertical chucking machine is a table which holds 6 to 12 rotating chucks. The work is loaded into the first chuck, and it (and each succeeding chuck) is indexed into position under 6 to 12 fixed machining heads. Thus, at every cycle, 6 to 12 operations are performed on the workpiece at once,

Fig. 18-12   Some of the quick-change tools which are used on a horizontal boring machine. Tapered shanks are similar to those for milling machine tools. (*Davis Tool Co., a division of Giddings & Lewis, Inc.*)

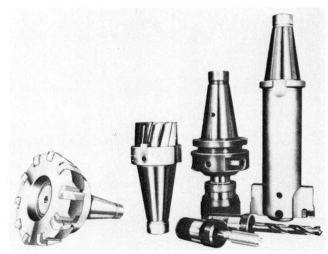

similar to the multispindle automatics described in Chap. 7.

Quite complex tooling, as shown in Fig. 18-15b, may be mounted at each station. As a result, the initial setup time for a new job may be two to four days, and the tooling is very expensive. However, by using throwaway carbide inserts, cutter changing is fairly simple.

These machines will often hold 0.001-in. (0.025-mm) tolerances, and a completed piece is produced every time the table indexes. If production is very high, automatic loading and unloading devices can be designed for this chucker. Either English or metric measurements may be used.

CONCLUSION

Each of the machines discussed in this chapter is made in many configurations to match the needs of the buyer. However, if these basic styles are understood, the capabilities of any variation can easily be mastered.

Review Questions and Problems

18-1. Describe a vertical boring machine and the kind of workpieces for which it is used.

18-2. How is a vertical boring machine different from and how is it similar to a double-housing planer?

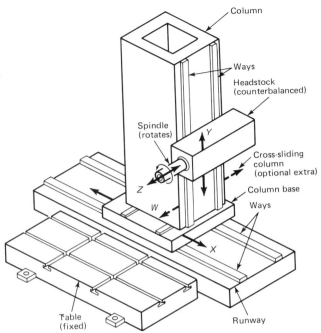

Fig. 18-14   The general construction and motions of one style of floor-type HBM. Table may be much larger than shown here.

18-3. What machining operations can be performed on a VBM?

18-4. How is the size of a VBM designated?

18-5. How does a vertical turret lathe differ from a vertical boring machine?

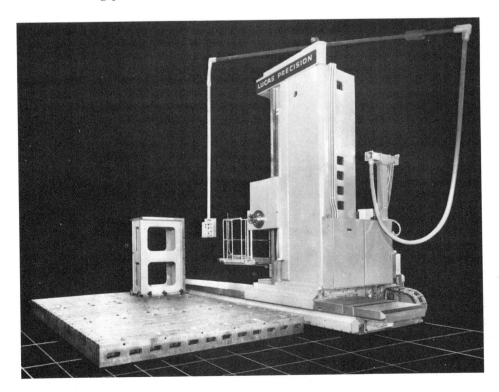

Fig. 18-13 A large floor-type HBM. Block on table is used sometimes instead of an angle plate. (*Lucas Machine Division, Litton Industries.*)

(a)

(b)

Fig. 18-15   (a) A "Mult-au-Matic" high-production, vertical 12-spindle indexing chucker. Tooling not shown at some stations. (b) A closeup of the tooling used at one station. Workpiece shown below in rotating chuck. (Bullard Co.)

18-6. What machining operations can be performed on a VTL?

18-7. How is the workpiece held on a VTL or VBM?

18-8. How does a vertical chucker differ from a vertical turret lathe in the type of workpiece it handles and in its construction? How is it similar?

18-9. Compare a VAC and a VNC.

18-10. Describe the two principal types of horizontal boring machines and the usefulness of each.

18-11. How is the size of these machines specified?

18-12. What machining operations can be performed on a HBM?

18-13. How is the workpiece held on a HBM?

18-14. How are the cutting tools held on a HBM?

18-15. A VTL is turning a stainless-steel casting at 35 fpm (10.7 m/min). If the casting is 2½-ft (760-mm) diameter, what rpm should be

used? If the feed is 0.005 in./rev (0.125 mm/rev), what feed in inches per minute should be specified? What feed in millimetres per minute?

18-16. A HBM is face milling a part using a 6-in. (150-mm) face mill having eight teeth. If the workpiece is to be milled at 90 ft/min (27.4 m/min), what cutter rpm should be used? If the feed is 0.004 inch per tooth (0.1 millimetre per tooth), what feed in inches per minute should be used? What feed in millimetres per minute?

18-17. A manufacturer wishes to purchase a 120-in. (3000-mm) vertical boring machine to machine both aluminum, at 250 fpm (76 m/min), and steel, at 60 fpm (18 m/min). The largest piece to be machined will be 8 ft (2400 mm) in diameter, and the smallest piece will be 3 ft (900 mm) in diameter. What are the maximum and minimum rpm that must be provided on this machine? Compute with both inch and metric units.

18-18. What machine would you specify to turn and face a 12-in.-diameter (300-mm) piece, 5 in. (125 mm) high, in lots of 500? In lots of 5000?

18-19. A VBM is being used to turn, bore, and face a casting as shown in Fig. 18-5a. The hole is to be bored to 3-in. (75-mm) diameter at a feed-rate of 0.015 in./rev (0.38 mm/rev) using a depth of cut of 0.025 in. (0.64 mm). The out-side is to be turned to 4-ft (1200-mm) diame-ter at a feedrate of 0.025 in./rev (0.64 mm/rev) using a depth of cut of 0.250 in. (6.35 mm). The top of casting is to be faced using a 0.125 in. (3.18 mm) depth of cut at a feedrate of 0.015 in./rev (0.38 mm/rev).

a. If the maximum cutting speed allowed is 60 fpm (18.3 m/min), what rpm should be used?

b. Compute the fpm (or m/min) for each surface at the rpm you calculated.

c. What is the total rate of stock removal in cubic inches per minute? Show all calcu-lations clearly.

d. What horsepower will be required if the hp/(in./min) rating is 0.80 at 75 percent efficiency?

e. How long will the machining operation take? Assume that turning and boring can be done along the full height of the workpiece (no allowance for clamping). Casting is 25 in. (635 mm) high.

References

The best sources of information are the manufacturers' catalogs as there is very little literature available on these machines. ASTME: *Tool Engineers Handbook*, Sec. 30 to 32, McGraw-Hill, New York, 1976.

broaching

Broaching is a machining process which is becoming widely used in both small and large manufacturing plants. Almost any shape can be broached, either on the inside or the outside of a part, as shown in Fig. 19-1. Broaches are made in both inch and metric sizes.

The major limitation is that broaches can cut only in a *straight line*. Thus, a broach can cut across a gap, but it cannot get beyond any projection. Within this limitation, broaches can cut keyways and splines; create flat, rounded, or gear-shaped surfaces; and make square, hexagonal, or irregularly shaped holes in practically any metal and in some plastics.

This variety of work is done while holding tolerances of ±0.002 to 0.0002 in. (0.05 to 0.005 mm) quite consistently during long runs of work.

TYPES OF BROACHES

The two major types of broaches are the *push* broach and the *pull* broach. A second division is *internal* and *external*. *Flat* broaches are a special class of external broaches.

The **push broach** must be relatively short since it is a column in compression and will buckle and break under too heavy a load. One company suggests that if the cutting length of the broach is more than 25 times the average root diameter, it should be designed as a pull broach. Figure 19-2 shows both push and pull types of broaches. These cost from $15 to $250 each.

Push broaches are often used with a simple arbor press if quantities of work are small. For medium- to high-volume production they are used in broaching machines. The keyway broaches in Fig. 19-3 are an example of simple push broaches. The cost is from $70 to $400 per set of 15 to 36 keyway and hole combinations.

Pull broaches are pulled either up, down, or horizontally through or across the workpieces, always by a machine. *Flat*, or nearly flat, broaches may be pull type, or the broach may be rigidly mounted, with the workpiece then pulled across the broaching teeth. Automobile cylinder blocks and heads often are faced flat by this method.

CAUTION: An internal broach will not make its own hole. A properly sized drill, bored, or cored hole must first be made in the location to be broached. If the holding fixture and the broach are rigid, the broach may make a few thousandths correction in the hole location, though it is not wise to count on this.

Pot Broaches

Cutting a shape completely around a workpiece by broaching has been done on a small scale for some time. However, about 1969 the process was developed to the point where it is now a very practical,

Fig. 19-1 (a) Some simple shapes which can be made by broaching. (*National Broach & Machine Division Lear Siegler, Inc.*)

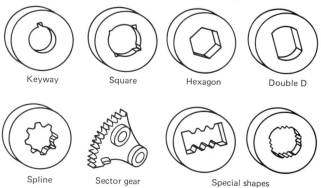

Keyway Square Hexagon Double D

Spline Sector gear Special shapes

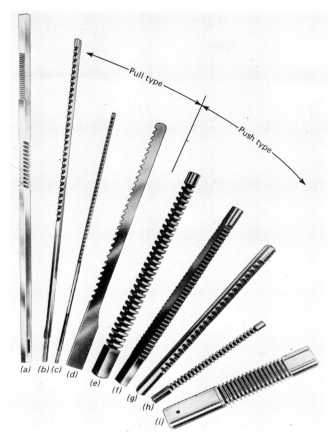

Fig. 19-2 A variety of broaches (inch and metric). (a) Sizes the width and depth of a slot. (b), (c) Keyway broaches. (d) Special keyseating broach. (e) Square broach (note round starting end). (f), (g) Special shapes. (h) Small square broach. (i) Waveguide sizing broach. (The du Mont Corporation.)

reliable method of rapidly producing gears, gearlike shapes, and irregular shapes around the circumference of a part.

Some of the work which is being done is shown in Fig. 19-4a. The circular broach with internal cutters is called a *pot broach*. These are made, as shown in Fig. 19-4b, with long "sticks" of cutters fastened into the body or by many "rings," each having the desired tooth size and shape. These assemblies are held securely in the housing shown and are mounted on special vertical broaching machines.

The workpiece is usually pushed up through the broach, though smaller parts may be pulled through. Tooth spacing, chip per tooth, etc., are about the same as in other broaches.

These broaches may cost $5000 to $15,000 each, but they can make an accurate, smoothly finished part in a few seconds with excellent repeatability. Thus, for quantities of over 25,000 pieces, this could be a very economical method.

THE BROACH

The broach is like a single-point tool with many "points," each of which cuts like a flat-ended shaper tool, though some broaches have teeth set diagonally, called *shear cutting*.

The principal parts of an internal broach are shown in Fig. 19-5.

The *pull end* is made in several shapes to fit different makes of broaching machines, including special snap-action ends for automatic machines.

The *front pilot* guides the broach into the hole, and also serves as a rough *no-go* gage.

The *rear pilot* is the size and shape of the finished hole, and guides the broach through the finished work.

The *follower diameter* is, for a pull broach, often shaped somewhat like the pull end so that it can be gripped by an automatic broaching machine and brought back to its starting position.

Broaching teeth must be made carefully because the teeth, and not the machine, determine the size and shape of the cut. The machine supplies only the power and some guidance for the broach.

The first tooth is usually sized to barely touch the prepared hole or the surface to be machined. Each successive tooth projects slightly more, until the last few teeth cut the finished shape.

Chip per tooth (which corresponds to the same term in milling) is the difference in height between two

Fig. 19-3 A set of HSS push-type keyway broaches complete with bushings. Made in inch and metric sizes. (The du Mont Corporation.)

Fig. 19-4   (a) Examples of work done by pot broaching. Top row is the blank, bottom row the broached product. (b) The sectional pot broach with inserted broach sections which was used to broach the left-hand part in (a). (*National Broach & Machine Division Lear Siegler, Inc.*)

TABLE 19-1   APPROXIMATE CUTTING SPEEDS AND ANGLES FOR HIGH-SPEED STEEL BROACHES

Material	fpm	m/min	Hook Angle
Cast iron and steel	10–30	3–9	Soft 15–20° Hard 8–120°
Aluminum	50 up	15 up	10–15°
Stainless steel	10–25	3–7.5	12–18°
Brass, bronze	30	9	0–10°

Tooth spacing (*or pitch*) *and tooth depth* are especially important because the chips cannot escape from the cutter on inside cuts. They will curl up in the space as shown in Fig. 19-6. Naturally, the longer the cut and the greater the depth of cut, the larger is the volume of chips made. To hold this larger volume of chips, the teeth must be deeper and the pitch must be greater. The *width* of the cut will not change the spacing needed.

CAUTION: Round internal broaches would cut a series of rings, and wide flat broaches would generate a big chip hard to get rid of. Thus *chipbreakers*, which are notches in the teeth, are ground into these types of broaches.

Large, flat surfaces are often cut by a series of small teeth, staggered so that together they cover the entire surface.

An approximate formula for tooth spacing is

$$\text{Pitch} = 0.35\sqrt{\text{length of cut}}$$

More frequently the designer refers to a table like Table 19-2. Not all manufacturers use the same data, but the differences are relatively small.

Broach size is widely variable. A small $\frac{1}{16}$-in. (1.6-mm) keyway broach weighs only a few ounces and costs about $18. A large push or pull broach for a 10-in.-diameter (250-mm) spline might weigh 2000 lb (907 kg) and cost thousands of dollars. The machines broaching automobile cylinder blocks have a fixed broaching surface about 10 ft (3000 mm) long.

The *cutting length* of a simple push or pull broach is easily determined once the pitch and chip per tooth are decided. The equation would be

$$N = \text{number of cutting teeth} = \frac{\text{depth of cut}}{\text{chip per tooth}}$$
$$(19\text{-}1)$$

$$\text{Cutting length} = \text{pitch}\ (N + \text{number of finish teeth})$$
$$(19\text{-}2)$$

adjacent teeth. This may be as small as 0.0005 in. (0.0125 mm) for semifinishing teeth or 0.002 to 0.006 in. (0.050 to 0.150 mm) for roughing teeth.

The *semifinishing teeth* (which are not always needed) will take a lighter cut than the roughing teeth. The three to six *finishing teeth* will all be the same size and shape as the finished cut.

Broach tooth angles are similar to those in all cutting tools. This face or hook angle varies as shown in Table 19-1. The *backoff angle* is really a clearance or relief angle to prevent excessive rubbing on the work. This is quite small, from $\frac{1}{2}$ to 3°.

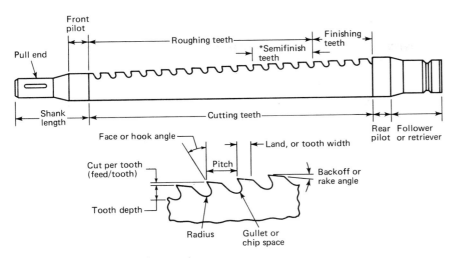

*Semifinishing teeth are not always used.
Note: Slope (rise per tooth) is exaggerated in these drawings

Fig. 19-5 The principal parts and angles of a broach.

EXAMPLE 19-1

It is desired to cut a ½-in. (12.7-mm) keyway through a 6-in.-long (150-mm) cast-iron gear hub (Fig. 19-7). The pitch (from Table 19-2) is 0.875 in. (22 mm). The chip per tooth will be 0.004 in. (0.100 mm). Four finishing teeth will be used. Keyway depth is half its width.

$$\text{Number of teeth} = \frac{0.250}{0.004} = 62.5 \text{ (use 63)}$$

$$\text{Cutting length} = 0.875 (63 + 4) = 58.625 \text{ in.}$$

In *metric*

$$\text{Cutting length} = 22 (63 + 4) = 1474 \text{ mm (58.0 in.)}$$

Cutting speeds are low as shown in Table 19-1. These speeds may be changed considerably depending on the exact material, depth of cut, size of chip, etc. Most broaches are made of HSS, such as T2 or T3. Carbide is used as inserts on flat broaching, but it is very expensive to make in most shapes.

THE BROACHING MACHINE

The two major divisions in broaching machines are the push or pull type and the surface broaching

Fig. 19-6 Illustration of how a chip fills the gullet of a broach.

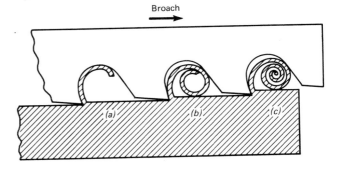

Fig. 19-7 A long keyway to be broached.

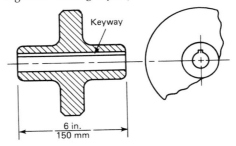

TABLE 19-2 TABLE FOR SIZING BROACHES (FOR MILD STEEL)

Length of Cut		Pitch		Tooth Depth	
in.	mm	in.	mm	in.	mm
½	12.7	¼	6.4	3/32	2.4
1	25.4	11/32	8.7	⅛	3.2
2	51	½	12.7	3/16	4.8
3	76	⅝	15.9	15/64	6.0
4½	114	¾	19.0	19/64	7.5
6	152	⅞	22.2	11/32	8.7
8	203	1	25.4	⅜	9.5

NOTE: These are not "standard" dimensions. Other sources show slightly different figures.
From "Broaching Practice," National Broach & Machine Co.

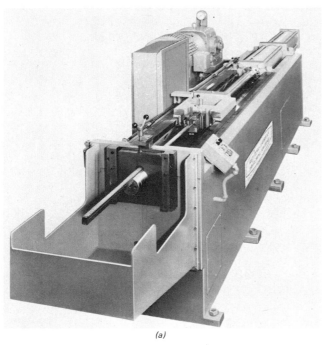

Fig. 19-8 (a) A hydraulic horizontal keyway broach. (*National Broach & Machine Division Lear Siegler, Inc.*) (b) A twin ram pull-down vertical broach with complex fixtures for high production. (*Detroit Broach & Machine Company.*)

types. Each one is made large and small, simple and complex.

Push or Pull Broaching

As mentioned in describing broaches, the choice of push or pull type is made according to the relative stiffness of the broach itself since the shape of the cut, the finish, and the accuracy is mostly in the design and sharpening of the broach.

Both push- and pull-type broaching machines may have either *horizontal* (Fig. 19-8a) or *vertical* (Fig. 19-8b) rams. Usually the vertical type is for smaller sizes of work. For splines and other shapes of about 4 in. (100 mm) or over, the horizontal broaching machine is used. Horizontal machines are also somewhat easier to use when broaches are 60 in. (1500 mm) or more in length.

steps in push or pull broaching

Whether one is using a simple keyway broach in a hand or hydraulic-arbor press or multiple-ram au-

tomatic broaching machines, there are five steps needed.

1. Load the workpiece into the fixture or other support.
2. Start the front end (pilot or pull end) through the work, or past the work if it is a surface broach.
3. Push or pull the broach through (or completely past) the work.
4. Remove the workpiece.
5. Return the broach to its starting position either by hand or automatically.

Steps 1 and 4 above, in large-lot work, are often done with automatic loading and unloading equipment.

This process may sound slow. However, a 36-in.-long (900-mm) broach cutting at 20 fpm (6 m/min) will take only 0.15 min (9 s) per stroke, plus handling time. Moreover, the resulting shape will often be the final shape, very accurately made, with a good finish. Thus, push or pull broaching can be a very economical manufacturing process. Production rates vary widely, but it is not unusual to broach 200 to 400 parts per hour on an automatic push or pull broaching machine.

Surface Broaching Machines

Surface broaching is most often used for large quantities of work and quite large, relatively flat surfaces. These machines may be *continuous* broaching machines (Figs. 19-9 and 19-10) or the *reciprocating* type (Fig. 19-11). In either case, the broaching surface must be rigidly mounted, and the workpiece must be solidly supported as it passes the broaching section.

Both types of machines are often equipped with automatic loading and unloading mechanisms. A good example of this is the automotive industry's finishing of cylinder heads and blocks in large automatic continuous broaching machines.

Capacities of Broaching Machines

Vertical broaching machines are made with single, double, or multiple rams. They can be purchased with strokes up to 100 in. (2500 mm) and up to 50 tons (45 metric tons or 400 kN) capacity. They may cost from $5000 to $70,000 each.

Horizontal broaching machines are made to handle broaches up to 90 in. (2300 mm) long and of 12-in.

(300-mm) diameter or larger. These have capacities of from 5 to 50 tons (4.5 to 45 metric tons or 44 to 445 kN). Depending on the type of work done, production may be from 10 to 300 pieces per hour. Cost is $10,000 to $100,000 depending on the size and the amount of automation used.

Continuous broaching machines are usually "standard-specials," that is, the basic machine is standard, but it is modified as to chain size, handling equipment, broach support, etc., to fit a particular job. Often they are equipped to handle a "family" of similar parts without excessive setup time.

The actual broach (fixed in position) may be from 30 to 250 in. (760 to 6350 mm) long, using a machine capacity of 40 tons (36 metric tons or 360 kN) or over. These machines, plus their broaches, may cost $80,000 or more. However, production rates of 300 to 800 pieces per hour have been obtained with continuous broaching machines.

SHARPENING BROACHES

Sharpening broaches is a fairly expensive job and may cost $100 to $400. The chip per tooth must be maintained as well as the shape. Sharpening is done on the *face* and the gullet of the teeth. This will gradually change the height of the tooth, which, on many broaches, changes the size. The finishing teeth will,

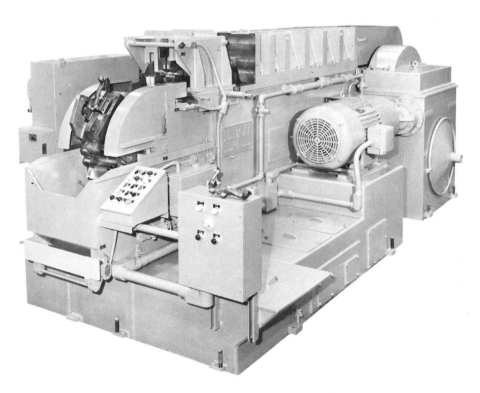

Fig. 19-9 A medium-sized continuous broaching machine, broaching steering knuckle pins at 30.6 fpm (9.3 mpm) removing 0.135 in. (3.43 mm) of stock. (*Detroit Broach & Machine Company.*)

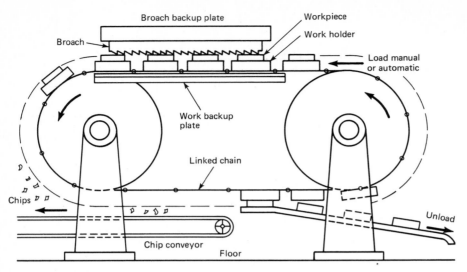

Fig. 19-10 Schematic of a typical continuous broaching machine.

for a few sharpenings, make up the difference. However, sooner or later the broach will cut under the minimum allowable size and must be discarded or recut for a different size.

The cost is not excessive when you consider that production is from 50 to 500 pieces per hour, broach life is 4000 to 7000 pieces per grind, and up to seven sharpenings can be done before the tooth becomes too weak to use. However, the cost of the large automatic broaching machines and $8000 broaches can be justified only when lots of from 10,000 to 100,000 pieces of one kind are to be machined.

The variety of shapes, styles, sizes, and construction of broaches and broaching machines is very large. However, they are all based on the principles just discussed.

Review Questions and Problems

19-1. What is the major limitation which applies to broaching?

Fig. 19-11 Schematic of a large reciprocating-type surface broaching machine. Broaching surfaces may be at any angle, may have many shapes, and may be on two or even three surfaces of the workpiece.

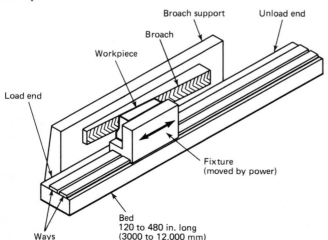

19-2. Compare push and pull broaches. What determines which should be used?

19-3. How does "flat" broaching differ from push or pull broaching?

19-4. Name the basic parts of a pull broach.

19-5. What determines the size of the chip per tooth? What factors determine the pitch?

19-6. Calculate the cutting length of a pull broach used to cut a $\frac{3}{8}$-in.-wide (9.5-mm) (depth = $\frac{1}{2}$ width) keyway in a steel gear hub 4 in. (100 mm) long if the chip per tooth is 0.003 in. (0.075 mm) and six finishing teeth are required. How long will it take to cut the keyway at 20 fpm (6.1 m/min)?

19-7. Determine the length of a broach required to cut a $\frac{3}{4}$-in.-wide (19-mm) keyway in a cast-iron gear hub 6 in. (150 mm) long if the chip per tooth for roughing is 0.004 in. (0.10 mm), and the broach has six semifinish teeth with a chip per tooth of 0.0005 in. (0.0125 mm) and four finishing teeth.

19-8. Name the five steps needed in a broaching operation.

19-9. How does broaching compare with other machining methods such as milling or shaping?

References

1. *Broaching Practice*, National Broach & Machine Co., 1953.
2. *Machining Data Handbook*, Machinability Data Center, Metcut Research Associates, Inc., 1973.
3. Manufacturers' catalogs and other literature.
4. *Metal Cutting Tool Handbook*, Metal Cutting Tool Institute, 1965.
5. ASTME: *Tool Engineers Handbook*, 3d ed., McGraw-Hill, New York, 1976.

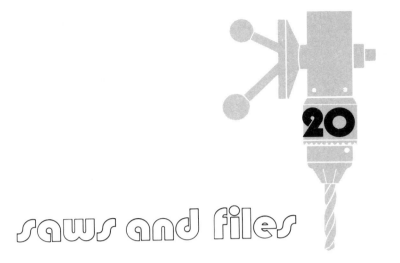

saws and files

Many parts for machines of all kinds are cut from round, square, flat, or hexagonal bar stock or tubing. Other parts may be made from structural shapes (such as I beams, channels, and angles) cut off to the desired length.

Cutoff may mean cutting various lengths of any material from aluminum to tough alloys, or it may mean cutting off "slices" 1 in. (25 mm) thick, from 4- to 10-in. (100- to 250-mm) diameter bars. This cutting operation should be done in the shortest possible time consistent with the desired quality. Often this is most efficiently done with some style of power saw.

CUTOFF SAWS

Cutoff saws are used almost exclusively for cutting stock to a specified length.

Power Hacksaw

The power hacksaw is the original and least expensive saw for the work. As shown in Fig. 20-1, these saws work the same as a hand hacksaw: they cut on the forward stroke and then lift slightly so that the blade does not drag on the return stroke.

The *size* of a power hacksaw is the cross section of the largest piece of stock which it can cut. Typical sizes are 6 × 6 in. (150 × 150 mm) to 24 × 24 in. (600 × 600 mm). The motors used will vary from 1 to 10 hp (0.75 to 7.5 kW).

The *speed* of these saws is in *strokes per minute*. This may be from 30 strokes per minute for large cuts with heavy saws on difficult materials, up to 165 strokes per minute on carbon steels and nonferrous materials. The hacksaw usually has four to six different speeds available.

Feed may be a positive advance per stroke or may be gaged by a friction or pressure drive. The smaller power hacksaws feed about 0.006 in. (0.150 mm) per stroke and the larger ones 0.012 to 0.030 in. (0.150 to 0.75 mm) per stroke. Feed pressures will be 450 to 750 lb (200 to 340 kgf) on the blades.

Work is held in a built-in vise, which may be hand or power operated.

Automatic power hacksaws (Fig. 20-2) will feed the stock a preset length, clamp the vise, cut off, and raise the saw for the next cut—all with preset gages and limit switches. These will cut accurate lengths to within 0.010 in. (0.25 mm) or less. They are, of course, expensive (from $5000 to $10,000), and so they would be used only if a large amount of work is to be done.

Fig. 20-1 A small hydraulically operated hacksaw. Made in capacities up to 12⅝ in. (320 mm) diameter. (*Kasto-Racine.*)

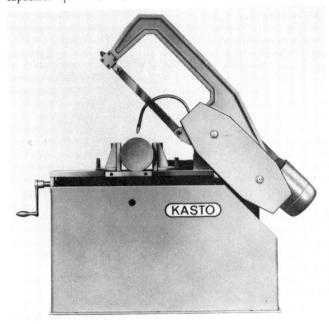

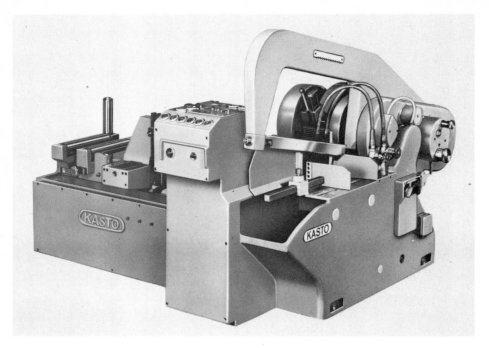

Fig. 20-2 A heavy-duty electro-hydraulic automatic power hacksaw. Length of work, feed, and clamping can be preset. Capacities to 15¼ in. (400 mm). *(Kasto-Racine.)*

Band-saw Cutoff Machines

Use a continuous band-saw blade as shown in Fig. 20-3. The blade is run between guide rollers which twist it into a vertical plane at the cutoff position. The cutting action is continuous, and thus faster than a hacksaw.

The *size* of a band saw is specified by the maximum size work which can be cut, the same as the power hacksaw.

The *speed* is in feet per minute (fpm) or meters per minute (m/min) and is indicated directly on the machine. There are usually four to six speeds available either with step pulleys or, on the larger machines, through a gear box. The speeds will run from about 60 to 300 fpm (18 to 90 m/min).

Feed is by the weight of the saw, which is modified by a limiting hydraulic valve. On the large high-production models a positive feed is sometimes available.

Work is held in a built-in vise, the same as with the power hacksaw.

Automatic band cutoff saws are available for high-production work. They operate much the same as the power hacksaws previously described.

Comparison of Hacksaws and Band Saws

The decision as to which type of cutoff saw to buy is often influenced by custom or habit. However, there are definite factors which can be considered.

cost
A hacksaw is much less expensive, often about half the cost of a band saw of equal size and power.

saw blades
The hacksaw blades may cost one-half to one-quarter the cost of a band-saw blade. However, the hacksaw will become dull in one-half to one-quarter the number of cuts that the band saw will make.

The hacksaw blade is almost unbreakable and is somewhat less likely to have its teeth stripped off by hard spots in the material being cut.

kerf (width of cut)
The band-saw blade is thinner than the hacksaw blade, especially for the larger sizes. Thus, less metal is wasted in the cut. However, this "saving" is often lost because of the 2- to 6-in.-long (50- to 150-mm) "stub end" which is thrown into the scrap bin when the bar of stock is used up.

speed
The band saw will cut off stock up to twice as fast as the hacksaw. However, it does take more care and more time to change blades, adjust saw guides, and regulate feeds. Thus, the plain hacksaw can be used by less experienced operators.

Other Cutoff Saws

If the material to be cut is very hard, an *abrasive wheel cutoff saw* is used. These abrasive wheels are from 6- to 30-in. (150- to 750-mm) diameter. The large-sized machines are also used for cutting extra large stock, I beams, etc., which may have to be handled with a hoist. The cut surface is quite smooth, and almost any material can be cut (refer to Fig. 15-20).

Cold saws, which use circular steel blades similar to

those used in wood saws, are used for both small and large work. They cut quite fast and give an excellent finish.

These do not run "cold," but with proper cutting fluids the temperature is low enough to avoid damaging the material. These saws are often designed as automatic cutoff machines, especially for mild steel and nonferrous materials.

CONTOUR BAND SAWS (VERTICAL BAND SAWS)

These are vertical band saws as shown in Fig. 20-4. They can be used for cutoff work but are not equipped to make the heavy straight cuts required as easily as cutoff saws. The principal use of these vertical band saws is for cutting irregular shapes, making beveled cuts on tubing or solid stock, and doing the slotting or slitting needed for clamping. They also are used sometimes for cutting off the sprues and risers from castings.

Tool- and diemakers use the contour saw for roughing out both interior and exterior shapes. By substituting special *file bands* or narrow abrasive belts, the tool- and diemaker can get the size, shape, and finish needed in accurate work.

Size of a contour saw is the distance from the blade to the column. This is the radius of the largest circle which can be cut and is referred to as the *throat size*.

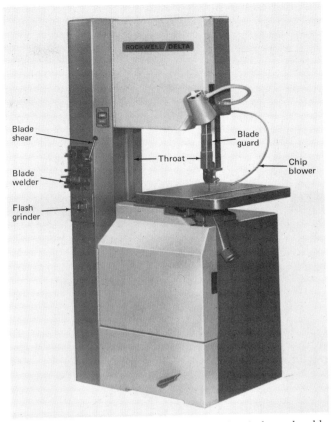

Fig. 20-4   A vertical-band contour saw. Blade grinder and welder shown on left side. (*Power Tool Division, Rockwell International.*)

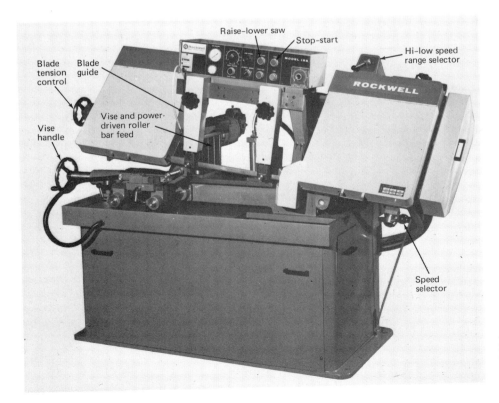

Fig. 20-3   A modern semiautomatic horizontal band saw for cutoff work. (*Power Tool Division, Rockwell International.*)

The sizes are usually from 15- to 36-in. (380- to 900-mm) throat, though machines with 60-in. (1500-mm) throat are available.

Speed is read directly from a dial or vertical scale in feet per minute (fpm) or metres per minute (m/min). Most contour saws have a gear box which gives two or three ranges of cutting speeds. A common speed range is from 50 to 5000 fpm (15 to 1500 m/min). Special high-velocity machines can reach 15,000 fpm (4500 m/min).

Feed is frequently by hand pressure, the same as when cutting wood. Mechanical and hydraulic feed attachments can be added, but there is no specific "inches per minute" control on any of the feeding devices.

Work holding is often by hand since the saw cuts downward, thus, holding the work on the table. Special angle jaws (Fig. 20-5) are convenient accessories for guiding the work.

SAW BLADES

All saw blades have certain common characteristics and terminology.

Rake angles are 0° (neutral rake) on most saw blades (Fig. 20-6a). Some have a positive rake angle as shown in Figs. 20-6b and 20-7c.

The **width** of a saw blade is its total width including the teeth, as shown in Fig. 20-6a.

The **set** of a saw blade means the offsetting of some teeth so that the *back* of the blade clears the cut as shown in Fig. 20-6c. Two principal types of set are used, as illustrated and described in Fig. 20-6d and e. The **raker set** is the most frequently used and is furnished with all hacksaws and band saws unless otherwise specified.

The **kerf** is the width of the cut made by the saw blade, i.e., the material cut away. The thickness of the blade is called the *gauge (or gage)*. Common gages for saw blades are: Hacksaw blades, 0.032 to 0.100 in. (0.8 to 0.25 mm) thick; band-saw blades, both cutoff and contour blades, 0.025 to 0.042 in. (0.6 to 1.0 mm).

The **pitch** of a saw blade is the distance between the tops of two adjacent teeth. This is specified in teeth per inch (sometimes abbreviated as tpi) as shown in Fig. 20-6f. The most used pitches are 4, 6, 8, 10, and 14 teeth per inch (approximately 6, 4, 3, 2.5, and 1.8 mm pitch). Occasionally larger or smaller pitches are used, as noted in Figs. 20-6 and 20-7.

The **length** of a power *hacksaw* blade is its end-to-end measurement. This depends on the size of the saw and may be from 12 to 30 in. (300 to 760 mm) long. The longer lengths are usually wider and thicker to give the needed strength.

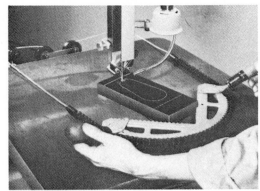
Workholding jaws

Hydraulic contour feed controls

Fig. 20-5   Special jaws and clamps which can be used with vertical (contour) band saws. *(DoAll Company.)*

The length of a *band-saw* blade varies with the size and make of the saw. These may be cut to length from 100-ft (30-m) coils and welded together on the butt-welding attachment which is supplied with every band-sawing machine, or they may be ordered "cut and welded" by the vendor. These blades average about 15 ft 6 in (4.7 m) long.

The *teeth* for band saws are made in the three styles shown and described in Fig. 20-7. Hacksaw blades are usually the standard tooth shape.

The *material* for all saw blades may be:

1. *Carbon steel*—general utility for small-lot, low-speed work. The least expensive blade, these may have a hard "back" for greater wear.

2. *High-speed steel*—this costs two to three times as much as carbon steel, but it is much longer wearing and is a necessity for the difficult-to-machine metals.

3. *High-speed edge*—a carbon-steel blade with a narrow strip with HSS teeth welded on. This is a tough blade, intermediate priced, and widely used for most materials.

4. *Tungsten carbide–tipped blades*—available in a few sizes. Used only on large, very rigid sawing

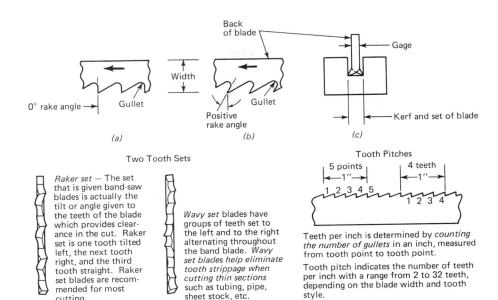

Two Tooth Sets

Raker set — The set that is given band-saw blades is actually the tilt or angle given to the teeth of the blade which provides clearance in the cut. Raker set is one tooth tilted left, the next tooth right, and the third tooth straight. Raker set blades are recommended for most cutting.

(d)

Wavy set blades have groups of teeth set to the left and to the right alternating throughout the band blade. *Wavy set* blades help eliminate tooth strippage when cutting thin sections such as tubing, pipe, sheet stock, etc.

(e)

Tooth Pitches

Teeth per inch is determined by *counting the number of gullets* in an inch, measured from tooth point to tooth point.

Tooth pitch indicates the number of teeth per inch with a range from 2 to 32 teeth, depending on the blade width and tooth style.

(f)

Fig. 20-6 *(a)*, *(b)* Rake angles. *(c)* Kerf and set. *(d)*, *(e)* Raker and wavy set. *(f)* Teeth per inch. *(Adapted from Armstrong-Blum Mfg. Co.)*

machines for high-production sawing of difficult materials.

Different manufacturers use special alloys of carbon and high-speed steels; thus a direct comparison cannot always be made.

Selecting a Saw Blade for Power Sawing

Choosing the blade material is relatively simple. If most material is low-alloy steel, aluminum, and similar materials, a carbon-steel blade is satisfactory for low production. Otherwise, the HSS or HSS-tipped blades would be used.

The tooth *pitch* is important. A saw cuts like a broach, with the advantage that it is not totally enclosed. The chips curl up in the saw-tooth gullets as the saw advances. Thus, if long cuts are made with teeth spaced too closely, the gullets will fill up and the teeth will stop cutting, causing a long cutting time.

Figure 20-8 illustrates the effect of too many or too few teeth. At least two teeth should contact the work.

This may mean that 24 or 32 teeth per inch (1.0- or 0.8-mm pitch) are required for sawing tubing.

The *size* of the blade to be used for cutoff saws is usually specified by the manufacturer. Contour saws can, by changing the saw guides, use a wide range of blade widths. Naturally, the thinner blades cut smaller circles or arcs. Examples of minimum radii which can be cut are:

a $\frac{1}{4}$-in.-wide (6.4-mm) blade: $\frac{5}{8}$-in. (16-mm) radius

a $\frac{3}{4}$-in.-wide (19.1-mm) blade: $5\frac{7}{16}$-in. (138-mm) radius

For economy of sawing time and blade life, use the widest blade allowable in your saw within the limits of the radius to be cut.

FILES

Though there are dozens of shapes and sizes of files, their direct use in the manufacturing process is quite limited. Files are used principally for burring (or deburring), i.e., removing the sharp rough edges which result from most machining processes. When burring

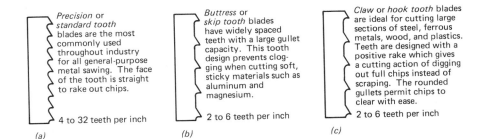

Precision or *standard tooth* blades are the most commonly used throughout industry for all general-purpose metal sawing. The face of the tooth is straight to rake out chips.

4 to 32 teeth per inch

(a)

Buttress or *skip tooth* blades have widely spaced teeth with a large gullet capacity. This tooth design prevents clogging when cutting soft, sticky materials such as aluminum and magnesium.

2 to 6 teeth per inch

(b)

Claw or *hook tooth* blades are ideal for cutting large sections of steel, ferrous metals, wood, and plastics. Teeth are designed with a positive rake which gives a cutting action of digging out full chips instead of scraping. The rounded gullets permit chips to clear with ease.

2 to 6 teeth per inch

(c)

Fig. 20-7 The three most used styles of saw teeth. *(Armstrong-Blum Mfg. Co.)*

Coarse pitch gives chip space for long cuts (2 to 8 teeth per inch)

(a)

Fine pitch — teeth will clog on wide cuts. Must be used for narrow stock (10 to 14 teeth per inch)

(b)

At least two teeth on material (14 to 32 teeth per inch for tubing)

(c)

Coarse pitch straddles narrow work, stripping off the teeth

(d)

Fig. 20-8   Relationship of tooth spacing to types of cutting being performed.

is done by hand, several styles of files and scrapers are used. Some of the many kinds are shown in Fig. 20-9.

Many files are also used by tool- and diemakers for sizing and smoothing the dies, jigs, and fixtures which they make.

As files are hardened from HRC 45 to HRC 50, they can be used to get a rough check on the hardness of a part before attempting to repair or rework it.

Some flat files are cut on all sides, but some have one or two smooth "safe" edges so that they can be used close to surfaces which should not be scratched.

The *kinds* (shapes) of files are numerous. The ones most used for burring or smoothing on a lathe are:

Mill file—rectagular tapered end, single cut

Flat file—like a mill file but double cut

Hand file—rectangular, tapers in thickness only, with one safe edge, double cut for fast, heavy filing

Round, half-round, square, and triangular files are used for burring and for final fitting of parts at assembly (Fig. 20-9).

A special *"long angle" lathe file* is made like the mill file, single cut, but with two "safe" edges. This file is often used to break sharp edges and to deburr a workpiece while it is still on the lathe. Occasionally, it is used for smoothing the work.

Most of these kinds of files appear in Fig. 20-9, as do the *curved* tooth files used for aluminum and copper, and some of the small *Swiss-type files* used by diemakers.

The *size* of a file is specified by its length. The other dimensions are standardized by the manufacturers according to the kind or style of file. The most used lengths are 6 to 14 in. (150 to 360 mm), though larger and smaller files are made.

Cut or *tooth spacing* of a file is specified as coarse, bastard, second cut, or smooth cut. Dead smooth and rough (extra coarse) are sometimes used also. The actual spacing of the teeth decreases with a decrease

in the length of a file. Thus, an 8-in. (200-mm) mill bastard file will have more closely spaced teeth than a 10-in. (250-mm) mill bastard file (see Fig. 20-10).

Figure 20-10 also shows *single-cut files*, which are used to produce smooth surfaces, and *double-cut files*, which are used with heavier pressure for fast metal removal with a rougher finish.

Safety requires that a file *never* be used without a handle firmly secured over the tang. Especially on lathe work, the bare tang can be driven right through your hand.

Do not throw files together in a heap. Remember that they are cutting tools, with shaped and hardened cutting edges. Treat them as you would any good cutting tool.

Do not use a file for burring (except maybe on lathe work) if other methods such as tumbling, sand blasting, or vibrating will do the job. Hand burring is usually quite expensive.

Fig. 20-9   Several styles of files. (*a*) Curved tooth, for soft metals. (*b*) Mill files. (*c*) Triangular files. (*d*) Half-round file. (*e*) Cantsaw, for saws. (*f*) Rasp, for wood or soft metal. (*g*) Round file. (*h*) Scrapers, for hand deburring. *(Nicholson File Co.)*

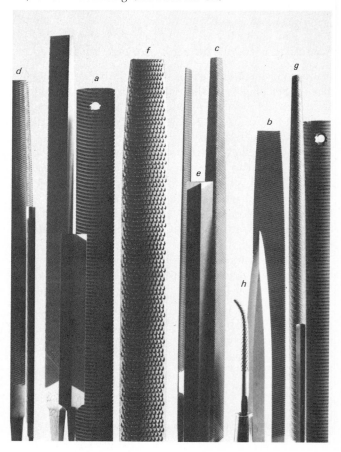

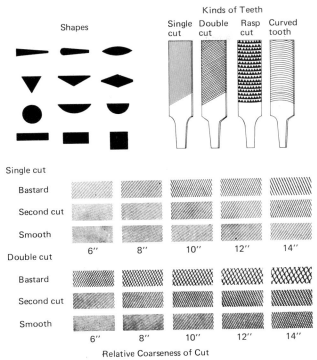

Fig. 20-10 The spacing of the teeth on a file varies with the coarseness (cut) and the length of the file. Some of the shapes and kinds of teeth used for files.

Review Questions and Problems

20-1. What are cutoff saws? What two basic types are available?

20-2. How is the size of cutoff saws specified? The speed? The feed?

20-3. List the factors to be considered when deciding which type of cutoff saws to purchase and tell how each factor affects your decision.

20-4. When are abrasive wheel cutoff saws used?

20-5. What are contour saws and when are they used?

20-6. What is the "set" of a saw blade and why is it important?

20-7. What is the "kerf"? What is the gage of a saw blade? What is tooth pitch?

20-8. What materials are used to make saw blades?

20-9. What should be considered when selecting a saw having the correct tooth pitch?

20-10. What is the principal use of files in manufacturing?

20-11. When are files economically used for deburring?

20-12. What is a "safe edge" on a file?

20-13. List some of the kinds (shapes) of files most commonly used.

20-14. What is meant by a single-cut file? A double-cut file?

20-15. How is the "cut" or tooth spacing specified?

20-16. What safety precaution should always be followed when using a file?

References

1. *File Filosophy*, Nicholson File Co., 1971.
2. Saw and saw machine manufacturers' catalogs.
3. *Technical Guide to the Selection and Use of Marvel Cutting Tools*, Armstrong-Blum Co., Chicago, Ill., 1971.
4. ASTME: *Tool Engineers Handbook*, Sec. 65 to 67, McGraw-Hill, New York, 1976.

welding and related processes

Welding (arc, gas, or resistance) is the joining of parts by fusion, that is, melting the material of two pieces so that the metal flows together, making a relatively homogeneous joint. Frequently metal is added to the joint to strengthen it. This definition is necessarily incomplete because there are over 50 variations of welding processes. However, five of these are used for possibly 80 percent of all the welding done, and a few others are used quite frequently. These are the methods which will be discussed in this chapter.

In some manufacturing plants, the "welding department" is a 5-m-square (16-ft) room with one or two welding machines in it. In another shop, the welding department may be over 100 m (328 ft) long by 50 m (164 ft) wide, with a wide variety of welding equipment available. Some shops use welding only for occasional repairs, while others will design many small and large parts as weldments.

There is a vast store of literature on welding processes, machines, rods, metallurgy, etc., available for further study if this becomes important in your future work. Some of the resources are listed at the end of this chapter.

USES OF WELDING

Although welding did not start to be widely used until 1940, it is now used in making bridges and buildings. The large and small tanks used with air and vacuum pumps, paper-processing tanks, and oil tanks are usually welded. Overland pipelines and farm and road equipment use large amounts of welding.

Machine bases and frames, brackets of many kinds, lever arms, and bearing supports are often designed as weldments. The automobile industry uses hundreds of welding machines. A few of the possibilities are shown in Fig. 21-1.

Advantages of Welding

1. A properly made weld can be stronger than the part on which it is used.
2. No holes are needed for bolts or rivets.
3. No patterns (such as used in castings) are needed.
4. The equipment is inexpensive, costing from $100 to $8000, with the average being from $300 to $2500.
5. The equipment can be portable and used indoors or outdoors.
6. The process allows considerable freedom in design.

Disadvantages of Welding

1. A good welding job requires a skilled operator. Even automatic welding needs a skilled operator to set it up, though new processes are decreasing the amount of manual skill needed.
2. Each part of a weldment must be cut to size and shape (and sometimes machined on the edges) before it can be welded.
3. Jigs and fixtures (sometimes quite large) are often needed to hold parts in position for welding.
4. There is a constant battle with the distortion which is a result of the very high temperatures and localized heating which is normal in most welding. The distortion can often be controlled, as will be described later in this chapter.

TYPES OF WELDS (Fig. 21-2)

There are a number of variations in types of welds

made, but the joints shown here are the most commonly used.

Fillet Weld

Figure 21-2a shows the fillet weld, used wherever two or more pieces come together at an angle, usually 90°. This is the most frequently used weld. No special joint preparation is needed except to fit the parts well.

Square Plain Butt Weld

This weld type (Fig. 21-2b) is used on material $\frac{3}{8}$ in. (9.6 mm) or lighter, though heavier material can be welded this way when using automatic welding. As no joint preparation except fit-up is needed, this is an inexpensive weld.

Vee Butt Weld

The vee butt weld (Fig. 21-2c and d) is used for joining flat plates or the seam on circular pieces, such as tanks and pipes. The *single vee* is used where welding can be done from only one side. It is used for metal thicknesses of from $\frac{1}{4}$ in. (6.4 mm) to $\frac{1}{2}$ in. (13 mm). The vee is cut at an angle of 45° to 60° by machining, rough grinding, or flame cutting. The vee is sometimes made in a U shape especially if the weld is deep.

The *double-vee* joint is preferred for material $\frac{1}{2}$ in. (12.7 mm) or over. It costs more for preparation but requires only about half as much welding rod as the single joint. Moreover, warpage can be reduced by alternating the welding, first on one side, then on the other.

Fig. 21-1   Some typical work done with arc welding.

Fig. 21-2   The principal kinds of arc welds.

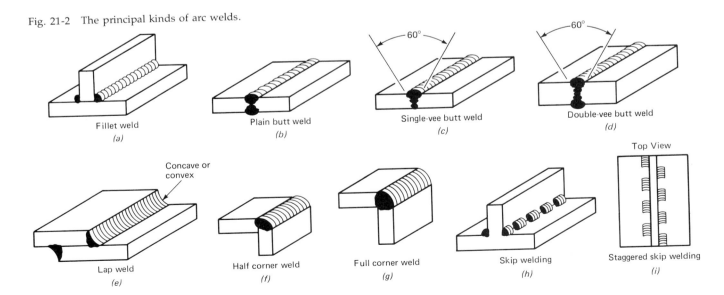

Fillet weld
(a)

Plain butt weld
(b)

Single-vee butt weld
(c)

Double-vee butt weld
(d)

Concave or convex

Lap weld
(e)

Half corner weld
(f)

Full corner weld
(g)

Skip welding
(h)

Top View

Staggered skip welding
(i)

Lap Joint

This weld consists of two (sometimes only one) fillet welds (Fig. 21-2e). This joint does not require any special preparation, so it is quite inexpensive. The single-fillet lap joint should be used only where light loads are to be held. For heavier loads, the double-fillet lap or the double-vee butt welds should be used.

Corner Joints

Corner joints are used at any angle, though the 90° angle shown is the most common. The *half-open* corner weld (Fig. 21-2f) is used for lightly loaded corners. It is fast, inexpensive, and fairly easily held in position for welding.

The *full corner* weld (Fig. 21-2g) is more expensive and more difficult to hold in position, but it is suitable for heavy loads. For extreme conditions a weld is also made on the inside of the full-open joint.

Corner joints on material thinner than $\frac{1}{8}$ in. (3.2 mm) require special care so that the weld does not burn through.

Tack Weld

A *tack weld* is a small weld, about $\frac{1}{8}$ to $\frac{1}{2}$ in. (3 to 13 mm) long, made at a few points on an assembly merely to hold it in position so that the operator can more easily do a complete job. Tack welds can also help to control warping.

Skip Weld

This type of weld consists of alternate welded and unwelded spaces along a joint (Fig. 21-2g). It is used where full strength is not needed, as it saves time and metal. Staggered skip welding (alternating on opposite sides) (Fig. 21-2h) can reduce warping.

SOME BASIC VOCABULARY

There are some abbreviations and terminology which will be used in this chapter. These are listed here to make it easier to understand the discussions.

ACHF—Alternating current welder, with a high-frequency current added for starting only or for welding of some types.

Arc blow—A condition sometimes present in dc welding where the arc wanders from the desired position as if it was being blown to and fro.

Duty cycle—The percentage of a *10-min* cycle during which a welding machine can be used without overheating. Thus, a 40 percent duty-cycle welder can be used at full capacity for only 4 out of every 10 min. There are many 100 percent duty-cycle welders made; in fact, they are a necessity for automatic welding.

DCRP—Dc welding machine used with *reverse polarity*, that is, with the electrode as the plus (positive) side of the circuit.

DCSP—Dc welder with the electrode *standard polarity*, that is, the electrode is the minus (negative) side of the circuit.

Fit-up—This refers to the accuracy or closeness with which parts to be welded are positioned before welding. A poor fit-up (parts held with gaps between them) increases the cost of welding.

Flux—A chemical added to the electrode (welding rod) coating to dissolve the surface oxides on the metal being welded.

Slag—A mixture of oxides and impurities which floats to the surface of a weld while the metal is molten. This is an acidic or basic material which has been added to the coating of the welding rod. Cellulose, silica, limestone, and rutile (TiO) are used.

The slag slows the cooling and protects the weld from absorbing harmful gases from the atmosphere. Slag must be removed before making a second weld over the first pass. This may be done by wire brushing or chipping.

The *speed* of welding is given in inches per minute or pounds of welding rod per minute. This is usually assuming 100 percent efficiency. Hand welding is seldom more than 60 percent efficient because of the need for changing electrodes as they are consumed, fitting up, chipping slag, etc. Automatic welding may be up to 80 percent efficient.

WELDING MACHINES

As electric-arc welding is used in about 85 percent of all the welding done, these machines will be considered first. An exception is the automotive industry, which uses a high percentage of resistance welding, a process which will be discussed later in this chapter.

A welding machine is a fairly simple piece of equipment. Its basic job is to generate a low-voltage [10 to 50 V (volts)], high-amperage [50 to 2000 A (amperes)] electric current, either alternating current (ac) [usually 60 Hz (hertz)] or direct current (dc). Many welding machines can supply either ac or dc by a flip of a switch. The basic circuit, plus ground clamps and electrode holder, is shown in Fig. 21-3.

AC Welders

Basically heavy-duty transformers, ac welders (Fig. 21-4a) can change the 110- or 220-V current to low voltage. Capacitors and resistors are added to raise the efficiency (power factor). Other refinements such as a high-frequency stabilizer are added, especially for heavy-duty work.

The different amperages needed either are tapped off the transformer, through plug-in outlets (as shown), or have a sliding coil to allow continuous selection of amperage to be used.

DC Welders

Dc welders (Fig. 21-4b) may use a motor-generator (MG) set. This is a diesel or gasoline engine or an ac electric motor driving a generator. Other dc welders use solid-state (semiconductor) selenium or silicon rectifiers added to an ac transformer. These rectifiers allow only half of the complete ac wave to pass through, creating a pulsating direct current. The desired amperage is selected from a continuous range dial.

AC-DC Welders

While more expensive, ac-dc welders allow a wide variety of work to be done with one welding machine. A selector switch connects the proper circuits.

Details of these circuits can be found in the references at the end of the chapter or by contacting the equipment manufacturers.

AC and DC Welding Compared

Most manual welding of $\frac{1}{4}$-in. (6.4-mm) and thicker steel is done with ac welders. The welding of thinner sheet metal is often more efficiently accomplished with a dc machine.

Usually dc welding is preferred when welding nonferrous metals. Dc welders are used most frequently with the gas-shielded welding methods (TIG and MIG) which will be described later. Direct current can also be used with the electrode (the welding rod or wire) having either positive or negative polarity.

Although dc welding is used with a greater variety of welding processes, the welding of steel is by far the largest use, so more ac welders are in use today.

ARC WELDING METHODS

The five most used arc-welding methods will be described in the order of their frequency of use. Follow-

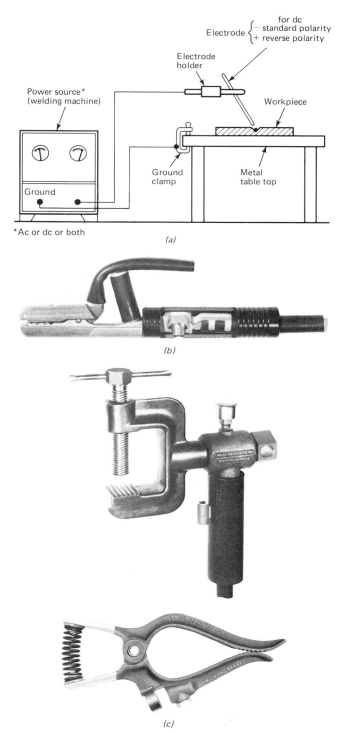

Fig. 21-3 (a) The basic arc welding circuit. (b) Hand-held electrode holder. (c) Types of ground clamps. (*Tweco Products, Inc.*)

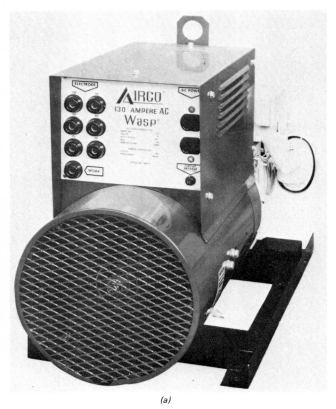

(a)

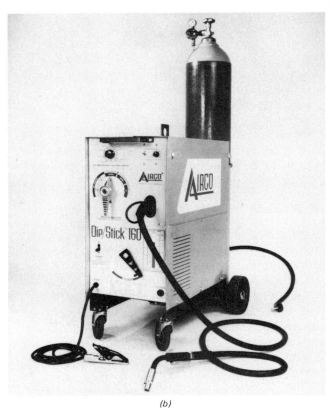

(b)

Fig. 21-4 (a) A gasoline-engine-driven, small welding machine with 35 percent duty cycle. (b) A 160-A dc welding machine, for stick or inert gas. Shown set up for tungsten inert gas. (*Airco Welding Products Division, Airco, Inc.*)

ing these are three specialty types of welds which are solving certain problems more economically than the standard welding procedures.

Shielded-Arc Welding (Stick Welding)

This, the first welding method ever used, is still used for more than 60 percent of the arc welding done today. This percentage is down from over 80 percent in 1962 and is still declining slowly as faster methods are developed.

Early in the development of welding, it was discovered that if the welding rod (electrode) was *coated* with the appropriate material, the weld was made more rapidly, the arc was more stable, the weld was more ductile, and the welded joint was less likely to corrode. These are called *shielded rods*, compared to *bare rods*. The basic process is shown in Fig. 21-5.

This *hand-held shielded-arc* or *stick welding* is the most frequently used method of welding. It is used mostly on steel (of any alloy) and is not often used on materials such as aluminum, stainless steel, and titanium.

The equipment for shielded-arc welding is inexpensive: the cost would be for the welding machine

plus an electrode holder and the proper electrodes. Electrodes for most of this type of welding cost only 15 to 30 cents per pound.

The welder strikes an arc which develops a temperature of 6000°F (3200°C) or over. This arc melts both the electrode and the base material. By moving the welding rod at the proper speed along the joint, the added metal and the base fuse together, making a joint which is often stronger than the original work-

Fig. 21-5 The action of the shielded arc (SMA) or stick welding process. (*Adapted from Lincoln Electric Co.*)

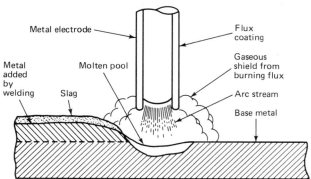

piece. An example of shielded-arc welding on large work is shown in Fig. 21-6.

The *electrode* (welding rod) is a coated metal rod from $\frac{1}{16}$- to $\frac{3}{8}$-in. (1.6- to 9.6-mm) diameter and 15 to 30 in. (380 to 760 mm) long and of approximately the same chemical composition as the material being welded.

The *coating* may be only a few thousandths of an inch thick but is more frequently $\frac{1}{64}$ to $\frac{1}{32}$ in. (0.4 to 0.8 mm) thick. It is made of materials such as cellulose, silica, lime, or rutile (TiO).

Welding rods are numbered and color-coded to identify the different types. Some typical rods and their uses are given in Table 21-1.

Iron-powder-covered electrodes have a very heavy (up to $\frac{1}{4}$ in., 6.4 mm) coating containing a large quantity of powdered iron. This type of welding rod gives very smooth welds and can often double the welding speed. It is a much more expensive electrode, but it is used especially when large welds on long sections must be made.

Shielded-arc welding is usually ac when welding steel in the flat or horizontal position. DCSP and DCRP are used for overhead and vertical welds (Fig. 21-7). These "out-of-position" welds (overhead and vertical) also require the use of different electrodes.

Gas-Metal-Arc Welding (GMAW)

Although *gas-metal-arc* is the official description of this welding method, the earlier name of *MIG* (metal inert gas) is still widely used, especially in the shops. Both names will be used here so that you will recognize them as the same process.

Some *advantages* of GMA (MIG) welding are:

1. No flux required
2. High welding speeds, often twice the rate of stick welding
3. Increased corrosion resistance
4. Easily automated welding
5. Welds all metals including aluminum and stainless steel

These factors have doubled the use of MIG welding since 1963.

Metals such as stainless steel and aluminum are difficult to weld because a very strong oxide coating forms on them very rapidly. This prevents bonding of the electrode material with the workpiece material.

In gas-metal-arc (GMA) welding (Fig. 21-8), the welding area is flooded with a gas which will not combine with the metal (an inert gas). The rate of flow of this gas is sufficient to keep the oxygen of the

Fig. 21-6 A student preparing to do vertical shielded-arc (stick) welding. Notice complete protection with goggles, helmet, long sleeves, heavy gloves, and leather jacket. (*Lincoln Electric Co.*)

air away from the hot metal surface while welding is being done.

gases used for GMAW (MIG)

Carbon dioxide (CO_2) is used for working with steel, as GMA is a clean, faster method for welding steel. Carbon dioxide is used principally because it is inexpensive.

For welding aluminum or copper, argon or argon-helium mixtures are used. Stainless-steel MIG weld-

Fig. 21-7 The four positions in which welding may need to be performed.

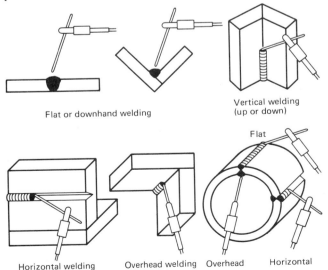

Flat or downhand welding

Vertical welding (up or down)

Horizontal welding

Overhead welding

Overhead

Flat

Horizontal

TABLE 21-1   THE MOST FREQUENTLY USED
ELECTRODES (WELDING RODS)
for WELDING STEELS

Electrode Number	Current Used	Use
6010	DCRP	Deep penetration, smooth arc, all positions x-ray quality welds, light slag.
6011	Ac or DCSP	Similar to 6010, but any type of current.
6012	Ac or DCSP	Excellent general-purpose rod. Medium penetration, all positions, will fill in poor fitups. Higher deposition rate than 6010 or 6011. Most used electrode.
6013	Any	Shallow penetration, good for thinner metals. All positions
6014	Any	Flat and horizontal welds only. About 30 percent iron powder, high deposition rate.
6018 7018	Any	Low hydrogen with iron powder. Good for heavy plate and large welds.
6024 7024	Any	About 50 percent iron powder for flat and horizontal positions. Mild penetration, fast deposition rate.

First two numbers = approximate tensile strength of the electrode material
Third number = permissible welding position (1, all positions; 2, horizontal and flat only; 3, flat only)
Fourth number = type of coating and current

ing is done with either argon-oxygen or helium-argon gas mixtures. Titanium requires pure argon gas shielding, and the copper-nickel and high-nickel alloys use argon-helium mixtures. GMA (MIG) welding equipment in use is illustrated in Fig. 21-9.

GMAW equipment

The *welding machine* is dc constant voltage, with both straight and reverse polarities available, and capac-

ities from 25 to 900 A, depending on the size of the work to be done. The weld metal is deposited as drops (short-circuiting) or spray (Fig. 21-10) depending on the amperage.

The *wire feeder* holds coiled wire (the electrode) of the proper alloy. The wire is pushed through the cable into the welding nozzle or "gun." Speed is controlled by the rate of "burn off" or melting of the wire and may be up to 200 ipm (5.1 m/min).

The *welding wire* (continuous electrode) is very often bare. Very lightly coated or flux-cored (a flux inside a tubular electrode) wire is also used. The wire is usually in diameters of 0.035 to $\frac{1}{16}$ in. (0.90 to 1.6 mm); however, sizes up to $\frac{1}{8}$ in. (3.18 mm) are made. It is shipped in spools or coils ready for use.

The *composition* of the welding wire is made for all alloys of stainless steels, aluminum alloys, and steel with some variations in the alloys to make the electrode wire work better with the different automatic welding processes.

The *gases* are stored in heavy walled cylinders and are metered through valves with pressure and flow gages to assist in controlling the proper percentage of each gas. Gas ranges from 15 to 50 ft³ [425 to 1400 l (litres)] per hour of actual welding are used.

The nozzle or "welding gun" (Fig. 21-11) has to bring the continuous wire electrode, the gases, and a trigger control mechanism into one compact lightweight unit which an operator can handle.

automatic GMA welding

MIG (GMAW) welding can be completely automatic. The welding nozzle can be mounted on a carrier which travels along a track for straight welds such as seams in large tanks. The welding equipment is sometimes held stationary while a cylinder is turned under it, thus welding all around automatically. These methods can result in considerable cost reduction if the quantity justifies the extra equipment.

Fig. 21-8   Simplified schematic of one type of GMA welding. Control wires omitted. Water cooling of the nozzle is added for 750 A or over.

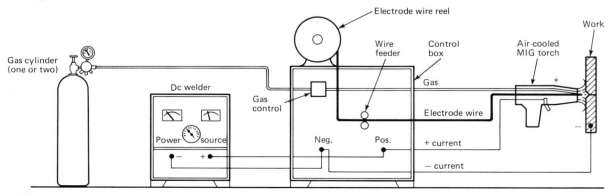

Gas-Tungsten-Arc Welding (GTAW)

This process, often called *TIG (tungsten inert gas) welding*, is similar to MIG in that it uses the same gases for shielding. It is used for welding the same metals: aluminum, stainless steel, and nonferrous alloys. Gas-tungsten-arc (TIG) welding is not used as often on plate over $\frac{1}{4}$ in. (6.4 mm) thick, but it is easier than MIG welding for thin plates and small parts.

A sketch of the *equipment used* is shown in Fig. 21-12. The gas flow and its controls are similar to those used with GMA welding. However, GTA uses a *permanent, nonconsumable tungsten electrode*. The tungsten electrode is used only to generate an arc. This arc does not melt the tungsten, which has a melting point of over 6000°F (3300°C). The end of the welding gun where the arc is created either is made of high-impact ceramic or is water-cooled.

Due to the gas shielding and the lack of filler metal, exceptionally clean welds can be made with no scale or alloying of the workpiece material. GTA is not as fast as GMA, but it makes very smooth welds of high quality.

When welding metal which is over 0.100 in. (2.5

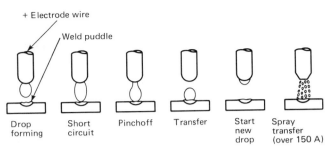

Fig. 21-10   The two ways in which filler wire melts and transfers in gas-metal-arc welding.

mm) thick, additional metal is fed into the weld area *by hand*, often using bare rods 36 in. (914 mm) long as shown in Fig. 21-13. This, of course, uses both the operator's hands, so the work must be held in position with clamps or fixtures. TIG welding requires a highly skilled operator. Automatic feeding of the filler wire into the TIG arc to speed up the process has been developing since 1972.

The welding machine is usually set for DCSP (negative electrode), though DCRP is very good for

Fig. 21-9   (*a*) A complete GMA welding outfit: machine, wire feeder, gas, and welding gun. (*b*) MIG welding stainless steel. (*Linde Division, Union Carbide Corp.*)

(a)

(b)

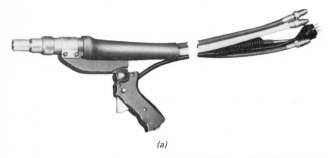

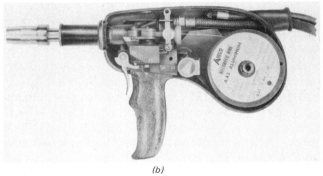

Fig. 21-11 (*a*) A 750-A, water-cooled MIG welding gun. (*b*) A small gun, with 1-lb (0.45-kg) spool of wire, for small work. (*Air Reduction Sales Company.*)

thin aluminum plate. As little as 3 A may be used for work on very thin metals and intricate parts.

Cleaning the weld area is important in both GMA and GTA. The surface to be welded must be free of rust, oil, grease, paint, and scale or the weld may have imperfections.

Submerged-Arc Welding

Submerged-arc welding, sometimes called *hidden arc or subarc welding,* is exactly that: the arc cannot be seen. There is no need for a large helmet or a leather apron, though safety glasses should be worn as routine protection. Use of this process is increasing as its advantages become known.

This is one of the fastest and most efficient processes for welding long flat joints such as for heavy welded pipe or long butt joints in plate or structural shapes. Subarc welding can also be used to join ends of pipe or tubing if the welding machine is kept stationary and the pipe rotated under it. Hand welding of shorter length joints is easily accomplished. The basic process is illustrated in Figs. 21-14 and 21-15.

The *granular flux* is fed, at a controlled rate, from the *hopper* just ahead of the electrode. Thus, the "working" end of the electrode is always hidden from view.

The flux may be made of silica, metal oxides, and other compounds fused together and then crushed to proper size. Another group of fluxes is made of similar materials "bonded" and formed into granules.

Much of the flux is melted, but the top layer remains unchanged and, since it is quite expensive, it is used again. The unfused flux is easily cleaned off the weld, often with a vacuum-type cleaner.

The bare *electrode wire* is fed from a reel down through the *gun* or nozzle. The operator can "inch" the wire (move it slowly, a little at a time) for the start of the arc, and then set the proper feedrate on the

Fig. 21-12   A simplified schematic drawing of a GTA (TIG) welding system.

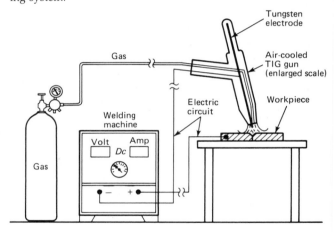

Fig. 21-13   GTA welding a Hastalloy X (high-nickel alloy) combustion chamber. (*Miller Electric Mfg. Co.*)

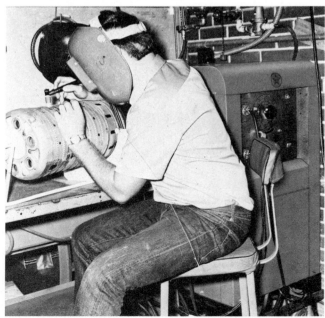

control box. Wire of several alloys for steel, stainless steel, copper, etc., is available in diameters from $\frac{1}{8}$ to $\frac{1}{4}$ in. (3.18 to 6.35 mm). Voltage used is from 25 to 40 V. *Current* used depends considerably on workpiece thickness and will vary from 250 A for thin sheet metal up to 1400 A for heavy plate [1 to $1\frac{1}{2}$ in. (25 to 38 mm)].

The principal *disadvantage* of submerged-arc welding is that it is not easily used if the workpiece is tilted more than 5°. Special *skirts* must be attached to hold the granulated flux in place for vertical welds or welds at any angle.

The *advantages* of subarc welding are listed below.

1. Partly because it is often automated, it is much faster than regular arc welding. Speeds up to 150 ipm (3800 mm/min) are possible on $\frac{1}{8}$-in.-thick (3.18-mm) steel at 100 percent efficiency.

2. The deep penetration of the submerged arc requires fewer passes on thick plate and may eliminate the need for edge preparation (cutting vees and grooves).

3. Less distortion occurs from high speeds and uniform heat input, especially when automated.

4. The operator can work more easily without the helmet and other safety equipment and is not exposed to the usual spatter.

Cored Wire Welding

An *inner-core-type wire electrode* in coils was introduced about 1970. This is an "inside-out" wire with the flux inside a tubular electrode. This welding wire is more expensive, so it must pay for itself through either higher speed or better quality or both.

The *equipment* needed for one company's rod is simple: a constant-voltage dc power source, a wire feeder, and a lightweight welding gun, which is like the MIG holder without the gas and cooling attachment (see Fig. 21-16).

The slag-gas shield coming out through the weld metal speeds up the welding and can (in the no-gas system) be used outdoors and in any position, which is especially desirable in structural steel and pipeline welding.

The cored wire electrode can tolerate considerable rust and scale on the work surface. The welds are strong and tough. The arc starts easily, and cored

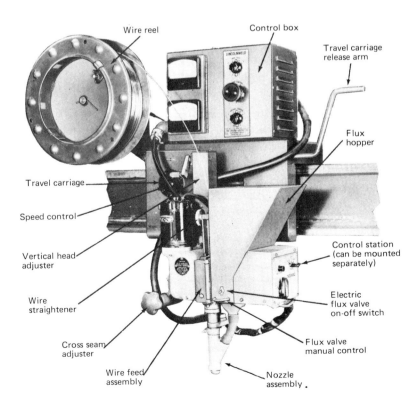

Wire reel
Control box
Travel carriage release arm
Flux hopper
Travel carriage
Speed control
Vertical head adjuster
Wire straightener
Cross seam adjuster
Control station (can be mounted separately)
Electric flux valve on-off switch
Flux valve manual control
Wire feed assembly
Nozzle assembly

Fig. 21-14 A 750-A automatic submerged-arc welder shown mounted on an I-beam rail. Power source not shown. *(Lincoln Electric Co.)*

wire produces high-quality welds at faster speeds than stick welding even when hand held. Though this is one of the newest welding methods, by 1973 it was being used for an estimated 10 percent of all arc welding.

Electron-Beam Welding (EBW)

Because an electron beam can be focused to a spot diameter as small as 0.001 in. (0.025 mm), it has been used for several years for drilling holes in diamonds and for soldering very fine wires in electronic work.

By 1968 work was started on electron-beam welders having power up to 50 kW and over, and high-production use of electron-beam welding began. By 1974, some automobile companies were placing orders for a dozen EB welders at a time.

Today automobile, airplane, aerospace, farm, and other types of equipment [including ball bearings over 4 in. (100 mm)] are being welded by the electron-beam process.

The basic theory of electron-beam welding is shown very much simplified in Fig. 21-17. The electrons are emitted from a filament (usually made of tungsten) and electrically accelerated to one-half to

two-thirds the speed of light. The stream of electrons is then electromagnetically aligned and focused, giving a highly concentrated beam. As this stream of electrons hits the workpiece, their kinetic energy is changed to thermal energy and the temperature is high enough to vaporize most materials.

A *high vacuum* is necessary around the filament so that it will not burn up. Thus, much EBW is done in a vacuum chamber. It takes 5 to 30 min. to evacuate these chambers, depending on their size, and this is nonproductive time. A *hard-vacuum* electron-beam welder is shown in Fig. 21-18. The welding head or the work is moved by numerical control or by hand,

Fig. 21-15 (*a*) A student using a hand-held submerged-arc welder. Notice the eye and hand protection. (*b*) Automatic submerged-arc welding of a long trough. (*Lincoln Electric Co.*)

(a)

(b)

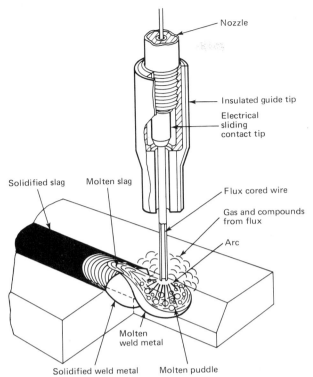

Fig. 21-16   The process used for cored wire welding. (*Adapted from Lincoln Electric Co.*)

monitoring the position by use of a television camera and screen.

Soft vacuum, *partial vacuum*, or *commercial vacuum* EBW machines have been developed and used in manufacturing since about 1970. These machines keep the filament under high vacuum, but the workpiece is located in a section of the machine which is under partial vacuum. This partial vacuum is easily maintained. A partial vacuum electron-beam welder is shown in Fig. 21-19.

Out-of-vacuum electron-beam welders are now being used in some quite high production welding jobs. The workpieces are kept within $\frac{1}{4}$ to $\frac{1}{2}$ in. (6.4 to 12.7 mm) of the vacuum chamber so that the electrons will not be scattered very much by passing through the air.

The *disadvantages* of EBW are the cost, often $60,000 to $150,000; the time and equipment needed to create the vacuum; and the precautions which must be taken to prevent eye damage and, in some machines, x-ray exposure.

The *advantages* of EBW are that the welds are clean, with no porosity since there is no air; no shielding gas is needed; and as the energy input is in a narrow, concentrated beam, distortion is almost eliminated.

The speed of EBW may be as fast as 100 ipm (2500

mm/min), and it will weld or cut any metal or ceramic, sometimes as thick as 6 in. (150 mm). Edge preparation for welding consists merely of closely aligning the edges. The beam can reach down inside a hole to weld a bushing or plug at the bottom. Some examples of electron-beam-welding jobs are shown in Fig. 21-20.

Laser Welding

The principles of operation and uses of the laser are discussed fully in Chap. 29. The laser welding process has certain advantages over electron-beam welding:

1. It operates in air, no vacuum required.
2. It can be "aimed" with simple mirrors.
3. Welds can be made *inside* transparent glass or plastic housings.

A disadvantage is that to generate the power needed for heavy welding requires a large, expensive laser, though progress is being made in decreasing the size.

Thus, lasers are used most often for work such as welding very small wires to electronic devices and similar work, called *microwelding*.

However, by the end of 1971, continuous gas lasers were being developed to the point where the power is now sufficient to enable it to compete with the electron beam on light work (such as welding automobile bodies) and some heavier work. Commercial applica-

Fig. 21-17   Simplified schematic drawing of a high-vacuum electron-beam welding machine.

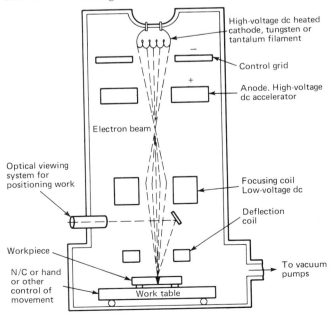

Fig. 21-18 A hard-vacuum electron-beam welder for titanium wing assemblies. Pumps to welding vacuum in 20 min. *(Sciaky Bros., Inc.)*

tions are now being developed, but the laser's use in heavy production is still limited.

Numerical control can be used to move the workpiece (or the electron beam or laser beam) along an irregularly shaped welding or cutting job. Thus, when the quantity of production justifies it, highly automated systems can be installed.

Inertia and Friction Welding

These two rather similar processes are used most frequently when one or both of the parts to be joined are cylindrical. The other part may be a flat plate, the "eye" of a piston rod, or another piece of pipe, rod, or tubing as shown in Fig. 21-21.

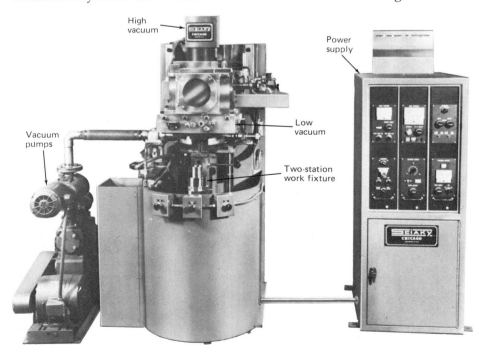

Fig. 21-19 A semiautomatic soft, or commercial, vacuum electron-beam welding machine. Electron-beam gun moves to make the weld. *(Sciaky Bros., Inc.)*

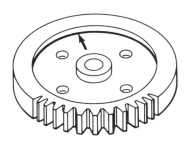 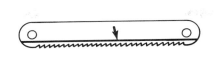 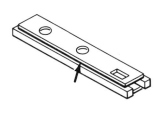 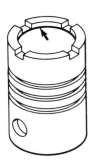

Fig. 21-20 A few examples of electron-beam welding. Arrows point to the welded joints.

The two processes are similar since they both use friction to develop heat. The temperatures developed are below the melting point of the metals being welded but high enough to create plastic flow and intermolecular bonding.

Friction welding consists of four steps:

1. One part is accelerated to the desired speed.

2. The nonrotating member is moved forward to meet the rotating member.

3. Pressure and rotation are maintained until the resulting high temperature makes the metals plastic.

4. The rotating member is quickly stopped, and the pressure is rapidly increased.

Inertia welding tends to be quicker than friction welding and depends on a heavy flywheel. Figure 21-22 shows one model of an inertia welding machine. The process consists of three steps:

1. The flywheel, with one part chucked in it, is accelerated until it stores enough kinetic energy to produce the weld.

2. The drive is disconnected from the power source, and the two parts are instantly brought together under high pressure.

3. The absorbed energy is changed to heat, bringing the two parts to the necessary plastic condition. Just before the flywheel stops, the pieces "freeze" in a permanent bond.

The *advantages* of friction or inertia welding include:

1. Many dissimilar metals can be securely joined

Fig. 21-21 Parts which have been inertia welded. (*a*) Roller bracket, tested 1.5 million cycles. (*b*) Automotive engine valve. Heat-resistant head welded to wear-resistant stem at 600 parts per hour. (*Production Technology, Inc., A Caterpillar Subsidiary.*)

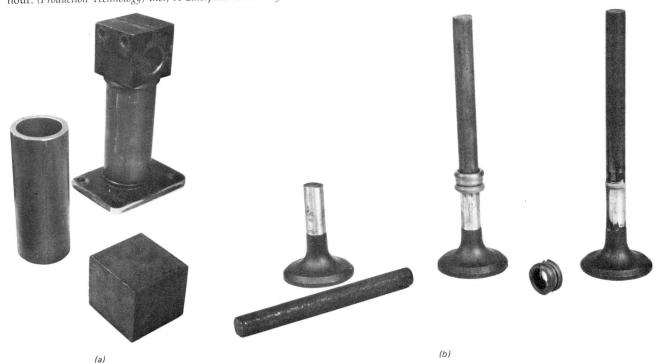

(a)   (b)

together. High-cost metals, such as stainless steel, need be used only in the section where needed.

2. Once the parameters of a job have been determined, the job is a fast one, sometimes only a few seconds.

3. Heat is localized at the weld and is quickly dissipated so that there is only a slight effect on the parts joined. The weld may not have to be heat treated.

4. Repeatability is reported as excellent, and several jobs have been fully automated.

5. No flux, gas, filler metal, or slag are present to cause imperfections. Also no smoke, fumes, or splatter is produced.

6. Smaller shafts can be welded to large plates, instead of turning them down or making a long, expensive forging.

The *disadvantages* include:

1. A sometimes quite heavy "flash" is forced out in all inertia and friction welds. If tubing is welded, the flash is also on the inside. This flash may have to be machined off, though sometimes it is not bothersome.

2. Lengths can be held to only about 0.010 to 0.020 in. (0.25 to 0.50 mm) tolerance, which may require added machining.

3. The flash from medium- and high-carbon steels will be hard and, thus, must either be removed while it is hot or annealed before it is machined.

4. Thrust pressures in inertia welding will range from 10,000 to 40,000 psi (69 to 276 MPa), which requires a heavy, rigid machine.

5. The use of either inertia or friction welding is restricted to flat and angular butt welds, where one part is normal (perpendicular) to the other part.

GENERAL CONSIDERATIONS IN ARC WELDING

The following discussions apply to all types of arc welding. Laser and electron-beam welding are included in some of these considerations. Exceptions were noted in the previous discussions.

Work Holding for Arc Welding

If the welding job is to connect the long seam in a tank formed by rolling the steel into a circle or to connect flat plates, the work may hold itself in place. However, in making brackets, wheels, machine frames, or bases, several different pieces of various shapes of metal must be held in the proper positions.

If the work is small, say under 24 in. (600 mm) on a side, the welder may position the parts and *tack weld* each one in place. However, it is difficult to position the parts with any accuracy unless a helper works with the welder.

A *welding fixture* is usually made if more than about 10 assemblies are needed. This may be fairly simple, with clamps used to hold the loose pieces in place until they are welded. Some welding fixtures, like those used in aircraft work, can be very complicated and expensive.

Fig. 21-22 An inertia welding machine, 4000 rpm, 175,000-lb (780-kN) force, for parts up to 2.625-in. (66.7-mm) diameter. *(Production Technology, Inc., A Caterpillar Subsidiary.)*

Welding positioners (Fig. 21-23) may be used to hold the work, and they allow the welder to move the work in several directions so that more welding can be done with less fatigue.

These welding positioners are made for any size work, from 10 lb [4.5 kg (mass)] up to 1,000,000 lb [450 000 kg (mass)]. The small- and medium-sized models can be tilted and rotated quite easily by hand. The larger ones are motor driven.

As can be seen in Fig. 21-23, positioners are made which will tilt the work left, right, vertical, or horizontal or which will roll the work around so that the welding is always done in one position. A welding positioner for handling up to 2000 lb [450 kg (mass)] will cost about $3500 and may, if production is 500 pieces or more per year, pay for itself in a few months.

The Cost of Arc Welding

Welding "machines" or welding current generators with the necessary minimum equipment may be purchased for as low as $200 for a 125-A ac outfit up to $2000 for a 600-A outfit. GMA equipment with higher amperages may cost up to $5000 complete. Thus, the equipment cost is not high compared to many metal-working machines.

Welding rods may cost as little as 15 cents per pound (33 cents per kilogram) for low-carbon steel up to about $1 per pound ($2.20 per kilogram) for stainless-steel electrodes or wire. A pound of regular No. 6013 wire will make about 5 ft (1500 mm) of $\frac{5}{16}$-in. (8-mm) fillet weld, which is quite inexpensive.

The greatest cost in welding is the labor cost, and this will vary a great deal. Variation can be due to the type of welding rod used, the size of the weld, the welding process used, the skill of the operator, and the "position" (overhead, flat, etc.) in which the welding is done.

A few examples of the rate at which a weld can be made are listed in Table 21-2. These rates are at 100 percent efficiency or operating factor. It is seldom that more than 50 to 60 percent of this speed will actually be reached in production.

Distortion due to Arc Welding

In all welding processes a large amount of heat is put into the welded joint. This heat will spread out through the welded metal. The amount of spreading will depend on the welding process, the speed and depth of the welding, and the coefficient of heat transmission of the metal being welded.

Some processes such as electron-beam and laser welding use such a narrow seam of weld that almost

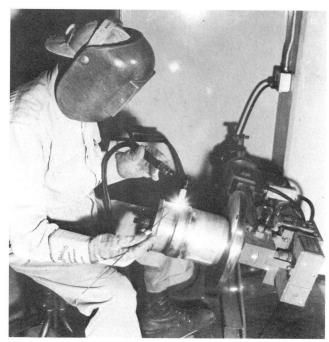
(a)

(b)

Fig. 21-23 Welding positioners. (*a*) Table model positioner for up to 200 lb (91 kg). (*b*) Universal Balance positioner for up to 4000 lb (1814 kg). (*Aronson Machine Co., a division of Airco, Inc.*)

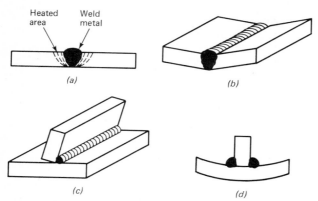

Fig. 21-24 Cause of and types of distortion in welding. (a) The reason for distortion; (b) butt weld; (c) single side weld; (d) double fillet weld.

no distortion occurs. In fact, electron-beam welding is being used to assemble some finished parts with no additional machining needed.

Most welding produces a spread of heated metal similar to that shown in Fig. 21-24a. The normal amount of expansion is often restricted by the stiffness of the surrounding metal. Thus, when this hot weld and the surrounding heated areas start cooling, the powerful contracting forces will bend or distort the welded assembly.

The *distortion* will almost always be *toward the weld*, as shown in Fig. 21-24b to d. In some jobs the distortion is relatively small, and can, if the work is not to close tolerances, be accepted. However, in most work, distortion from welding must be straightened out, or machined off, and this is an added expense.

prevention of distortion

While distortion in welding is seldom completely eliminated, it can be greatly diminished by several methods.

1. Do not overweld. A large, convex bead does

not add any strength but does increase the shrinkage forces. This also means as few passes as possible should be made by using large electrodes.

2. Locate parts out of position to allow for shrinkage. In Fig. 21-25c, the shrinkage will pull the two arms into the correct spacing.

3. Balance the forces by alternate side and skip welding as shown in Fig. 21-25a.

4. Design braces and supports so that they help control the forces causing distortion.

5. Clamp or bolt the work securely to a rigid table or fixture (Fig. 21-25b). If the clamps are left on during cooling, the work will "move" much less.

6. Often it is desirable to stress-relieve a weldment by proper heating and cooling in a furnace. Even very large, complex welded parts may be heat treated this way in furnaces made especially for handling parts over 20 ft (6000 mm) long and 10 ft (3000 mm) in cross section.

These same forces can be used to *straighten* a section of, for example, slightly bent steel pipe. By proper heating of the high (convex) side, sometimes combined with water cooling the surrounding area, the contraction will pull the pipe straight. This method has been used on I beams, bulkhead assemblies, and similar work. Repeated applications of the procedure may be needed to fully straighten heavy work.

Equipment for Arc Welding Safety

The welding arc gives off infrared (heat) rays, ultraviolet (sunburn) rays, and very bright visible light rays.

A *head shield*, or helmet, as shown in previous illustrations, must be worn at all times when welding. The *window* is covered by a very dark glass made so that it absorbs the rays. The operator can barely see anything through this until the arc is struck. The helmet is made so that a slight nod of the welder's

TABLE 21-2 TYPICAL WELDING SPEEDS*

Type of Weld†	Welding Position	Welding Process	ipm	Rate, per Pass (at 100% Efficiency) mm/min
$\frac{3}{8}$ fillet	Flat, horizontal	Shielded metal arc	13	330
$\frac{3}{8}$ fillet	Vertical or overhead	Shielded metal arc	3	75
$\frac{3}{8}$ butt	Flat, horizontal	Shielded metal arc	10–20	250–500
$\frac{3}{8}$ fillet	Flat, horizontal	Submerged arc	20–28	500–700
$\frac{3}{32}$ butt	Flat, horizontal	GTAW (TIG)	12	150
$\frac{1}{2}$ butt	Stainless steel, flat	Electron beam	40–63	1000–1600

*Averaged from several sources.
†On mild steel, except when noted.

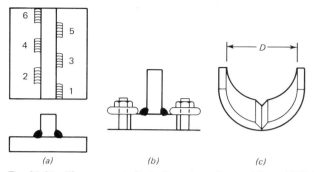

Fig. 21-25 Three ways to limit distortion when welding. (a) Welding sequence to limit distortion. (b) Clamps will limit distortion. (c) Distance D is larger than desired final dimension.

head causes the helmet to drop down over the face just before the arc is struck.

CAUTION: Never look at welding in process unless you are wearing a *head shield*, using a *hand shield*, or are 40 ft (12 m) away. *Eye burn*, an extremely painful sensation in the eyes, can result from even a few seconds exposure. This sensation does wear off in 12 to 48 h, and the eyes are seldom permanently injured.

Because of the actual sunburn which the welding rays will cause on any exposed skin and because of the spatter of molten metal, a welding operator keeps the entire body carefully covered as shown in previous illustrations.

The *gloves* are often made of heavy leather and extend up over the shirt's cuffs.

The *apron* is also leather to prevent sparks from burning through the operator's clothing.

The *pants* are heavy, with no cuffs into which hot sparks could fall.

The *shoes* are "high," that is, come up above the ankles, with heavy soles to protect the feet from pieces of metal on the floor.

All *wiring* to the welding equipment must be large enough to carry the maximum current available from the welding machine. The wiring should also be checked carefully to be certain that there are no breaks in the insulation.

GAS WELDING

By burning pure oxygen in combination with other gases, in special torches, a flame temperature up to 6000°F (3300°C) can be attained. The most frequently used gas is acetylene (C_2H_2). MAPP gas (methylacetylene propadiene, C_3H_4) is also quite common, and propane, butane, and hydrogen are used occasionally.

The heat of the gas flame melts the work metal and the hand-fed filler rod in much the same way as the shielded electric arc. The equipment is inexpensive ($250 will buy the basic equipment), and it is portable, so oxyacetylene welding is often used in small shops. It is especially useful for welding light metal (such as automobile bodies) and for repair jobs. Gas welding can be used with all the common metals.

The disadvantages are that it is much slower than electric-arc welding and does not concentrate the heat close to the weld. Thus, the heated area is larger, which causes more distortion. If electric welding is available, gas welding is seldom used for work over $\frac{1}{8}$ in. (3.2 mm) thick.

The gas is purchased in cylinders and connected through regulating valves and pressure gages into flexible hoses attached to the nozzle. A typical arrangement is shown in Fig. 21-26.

The welding torch mixes the oxygen and another gas inside in the proportions set by the pressure regulators. Several sizes of *tips* are available for the torch. Larger diameter tips (up to 0.150-in., 3.8-mm diameter) are used for greater thicknesses of metal.

Steel is easily gas welded. *Aluminum* requires the use of a flux brushed onto the work and the rod. *Stainless steel* is not usually gas welded if corrosion resistance must be maintained.

The *acetylene gas* must be handled very carefully, as it is explosive at 25 psi (172 kPa). Therefore, the acetylene is stored by dissolving it in acetone which is held in asbestos, kapok, etc. In this way it can be stored at 200 psi (1.4 MPa). All fittings connecting acetylene gas are lefthand threaded, and the hoses are colored red.

Oxygen is stored in cylinders under 2000 psi (140

Fig. 21-26 Oxyacetylene welding and cutting torch arrangement.

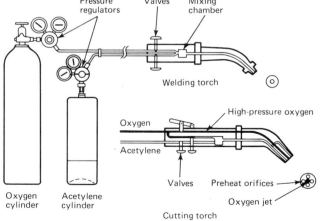

Note: Torches are at a larger scale

MN/m². The hoses are green or black with righthand-threaded fittings.

MAPP gas, though nearly 1000°F (538°C) cooler, is satisfactory for much work and is much safer to handle.

Safety Precautions in Gas Welding

1. The operator should wear safety, medium-dark tinted glasses but does not need the shield used in arc welding. A hat and coveralls should be worn to protect the operator from flying sparks. Gloves may be worn.

2. Gas cylinders (especially acetylene) should not be allowed to drop and should always be stored in an upright position.

3. Oil and grease should not be allowed to accumulate on or around the oxygen tanks, as they may ignite spontaneously in the presence of pure oxygen.

FLAME CUTTING (TORCH CUTTING)

When workpieces such as 6-in.-wide (150-mm) steel gears, cams, or sprockets are to be made or when heavy steel plate must be cut to size or cut to a specified shape, one of the most frequently used processes is *flame cutting*, sometimes called *torch cutting*. An example of some of this work is shown in Fig. 21-27. Even steel plate 15 in. (380 mm) and over can be flame cut.

The *oxyacetylene welding torch*, previously described, is the usual cutting tool, especially if steel is being cut. The flame from the oxyacetylene torch first heats a

Fig. 21-27   Example of flame-cut work. The gear is 8 in. (200 mm) thick. *(Joseph T. Ryerson & Son, Inc., a subsidiary of Inland Steel Company.)*

spot on the metal to its ignition temperature (about 1600°F, 871°C for steel), and then an added stream of high-pressure oxygen causes it to "burn," or oxidize, and blows the melted metal out of the kerf. The torch is moved along the desired path and the metal is cut.

The *accuracy* of a torch cut is about $\pm\frac{1}{16}$ in. (1.6 mm) for cuts up to 6-in.-thick (150-mm) steel. On thinner plate, or with special care, $\pm\frac{1}{32}$-in. (0.8-mm) tolerance can be maintained.

The *finish* from torch cutting, as can be seen in Fig. 21-27, can be close to 250 μin. (6.4μm), though it is often much rougher.

Distortion is not a problem when 1-in. (25-mm) or thicker plate is being cut. Narrow sections and plate $\frac{1}{2}$ in. (13 mm) or less may require clamping or the use of the methods used to prevent distortion in welding.

The effect of heat on steel with less than 0.30 carbon is not harmful. The edges can be machined readily, and the maximum penetration of the heat-affected zone is only about $\frac{1}{8}$ in. (3.2 mm) in a 6-in.-deep (150-mm) cut. A hardened edge will result when higher-carbon steel is cut.

Methods of Torch Cutting

When small lots are to be cut or when steel I beams are to be cut off or shaped, the torch is used *by hand* following chalk or soapstone marks on the metal.

For more accurate cutting, a pantograph moved by a manually guided roller or stylus can be used to follow a pattern made of wood or masonite. This arrangement can be used to guide multiple cutting torches as shown in Fig. 21-28. *Electric-eye* tracers and numerical control tapes are used to completely automate the cutting when the quantity needed is large.

Stacking of several layers of $\frac{1}{4}$-in. (6.4-mm) or thinner plate is another way of speeding production of torch cut parts.

Metals which Can Be Cut

All low-alloy steels and aluminum alloys are easily cut with oxyacetylene torches. Some stainless steels can be cut this way, but many of the most used alloys cannot be easily cut. Cast iron cannot be flame cut by the standard method.

Powder cutting is a flame cutting process in which a finely powdered mixture with cast iron in it is blown into the torch stream. The heat supplied by the burning iron powder raises the temperature enough so that the chromium and nickel alloys and stainless steel will melt. This is a more expensive process and leaves a somewhat wider kerf than regular flame cutting.

Plasma-arc cutting can be used on any metal. With

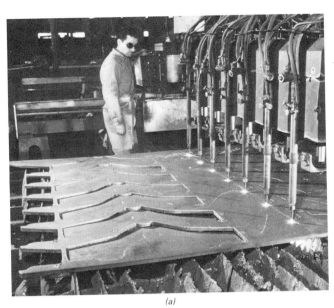

Fig. 21-28 (a) A seven-torch flame-cutting machine following a template. (b) Partial view of a 20-torch flame-cutting machine with an 80-ft. wide (24.4-m) cutting bed, N/C or electric eye control. (*Joseph T. Ryerson & Son, Inc., a subsidiary of Inland Steel Company.*)

this process, cutting speeds are fast, and distortion, even in thinner plates, is minimal.

A tungsten electrode cathode in the torch either completes an arc with the side of the torch (non-transferred arc) or uses the workpiece metal as the anode (transferred arc). An inert gas such as argon, helium, or nitrogen is passed through the arc and is ionized. When these particles reform into atoms, they generate a temperature as high as 30,000°F (16,500°C).

This high-temperature flow can be kept in a narrow beam, thus preheating only a small kerf which is oxidized and blown away by the stream of oxygen which flows through another part of the torch nozzle.

Plasma-arc cutting is expensive, but it is often economical, especially for cutting stainless steel and aluminum, because of the speed of cutting, the smooth edges, the small kerf, and a very narrow heat-affected zone.

HARD SURFACING

There are many times when only a portion of a part really needs special hardness, toughness, wear resistance, or chemical resistance. Examples are the working edges of oil drills, a section of a shaft on which seals must run, and some parts of tools and dies.

Two methods are most frequently used for applying a hard or other special metal to selected areas of a workpiece.

Arc Welding

All arc-welding processes can be used to bond special metals to the original equipment or to build up worn-out areas. These metals may be high-carbon steels, high-speed steels, stainless steels, tungsten carbide, or some nonferrous metals. The desired metal in the form of rod or wire is welded, row after row, over the entire area which is to be surfaced. Often more than one layer is needed, especially since the first pass is "diluted" by the base metal.

The resulting surface is rough and may be ground smoother as required. Repairs of worn areas on shafting may either be turned or ground to size and finish.

Metalizing or spray coating is used where a thinner, less well-bonded layer is sufficient. Almost any metal (including steels, stainless steels, aluminum, and zinc) can be sprayed with a special gun. Ceramics, such as aluminum oxide, may also be sprayed onto metals.

The *thermal spraying*, as it is referred to in the American Welding Society Handbook, uses relatively simple, inexpensive equipment. The source of heat for melting the metal is frequently an oxy-fuel gas torch using acetylene (5660°F, 3127°C). For lower-melting-point metals (such as zinc and lead), hydrogen or propane gas occasionally is used. Electric-arc and plasma-arc welding also are used in some systems.

A complete oxyacetylene metalizing system is

shown in Fig. 21-29. The wire of the alloy to be used is pulled from the reel. The oxyacetylene mixture and compressed air and wire are brought together inside the torch, or gun as it is often called.

The hot flame melts the wire; the pressure of the air atomizes (makes into fine particles) the melted metal and drives it onto the workpiece. Sometimes powdered metal is used instead of wire; then the process requires a somewhat different gun.

The result of spray coating is principally a *mechanical bond*. There is a slight fusion with the workpiece, but the layer of metal can usually be peeled off without much force. Thus, it is necessary to prepare the surface which is to be covered.

Surface preparation consists in first getting a perfectly clean surface, and then roughening the surface. Quite frequently this is done on round work by cutting a 20 pitch or coarser thread in the area to be coated. Sometimes a knurling tool is run over the threads to give an extra "tooth" to the surface.

If flat surfaces are spray coated, they must be roughened (sometimes with grit blasting), and the coating should be brought down over the edge so that it will not peel off.

The coating may be from 0.010 to 0.125 in. (0.25 to 3.2 mm) thick. An allowance usually must be made for turning or grinding to the needed finish and size.

RESISTANCE WELDING

Resistance welding is used especially with sheet metal from 0.02 to 0.125 in. (0.5 to 3.18 mm) thick and with steel pipe and tubing. Many metals, including most grades of steel, stainless steel, and some aluminum, can be resistance welded.

Spot Welding

Millions of spot welds are used in making automobiles, large and small kitchen appliances, steel furniture, etc. This process is the most frequently used resistance-welding process. Basically, the welding occurs due to pressure and heat. The heat is developed because a high current density going through the resistance at the surface of contact of the metals in the workpiece creates a temperature just under the melting point.

The pressure is applied by *rocker-arm* or *press-type electrode holders*. The pressure may be developed by a foot lever or by air or hydraulic cylinders. The basic setup is shown in Fig. 21-30.

The welding *tips* are made of copper alloys which must be good conductors and must be strong enough to stand considerable pressure. The contact surfaces are shaped, often to a flat circle of from $\frac{3}{16}$- to $\frac{7}{8}$-in. (4.8- to 22-mm) diameter. The welding tips may be hollow part way down so that cooling water can flow into them.

The spot welding cycle is:

1. Apply pressure to the work from 200 to 3200 lb (0.9 to 14.2 kN).

2. Current turns on and makes the weld, using 1 to 10 V and 1000 to 50,000 A. Time may be from two

Wire feed speed control

Wire from reel

Gas hoses

Fig. 21-29 A spray coating being applied to rebuild the bearing surfaces on a drive shaft. (*Metallizing Company of America, Inc.*)

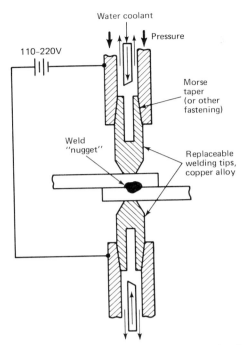

Fig. 21-30   Spot-welding tips and basic circuit.

percent of the metal thickness, and a maximum of 80 percent. Deeper welds create "dimples" in the work and damage the electrodes.

Aluminum can be spot welded, but it requires high current for a very short time and lower pressures than are used for steel.

Fig. 21-31   (*a*) A semiautomatic rocker-arm-type spot welder. Can also be supplied without the controls. (*b*) An eight-head automatic spot welder with fixture for welding reinforcing wire screen. (*Alphil Spot Welder Mfg. Corp.*)

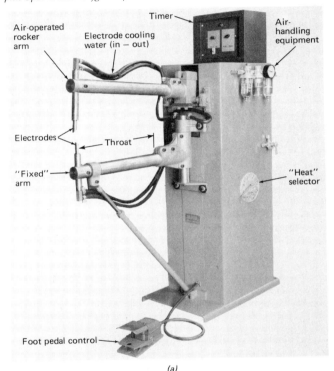

(a)

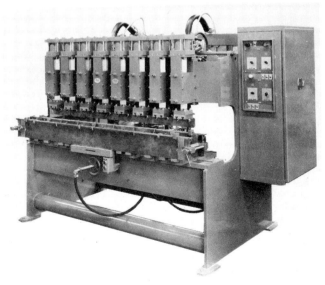

(b)

cycles ($\frac{1}{30}$ s) to 3 or 4 s. In production, welding time is usually less than 1 s.

3.   Current turns off, while pressure stays on for a brief time to allow the weld to "freeze."

4.   The tips open, and the part is moved or removed.

Figure 21-31 shows two styles of spot welder. The welding current is adjustable up to the capacity of the welder. In many machines, the pressure and the welding time can also be preset so that the cycle is automatic. In production work, special fixtures are often fastened to the welder to speed up the loading, positioning, and unloading of the work.

Spot-welding machines are made in capacities from 10 to 150 kVA and cost from $700 to $5000. Added controls and tooling and multiple welding heads will increase these costs.

considerations in spot welding
Cleanliness of the workpieces is important. Oil and dirt can cause defective welds. The electrodes must also be kept clean and properly shaped.

Welding tips of smaller diameters create greater current density. However, for thicker metals, larger tip diameters with high amperage are needed in order to generate enough heat to make the weld.

Metals with higher resistance (such as high-carbon steels and stainless steels) will heat up faster and thus require less current.

The *depth* of the weld should be a minimum of 25

Galvanized steel is somewhat difficult to spot weld, as the zinc tends to alloy with the welding tips. However, very secure welds can be made by using somewhat more current, time, and pressure than would be used with uncoated steel.

Copper, because of its high conductivity, is very difficult to spot weld. However, several copper alloys can be successfully spot welded to themselves and to other metals.

Seam Welding

Seam resistance welding is a variation of spot welding which uses rotating copper disks instead of welding tips. The general setup is shown in Fig. 21-32. Higher pressure and current are needed, and the welding disks must be power driven. Thus, a seam welder can cost $15,000 or more.

The closely spaced spots are made by automatically switching the welding current on and off as the rolling disks continue to move along the seam.

Seam welding can be used to form tubular shapes out of thin sheet metal or to weld two flat pieces together. It is not used to make steel or copper tub-

Fig. 21-32   Two uses of seam welding. (*a*) Seam weld on flat stock. (*b*) Roll spot weld on cylindrical work.

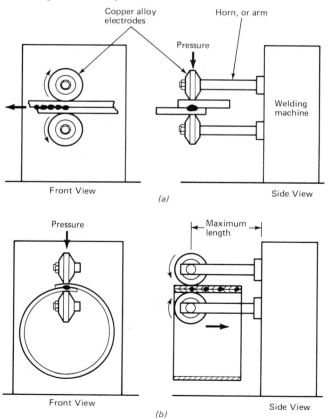

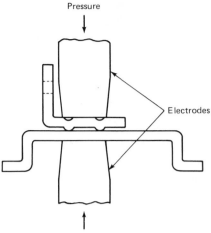

Fig. 21-33   A typical projection welding job.

ing. Seam welding of pipe and tubing is discussed in Chap. 25.

Projection Welding

A specialized form of spot welding is projection welding, and it uses the same press-type machine with wide, flat electrodes. As shown in Fig. 21-33, one part has specially shaped projections on the bottom. The current almost melts these projections which, under pressure, spread out and create a strongly welded connection. The projections are often made during punch-press forming. They must be heavy enough to make a secure weld but not heavy enough to burn through or slow down the cycle.

Stud Welding

This is a specialized form of projection welding which uses specially made welding guns (Fig. 21-34a). The studs are available in hundreds of different forms and are often welded to flat surfaces as attachment bolts or to the inside of tanks to hold insulation or support wire mesh, studs for engine mounts, and nameplate fasteners. The same basic procedure of heat-press-cool is used (Fig. 21-34d), and the weld will be as strong as the metal used. Typical jobs are shown in Fig. 21-34b and c.

The base to which the studs are attached must be heavy enough to hold the studs, usually $\frac{1}{8}$ in. (3.2 mm) or over, except for tiny studs which are applied with a specially made gun.

Welding current needed may be from 400 A (dc) for a $\frac{7}{16}$-in.-diameter (11-mm) stud to 4000 A for a 1-in.-diameter (25.4-mm) stud. Most steels and stainless steels can be stud welded. Aluminum stud welding requires the use of argon or helium gas.

(a)

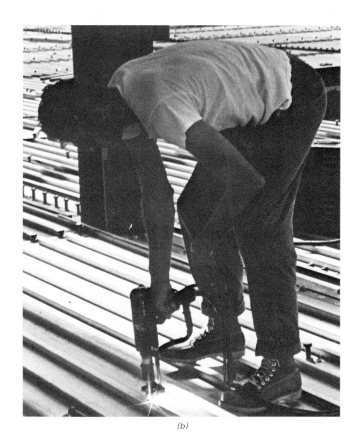

(b)

(c)

Fig. 21-34 (a) A complete stud welding outfit. (b) Welding studs for special floor construction. (c) Welding ⅞-in.-long (22.3-mm) studs to powdered metal parts, 720 per hour, located ±0.005 in. (0.125 mm). (d) The basic stud welding cycle. (*Nelson Stud Welding Co., a division of TRW Inc.*)

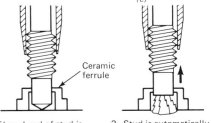
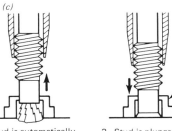
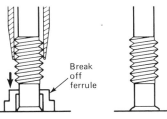

Ceramic ferrule

Break off ferrule

1. Fluxed end of stud is placed in contact with work.

2. Stud is automatically retracted to produce an arc.

3. Stud is plunged into pool of molten metal.

4. Weld is complete.

(d)

A ceramic ferrule is used to hold the molten metal in place and shield the arc. These are very inexpensive and are thrown away after use. Special capacitor discharge systems may be used, especially for aluminum and brass.

Stud welding is often automated. A single gun can make 6 to 30 welds per minute, and equipment can be designed to use multiple guns.

BRAZING

Brazing is the joining of like or dissimilar metals by flowing a nonferrous molten filler metal between them. The filler metal always has a lower melting point than the base metals; thus, the heat required will seldom cause distortion. Brazing is done at temperatures *over* 800°F (427°C).

Advantages and Use of Brazed Joints

1.   Properly brazed joints are pressure tight.
2.   Joining two thin pieces or thin-to-heavier sections and joining dissimilar metals can be done without warping and with adequate strength.
3.   The process is easily automated for long runs, or semiautomatic setups can be used to speed shorter production runs.

For small [under about 4-in.-diameter (100-mm)] work (such as fastening tubing, studs, plugs, or sheet metal parts to thick or thin sections of almost any shape), brazing should be considered a possible process to use. A well-known brazing job is fastening carbide tips to steel holders for cutting tools of many uses.

Some large brazing jobs, such as the honeycomb construction of airplane parts, go beyond these limits. However, these are special applications, and the tooling and heating equipment is very expensive.

The *strength* of the joint depends on the brazing metal used, the penetration of the brazing metal into the grain boundaries of the base metal, and the area of the brazed joint. The brazing metal is not as strong as the base metal; however, brazed joints can be designed which are so strong that the work will fail instead of the joint. Figure 21-35 shows a few examples of the uses of brazing.

Disadvantages of Brazed Joints

1.   Since the entire area to be brazed must be heated, large cast sections or large heavy plates cannot easily be brought up to temperature.
2.   The close-tolerance fit-up of the joints may require expensive machining.

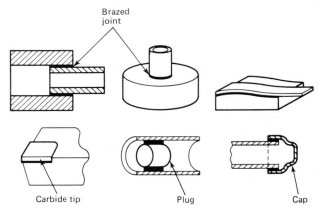

Fig. 21-35   Some examples of the use of brazed joints.

3.   Flux must be thoroughly cleaned off after brazing, as it is often corrosive.

Making a Brazed Joint

The simplest brazing, used for small lots, is done by hand using a gas-flame torch and brazing metal in wire form.

1.   The surfaces must be thoroughly *clean*.
2.   *Flux* must be applied to dissolve the oxides present and prevent the formation of oxides during heating.
3.   *Heat* is applied to the *workpieces* in the area which is to be brazed. The entire area must be brought to the melting temperature of the brazing material.
4.   The brazing metal, in the form of a wire, is touched to the joint. It melts and is pulled into the entire area by *capillary action*. Or a preformed ball, ring, or other shaped piece of brazing metal is placed in the joint, and this flows rapidly when the right temperature is reached.

CAUTION: As capillary action works only in narrow gaps, the parts to be brazed should be fitted so that there is only 0.0005 to 0.003 in. (0.013 to 0.075 mm) clearance between them. Smaller or larger clearances can, if necessary, be used, but they require special planning.

The *brazing filler metals* most frequently used are copper alloys and silver alloys. A wide variety of formulations is available. Selection is on the basis of cost, strength required, corrosion resistance needed, and the melting temperature as it might affect the workpiece. Some nickel-chromium brazing filler alloys (called *heat resistant*) are used when operating

(a)

(b)

(c)

Fig. 21-36 (a) Brazing tubing onto bellows using fixtures. (b) Furnace brazing studs onto plates. (c) Induction coil brazing six pieces at once. Fixture raises and lowers the work. *(Handy and Harman.)*

temperatures are over 600°F (315°C). Special aluminum-silicon alloys are used for brazing aluminum.

Methods of Brazing

Hand-operated oxyacetylene torch used with wire filler was mentioned previously. This method is quite quick, but it is not efficient for large quantities, although fixtures, as shown in Fig. 21-36a, can speed up the work.

Furnace brazing (Fig. 21-36b) is widely used. If a number of similar items are to be brazed, the brazing preform is inserted, and the parts are lightly clamped together. In the batch method, one or more trays full of parts are put into the furnace long enough for the brazing metal to melt.

A continuous furnace brazing setup may involve open burners or a small partially enclosed furnace. The parts to be brazed are passed through the flame or furnace on a conveyor. Brazing paste or preforms are inserted by hand or automatically.

Controlled atmosphere furnaces, using nitrogen or other inert gases and occasionally a vacuum, do very high-quality brazing as no oxides are formed during the heating cycle. Large air-frame parts are welded this way.

Induction brazing (Fig. 21-36c) is a semiautomatic process in which the heat is supplied by high-frequency induction coils. The equipment for this is relatively inexpensive, and the heating coils are easily made from copper tubing to localize the heat in the desired areas.

The induction heating cycle may take only 2 to 10 s, and it can be controlled accurately to give uniformly sound joints. The joint is preloaded with brazing filler metal.

Other brazing methods such as salt-bath and electric resistance heating are used for specialized types of work.

SOLDERING

Soldering, except for making electrical connections, is not a widely used manufacturing process. One use is in assembling low-pressure air and water lines, though, even for this work, special flare or compression-type fittings are more frequently used.

Soldering is done mostly with lead-tin alloys which melt at 365 to 450°F (185 to 232°C). The percent of tin varies from 25 to 60 percent. The higher percent is more expensive but has greater fluidity. The popular 60 percent tin–40 percent lead is close to the eutectic (lowest melting temperature) point, melts at about 365°F (185°C), and makes excellent joints.

Soldered joints are made by heating the body of the

work until the solder melts on contact and fills the joint by capillary action. Thus, the joints should have only 0.002 to 0.008 in. (0.05 to 0.2 mm) clearance. Dip soldering, immersing an assembly in molten solder, is used for large surfaces. Soldering irons of many sizes are used for electrical wiring.

Review Questions and Problems

21-1.   List the advantages and disadvantages of welding.

21-2.   What are the three basic types of welding machines. Where is each type used?

21-3.   Describe eight types of welds.

21-4.   What safety equipment should be used when welding?

21-5.   Describe shielded-arc welding. What are its advantages?

21-6.   Compare gas-metal-arc welding and gas-tungsten-arc welding and list the advantages of each.

21-7.   Describe submerged-arc welding and list its advantages and disadvantages.

21-8.   Describe electron-beam welding. What are its advantages and disadvantages?

21-9.   Compare friction welding and inertia welding. What are the advantages and disadvantages of each?

21-10. What advantage does laser-beam welding have over electron-beam welding?

21-11. What steps should be taken to prevent excessive distortion in welding?

21-12. What is spot welding and what must be considered when doing spot welding?

21-13. What is cored wire welding? What are its advantages?

21-14. How is seam welding done?

21-15. Describe projection welding.

21-16. How is gas welding done, and what are its advantages and disadvantages?

21-17. What safety precautions should be taken when doing gas welding?

21-18. Describe how flame cutting is done.

21-19. What is the principal difference between brazing and welding?

21-20. What are the advantages and disadvantages of brazing?

21-21. Define soldering.

References

1. Courses, standards, books and booklets of the American Welding Society (AWS).

2. Fundamentals of Gas Metal—Arc (MIG) Welding, and other booklets of the Miller Electric Mfg. Co., 1969 and others.

3. Giachino, Weeks, Brune: Welding Skills and Practices, 3 ed., American Technical Society, 1967.

4. "How To . . . . ." pamphlets on several welding topics, Linde Div., Union Carbide Co., New York, 1970, 1971.

5. Little, Richard L.: Welding and Welding Technology, McGraw-Hill, New York, 1973.

6. Patton, W. J.: The Science and Practice of Welding, Prentice-Hall, Englewood Cliffs, N. J., 1967.

7. Procedure Handbook of Arc Welding Design and Practice, The Lincoln Electric Co., 1973.

8. Welding Handbooks (five sections), American Welding Society, 1969 and others.

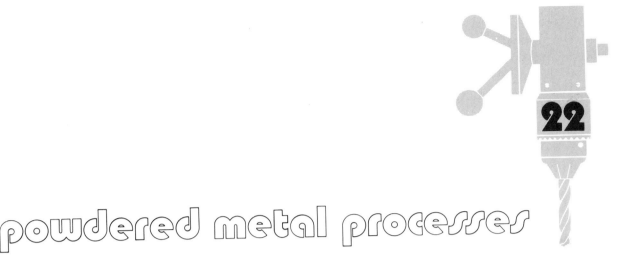

powdered metal processes

Real progress in powder metallurgy began about 1930. Some of the earliest products were tungsten carbide tips for cutting tools and "self-lubricating" porous bronze bearings.

Today thousands of powdered metal parts are made for automobiles, farm equipment, computers, electric appliances, and electronics. These products include gears, levers, filters, cams, pinions, and previously impossible combinations of these. A few of these parts are shown in Fig. 22-1. The parts may weigh from a few ounces up to 70 lb [a few grams to 32 kg (mass)].

P/M (powdered metal) parts can be made from many metals. Iron, iron-nickel alloys, steel of several alloys, stainless steel, copper, bronze, and aluminum alloys account for most of the 500,000,000 lb (227,000,000 kg)* of powder used in 1974.

Some of the reasons for the tremendous growth in this field are:

1. The process has "built-in" automation. Once set up, a press can run practically unattended. Production rates up to 1000 parts per hour and over are being achieved.

2. Frequently no additional machining is needed, though the parts can easily be drilled, milled, ground, or tapped when necessary.

3. Tolerances on diameters and width dimensions are easily held to 0.002 in. (0.05 mm). Length dimensions can be held to 0.005 in. (0.125 mm) without repressing.

4. There is almost no wasted metal, as the powder is automatically "measured" for each part. Moreover, very little scrap is produced.

5. Complex parts (such as internal keys, combination gears, and odd-shaped cams) are produced (once the dies are made) almost as easily as very simple parts.

*Metric units are often given with space used instead of commas, for instance 227 000 000 kg.

The basic steps in making a powdered metal part as well as some of the most used variations of the process are shown in Fig. 22-2. These will be explained in the following pages.

PRODUCTION OF POWDERS

The powders used in the manufacture of P/M parts cost from 20 cents per pound for iron to over $2 per pound for titanium and are produced by combinations of chemical and mechanical methods.

Atomization

Atomization can be used to produce powders from any metal which melts easily. These include lead, zinc, copper, aluminum, iron, steel, and stainless steel.

The pure metal or the desired alloy is melted. The stream of molten metal is atomized and cooled by a 100-psi (690-kPa) stream of air or water.

Fig. 22-1 A few of the wide variety of parts which can be made by powder metallurgy. Some shapes simply could not be made by machining. (*Metal Powder Industries Federation.*)

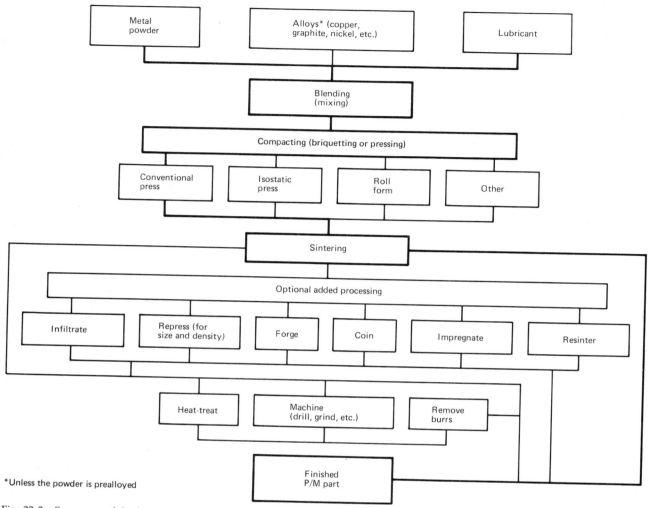

Fig. 22-2   Summary of the basic steps in producing powdered metal (P/M) parts.

Reduction

Reduction means removing the oxygen from a metal's oxide. The process can be used for most metals. The hydrogen reduction process is basically:

$$\text{Metal oxide} + \text{hydrogen} \rightleftarrows \text{metal} + \text{water}$$

The actual process is much more complex than this equation seems to indicate. Hydrogen reduction is used especially for producing powders of tungsten, molybdenum, copper, cobalt, and iron. It is a relatively slow process which produces pure, irregularly shaped powders.

Sponge iron is a spongy, brittle mass produced by hydrogen reduction of iron oxide. The oxide used may be iron ore, or it may be made from the hundreds of tons of scale which comes off the billets in the steel mills. Nickel, molybdenum, and carbon (graphite) are added as desired for strength and hardenability.

Electrolytic Deposition

The electrolysis of metals from solutions of their salts is much the same process as electroplating. Copper powder, especially, is made by this process. In one plant, pure copper anodes are suspended in an electrolyte with lead-alloy cathodes which measure about $24 \times 36 \times \frac{3}{8}$ in. ($600 \times 900 \times 9.5$ mm) in size. Thousands of these plates are used in a single plant.

By controlling the direct current, temperature, circulation, etc., pure copper powders of various sizes and densities are deposited on the cathodes. This powder is flushed off the cathodes, washed, dried, ground to size, sieved, and blended, ready for sale.

Mechanical Crushing

Mechanically crushing, or pulverizing, a chemically combined, hard, brittle mass is the process used for hard alloys such as tungsten carbide (WC) and titanium carbide (TiC). When used for cutting tools, cobalt powder is added in the mixing (blending) process. Mechanical crushing is also frequently part of the production cycle of other powders. Some powdered metal parts and the metal powder used are shown in Fig. 22-3.

Other Methods

Additional means used for producing metal powders include precipitation from liquid or gas and decomposing iron or nickel carbonyls, $Fe(CO)_5$ and $Ni(CO)_4$, by low-temperature heating.

ADDITIONAL STEPS IN PREPARING P/M POWDERS FOR USE

Annealing is often necessary after processing, as P/M powders must be soft so that they can be compressed more easily.

Sizing and Blending

Sizing through various sized sieves is necessary because the mix of grain sizes affects the final properties. Powder is sold with a specified mix.

Blending (mixing) is used:

1. To combine two or more metal powders to create an alloy
2. To add carbon (graphite) powder to achieve the desired carbon content of iron or steel
3. To add and distribute the lubricant

The blending is done in large mixing barrels. Often about 30 min is sufficient.

P/M powders may be purchased premixed, or they can be mixed at the user's plant. Even if purchased premixed, a "stirring" is often done to avoid segregation which may have occurred in shipping.

Lubricating the Powder

Powdered metals do not flow like water when they are being pressed. There is friction between the particles of powder and at the surfaces of the punches and the die. There is also considerable friction between the part and the die as the part is ejected (stripped).

To help overcome these frictional forces, from 0.5 to 1.5 percent of a lubricant is added to the powder mix. Lithium stearate, zinc stearate, and occasionally wax are the lubricants used. These disappear as gases during the sintering (heating) operation.

CHARACTERISTICS OF THE P/M POWDER

In addition to the chemical composition and the grain sizes of the powder, two other characteristics are important.

Apparent density is the mass of the loosely piled powder in grams per cubic centimeter (g/cm^3). The powder flows into the mold by gravity, and thus, if the desired density of the finished product is known, the compacting ratio (ratio of fill size to finished size) is:

$$\text{Compacting ratio} = \frac{\text{density of part (g/cm}^3)}{\text{apparent density (g/cm}^3)}$$

This is often close to 2:1.

The *flow rate* of a P/M powder is the time in seconds that it takes for 50 g of the powder to flow through a standard small orifice. Flow rates range from 15 to 35 s. This figure is used only as a comparison, since in actual use the powder flows into the cavity through a large nozzle, called *the shoe*. The rate of flow from the shoe into the die is a limiting factor in rate of production.

COMPACTING (PRESSING OR BRIQUETTING) THE POWDER

Compacting forces may be listed in different terms. The following all list the *same* force:

English units: 20 tons/in.² (also abbreviated as tsi)

Some European countries: 2.8 tons/cm² (metric tons)

Metric, SI units: 276 MN/m² = 276 MPa

Some powders (such as brass, bronze and aluminum) are efficiently compacted (pressed) at the lower pressures of 10 to 25 tons/in.² (138 to 345 MPa) and are not too abrasive. However, iron and steel

Fig. 22-3 Some P/M parts and the metal powder from which they were made. *(Hoeganaes Corp., a subsidiary of Interlake, Inc.)*

powders, nickel alloys, and the like often require compacting pressure of 35 to 50 tons/in.² (480 to 690 MPa). The limit to permissible pressure is usually the strength of the tooling.

Many P/M parts are small and can be made with 4-ton (35.6-kN) presses. For example, a $\frac{1}{4}$-in-diameter (6.3-mm) brass part will require less than 1 ton (8.9-kN) pressure. By contrast, a 2-in.-diameter (50-mm) steel or iron part may need more than 100 tons (890 kN) pressure to obtain the required density. Maximum powdered metal part size is limited by the capacity of the press.

Compacting presses (Fig. 22-4) are available from 4 tons (35.6 kN) to 600 tons (5.34 MN) capacity. A few 1000-ton (8.90-MN) presses are in use. These are used for larger parts, such as a 7-in. (180-mm) steel cam which has more than a 25 in.² (160 cm²) area. Rated stripping (ejection) pressures are 40 to 50 percent of the compaction rating. Presses may be of either mechanical or hydraulic operation, with a few "high-energy" presses (see Chap. 30) in use. Isostatic presses do special work, as will be described later.

Density of the part, in grams per cubic centimeter, will increase with greater pressure. Greater density also means higher strength, higher apparent hardness, and improved elongation. Density is also expressed as a percent of the weight of an equal mass of melted metal (see Table 22-1). Densities of 80 percent are commonplace, and 90 to 95 percent can sometimes be attained in one pressing. Higher densities are possible by repressing after sintering (see Fig. 22-9).

Fig. 22-4 (*a*) 12-ton (10.9-t) Stokes press producing bronze bushings at Pitney Bowes Co. (*b*) A 500-ton (454-t) double-action P/M press with OSHA guards. (*Cincinnati, Incorporated.*)

(a)

(b)

TABLE 22-1 MELTING POINTS AND DENSITIES OF SOME METALS USED FOR POWDER METALLURGY

Metal	Approx. Melting Point*		100% Density, g/cm³
	°F	°C	
Aluminum	1220	660	2.70
Brass (65-35)	1700	927	8.47
Cobalt	2720	1490	8.90
Copper	1980	1083	8.94
Iron (pure)	2800	1535	7.87
Nickel	2650	1453	8.9
Steel (Av.)	2500	1370	7.9
Titanium	3200	1820	4.5
Tungsten	6170	3140	19.3

*Melting points will vary for various alloys.

Green density is the density of the part as it comes from the press before sintering. This is quite close to the final density. The *green strength* of a part is quite low but strong enough for all usual handling. This "green" strength is due to a mechanical interlocking of the irregularities of the surfaces of the powder particles.

The *size* of the compacted part will increase from 0.5 to 1.5 percent when it is stripped (ejected) from the die (sometimes this is called *pop-out*). This must be taken into account when planning the size of the tooling.

Tooling for Compacting

As shown in Fig 22-5, the principal parts of a set of powdered metal tools are:

1. The die, which may be carbide lined.
2. Upper and lower punches. There may be more than one of either or both of these.
3. Core rods, if holes are required in the part.

The die parts are made of tool steel, hardened to 40 to 60 HRC, ground and fitted to very close tolerances. The die, if carbide lined, will last for 1,000,000 to 5,000,000 parts. The punches and core rods may last from 10,000 to 100,000 parts before they wear out of tolerance.

The *cost* of a set of tooling for making simple round bushings may be less than $500; most tooling costs under $2000. However, die sets costing over $10,000 have proven to be economically justified for special parts.

The *setup time* to remove a set of tools and replace it may be only an hour, for simple round bushings. More complex, multilevel tools may take up to 8 h, and the first run of a complicated new job may take two days before it is in production.

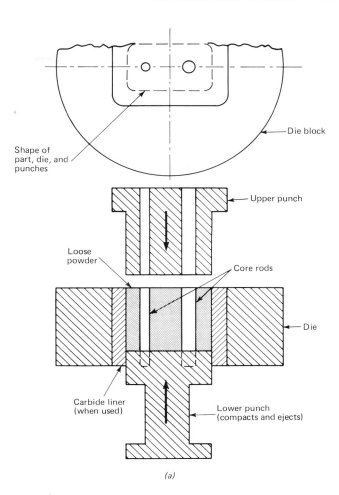

(a)

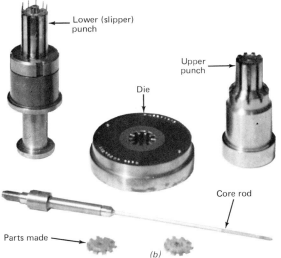

(b)

Fig. 22-5 (a) Sketch of principal parts of a simple P/M die set, or "tooling," for compacting metal powders. (b) Actual tooling for one P/M job. (*Pitney-Bowes, Inc.*)

Fig. 22-6   Three types of compacting cycles for powdered metal parts.

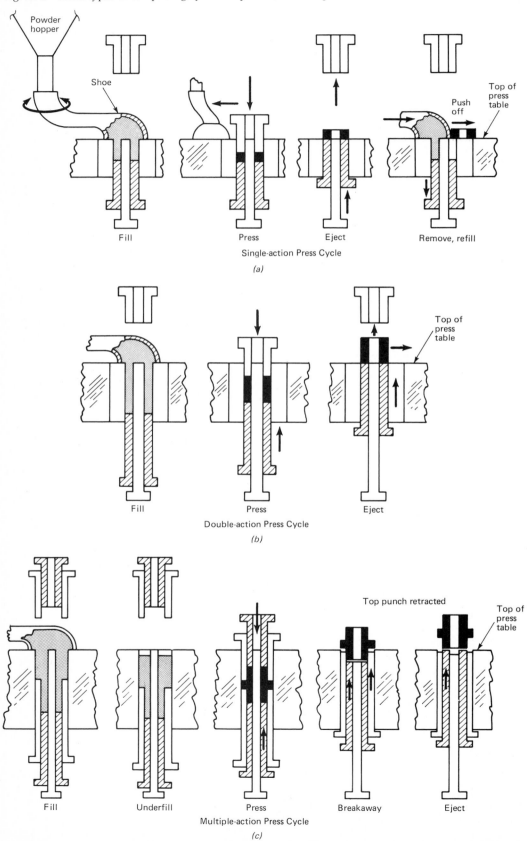

Powder hopper

Shoe

Top of press table

Push off

Fill   Press   Eject   Remove, refill

Single-action Press Cycle

(a)

Top of press table

Fill   Press   Eject

Double-action Press Cycle

(b)

Top punch retracted

Top of press table

Fill   Underfill   Press   Breakaway   Eject

Multiple-action Press Cycle

(c)

Methods of Compacting

A surprisingly large number of P/M parts can be made by the simplest method, shown in Fig. 22-6a. If the part is not more than about ¼ in. (6.35 mm) thick, the *single action* of the upper punch will make a uniformly dense part. The lower punch is then used only for ejection of the piece.

If the part is longer, one compacting action will not compress the powder farthest away. Then a *double-action* press is needed, as shown in Fig. 22-6b. If the length-to-diameter ratio is over 4:1, the part probably should not be made by powder metallurgy.

A part like the one shown in Fig. 22-6c may require double punches top and bottom, with the motion of each punch separately controlled. For complex, multilevel parts, presses may be equipped to handle up to five different levels on the part. These presses and their tooling are very expensive. However, they may make a P/M part which cannot be made any other way, or a P/M part which combines several previously separate parts, and make them to close tolerance more consistently.

Other methods, such as using movable dies and sloping punches to push the powder around, are opening up an even wider field of possibilities for P/M parts.

SINTERING P/M PARTS

Sintering means heating the powdered metal "green compact" to a temperature below its melting point in a controlled atmosphere. Iron, steel, stainless steel, and similar alloys are often sintered at 2050°F (1120°C). Copper and copper alloys are sintered at about 1600°F (870°C).

The *atmosphere* is controlled to prevent oxidation (rusting) of the work. The most frequently used atmospheres are dissociated ammonia, hydrogen, and Exogas. The relatively simple equipment needed to generate these gases is a part of the sintering equipment.

CAUTION: Hydrogen is highly explosive, so the proper safety precautions must be used.

The *temperature* is generated by electricity or by gas burners around the furnace and is closely controlled.

The *changes in the metal* during sintering can be quite complex. Basically, the powder chemically bonds together due to the diffusion of atoms at the points of contact of the metal particles. These small *necks* of contact grow larger and develop into grain boundaries. The continued movement of the atoms tends to eliminate the surface area of the original par-

ticles, thus tending to form (but never quite attaining) a part without pores. In low-density, lightly compacted, porous parts, the necks cannot enlarge into close grains because the pores are too large to be bridged by the interatomic action.

During sintering, solid solutions [such as copper or graphite (carbon) becoming dissolved in iron] are also being formed. Some of these are precipitated out of solution as the part cools.

The **furnace** (Figs. 22-7 and 22-8) for sintering has three sections:

1. A preheat section at 600 to 1000°F (315 to 482°C) which burns off the lubricant and preheats the part.

2. The high-heat furnace section, where the sintering occurs.

3. A cooling section, with a water-cooled jacket to bring the parts out at a temperature low enough so that they can be handled and also to decrease oxidation. Some iron and steel parts are dipped in oil immediately for corrosion protection.

The *time* actually in the high-heat section is usually 15 to 40 min, which means about 1 to 2 h total time through the process.

A mesh-type, high-nickel-alloy steel belt moving at 5 to 30 ft (1.5 to 9 m) per hour is most often used to carry P/M parts through the sintering process. Some parts may be placed directly on the belt. However, small parts must be placed in carbon or ceramic trays or in wire baskets which then are placed on the belt. The belt or trays are usually loaded by hand.

Furnaces with driven rollers instead of belts, while more expensive, give longer life with heavy loads. *Unloading* may consist of letting the parts drop off the end of the belt into containers or of emptying the trays as they come past the operator.

The *size* of the part is often affected by the sintering process. Some mixtures of iron, copper, and carbon will not change size at all. Other mixtures may grow or shrink up to 2 percent depending on the time, temperature, density, and alloy used. See Fig. 22-9 for one example of this variation.

CAUTION: If final part size is to be held to close tolerance, the die, punches, and core rods must be undersized or oversized to compensate for both the pop-out and the size effect of the sintering.

ALUMINUM P/M PRODUCTS

Aluminum powdered metal parts were, until about 1969, considered very difficult to make. However, powders, lubricants, and processes have now been developed which make it possible to take advantage

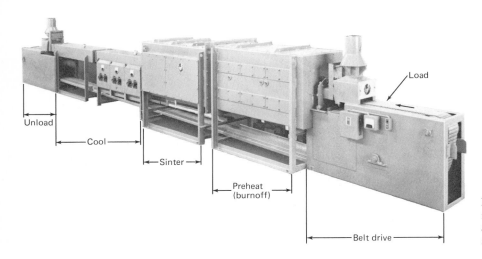

Fig. 22-7   A mesh-belt sintering furnace with 24-kW high-heat chamber and variable-speed drive. *(Harper Electric Furnace Corp.)*

of the light weight and corrosion resistance of aluminum.

The *aluminum powders*, somewhat coarser than iron powders, are usually supplied as preblended alloys, with 1.5 percent of a special wax lubricant already blended in. After sintering, and sometimes plastic impregnation, the parts can be heat-treated and machined much the same as any cast aluminum part.

Compacting can be done as low as 7 tons/in.^{2} (96.6 MPa), though twice that is more usual. This means that a press can produce parts with twice the area which it can press in iron.

Sintering is very critical. Studies by the Aluminum Company of America show that a furnace atmosphere of nitrogen or a vacuum of 50 to 200 micrometers of mercury give the best results. Argon and disassociated ammonia can be used. All gases must be very dry. The temperature must be between 1100 and 1150°F (593 and 621°C). Sintering time is 15 to 40 min depending on the thickness of the part.

INFILTRATION AND IMPREGNATION

Practically all iron and steel powdered metal parts have some porosity. This is very desirable in self-lubricated bearings or filters but not for other en-

Fig. 22-8   A roller-hearth-type heavy-duty sintering furnace, 472-kW capacity. *(Harper Electric Furnace Corp.)*

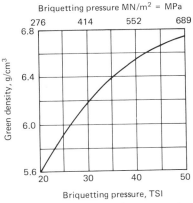
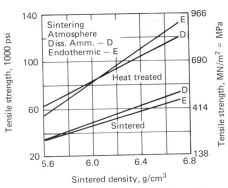
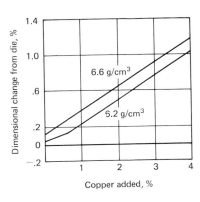

Fig. 22-9   The effects of pressure, density, and composition on one type of powder. *(Hoeganaes Corp., a subsidiary of Interlake, Inc.)*

gineering uses. This porosity also makes it difficult to apply nickel plating or similar finishes to P/M articles.

Metal infiltration, usually with a special copper alloy, can fill the pores. The proper size piece of copper is placed on either the top or the bottom of the part, and when the copper melts, during sintering of the iron, capillary action causes the copper to fill the pores. This copper infiltration gives the P/M parts:

1. Higher tensile strength, fatigue strength, and hardness
2. Higher and more uniform density
3. Sealed surface porosity so that secondary operations such as plating can be done

Plastic impregnation will also fill the pores and, as a secondary benefit, improve the machinability of the steel. This impregnation is done with vacuum and pressure cycles applied to fairly large batches of parts at one time.

Oil impregnation of low-density powdered metal bearings to make "self-lubricating" bearings was one of the earliest uses of powdered metal. The impregnation is usually done by use of a vacuum applied to the parts while they are immersed in oil.

OPERATIONS AFTER SINTERING

It is almost correct to say that a sintered powdered metal part can be treated like any other metal part of the same alloy.

Coining is a repressing of the P/M part in the same (though readjusted) dies or a separate set of tooling. Since the sintering operation has soft annealed the part, a slightly higher coining pressure can cause considerable plastic flow. This increases the density, hardness, and tensile strength.

Coining can also be used, as the word implies, to create patterns, lettering, or small notches on the P/M part.

CAUTION: This, of course, changes the size of the part, especially in the direction of pressing. Thus, the original tooling dimensions must allow for this change.

Sizing is often considered a form of coining. However, sizing dies are usually specially made tooling which can create dimensions to within 0.0005 in. (0.0125 mm) or less of exact size.

Both sizing and coining operations are often hand fed, one at a time operations. When large quantities are involved, this can be automated economically.

Heat treatment for strength and hardness of P/M parts can be done much the same as with solid metals. The proper carbon, molybdenum, manganese, and nickel alloys can be added to the powder before compressing.

CAUTION: Hardness of P/M parts is referred to as *apparent hardness* since the parts are, in varying degrees, porous. Thus the microhardness, which is important in sliding wear, will not be indicated accurately by standard hardness testing procedures. Microhardness tests should be used.

Tumbling or other barrel finishing methods may be needed to remove the sharp edges and fine *flash* produced at the junction of the die and punches in some P/M parts.

Machining (such as turning, drilling, tapping, grinding, etc.) can be done the same as with solid metals. Quite sharp tools with positive rakes should be used. In lower-density materials (below 90 percent), the porosity creates a minute "interrupted cut" condition. This sometimes makes cutting tools wear faster, so carbide tools should be used for large quantities.

LIMITATIONS OF POWDER METALLURGY

1. Strength of the part on some article, such as heavy-duty gears, may be a limitation, though repressed P/M parts may equal solid metal strengths.

2. It must be possible to eject the part from the die. Thus, grooves around the circumference, double or reentrant tapers, standard threads, crosswise holes, and diamond knurls cannot be made (see Fig. 22-10a to c).

3. Wall thickness should be 0.060 in. (1.5 mm) or thicker, so the powder can flow into the space, though thinner walls have been made. The length/thickness (L/t) ratio should be less than 4, or powder may not fill the cavity (see Fig. 22-10d).

4. The length/diameter ratio of solid parts should be less than 10 (see Fig. 22-10e).

5. The part should be designed with as few levels (diameters) as possible. Multiple levels can be compacted, but the tooling is more complex (see Fig. 22-10f).

SPECIAL P/M PROCESSES

Isostatic Pressing

Isostatic means that pressure is uniform on all sides. This can be attained by surrounding the part with either liquid (usually oil or water) or gas, under pressure. This process creates substantially equal density and strength through the P/M part. It is used for both metal and ceramic powders.

Isostatic pressing is used especially for making very large powdered metal parts such as a 24 × 70 in. (600 × 1780 mm) tungsten crucible and large carbide tools. The pressure may be from 5 to 50 tons/in.² (69 to 690 MPa), and it is uniform over the entire part. Thus, the larger areas do not require heavier pressures as in conventional pressing. Large hollow tubes and irregularly shaped parts of all sizes can also be made. Large *preforms* for forging are sometimes made this way.

This process requires no lubricant, the tooling is inexpensive; the equipment is quite simple to use; and close to 100 percent density can be achieved. However, the production rate is not high; tolerances are not as close as with regular pressing; and a certain amount of experimenting is often necessary before a successful run can be made.

The *molds*, or *bags*, are made of rubber or plastics such as polyurethane and may be formed over a solid

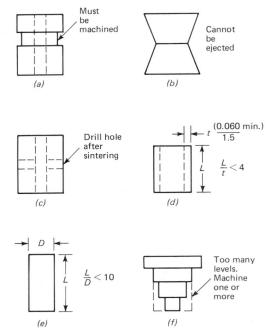

Fig. 22-10   Some design limitations of powdered metal parts.

pattern to the required shape. A solid core may be used to control the inside deminsions as shown in Fig. 22-11b.

In *dry bag pressing*, the bag is permanently fixed into a small pressure container. It is filled with a carefully measured amount of powder, sealed, and subjected to hydrostatic pressure. The dry bag system is used more often for smaller, less complex P/M parts. Solid cores can be used, and this system can be automated.

Hot isostatic pressing is sometimes used to make special items such as large solid-carbide boring bars or rolls for metal processing. The part is placed in a pressure vessel which can be heated to 2000°F (1090°C), and the pressure is supplied by gas.

Forging (Hot Forming) of P/M Parts

When forgings are made from hot bar stock, several *strikes* must be made, large burrs or flash are created, and much metal is wasted. However, high strengths and fairly complex shapes are easily produced by forging. (See Chap. 27 for a discussion of forging.)

If a P/M *preform* of either a measured *slug* or an approximation of the needed shape is made by mechanical or isostatic pressing, the part can be pressed, sintered, heated to forging temperature in a furnace or by induction heating, and then forged to its final shape and size. Very close to 100 percent density is often achieved in one stroke of the press.

This process was first tried in 1969 and was first

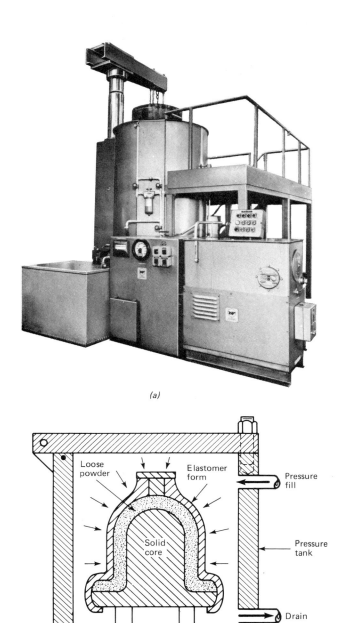

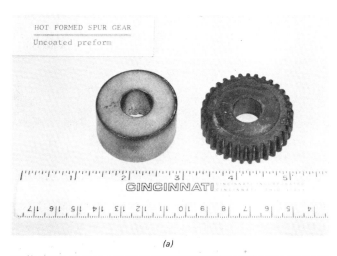

Fig. 22-11 (a) A 30 × 36 in. (760 × 910 mm) cavity, 10,000-psi (69-MPa) capacity isostatic press. Note heavy breech sealing cap at the top. (*National Forge Co.*) (b) Diagram of a simple *wet bag* method of isostatic pressing. Solid core is shown.

Fig. 22-12 Powered metal forging press and parts made with it. (*Cincinnati, Inc.*)

used for automobile gears. Equipment is expensive, but economically feasible uses are fairly rapidly being developed by both powder metallurgy and forging companies. Figure 22-12 shows a press and the first parts made.

Closed dies are used, and relatively minor sideways flow is usually planned for. Most of the compacting is done in the direction of the press stroke.

Fairly close tolerances are achieved, no metal is wasted, the production time is fast, and strengths are equal to those of bar-stock forgings. If isostatic pressing or the very large P/M presses are used for the preforms, quite large forgings can be made, with the possibility of considerably reduced costs.

High-energy-rate forming (HERF), sometimes called *high-velocity forming (HVF)*, is described in Chap. 27.

This is being used to some extent to obtain high-density compacting of the superalloys.

Rolling of strips of metal, either directly from the powder or from isostatically pressed P/M billets, is being done on a limited scale. Nickel strip for coins is being produced in large quantities by both methods.

Review Questions and Problems

22-1. What are the advantages and disadvantages of making parts by powder metallurgy?

22-2. Define "apparent density" and "flow rate."

22-3. Name the principal parts of the tooling for powdered metal.

22-4. The apparent density of a bronze powder is 3.5 g/cm^3 and the final density is to be 7.6 g/cm^3.
  a. What is the compacting ratio?
  b. How deep a fill will be needed for a 25-mm-long bushing?

22-5. A bronze bushing 1 in. long, $\frac{3}{4}$-in. OD, and $\frac{1}{2}$-in ID is to be pressed at 10 tons/in.^{2}. How many tons of pressure are needed?

22-6. Should the bushing in Prob. 22-5 be made on a single- or double-acting P/M press? Why?

22-7. Convert Prob. 22-5 into SI metric units and solve it.

22-8. a. How is the die filled with loose powder?
  b. How is the ejected part moved off the machine table?

22-9. The sintering section of a furnace is 45 ft (13.7 m) long. Sintering time is to be 30 min. At what rate should the mesh belt be run? Solve in U.S. and metric units.

22-10. The total length of the sintering unit (preheat, sinter, cool) is 80 ft (24,400 mm). How long will it take a part to go through the entire cycle at the rate determined in Prob. 9?

22-11. What is meant by infiltration of a P/M part?

22-12. Describe coining. Why is sizing sometimes called a coining process?

22-13. Two rectangular blocks 100 × 250 × 100 mm and 150 × 200 × 25 mm are to be isostatically pressed at 400-MPa pressure. How much stronger a chamber is needed for isostatically pressing the two at once as compared to pressing them one at a time?

22-14. What are the advantages and disadvantages of isostatic pressing?

22-15. Why is it sometimes advantageous to make P/M preforms for forged parts?

References

1. Clark Francis: *Advanced Techniques in Powder Metallurgy,* Rowan & Littlefield, Inc., 1963.
2. *Creating with Metal Powders,* Hoeganaes Corp., Riverton, N.J. 08077.
3. DeGroat, George H: *Tooling for Metal Powder Parts,* McGraw-Hill, New York, 1958.
4. Hurschhorn, Joel S: *Introduction to Powder Metallurgy,* American Powder Metallurgy Institute, 1969.
5. Kunkel, Robert: "P/M—Profitable Tooling for Powder Metal Parts," *Manufacturing Engineering and Management,* p. 37, May 1970.
6. Kunkel, Robert: *Tooling Design for Powder Metallurgy Parts,* Society of Manufacturing Engineers, Detroit, Mich., 1968.
7. Metal Powder Industries Federation, 201 East 42nd St., New York, N.Y. 10017. Many publications available.

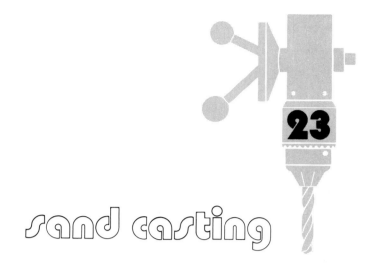

sand casting

Sand casting is pouring melted metal into a hollow mold which has been made in especially prepared sand. Of the several casting processes available, sand casting is by far the most widely used.

Although many materials are sand-cast, iron (such as gray iron, white iron, ductile or nodular iron, and malleable iron) is the most commonly used. *High-alloy irons* (over 3 percent alloy) are also being cast. Examples are: a nickel iron which is nonmagnetic, corrosion-resistant, wear-resistant, and the heat-resistant alloys. The most frequent alloying elements are silicon, chromium, and copper.

Steel (iron which has less than 2 percent carbon) is also sand-cast. Many low and medium alloys of steel, including acid-resistant types, are cast. Aluminum alloys, brass, bronze, and magnesium alloys are also sand-cast.

ADVANTAGES AND DISADVANTAGES OF SAND CASTING

The *advantages* of sand casting are:

1. Sand castings are made in any size from 8 oz (227 g) to over 300 tons [272 t (mass)].
2. A great variety of shapes can be made. Some of these shapes would be impossible to make any other way.
3. Production of many sizes and types of castings can be fully or partially automated.

The *disadvantages* are:

1. The necessary allowance for size variations is high compared to some other methods of casting. Typical tolerances are shown in Table 23-1.
2. The *skin* of a casting may contain sand, which causes rapid cutting tool wear.
3. The pattern and cores are an added cost which must be justified.

THE STEPS IN MAKING A SAND CASTING

The designer must know something about the casting process so that the engineering drawing shows a part which can be cast efficiently. It is possible to design a part which cannot be made in a sand mold, not because of its size, but because of its shape.

The process of making a sand casting is diagramed in Fig. 23-1. The melting and alloying of the iron and steel used is considered briefly in Appendix A.

The Pattern

The pattern is made to the shape of the part shown in the drawing. It is made of clear pine, mahogany, aluminum, brass, or plastic. If more than 500 parts are to be cast from one pattern, the pattern should be made of metal, usually aluminum. A few of the great variety of patterns which can be made are shown in Fig. 23-2.

The patternmaker must consider several items in

TABLE 23-1 TOLERANCES TO BE EXPECTED FROM SAND CASTINGS*

Metal	Major Dimension	Tolerance	
		±in.	±mm
Gray cast iron	To 10 in. (250 mm)	$\frac{1}{16}$	1.6
	10–in. (250–460 mm)	$\frac{1}{8}$	3.2
	Over 18 in. (460 mm)	$\frac{1}{4}$	6.4
Steel	To 10 in. (250 mm)	$\frac{3}{32}$	2.4
	10–36 in. (250–910 mm)	$\frac{1}{4}$	6.4
	36–120 in. (910–3050 mm)	$\frac{5}{16}$–$\frac{1}{2}$	8.0–13.0
Aluminum alloys and malleable iron	To 24 in. (610 mm)	$\frac{3}{32}$	0.8
	Over 24 in. (610 mm)	$\frac{5}{64}$	2.0
Brass and bronze	To 18 in. (460 mm)	$\frac{3}{32}$	2.4
	Over 18 in. (460 mm)	$\frac{1}{8}$	3.2

*Closer tolerances can be held with careful planning, but tolerances may be larger across the parting line.

303

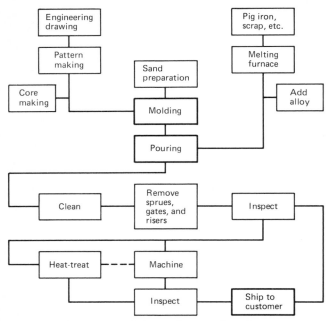

Fig. 23-1   The steps needed to make a sand casting.

(a)

(b)

Fig. 23-2   (a) A two-piece (split) pattern for a valve. (b) Loose, solid patterns for machine parts. (*Photograph by Samuel Lapidge.*)

addition to the dimensions shown on the drawing. Figure 23-3 shows the shape as designed, and the changes the patternmaker must make so that the casting will finally emerge at the correct size.

A **shrinkage allowance** must be made for any object which is heated and then cooled. This ranges from $\frac{1}{8}$ in./ft (10.4 mm/m) for gray iron, to $\frac{1}{4}$ in./ft (20.8 mm/m)

for cast steel and some bronze. The patternmaker uses *shrink rules* which are 18 in. (450 mm) or longer scales which have divisions larger than on standard scales so as not to have to compute every little variation.

If a metal pattern is used, this is often made from a poured pattern, and *double* shrinkage must be allowed for its mold.

A **machine finish allowance**, or *machining allowance*, must be made on any surface which is to be machined after casting. This can vary considerably according to casting size, but typical values are cast iron, $\frac{3}{32}$ to $\frac{3}{16}$ in. (2.4 to 4.8 mm); brass and aluminum, $\frac{1}{16}$ to $\frac{1}{8}$ in. (1.6 to 3.2 mm); and cast steel, double those for aluminum.

A **draft** of about $\frac{1}{8}$ in./ft (10.4 mm/m) must be allowed so that the pattern can be pulled up out of the sand without breaking the edges of the sand mold. In shallow castings and for metal patterns, the draft can be less, and more draft may be needed on castings over 12 in. (300 mm) deep. The draft is kept as small as possible, as it is often extra metal which must be machined off later.

Fillets (rounded inside corners) must be used or the casting may crack when it cools. These may be from $\frac{1}{8}$ to 1 in. (3 to 25 mm) radius. On round patterns the fillets can be cut on the lathe. For flat and irregular shapes, the fillets are made of beeswax, leather, or plastic. The fillets are purchased in strips or rolls, already cut to size and shape. They are fastened into place by heating the beeswax or by gluing the other types.

All outside corners should also be rounded, and this is easily done by forming the pattern in the lathe or by sanding the straight edges.

Patterns are seldom made of one piece of wood because they would warp while in storage. They are made like plywood, with the grain reversed as the thickness is built up.

The **cost** of a pattern may be as low as $25 to $50; however, a 36-in.-diameter (900-mm) pattern may take two to four days to make, depending on its complexity, and may cost up to $500. Metal matchplate patterns for complex parts may cost even more.

Cores

A core is a preformed piece of material that is put into a mold so that the metal will flow around it and leave an opening in the casting. Figure 23-4 shows some examples of cores and how they can be used. A core can be used to form quite complex shapes and passages inside and outside a casting. Sometimes several separate cores are needed.

The core must be supported in the mold, so the pattern is made with extended pieces which will leave an impression in the sand into which the core can be

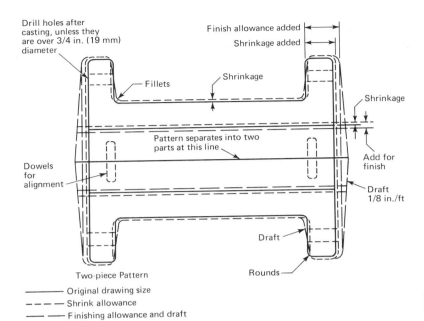

Drill holes after
casting, unless they
are over 3/4 in. (19 mm)
diameter

Finish allowance added

Shrinkage added

Fillets

Shrinkage

Shrinkage

Pattern separates into two
parts at this line

Add for
finish

Dowels
for
alignment

Draft
1/8 in./ft

Draft

Rounds

Two-piece Pattern

———— Original drawing size
- - - - Shrink allowance
—— —— Finishing allowance and draft

Fig. 23-3 The allowances which must
be considered when making a pattern.

placed. These are called *core prints* and are shown in Fig. 23-4.

The "standard" type of core, called an *oil-sand core*, is made of sand with a binder, or *glue*, made of oil or resins and some cereal. This combination, after being thoroughly mixed or *mulled*, is pressed into a *corebox*.

The corebox is actually a reverse image pattern of the core, made of wood, metal, or plastic. Examples of coreboxes are shown in Fig. 23-5.

The "green" (unhardened) core is removed from the corebox and baked in a gas or electrically heated oven.

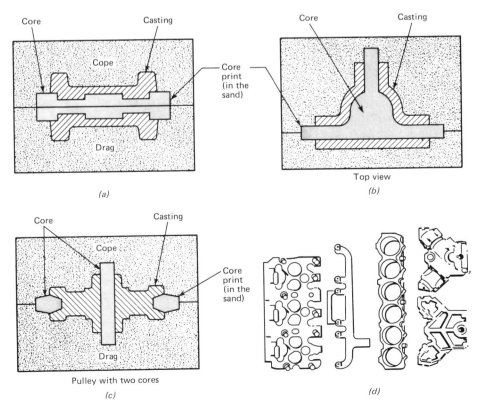

Core

Casting

Cope

Drag

(a)

Core

Casting

Core
print
(in the
sand)

Top view

(b)

Core

Casting

Cope

Core
print
(in the
sand)

Drag

Pulley with two cores

(c)

(d)

Fig. 23-4 (*a*) to (*c*) Some examples of simple cores in place. (*d*) Examples of shell-molded cores. (*Shalco Systems.*)

The dried core must be strong enough to resist the force of the hot metal as it rises in the mold but fragile enough to be crushed by the shrinkage of the metal or to break when the casting is shaken out after it is removed from the sand. Very long slender cores are reinforced by a steel rod in the center. The rod drops out when the core breaks up.

modern core-making processes

The CO_2 process is used especially for medium- and large-sized cores. It is important to realize that cores 6 and 8 ft (1800 to 2400 mm) in length or diameter are not unusual.

In the *carbon dioxide process*, dry core sand is mixed with sodium silicate (water glass) and packed into a corebox. Carbon dioxide gas (CO_2) is passed through the mixture, and the chemical reaction hardens the core. The core is stripped from the corebox and is ready to use. Complete sand molds are also made by this process.

Fig. 23-5   (*a*) Simple pressed core. (*b*) Corebox for one-half of core shown in (*c*). (*d*) Corebox for one-half of the core in Fig. 23-4*a*. (*e*) A precision-cast corebox. (*Unicast Development Corp.*)

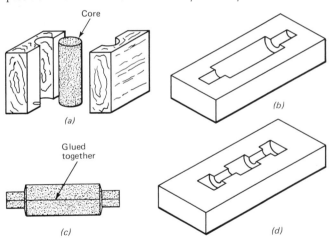

(*e*)

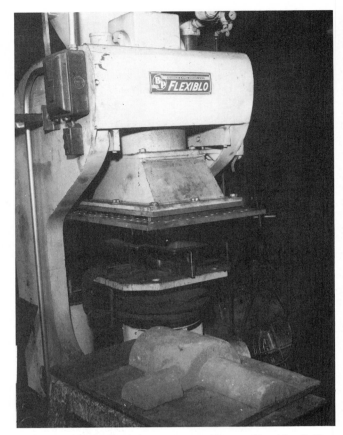

Fig. 23-6   A 55-lb (25-kg) capacity coreblower. (*Beardsley & Piper Division, Pettibone Corp.*)

Furan resin cores are widely used because they do not require heat, hold dimensions well, and are cured in the corebox so they do not warp or sag.

The sand is mixed with 1 to 2 percent of furan resin. This resin is made up of furfuryl alcohol, urea, formaldehyde, and an acid catalyst such as phosphoric acid.

The corebox is filled with this sand-furan mixture and, as the mixture is exothermic (gives off heat), it starts to cure immediately. Cores are ready for use in 4 to 8 h, or curing can be accelerated by heating for 15 min in a furnace at about 300°F (150°C).

This No-Bake process can be used to make cores weighing several hundred pounds. They are simply allowed to cure longer. These can be "cast" in wooden coreboxes. The centers of large cores may be filled with coke or other light materials to reduce cost and weight.

Furan hot-box cores use a similar sand-resin mix, blown into heated cast-iron coreboxes in a special core-making machine. The sand is blown into the hot box, cured for 5 to 60 s, and then ejected onto a moving belt. The center of larger cores hardens within a few minutes as the reaction continues. This process can be completely automated.

The disadvantages of furan cores are: a higher cost of additives, a limited bench life before the sand sets and cannot be used, and the presence of some odor.

Shell molding, described in Chap. 24, is used for making small- and medium-sized cores, especially if they are of intricate shape.

Coreblowers (Fig. 23-6), which automatically fill the coreboxes under pressure, will give high production rates for small- and medium-sized cores.

Extrusion of round or other solid shapes with a simple press and die is a fast method for making cores up to 3 in. (75 mm) in diameter. The extrusions are cut to the required length and then baked.

Fig. 23-7 (a) A hinged wooden flask (cope and drag). *(Photograph by Samuel Lapidge.)* (b) Small aluminum hand flask, and 18 × 48 in. (460 × 1200 mm) machine flask. *(The Hines Flask Co.)*

(a)

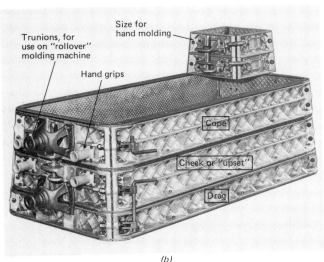

(b)

The Molding Sand

Molding sand is composed of sand, clay, water, and additives such as cereal. The basis for molding sand is clean *silica sand*, which is found in many parts of the world, though few deposits are found west of the Great Lakes in the United States. The sand is sifted, and particles finer than 270 mesh (53 μm) or coarser than 30 mesh (595 μm) are seldom used.

As this natural sand will seldom stick together, 4 to 15 percent of clay is added as a *binder*. The most commonly used clay is *bentonite*, derived in nature from the decomposition of volcanic ash. Chemically, bentonite is made up mostly of montmorillonite, $(MgCa)O \cdot Al_2O_3 \cdot 5SiO_2 \cdot NH_2O$.

Another binder used is *kaolinite*, most often called *fire clay*, $Al_2O_3 \cdot 2SiO_2 \cdot 2H_2O$. This same clay, as white as china clay, is used for porcelain dishes.

Cereal flours (such as corn, wheat or rye flour) produce a sticky, starchy material and also serve as a cushion by burning up and leaving voids as the sand expands from the heat of the poured metal. From 0.5 to 2 percent cereal is often added to the sand. Wood flour (ground wood fibers) and lignin (left over from making wood pulp) are sometimes used for the same purpose.

Other additives are occasionally used. *Sea coal* (finely ground soft coal) may be used to improve the surface on gray iron. Resins, oil, and pitch may also be used.

The *water content* of the molding sand is fairly critical. It is usually 3 to 4 percent by weight. Too little water makes a weak mold, and too much water can cause "scabs" on the casting due to too much gas pressure.

Making a Green Sand Mold for a Casting

The word "green" does not refer to the color of the sand. It means that the sand is a damp, plastic mixture as compared to a dry or hardened sand mixture. Sand molds may be for small 2 in. (50 mm) parts or for large 10 to 20 ft (3000 to 6000 mm) castings for turbines, large boring mills, etc. The largest volume of castings are in the range of up to 5 ft (1500 mm) maximum dimension. Within this range of sizes the following process is typical. The basic steps are illustrated in Fig. 23-8; more automated processes will be described later.

1. A *molding board* with a smooth surface is placed on a table or on the floor. The bottom half of a molding flask (Fig. 23-7), called the *drag*, is placed on the board. The half pattern is then placed somewhat to one side of the drag (Fig. 23-8a).

2. First fine sand and then regular green sand is

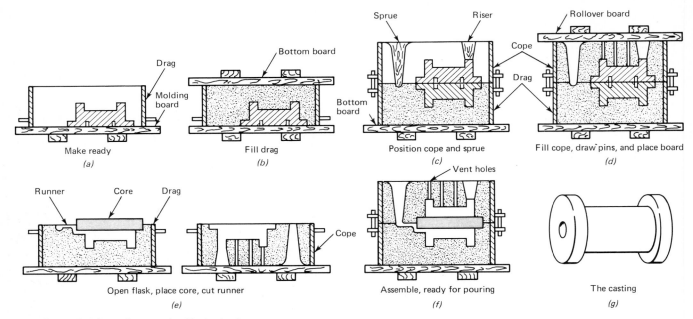

Fig. 23-8   The principal steps in making a mold for sand casting. Hand work shown.

poured into the drag and pounded firmly into place, compacted around the pattern. The moldmaker uses a wooden *rammer* for molds up to about 24 in. (600 mm) square. For large molds, air-operated rammers are used for manual compacting. The sand is then smoothed off level with the edge of the drag, called *striking off* (Fig. 23-8*b*).

3.   The *bottom board* (sometimes a rougher surface) is placed on top of the drag (Fig. 23-8*b*). It is held firmly by hand, clamps, or machine and the drag is turned over.

4.   The *cope* (top half of the flask) is placed on top of the drag and aligned by pins, and the top half of the pattern is positioned by pins on top of the lower half (Fig. 23-8*c*).

5.   Wooden or metal pegs for sprue and riser are positioned (Fig. 23-8*c*), and the cope is filled as in item 2 above. The sprue pin and riser pegs are pulled, and the pouring basin is cut out. The rollover board is placed on top of the flask (Fig. 23-8*d*).

6.   The cope is lifted off and turned upside down. The patterns are removed, and the runners and gates are cut into the drag (Fig. 23-8*e*).

7.   The cope is again placed on the drag, carefully aligning it on the core prints guided by the pins in the flask (Fig. 27-8*f*).

8.   The flask is removed, a steel or wooden retainer is placed around the mold, and it is ready to be filled with molten metal. If the mold is deep, weights are placed on the top of the cope so that the liquid pressure will not lift it during pouring. Often a steel

pin, about $\frac{1}{16}$-in. (1.6-mm) diameter, is jabbed into the mold several times from the top to make venting holes through which air and gas can escape. Figure 23-9 shows a green sand mold ready for filling and gives the names of the various parts.

Making the Casting

The prepared mold is often guided on roller conveyors to the pouring area, since many foundries pour hot metal only once or twice a day.

The molten metal is tapped from the furnace into large transfer ladles supported by overhead cranes. These may be used to pour directly into the large molds, or they may be used to fill hand-carried ladles or smaller ladles on rails or cranes.

The metal is poured through the sprue until both it and the riser are full. The metal is cooled until it is solid (frozen) in the sand mold.

The riser is important since it first vents the air as the mold fills, and then serves as a reservoir of metal to supply the casting as it shrinks due to cooling. On large molds, two or more risers may be needed.

Shakeout is the next step. The band is removed, and the mold is placed on a vibrating conveyor. The rather heavy vibration shakes off the sand and shakes out the core, leaving the casting to continue along the conveyor.

The sand drops into a pit and is later reconditioned by adding about 10 percent new sand plus binders if needed. This sand is then reused for more molds.

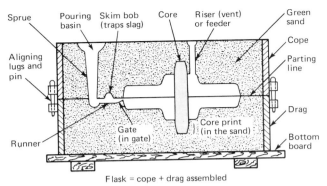

Fig. 23-9 A complete sand mold ready for pouring, and the nomenclature used.

The casting now still has sprue, risers, and runners attached. In fact, it is sometimes difficult on 10- to 100-lb [4.5- to 45-kg (mass)] castings to see the real shape. These appendages are now *trimmed off* with band saws, cold saws, or abrasive saws. Next, the largest rough spots and the nubs from the sprues, etc., are snagged (rough ground) to the shape of the casting. The casting is now ready for delivery to the customer.

OTHER TYPES OF MOLDS FOR CASTING

The process described so far is very basic and is used in many foundries. However, the need for faster, more accurate, or less expensive methods has been met by several improvements on the basic process.

Dry Sand Molds

The term *dry sand* means that no water is in the sand which is next to the molten metal. The dry part may be the entire mold or only about $\frac{1}{2}$ in. (13 mm) deep. Dry sand molds are more expensive but give better finishes and closer size control. They are usually used for steel castings and, sometimes, for iron castings.

The CO_2 and *furan* processes used for cores can also be used to make a complete mold. As these processes are more expensive than green sand molding, they are used principally when greater accuracy or better finishes are required.

Making Large Sand Molds

When a casting is very large, over 10 ft (3000 mm) in length or diameter and about 4 ft (1200 mm) deep, a *pit* casting is made. A hole is actually dug in the foundry floor and a form of the approximate size is built from bricks or wood.

The entire mold for the casting, sidewalls, base, and cores may be made from furan-process dry-sand

pieces fitted together. Of course, the mold may also be made with green sand, though this is less frequently used now.

For many medium-sized castings (such as machine tool bases), a *skin-dried mold* may be used. These are

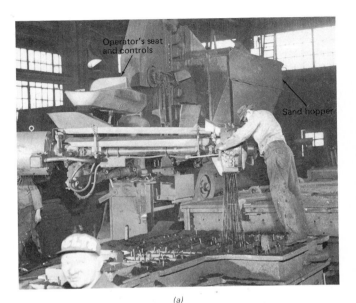

(a)

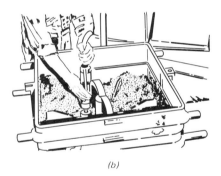

(b)

(c)

Fig. 23-10 (a) A mobile sand slinger filling large steel flasks. (*Beardsley & Piper Division, Pettibone Corp.*) (b) Using an air hammer to ram the sand in the drag. (c) A mold for a 700-lb (318-kg) casting being poured using two ladles at once. Note weights on top of flask.

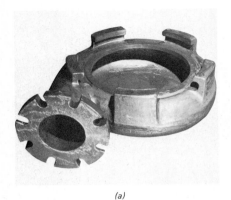

(a) *(b)* *(c)*

Fig. 23-11   Typical sand castings. (*a*) A 36-in. (914-mm) diameter, 200-lb (90-kg) pump part. (*b*) The core casting of a hydraulic valve. (*c*) Cast-iron gear housings for boring mill. *(Bullard Castings, Inc., one of the White Consolidated Industries.)*

made of green sand, usually in large metal flasks. A refractory material plus a bonding agent such as bentonite clay is sprayed or painted on the sand. The surface may be dried with a torch, or alcohol may be mixed with the clay, set on fire, and burned off.

The sand for medium and large molds is often put into the mold by a machine called a *sand slinger* (Fig. 23-10). This is a mobile piece of equipment which throws the prepared sand into the flask at such high speeds that the sand is compacted when it lands. A large sand slinger can fill a mold at over 100 lb/min [45 kg (mass)/min]. Some typical sand castings are shown in Fig. 23-11*a* to *c*.

OTHER TYPES OF PATTERN EQUIPMENT

The process described in Fig. 23-8 is still used for thousands of castings because many orders call for only 1 to 100 pieces. For these small quantities, it is still an excellent, and fairly quick, process, as many simple molds can be made complete in 5 min or less. For larger quantities, faster methods are available using pattern plates and match-plate patterns.

Mounted Patterns

The filling and compacting of the sand mold and the removing (drawing) of the patterns can be automated

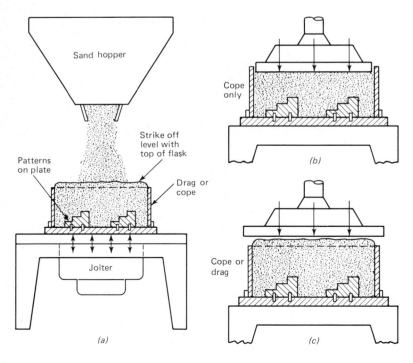

Fig. 23-12   Using a mounted pattern for hand molding, and three methods of compacting the sand by machine. (*a*) Using a jolting unit. (*b*) Squeeze and leave underfill. (*c*) Overfill, squeeze, and strike off level.

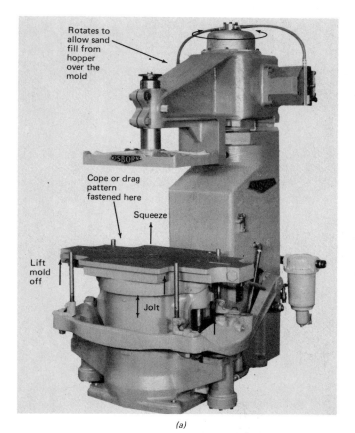

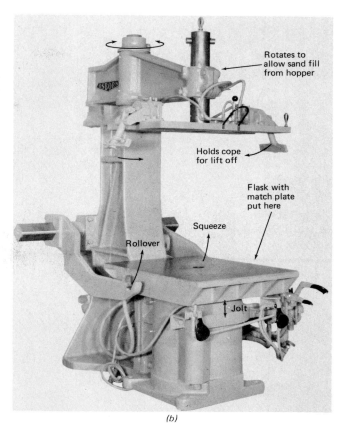

Fig. 23-13 (a) A ram-jolt-squeeze-stripper mold-making machine. (b) A ram-jolt-squeeze-turnover-stripper mold-making machine. A complete mold is produced at each cycle of 3 to 5 min. (*The Osborn Manufacturing Co.*)

if the loose patterns are mounted on a wooden or aluminum board. Sprues, gates, risers, and runners are also put on the board.

If the halves of a split pattern are aligned and mounted on opposite sides of a single plate, it is called a *match-plate pattern*. If each half of the pattern is mounted on a separate board, it is called a *double match plate* or a *cope-and-drag set*. In both types, several small patterns can be fastened to one board connected by suitable runners.

As shown in Fig. 23-12, the mounted cope or drag can speed up even hand molding. However, the biggest saving is when the mounted patterns can be used with semiautomatic moldmaking machines.

Figure 23-13a shows a ram-jolt-squeeze-stripper machine which uses a mounted cope or drag pattern fastened to the table. Figure 23-13b shows a ram-jolt-squeeze-turnover-stripper machine which uses a match plate held between a cope and drag. The drag is filled, jolted, and squeezed, then the assembly is lifted up and the operator turns it 180° so that the cope is on top. This is filled, jolted, and squeezed,

and then lifted up so that the match plate can be removed (stripped) and the cores inserted. The cope is lowered, and the mold is taken away by a hoist. It is ready for pouring.

Figure 23-14a to c shows a slightly different process for making a cope or drag half of a mold. Notice that several different parts are being made in one mold. In this method, the use of many flexible rams distributes the pressure more evenly than a single plate can. It is claimed that this eliminates soft spots in the mold and, thus, produces a more accurate casting. It also eliminates the noisy process of jolting.

CAUTION: If you visit a foundry, you will see the seemingly simple process going on quite casually. Do not underestimate the amount of attention which has been given to the proper sand mixture. It is important to realize that the simple pounding of sand in the flask requires the "feel" of not too hard or too loose.

There is also a lot of knowledge (and some mathematical formulas) and experience used in de-

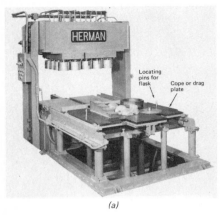
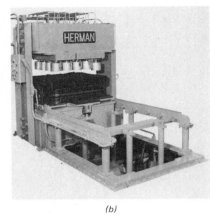
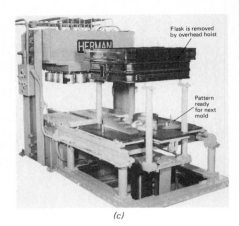

(a)   (b)   (c)

Fig. 23-14   (*a*) The cope or drag plate pattern is fastened to the machine table. Half of the flask is placed over the pattern and filled with sand from an overhead hopper. (*b*) The filled mold moves under the rams and is compacted. (*c*) The mold slides out, is lifted from the pattern, stripped, and removed by a hoist. Cycle can now start over. (*The Herman Corp.*)

ciding the size and position of sprues, runners, gates, and vents. Pulling the pattern requires a trained, steady hand, and pouring at the right temperature and speed is an important part of getting a good casting. It is all quite simple, when you know how.

Full Mold Process (Patented)

Polystyrene foam in blocks and sheets can be sawed, carved and glued, or even taped together into many varied shapes. As it weighs only 1 lb/ft³ (16 kg/m³), very large patterns can be made and handled easily. Some medium-size foam "patterns" are shown in Fig. 23-15. Each is actually a model of the part to be made.

The polystyrene pattern is packed in sand the same as the wooden pattern, though less firmly packed. However, when the metal is poured in, it vaporizes the polystyrene and fills the space left. Thus, there is no "cavity" at any time.

This process is especially useful for repair work and experimental or prototype work when only one or two castings are needed and a wood pattern would be too expensive. It is also advantageous where very large machine bases or automobile body dies are to be cast. The process is licensed by Full Mold Process, Inc., which holds the patents.

The special advantages of this process are that no draft is needed; rounded corners are not necessary; little, if any, binder is needed in the sand; fewer cores are needed; and less skilled help can make the "patterns." Foam models can often be made in one-quarter to one-half the time needed for making a wooden pattern. The disadvantages of the method

are that the finish is rougher than sand casting unless a smooth type of polystyrene is available, though this is not serious in medium- and large-sized castings. Venting the gas generated is sometimes a problem.

This system requires a new shape or pattern for every casting. However, simple forms can be sawed several at a time, and if a large number of relatively small castings are needed, the "patterns" can be made cheaply in automatic foam molding machines.

Centrifugal Casting

Not only circular shapes, but symmetrical and unsymmetrical shapes, as shown in Fig. 23-16, are centrifugally cast in steel, nickel alloys, copper and bronze alloys, and aluminum.

These centrifugal castings are usually from 6- to 60-in. (150- to 1500-mm) diameter, though larger and smaller castings are made. Castings as long as 34 ft (10,000 mm) can be made, though 10 to 20 ft (3000 to 6000 mm) is more often cast. Tubular and other shapes can be cast in alloys which cannot be easily rolled, forged, or extruded.

The axis of rotation may be either horizontal or vertical. Machines with a horizontal axis are used for long cylindrical (or approximately cylindrical) shapes as in Fig. 23-16. The process is to pour a predetermined amount of metal into a rotating mold. The rotation is fast enough to develop 50 to 100 g. Speeds of 200 to 1000 rpm are commonly used.

As the mold continues to spin, solidification of the metal progresses from the outside surface of the casting toward the inside surface. Any impurities and gases, being lighter than the metal, are forced to the inside of the casting. The final inside diameter de-

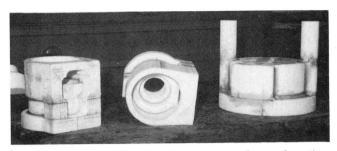

Fig. 23-15 Some polystyrene foam patterns for sand casting. About 24-in. (610-mm) diameter. Much larger ones are also made. *(Bullard Castings, Inc., one of the White Consolidated Industries.)*

general idea of this type of machine and some of the shapes made. This type of casting requires that molds made of dry sand-type material be placed in the rotating flask to form the desired shape. The solidification is not entirely unidirectional as cooling takes place from all surfaces of the casting. A central core may be used.

Many of these parts could be made as ordinary sand castings, so centrifugal casting would be chosen only when the greater uniformity of grain structure, freedom from defects, and higher mechanical properties are necessary for the end use of the product.

pends on the amount of metal originally poured into the mold.

The inner metal containing gases and impurities must be bored out. This may sometimes require removing more than an inch thickness. The metal that is left is free of defects, with uniformly high, nondirectional mechanical properties, with no sprues or gates to be removed. The outside of the casting is turned to the final diameter and finish required.

The mold for aluminum can be of iron; but for iron or steel castings, the mold is lined with sand or refractory-type materials or graphite.

Semicentrifugal Casting

When the axis of spinning is vertical, many shapes can be cast (such as wheels, bushings, very large bearing races and enclosures, and large venturis). If cores and sprues are used, this is sometimes referred to as semicentrifugal casting. Figure 23-17 shows the

Centrifuging Casting

If a number of small, high-quality castings are needed, the centrifuging method is occasionally used. In this system, as shown in Fig. 23-18, one or more molds are mounted in a rotating machine. The rotation flings the metal into the mold, which gives it better grain structure than would be obtained from a sand casting.

Some of the machines for centrifugal, semicentrifugal, and centrifuging casting are small and inexpensive, ranging from about $1500 to $20,000 for large machines. Tolerances as low as 0.010 in. (0.25 mm) can be held, though 0.015 in. (0.375 mm) is easier to maintain.

Flaskless Molding—Automated

For castings which will fit in flasks not over 35 × 22 in. (890 × 560 mm), a totally automatic system is being used in about 100 foundries. The system will cost from $800,000 to twice that amount, but the cycle

Fig. 23-16 (a) Vertical section through a centrifugal casting machine and some of the shapes which can be made. (b) Furnace rolls made by centrifugal casting. They have had shafts welded on. *(Wisconsin Centrifugal Inc.)*

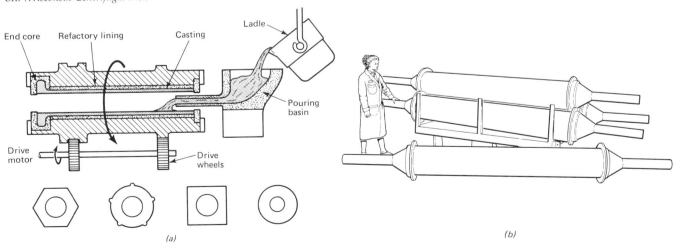

(a)

(b)

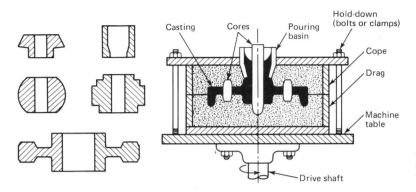

Fig. 23-17 Semicentrifugal (vertical-axis) casting and some of the shapes which can be made.

is fast, and practically no manual labor is needed. It is illustrated in Fig. 23-19.

Instead of flasks, a strong mold chamber is used. In the *vertical parting system*, one side of the chamber holds the cope and the other side the drag pattern. As shown in Fig. 23-19*b*, the cope of one mold and the drag of the next mold are made at the same time, facing *away* from each other.

A highly bonded sand is blown into the mold, then squeezed to high density. The sides of the mold chamber are pulled back and the sand mold, no flask or frame with it, is pushed out of the mold chamber.

Cores may be automatically inserted (not shown in the illustration), and the mold is pushed forward against the previous mold to form a complete casting opening.

No metal supports are needed, as each sand mold supports the others pressed against it. The molds are pushed under an automatic pouring station, then allowed to cool. They are then fed into a knockout barrel which separates the sand and cores and sends the castings on to be cleaned up.

To change to a different product, only replacement of the mold chambers and resetting of the sand and metal metering systems are required.

The economies are:

1. Elimination of flasks and flask-handling equipment
2. Minimum personnel requirements
3. No clamps or weights needed to hold flasks together
4. Fast molding cycle
5. Consistent dimensional accuracy and excellent parting line match

Horizontal parting line flaskless molding machines are also made. These will handle larger molds. This machine uses cope and drag plates similar to the usual setup. However, the patterns are reused after they have been withdrawn from the sand mold. This machine takes considerable floor space, is expensive, and could be economical only if over 10,000 fairly large identical castings are needed.

Fig. 23-18   (*a*) Schematic drawing of a centrifuging casting machine and some possible arrangements of the molds. (*b*) A small machine for vertical centrifugal or centrifying casting. Capacities up to 12,000 lb (5440 kg) are made. (*The Centrifugal Casting Machine Co.*)

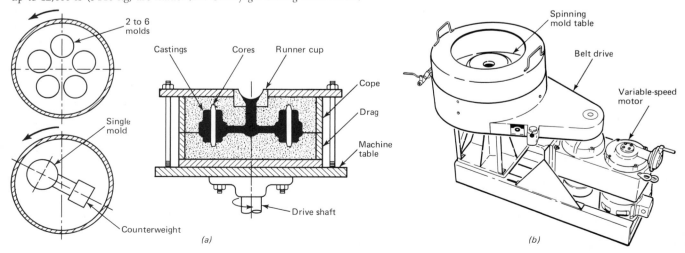

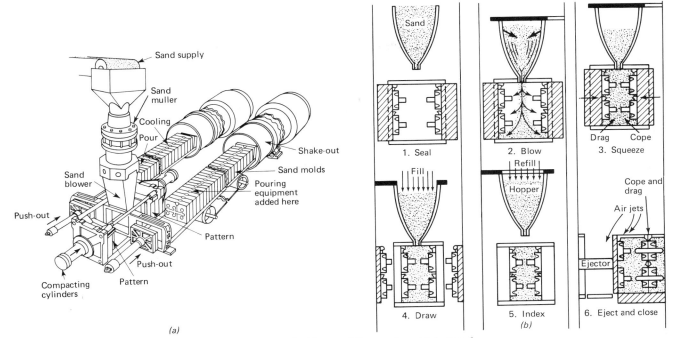

Fig. 23-19 (*a*) The Herman-Wallwork, vertical parting, flaskless molding system: high speed, high pressure. (*b*) The six steps in making the mold. (*The Herman Corp.*)

AUTOMATION OF SAND CASTING

Although jolt-squeeze-turnover machines are made as large as 8 ft (2400 mm) wide, a high percentage of the castings made will fit into a mold about 24 × 36 × 20 in. (600 × 900 × 500 mm), and thousands of smaller castings are made every day.

Systems for handling the automatic molding of work in this size range are made by several companies. These consist of power-driven conveyor systems connecting a casting removal section, sand filling stations, jolt-squeeze machines, turnover units, automatic pattern withdrawal, and some specialized units.

Many configurations of the "automated foundry" are possible. Figure 23-20 shows one possibility. The molding line shown would be up to 300 ft (90 m) long. However, it could easily be arranged in an L or U shape to fit available space in a factory, and it can produce up to 300 molds per hour.

The pattern boards can easily be changed, so these systems are quite flexible. Cores must, in most systems, be placed in the molds by hand.

The completed mold may be conveyed to the pouring area, filled, and cooled while returning to the beginning of the molding cycle, and then the flask is started through the line again.

This type of automated molding system will cost $500,000 or over, so it is practical only for large production work in places such as automobile companies, valve manufacturers, plumbing equipment companies, etc. It would not usually be practical for large, few of a kind jobs such as turbines, machine bases, and large pump bodies.

MACHINING OF SAND CASTINGS

Cast iron, in the grades most frequently cast, is easy to machine. The distribution of carbon in the iron causes it to break up into short chips when being cut so that no long tangles occur. On large iron castings, turning cuts $\frac{3}{8}$ in. (9.5 mm) and deeper are often made at cutting speeds of 200 fpm (60 m/min) with either positive or negative rake carbide tools. Other cast materials (such as aluminum, brass, steel, and stainless steel) are cut at about the same speed as the wrought materials.

Difficulties in machining sand castings may be caused by sand which is trapped in the outer layer; this is very abrasive. Thus, first cuts must be deep enough to go below this layer. This is especially true with the larger castings.

Hard spots do occur unless the casting is properly made. Actually these are also found in hot- and cold-rolled steel.

Dimensional changes take place when machining

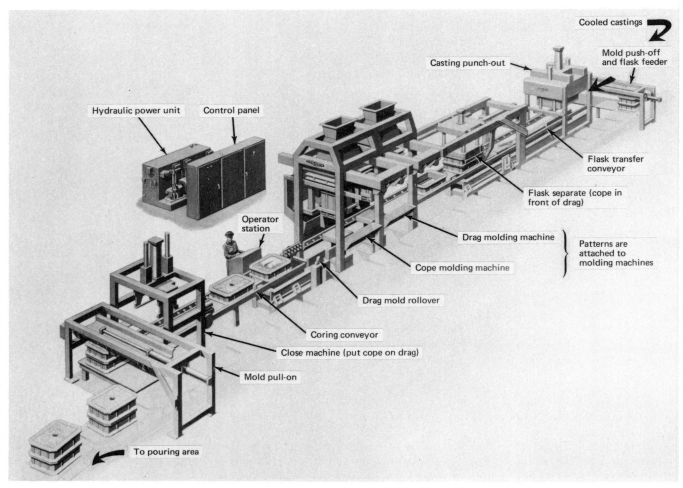

Fig. 23-20   A completely automated molding line. This can be connected to automated pouring and, as shown, the flasks are brought back and used again. *(The Osborn Manufacturing Co.)*

castings, especially those over 30 in. (760 mm) wide. The surfaces of the casting which touch the mold sand will freeze (harden) first. As the rest of the casting hardens, internal stresses are set up. When the *skin* is removed by machining, these stresses readjust, and in doing so, they change the shape and size of the casting. This change in size may be only 0.005 in. (0.125 mm) or it may be as much as 0.040 in. (1.0 mm). Thus, to finish a round casting inside and outside, it is usually necessary to rough turn the OD, bore the ID, and then turn the OD a second time to correct the *walking or moving* as it is sometimes called.

A stress-relieve anneal will considerably decrease this movement, though it seldom seems to eliminate it completely.

DESIGNING CASTINGS

The design of any piece of machinery should be as simple as possible, and this applies especially to sand castings. One of the first items to check on a casting design is: *Can the pattern be easily removed from the sand?*

With the skill of today's patternmakers, almost any shape can be made so that the pattern can be removed, but loose pieces and special cores are expensive. Many times a small change in design, as in Fig. 23-21a, will save pattern and molding costs.

Fillets and rounds must be used at all corners, as shown in Fig. 23-21e. If fillets are not used, the casting will often crack at the angle. If possible, the minimum size of fillet and rounds should be as shown.

Avoid slope openings, as shown in Fig. 23-21b, because the pattern cannot be pulled from the sand, necessitating cores or special methods.

Use webs instead of heavy cast sections to get the needed strength, as shown in Fig. 23-21c.

Avoid *hot spots*, heavy sections which will cool slowly and may be unsound (have shrinkage holes).

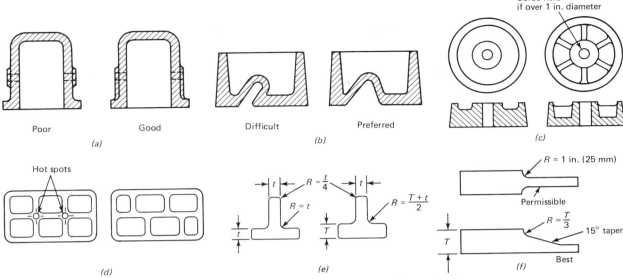

Fig. 23-21   Some suggestions for good design of sand castings. (a) Extend external bosses. (b) Do not use re-entrant details. (c) Use ribs instead of thick sections. (d) Cross ribs create possible "hot spots" and shrinkage cracks. (e) Radii for fillets and rounds. (f) Reduce different plate thickness on one side only if possible.

Fig. 23-21d shows one solution of this problem.

Use *cored holes to reduce the weight* and save on machining of castings. The smallest cored hole should be of 1-in. (25-mm) diameter. Cores longer than three times their diameter may require extra support or reinforcing. Cored holes should be $\frac{1}{8}$ to $\frac{1}{2}$ in. (3.2 to 13 mm) smaller than the finish bore size to allow for the core shifting position.

If *unequal sections* must be joined, use large radii or tapered transitions as in Fig. 23-21f.

See the references at the end of this chapter, or talk with your patternmaker or foundry superintendent for many more money saving ideas.

Review Questions and Problems

23-1.   What are the advantages and disadvantages of sand casting?

23-2.   What are the major (basic) steps in making a sand casting?

23-3.   What is a "shrink rule"?

23-4.   If the part shown in Fig. 23-3 is to be made using mild cast iron, what dimension should the hole in the pattern be to produce a *finished* hole in the part 1.500 in. (38.1 mm) in diameter?

23-5.   If the part shown in Fig. 23-3 is to have an unfinished outside diameter of $4\frac{3}{4}$ in. (120.65 mm), what will be the OD of the pattern?

23-6.   If the part shown in Fig. 23-3 is to be finish machined to a length of 6.250 in. (158.75 mm), what is the maximum length of the pattern including draft?

23-7.   What are fillets and rounds and why are they used?

23-8.   What is a core and why is it used?

23-9.   How are cores made? What properties must they have?

23-10.   Describe how a green sand *mold* for casting is made.

23-11.   Describe three types of patterns and list the advantages of each.

23-12.   Compare centrifugal and semicentrifugal casting. What types of parts are best suited for these types of casting?

23-13.   Is flaskless molding used for high or low production quantities? Why?

23-14.   What is the Full Mold Process of casting?

23-15.   What are the advantages and disadvantages of the Full Mold Process?

23-16.   List the difficulties which may be encountered when machining castings.

23-17.   What design principles must be considered when designing parts to be cast?

23-18.   Draw a sketch of a typical sand mold and name the principal parts.

23-19.   What are core prints?

23-20.   What are the advantages and disadvantages of the CO_2 and furan processes of making cores as compared to green sand cores?

References

1. *Casting*, 2d ed., Kaiser Aluminum and Chemical Sales, Inc., Oakland, Calif., 1965.
2. *Cast Metals Handbook*, American Foundrymen's Society, Gulf & Wolf Road, Des Plaines, Ill. 60016, 1957.
3. Cook, Glenn J: *Engineered Castings*, McGraw-Hill, New York, 1961.
4. *Foundry Sand Handbook*, 7th ed., American Foundrymen's Society, Des Plaines, Ill., 1963.
5. *Gray and Ductile Iron Handbook*, Gray and Ductile Iron Founder's Society, National City–East Sixth Building, Cleveland, Ohio 44114.
6. Heine, R. W., C. R. Loper, and P. C. Rosenthal: *Principles of Metal Casting*, 2d ed., McGraw-Hill, New York, 1967.
7. *Steel Castings Handbook*, Steel Founders' Society of America, 21010 Center Ridge Road, Rocky River, Ohio 44116.
8. "A Glossary of Foundry Terms," Section 13 of *Steel Casting Design Engineering Data File*, 1971. Also Sec. 1, *Design Rules and Data*, Steel Founders' Society of America, Rocky River, Ohio.
9. Sylvia, J. Gerin: *Cast Metals Technology*, Addison-Wesley, Reading, Mass., 1972.

additional casting processes

Many castings are made which do not require the strength of iron or steel. For these parts, metals with lower melting points (such as aluminum, zinc alloys, or brass) are sufficiently strong. For these materials permanent-mold and die-casting methods are often used.

PERMANENT-MOLD CASTING

Even with the help of automation, it is time-consuming to make a new sand mold for every part produced. To eliminate this necessity, especially when 500 or more of the same part are to be cast, permanent molds have been developed. These are most frequently used for aluminum, zinc, or brass castings, though in 1971 steel began to be commercially cast in special molybdenum and similar permanent-mold dies.

The Molds

The molds are made from steel or alloyed cast iron. These are expensive because the entire shape of the part plus the sprue and gating system must be cut or cast into the mold. The dimensions must allow for shrinkage of the metal when it cools.

The mold must be designed so that the part can easily be removed without distortion. Thus, the mold may be in two to four parts, with some sections moving outward away from the casting. This motion may be achieved manually, mechanically, or hydraulically. Figure 24-1 shows a simple example of a two-part permanent mold. This style can be used for a wide variety of parts.

There is really no "standard" permanent mold, since each mold is designed and built to cast a specific part. The basic principles described here apply to all permanent molds, but the mold itself is often the result of the experience and ingenuity of the mold designer.

Cores are made of steel if they can be easily withdrawn before the mold opens. This, of course, requires extra mechanisms. Sand cores, as used in sand casting, can be placed in the mold by hand. Studs, bushings, and other inserts may be cast directly into the part.

Venting of the air is through natural leaks in the mold, through 0.005- to 0.010-in.-deep (0.125- to 0.25-mm) slots, or through small drilled holes made to allow the air to escape.

A metal permanent mold will usually make 1000 to 10,000 parts before it must be repaired or built back to correct size. The parts made may weigh only a few grams or up to 55 lb [25 kg (mass)]. The smaller parts are often cast in a multiple mold which will cast several parts with each pouring.

The Process

Permanent molding is frequently a hand process. Close beside the mold is a 250- to 600-lb [110- to 270-kg (mass)] holding or melting furnace filled with

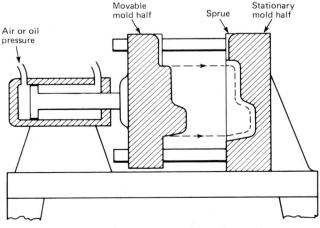

Fig. 24-1   One style of permanent-mold casting equipment.

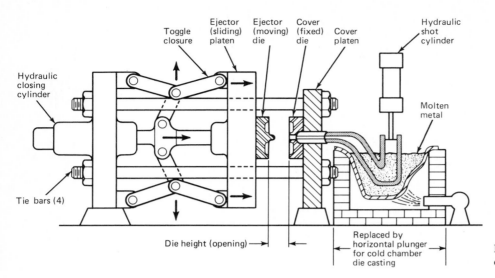

Fig. 24-2 Basic operation of a hot-chamber die-casting machine.

the alloy being used. This is kept at constant temperature by gas or electric heat.

Once a day or more often, the mold is sprayed with a *parting agent, or mold coating*, made of graphite, aluminum oxide, or other materials mixed with sodium silicate and water. This coating prevents thermal shock to the mold, and avoids occasional "soldering" of metal to the mold.

The cold permanent mold is, at start-up, preheated to 400 to 600°F (200 to 315°C). After pouring begins, the mold may be kept hot enough by the pouring cycle, or it may require a constant supply of additional heat, frequently from gas burners.

Cores, if any, are placed in the mold; it is closed, and the operator brings a ladle of metal over and pours it just the same as for a sand mold. After a wait of from 15 s to 1 or 2 min, the mold is opened and the solidified (frozen) hot casting is carefully removed. Total time may be from $\frac{1}{2}$ to 4 min. The casting is allowed to cool completely, then the sprue, gates, and flash are trimmed off. *Flash* is a thin "skin" of metal that has run between the faces of the dies.

Advantages and Disadvantages of Permanent-Mold Castings

1. Permanent-mold castings chill (solidify, or freeze) much faster than sand castings, so they have a finer grain structure, better strength, and are pressure-tight.
2. The finish is from 64 to 200 μin. (microinches) (1.6 to 5 μm) AA.
3. Production time is much shorter than for sand casting because the "one mold per casting" is eliminated.
4. Tolerances can be held to ±0.015 in. (0.38 mm) on the average, though not this close on larger castings.

However:

1. Only certain alloys can be cast in permanent molds.
2. Many complex shapes cannot be made because they cannot be removed from a rigid mold.
3. The dies may cost from $500 to $4000 or more.
4. Castings are usually under 30 lb [13.6 kg (mass)]. Larger ones are made, but the mold and handling equipment is expensive. These might more economically be made as sand castings.

Automation

There is no commercial, completely automatic, permanent-mold system, since the part is usually removed by hand even though the casting process may be automatic. Of course, a robot could be used if production were sufficient to warrant the $15,000 cost.

However, there are systems which close the mold, tilt up the ladle, feed the metal, time the cooling, and open the mold automatically. One of these is called the *Wessel process*.

There is also a low-pressure method in which molten metal is forced upward out of an airtight furnace into the mold.

DIE CASTING

Another casting process, which uses steel dies and is used extensively for aluminum, zinc alloys, and somewhat for brass and steel is die casting. The definition of die casting is producing castings by forcing molten metal under *high pressure* into reusable metal molds called *dies*.

This process is faster and gives better finishes and closer tolerances than permanent-mold casting. However, the dies are more expensive; the machines

Fig. 24-3 A 400-ton (363-t) hot-chamber die-casting machine. Hot chamber will be attached at left end. (*Lester Engineering Co., a subsidiary of Todd Shipyards Corp.*)

cost $5000 to $200,000; die casting cannot ordinarily handle large parts; and the machines take longer to set up.

Thus, the die-casting process is especially valuable for making quantities of 5000 to 100,000 parts to close tolerances and excellent finish so that they require only a minimum of additional machining.

The Die-casting Process

Basically a die-casting machine is fairly simple, as shown in Fig. 24-2. However, the large linkages, hydraulic system, cooling system, temperature controls, and safety interlocks used make the actual machine shown in Fig. 24-3 much more complex.

The *dies* are made of hot-work tool steels such as H13. The form of the part and the runners, gates, and vents are cut into the dies (Fig. 24-4). Today many of the die cavities are made by electrical discharge machining (EDM), described in Chap. 29. Skilled diemakers are required for much of this work.

Dies are made as *single cavity*, with only one part produced per stroke, or as *multiple cavity*, with 2 to 10 identical parts produced with each *shot* or cycle. It is not unusual to make two or more different parts in one *combination* die. The parts produced may be only 0.20 in. (5 mm) on a side, or may be as large as 24 × 48 in. (600 × 1200 mm). The dies for the large parts may weigh 4 tons [3600 kg (mass)].

Cores may be steel pins which are part of the die. They may be used on the sides of the die, with additional mechanism to withdraw them before the die opens.

A *draft* should be included on all vertical sides of the die and on all core pins so that the molded part can be ejected without bending or breaking. The draft is from 1° to 2°. The American Die Casting Institute, New York City, publishes tables of the required drafts.

Shrinkage, varying with the metal used, must be allowed for in dimensioning the dies.

Ejector pins are placed in the moving die so that as

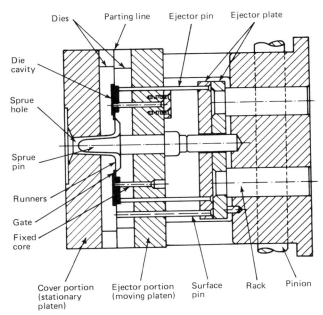

Fig. 24-4 Detail of the die section of a die-casting machine. (*Adapted from American Die Casting Institute.*)

the die opens, the pins push the part out of the die (Fig. 24-4).

A *trimming* die for removing the flash which forms around the edges is often made at the same time. This die is used in a simple vertical-action press.

Cooling coils are built into the die block to circulate water to solidify the molded metal as rapidly as possible. The flow is regulated so that the die does not become too cool while it is open.

Handling the Metal

There are two general methods of forcing the metal into the die. They are called *cold chamber* and *hot chamber*.

The **cold-chamber process** (Fig. 24-5a) is used for aluminum and brass alloys. As these metals melt at 1220 and 1700°F (605 and 933°C), they intereact with the iron of the system if left in contact too long. In most cold-chamber operations, the molten metal is kept in a heated holding crucible close to the machine. The operator ladles out approximately the correct amount of metal, pours it into the chamber, and presses a button, and the plunger drives the metal into the mold. After waiting 5 to 25 s for the casting to cool, the machine opens, and the part is ejected. The operator holds it with gloves or tongs, and puts it aside carefully because it is not fully hardened.

Pressures from 1000 psi (6.9 MPa) to over 12,000 psi (82.8 MPa) are exerted by the plunger. Modern machines have a "step-up" pressure at the end of the stroke to give the casting greater density.

Automation of cold-chamber die casting is increasing, since an automatic ladling system is made by some manufacturers. The ladle swings under a raised furnace nozzle, receives a measured amount of metal, swings, and tips the metal into the chamber. As it moves back, the cycle starts.

Ejection can be automatic, with the part falling through the bottom of the press into water (which cools it, and breaks its fall), and then onto a belt. Robots can also reach in and bring out the completed casting.

The **hot-chamber method** (Fig. 24-5b) is used for zinc and zinc alloys, such as Zamak. More tonnage of these alloys is die cast than all other alloys combined. The zinc alloy melts at about 800°F (427°C), and yet produces a casting with sufficient strength for car door handles, appliance covers, grills, small motor housings, and many lightly loaded machine parts, with a 25- to 100-μin. (0.6- to 2.5-μm) finish and tolerances and close as 0.003 in. (0.075 mm).

The hot-chamber process is basically automatic. During setup, the timing of the casting cycle is adjusted, the dies are aligned and fastened, and the fill stroke length and pressure are set. As long as there is enough metal in the melting furnace, cycling can be automatic.

The operator removes the castings from the dies, and often does some burring operations. Sometimes a trimming press is located close to the molding machine so the operator can trim one piece while the next piece is being cast.

Die-casting Machines

Die-casting machines for handling parts weighing 2 oz (56.7 g) will almost fit on a large table, and large machines may weigh over 150 tons (136 metric tons). Die-casting machines are rated by their *clamping tonnage* or by the tons of pressure the movable die will resist before it will be forced open. For example, if a die is for a part which has 100 in.² area and the shot pressure is 12,000 psi, or 6 tons/in.², the clamping pressure needed is:

Fig. 24-5   (*a*) Basic drawing of the cold-chamber method of die casting aluminum and copper alloys. (*b*) Basic system for the hot-chamber method of die casting zinc alloys.

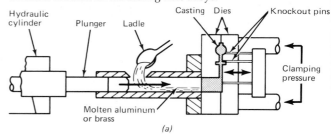

(a)

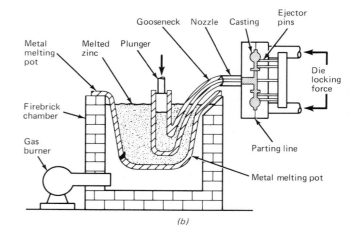

(b)

Clamping pressure = 6 tons/in.² × 100 in.²
= 600 tons

In metric units:

Die area = 645 cm² = 0.0645 m²

Clamping pressure = 82.7 MN/m² = 82.7 MPa

Clamping pressure = 0.0645 m² × 82.7 MN/m²
= 5.33 MN

Die-casting machines are commonly made with 5 to 600 tons (44.5 kN to 5.34 MN) clamping pressure, though several large 2500-ton (22.25-MN) machines are in use. These have a possible max-imum die size of 92 × 65 in. (2300 × 1600 mm). A few larger machines have been made. Figure 24-6 shows some die-cast parts and the dies used to make them.

vertical die-casting machines

An old idea which has had a rebirth since about 1968 is the vertical-action die-casting machine, shown in Fig. 24-7. This machine costs about the same as the horizontal machine of equal shot capacity and has several advantages.

The metal is brought into the cavity by vacuum, which eliminates mixing of air into the molten metal. As a result much lower pressures are needed

Fig. 24-6 (a) The die set, raw casting, and finished panel. Knockout pins outside the part. (b) A deep die-cast part with special venting system. (*Paramount Die Casting.*)

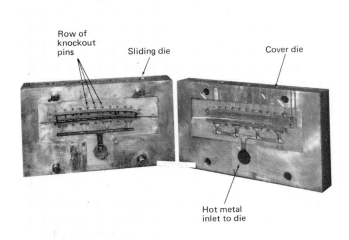

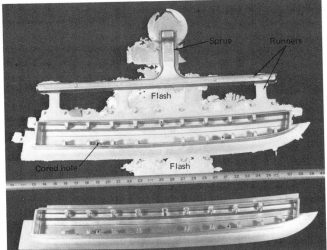

(a)

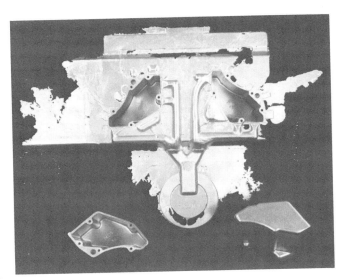

(b)

to make a sound casting, and venting the dies is no longer as much of a problem.

The very fast vacuum fill is especially advantageous when compared to the ladling required in cold-chamber casting of aluminum. Production rates may be doubled in some instances.

Other advantages are that it is much easier to place inserts accurately in a horizontal die, and alignment of the dies is easier than with the horizontal machine, which uses vertical dies. In addition, the vertical die-casting machine takes about one-third the space required for the horizontal models, and automatic unloading is greatly simplified.

The dies for the vertical injection-molding machine must have silicone or similar rubber seals, and vacuum runners to each die cavity, so the dies may be slightly more expensive.

For automatic unloading, a relatively simple shuttle plate is available which moves into the die space from the rear of the machine as the die opens. The casting is ejected onto the plate and quickly moved out for cooling.

Vertical injection-molding machines are most frequently made from 200 to 800 tons (178 kN to 7.12 MN) locking capacity, though both larger and smaller machines are available. They are used for automatic die casting of zinc, aluminum, and magnesium alloys.

SHELL MOLDING

Shell molding, sometimes referred to as *resin-shell casting*, uses a fairly fine sand mixed with 5 to 7 percent of thermosetting phenolic resins.

Fig. 24-7   (*a*) Vertical die-casting machine shown without dies in place. (*b*) Soleplate with heating element cast in. (*c*) Coffee pot heating unit die cast in vertical machine. (*Kux Machine Division, The Wickes Corp.*)

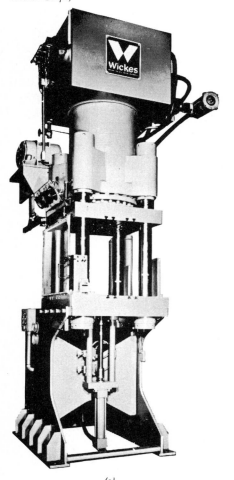

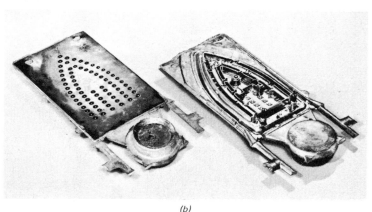

(b)

(a)

(c)

Advantages of Shell Molding

1. Parts can be cast to tolerances of ±0.005 in. (0.125 mm). This means that less metal needs to be removed for finishing. Machining allowances of 0.040 to 0.060 in. (1.0 to 1.5 mm) are enough.

2. Shell-molded parts may have finishes of 125 μin. (3.2 μm) or better.

3. Cored holes are often more accurately sized and positioned because the core can frequently be part of the pattern. Cored holes as small as $\frac{3}{8}$ in. (9.6 mm) are practical.

4. Draft of only $\frac{1}{2}°$ to 1° is sufficient, which leaves less stock to be removed by boring or turning.

5. Sections as thin as 0.10 to 0.20 in. (2.5 to 5 mm) can be cast, and smaller fillets and rounds may be used if they do not affect the strength of the part.

6. Any castable metal can be used in shell molding, in sizes up to 48 × 60 in. (1200 to 1500 mm), though most castings are about half this size or smaller.

Disadvantages of Shell Molding

1. Pattern cost is greater because they must be made of metal, usually aluminum or cast iron. Moreover, the pattern must not have deep tool marks, pinholes, or undercuts as these would prevent removal of the shell. The resin-sand mixture also is fairly expensive and more difficult to store and handle.

2. The size of the casting, compared to sand casting, is limited. However, sizes produced are, in general, larger than those made as die castings.

3. This method produces a strong, mildly unpleasant odor, though compounds can be added to decrease this smell.

The Shell-molding Process

The basic process, as shown in Fig. 24-8, is quite simple. However, to automate the process requires a machine with several very large motions, as shown in Figs. 24-9 and 24-10. The process is used to make molds for complete castings, and the same process can be used to make dry sand cores.

The Pattern

The pattern is carefully machined out of cast iron or aluminum, and it may include the cope and drag side by side; for larger pieces, cope and drag are made on separate machines. All sprues, runners, and gates are included in the pattern. Multiple cavities are often used. Four to ten patterns of the same part may be on one board. Standard shrink rules are used.

Knockout pins are also part of the pattern. They may be spring- or mechanically actuated and attached to a plate below the pattern.

The pattern is kept heated, by gas burners or electric coils, to between 350 to 500°F (177 to 260°C). It may be used for 500 to 2000 molds before rework is needed.

The Sand

A clean, moderately fine sand is coated with 5 to 7 percent of phenol formaldehyde or similar resin. This is done in large blenders. The sand is stored in a dump box on the molding machine (Figs. 24-9 and 24-10).

Forming the Mold (Fig. 24-8)

The dump box is tipped 180° so that the sand falls on top of the hot plate. The heat causes the resin to melt and "glue" the sand together. The longer the sand

Fig. 24-8 Four of the steps in making a shell mold by the hand-operated dump-box method.

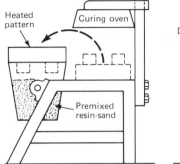

1. Invert pattern onto sand tank.

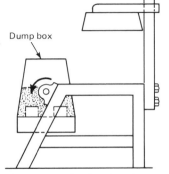

2. Invert sand tank so heated pattern is covered with resin.

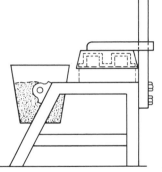

3. Resin-coated pattern is returned. Oven is positioned over the shell.

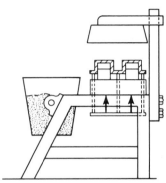

4. The shell is stripped from the pattern.

remains, the thicker the bonded coating. Usually a coating $\frac{3}{16}$ to $\frac{3}{8}$ in. thick is sufficient. This takes 15 to 60 s.

The dump box is tipped back, the mold is unclamped, and the entire pattern and mold is baked at about 600°F (315°C). In an automatic machine, the oven moves down over the mold. The mold is then pushed off (ejected) from the pattern and is ready to use.

Casting in Shell Molds

After any needed cores are placed, the cope and drag halves of the shell mold are clamped or fastened to-

Fig. 24-9   A 15 × 20 in. (380 × 510 mm) semiautomatic dump-box shell-molding machine. (*Dependable Shell Core Machines, Inc.*)

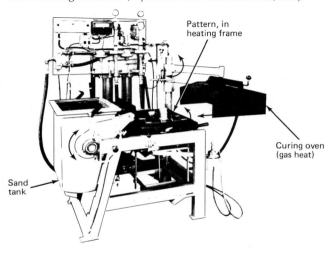

Fig. 24-10   Cope and drag shells made at the same time. Sand tank in the back, automatic cycle. *Left,* the patterns; *right,* the shell molds being pushed off. (*Shalco Systems.*)

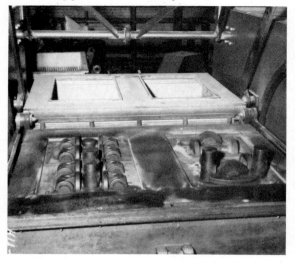

gether, often by gluing them in a semiautomatic machine like the one shown in Fig. 24-11. The glue squirts down onto the drag; the operator places the cope on the drag. The carrying plate then swings around under the pressing plungers which press the two molds together until the glue is dry, which takes 20 to 45 s.

The shells are strong enough so that the metal can be poured directly into them if the parts are small. However, for larger parts, several glued molds are often placed in a mold box on edge. The space between molds is filled with sand or small steel shot to support the shell's walls.

The mold box is moved to the pouring area, the metal is poured, and as soon as it has solidified, the mold box tips the entire contents onto a conveyor. The shot (or sand), the mold material, and the casting are separated by automated equipment.

Shell-molded Cores

Dry sand cores for any casting process may be made by the shell-mold process. The principal difference is that the sand-resin mix is usually *blown* into the mold under 50 to 80 psi (340 to 550 kPa) pressure. The machine has a horizontal closing and opening motion instead of the rotary motion just described.

Many cores are made in one piece, so the gluing operation is not needed. In large foundries, shell-molded cores are widely used, as they have better finish and accuracy than most other methods. Moreover, in volume production, the process can be completely automated, so these molds are often less expensive in spite of the costs of resin and equipment. This method also eliminates the need for any core ovens.

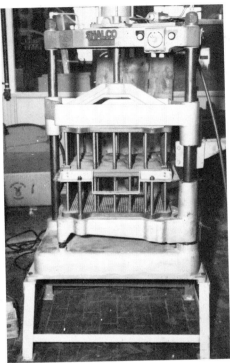

Fig. 24-11 Liquid adhesive applicator and shell-bonding machine. Shown pressing cope and drag halves together after applying the glue. (*Shalco Systems.*)

INVESTMENT CASTING

If a small casting, from ½ oz to 20 lb [14 g to 9.1 kg (mass)] or today even over 100 lb (45 kg), with fine detail and accurate dimensions is needed, investment casting should be considered. This process is used to make fuel pump parts, levers, nozzles, valves, cams, medical equipment, camera parts, and many other parts. It is sometimes called the *lost wax process*, and it has been used for hundreds of years by jewelry makers. Some foundries refer to this as *ceramic shell casting* because the metal is poured into a ceramic shell.

Any castable metal can be used, though aluminum and zinc parts can often be less expensively made by die casting. Both methods give about the same finish and accuracy, though the die-cast products may be stronger. Thus, investment casting is especially valuable for casting difficult-to-machine metals such as stainless steel, high-nickel alloys, and beryllium copper.

Investment casting is often profitable for pilot runs of as little as 12 pieces and is being used for quantities of over 100,000 parts per month. Most often the quantities ordered would be from 500 to 10,000 pieces. However, the process is slow and is one of the most expensive casting processes. If a design is changed, it may require expensive alterations to a metal die (as it would in die casting also).

The Pattern

A pattern of wax or polystyrene must be made for every piece cast. The wax is made in several different grades, out of beeswax, carnauba wax, paraffin, and other materials. The wax pattern may be made in steel, wood, plastic, or rubber molds (Fig. 24-12). The polystyrene patterns are made by injection molding into multiple-cavity steel dies (see Chap. 30). These dies may cost from $1000 to $20,000, so they are used only if large quantities of parts of small size are to be cast.

The Tree or Cluster

The wax or plastic patterns are next fastened, by heat or adhesive, to a common feed sprue and several sprues are fastened to one base. This is a hand operation, called *clustering*, as shown in Figs. 24-12 and 24-13.

Investing

Investing refers to the covering of the patterns with a heat-resistant slurry. Clusters for low-temperature metals, such as aluminum, can be put into a suitable size flask. Then a mixture of plaster of paris, silica, and talc, or similar materials, mixed with water is poured around the cluster. A vacuum is used to draw out the air so that plaster flows around every detail. The flask is then cured (dried) for several hours.

Shell molds for high-temperature metals, such as steel, and for large-sized castings are usually given an initial investment by dipping them several times into a slurry made from silica flour, magnesia, clay, or similar products mixed with liquid hardening agents such as ethyl alcohol, or ethyl silicate, and acid. This refractory coating is allowed to dry, and then the mold is dipped and dried several more times until a strong shell is formed.

Burning Out

The wax or polystyrene is now removed from the mold. This is called *dewaxing* or *burning out* the patterns. The molds are tipped upside down and heated. Some of the wax runs out and may be reused. The polystyrene is vaporized. The heat is then raised to about 1200°F (644°C) for aluminum and 1900°F (1040°C) for ferrous alloys so that the last bits of pat-

Both Methods

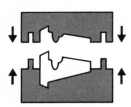

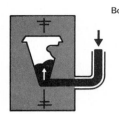

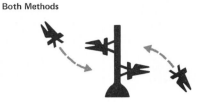

Master artisans construct the die.
The accuracy of the desired part depends upon these two steps.

A pattern is formed.
Wax or plastic is injected into the die to make a pattern in the shape of the desired part.

Then assembled into clusters.
As many patterns as is practical are assembled into clusters to facilitate mass production.

And coated with silica.
A fine silica coating on the pattern gives the cast part a smooth surface. From here, molds can be made in two ways.

Ceramic Shell Method

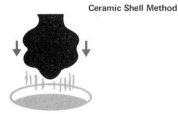

The pattern is coated with a ceramic shell.
By continual immersion, a thick coating of ceramic mold material is built up around the cluster.

Finally the thick shell is built.
After the ceramic coating has built up to sufficient thickness, it is ready for use as a mold.

Burned out of shell.
The shell is passed through a burn-out furnace to remove the patterns.

Poured into shell.
Molten metal is poured into the inverted shell in a manner similar to that shown.

Mold Flask Method

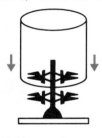

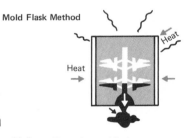

Or sealed in a container.
The coated clusters are mounted on a square steel base and enclosed in a metal flask.

And investment material added.
The metal flask is filled with an investment material which "sets up" tightly around the patterns.

Burned out of flask.
The flasks are carried through a furnace where the pattern material is burned out.

Poured into flask mold.
Molten alloy is forced under pressure into the cavities of the hot mold.

Both Methods

The shell is removed.
Once the casting is solid, the ceramic shell is blasted away.

The castings are cut off.
Gates and risers are removed with an abrasive wheel. Then the castings are cleaned.

Fig. 24-12   The principal steps of the investment casting process. (*Adapted from Stellite Division, Cabot Corp.*)

tern material are burned out (vaporized) and the mold is at the proper temperature for pouring.

Pouring the Metal

The metal may be poured by gravity, as in sand casting. However, a vacuum is used to pull air out of the mold so that no air bubbles will remain. Centrifugal casting and pressure casting are also used.

After the metal has cooled, the molds are broken away by cracking them under a hydraulic press or by sand blasting. The individual castings are then cut off the sprue, trimmed, and chemically cleaned or tumbled as required. Figure 24-14 shows an example of investment casting.

CERAMIC-MOLD PROCESSES

If long-wearing, accurate castings of tool steel, cobalt alloys, titanium, stainless steel, and nonferrous alloys, including beryllium, are needed, ceramic molds are often used instead of sand molds.

The two major ceramic-molding processes are the *unicast process*, licensed by Unicast Development Corporation, and the *Shaw process*, licensed by Avnet Shaw Division of Avnet, Inc.

Both processes use conventional patterns of wood, plastic, or metal set in cope and drag flasks. Instead of sand, a refractory slurry is used. This is made of a carefully controlled mixture of ceramic powder with a liquid catalyst binder (an alkyl silicate). Various blends are used for specific metal castings.

The ingredients are mixed just before using and are quickly poured over the pattern. In about 3 to 5 min, the slurry sets to a solid but flexible gel-like substance which can be stripped off the pattern.

In the Shaw process (Fig. 24-15), the mold is set on fire, and the alcohol in the catalyst burns off. This leaves a mesh of fine cracks throughout the mold, called *crazing*, which makes the hard ceramic permeable so that air and gases can escape during pouring.

The Unicast process "stabilizes" the green mold by spraying or dipping it in a chemical bath. This causes a catalytic interaction which creates a cellular or spongelike permeable mold structure. Some parts made by this process are shown in Fig. 24-16.

The molds are then dried, or cured, at temperatures up to 1800°F (982°C) for up to 1 h. They are then assembled, with cores in place, and poured in the usual manner. Castings up to 2000 lb [907 kg (mass)] have been made, though most are from 10 to 200 lb (4.5 to 91 kg).

These ceramic molds have a coefficient of expansion of practically zero, adequate venting all over, and walls that will not bulge under pressure. Thus, very accurate castings can be made.

These processes are being used to make castings for forging dies, die-casting dies, extrusion nozzles, tire

Fig. 24-13   (a) A complex wax pattern being "glued" together. (b) Investments ready to be heated and then have metal poured into them. (*Stellite Division, Cabot Corp.*)

(a)

(b)

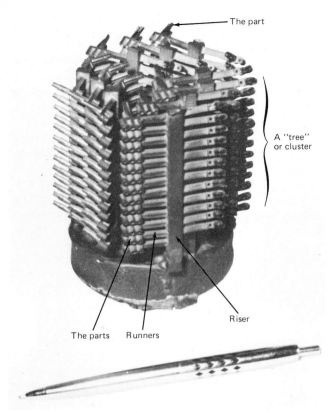

Fig. 24-14   An investment casting as it came out of the mold, 226 small connectors made of berillium copper.

molds, some cutting tools, and many parts for machines. The accuracy of the castings is from ±0.005 in. (0.125 mm) for small castings to ±0.045 in (1.14 mm) for castings over 15 in. (380 mm) on a side. Finishes of 75µin. (1.9 µm) are reported as average.

Ceramic-mold cast dies are finished to size by EDM (Chap. 29) or hand polishing. The materials for ceramic-mold casting are expensive, but the properties of the castings are considerably better than those of sand castings and much less machining is needed. Thus, for many uses, a net saving is accomplished.

PLASTER-MOLD CASTING

Plaster-mold castings have excellent surface finish, 30 to 90 µin. (0.75 to 2.3 µm), and consistent dimensional accuracy of ±0.005 in. (0.125 mm) and sometimes closer. They easily reproduce fine details and can be made in very complex shapes. Walls as thin as 0.020 in. (0.50 mm) can be cast, though 0.060-in.-thick (1.5-mm) walls are easier to cast. However, only nonferrous metals may be used. A typical plaster-mold is shown in Fig. 24-17.

Castings made this way may be as small as 1 oz (28 g) or over 1000 lb [453 kg (mass)], though most are from 3 to 30 lb [1.4 to 13.6 kg (mass)].

Sequence of Process Operations

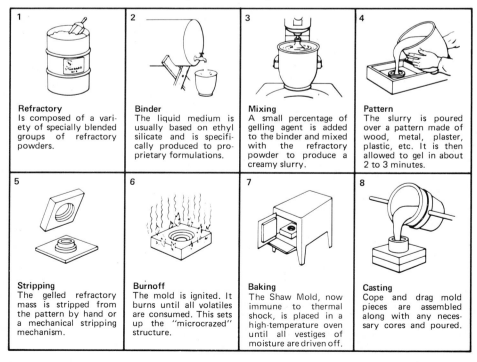

1	**2**	**3**	**4**
Refractory Is composed of a variety of specially blended groups of refractory powders.	**Binder** The liquid medium is usually based on ethyl silicate and is specifically produced to proprietary formulations.	**Mixing** A small percentage of gelling agent is added to the binder and mixed with the refractory powder to produce a creamy slurry.	**Pattern** The slurry is poured over a pattern made of wood, metal, plaster, plastic, etc. It is then allowed to gel in about 2 to 3 minutes.
5	**6**	**7**	**8**
Stripping The gelled refractory mass is stripped from the pattern by hand or a mechanical stripping mechanism.	**Burnoff** The mold is ignited. It burns until all volatiles are consumed. This sets up the "microcrazed" structure.	**Baking** The Shaw Mold, now immune to thermal shock, is placed in a high-temperature oven until all vestiges of moisture are driven off.	**Casting** Cope and drag mold pieces are assembled along with any necessary cores and poured.

Fig. 24-15   The eight steps in the Shaw process. (*Avnet Shaw Division, Avnet, Inc.*)

(a)

(b)   *(c)*

Fig. 24-16 Parts cast to shape by the Unicast process (*a*) Tool steel, titanium extrusion die. (*b*) Precision-cast milling cutter. (*c*) Cast-steel injection-molding die. (*Unicast Development Corp.*)

Making the Plaster-Mold Castings

A *pattern* must be made of sealed wood or plastic, though usually of aluminum. If an exact part is available, flexible thiocol or similar rubber compound may be used as a flexible mold. The finish of the pattern will be reproduced closely in the final casting.

The pattern must have some draft, though 1° or 2° is enough. All sprues, gates, and runners are also made as part of the pattern. Shrink rules are used, as in all casting processes, with added allowance for plaster shrinkage. *Coreboxes* are also made of metal or plastic so that the cores will be smooth.

The *plaster* is a special casting plaster made from plaster of paris, gypsum, talc, sand, and water. Several different formulas are used. A *permeable plaster* is made by mixing air into the casting plaster or by blending in special foaming agents. This permeable plaster allows gases and air to escape more easily during the pouring.

Molding is done much the same as with sand castings. The pattern is placed in the drag, and plaster is carefully poured in and allowed to set (setting is the first hardening which makes the mold firm enough to handle). The drag is turned over, parting powder is applied, and the cope is then poured, with sprues, etc., in place. After this sets, a small hole is made, and air pressure is used to separate cope, drag, and pattern.

Drying in an oven at 400 to 500°F (200 to 260°C) for 12 to 16 h, or at just under 1000°F (538°C) for $\frac{1}{2}$ to 2 h, is the next step. This drives out all the moisture; the mold shrinks about 1 to 1.5 percent when drying. Molds should be poured soon after drying, as they start to shrink and crack.

The plaster cores are set in place, the cope and drag are assembled and clamped together, and they are ready to be poured.

Pouring is done as in all casting processes. Steel and iron cannot be cast, as the high temperature will destroy the plaster mold. However, aluminum and brass alloys are easily cast.

When the metal has frozen, the plaster is broken off, sprues and runners are cut off, and the casting is cleaned and ready for use.

Plaster-mold castings are more expensive than sand or permanent-mold castings and have a larger grain size (unless cooled with chills), but the slow cooling leaves few internal stresses. They can be pro-

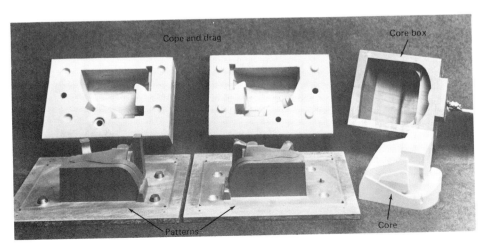

Fig. 24-17 The cope, drag, and core for a medium-size plaster casting. (*Ohio Precision Castings, Inc.*)

duced with less delay than die castings, yet they have properties nearly as good. Thus for quantities of 50 to 1000 pieces, they may be the most economical way of making high-quality nonferrous castings.

Fig. 24-18   Some of the shapes which have been continuous cast in brass alloys. The large piece is about an 8-in. (200-mm) diameter. (*American Smelting and Refining Company.*)

CONTINUOUS CASTING (C-C)

Continuous casting (C-C) is not a process for making individual parts. It is a method of getting a continuous length of square, rectangular, hexagonal, or other shapes of metal. It has been used with copper and brass and aluminum alloys since 1938, but only since about 1969 has it been used to any extent in the United States for casting steel. Some steel C-C machines were in use in Europe before this. Some shapes which have been continuous cast in aluminum and copper alloys are shown in Fig. 24-18. Many of these bars, such as the gear and cam shapes, are cut into $\frac{1}{4}$- to 1-in.-thick (6.35- to 25-mm) "slices" to make individual parts.

The Basic Process

Basically, continuous casting consists of pouring hot metal into a heated crucible directly from the furnace ladles. The hot metal flows through water-cooled dies made of graphite or copper. The partially cooled (but

Fig. 24-19   (*a*) The vertical casting machine for nonferrous alloys. (*b*) A vertical casting machine for steel. (*United States Steel Company.*)

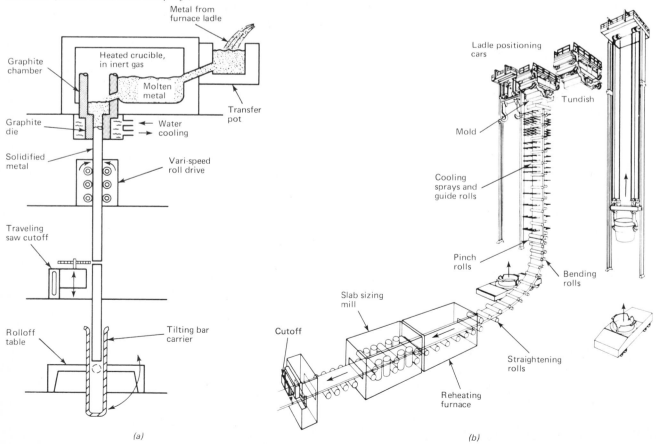

(a)   (b)

solid) shape is gripped by rollers and pulled forward at a rate which corresponds with the cooling and solidifying rate of the metal.

After the metal has gone through more rollers and cooled further, it is cut off by traveling oxyacetylene torches or circular saws, and pushed aside for further processing, or led directly into reheating and reforming rolls or onto the rolling mill. This *strand* of metal will continue to be produced as long as there is metal in the crucible.

Continuous-casting Machines

As shown in Figs. 24-19 and 24-20, both horizontal and vertical C-C machines are in use. Most U.S. C-C machines and all steel-casting machines are the vertical type. Figure 24-19*b* shows a large C-C steel-casting machine which bends the strand of metal before it is cut off. This machine is a continuous-continuous type, since metal is immediately processed in the rolling mills. This machine, in 1973, poured a continuous steel slab over a mile long— 10,000,000 lb [4,500,000 kg (mass)] of steel.

Uses of Continuous Casting

Steel and ductile iron which are continuous cast have less scale, fewer impurities (these tend to come to the top of the crucible), and a higher yield. The yield is the percent of metal poured which can actually be used. In the "old" method, a large piece had to be cropped (cut off) from the end of every billet because it had slag and air in it. With continuous casting, these impurities do not get into the metal.

A major advantage of continuous casting is that the metal is cast into the desired shape. Steel will come out in large slabs up to 76 in. (1930 mm) wide by 9 in. (230 mm) thick or in square billets all ready to be used in large rolling mills. Other shapes (as shown in Fig. 24-18) have been continuous cast in ductile iron, aluminum, and brass alloys.

The final advantage is the speed with which the metal is formed. The large C-C machine in Fig. 24-19*b* can cast more than 6 ft/min (1800 mm/min) of high-quality steel. Other machines, of course, go more slowly.

SUMMARY

This chapter, and Chap. 23, have described what may at first seem a confusing array of casting methods. There are many instances when a part may be made by any of two or three different casting methods. In this case, the quantity needed, the accuracy required, or the finish required may govern the

Fig. 24-20 A two-strand horizontal casting machine which can be used for both iron and nonferrous bars. Ductile iron bars are shown. (*Quebec Iron & Titanium Company.*)

selection. The final cost, including any necessary machining and finishing, may be the deciding factor.

Table 24-1 shows some general comparisons of the major processes described in the two chapters. Many variations on each process are in use, and a machine designer, manufacturing engineer, or purchasing agent should call in the experts in each field before making a final decision.

Review Questions and Problems

24-1.   Describe permanent-mold casting and list its advantages and disadvantages compared with sand casting.

24-2.   What is die casting?

24-3.   What are the advantages and disadvantages of die casting?

24-4.   Describe a cold-chamber die-casting machine. Make a sketch.

24-5.   Describe a hot-chamber die-casting machine. Make a sketch.

24-6.   A round plate of 8-in. diameter with a 2-in.-diameter hole is to be die cast at 1500-psi pressure. What should be the capacity of the machine in tons of locking pressure?

24-7.   How is shell molding done? What are its advantages and disadvantages?

24-8.   Describe vertical die casting and list its advantages and disadvantages.

24-9.   What is investment casting and what are its advantages?

TABLE 24-1 COMPARISON OF CASTING PROCESSES

Factors	Processes						
	Sand Casting	Permanent Mold	Plaster Mold	Shell Mold	Die Casting	Investment Casting	Unicast and Shaw Mold
Sizes	0.5 kg–tons	0.5–150 kg	0.5–10 kg	0.5–15 kg	0.2–75 kg	0.1–5 kg	1–1100 kg
Metals cast	All	Nonferrous	Brass, bronze, aluminum	All	Zinc, brass, aluminum, magnesium	All	All
Pattern or mold cost	Low $25–$1000	Medium $300–$3000	Medium $200–$1500	Fairly low $50–$1000	High $300–$5000	Medium to high $100–$3000	Medium $200–$2000
Thinnest section, mm	3–5	2.5–4	0.8–2	1.6	0.6–1.0	0.5–1.0	1.0–3.0
Usual lot size, pieces*	1–1000	500–5000	100–2000	500–5000	1000 to 100,000	20–1000	1–500
Usual tolerance ± first in./in. or mm/mm†	0.03–0.06	0.04–0.06	0.005–0.010	0.003–0.005	0.001–0.003	0.002–0.005	0.002–0.005
Commercial finish, $\frac{\mu in.}{\mu m}$	$\frac{250–1000}{6–25}$	$\frac{150–700}{4–18}$	$\frac{16–100}{0.4–2.5}$	$\frac{60–200}{1.5–5}$	$\frac{20–125}{0.5–3.2}$	$\frac{16–60}{0.4–1.5}$	$\frac{50–125}{1.3–3.2}$
Production rate, average per hour	1–300	20–100	4–40	10–50	100–1000	10–300	10–30

*In the automobile industry these quantities will be much higher.
†Additional inches or millimetres approximately one-third of first inch.

24-10. Compare permanent-mold casting and investment casting with respect to
  a. Metals which can be cast
  b. Tolerances which can be held
  c. Finish of parts *as cast*
  d. Steps required in the complete process

24-11. How is continuous casting done? What are its advantages?

24-12. The part shown in the figure below is to be manufactured using one of several manufacturing processes. Consider sand-mold casting, shell-mold casting, permanent-mold casting, die casting, investment casting, or machining from bar stock. Material is aluminum.

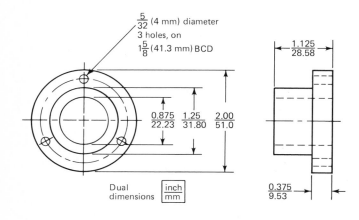

24-12 figure.

  a. Which process would you use to make 100 pieces with tolerance of ±0.005 in. (0.13 mm) on all dimensions? Why?
  b. Which process would you use to make 2000 pieces with tolerance of ±$\frac{1}{64}$ in. (0.4 mm) on all dimensions? Why?

Note: Fillets allowed as needed.

References

1. *An Introduction to Die Casting*, American Die Casting Institute, New York.
2. Broad, Edward R.: "Investment Casting Metals," *Machine Design*, p. 177, June 23, 1966.
3. *Casting*, Kaiser Aluminum & Chemical Sales, Inc. Oakland, Calif., 1965.
4. Cook, Glenn Jr.: *Engineered Castings*, McGraw-Hill, New York, 1961.
5. *Durez Guide to Shell Molding*, Durez Division, Hooker Chemical Corp., North Tonawanda, N.Y. 14120.
6. Heine, R. W., C. R. Loper, and P. C. Rosenthal: *Principles of Metal Casting*, McGraw-Hill, New York, 1967.
7. *Investment Casting Handbook*, Investment Casting Institute, 1717 Howard St., Evanston, Ill. 60202.
8. *Metalcaster's Reference and Guide*, American Foundrymen's Society, Des Plaines, Ill. 60016, 1972.
9. *Steel Castings Handbook*, Steel Founders' Society of America, Rocky River, Ohio, 1970.
10. *Product Standards for Die Castings* (specifications of tolerance, finish, testing, etc.), American Die Casting Institute, Inc. (ADCI), 366 Madison Ave., New York, 10017.

rolling, drawing, bending, and extrusion

Not many graduates will be actively involved in the large steel-making industry, and the study of the details would be of little real value. However, a brief look at how we get our raw shapes could be helpful.

Both rolling and drawing are squeezing processes, and the finished shapes (such as structural steel) or the shapes such as plate and rod which are often used for further machining are called *wrought* products as compared to castings.

MAKING STEEL SHAPES

By combining iron ore, coke, and limestone in the proper proportions in a blast furnace, pig iron is produced. This pig iron, plus scrap, is then remelted in open hearth furnaces, basic oxygen furnaces (BOF), or electric or vacuum furnaces (see Fig. 25-1 and Appendix A). During this process, the carbon content is controlled and the proper amounts of alloys are added to make the many basic steel compositions.

The molten steel from these furnaces is treated in either of two ways. The most modern method is to hoist it to the top of a continuous-casting (C-C) mill and cast it directly into square blooms (large) or billets (small) or rectangular slabs as described in Chap. 24.

Without the C-C setup, the steel is poured into molds which make very large *ingots*, up to 27 in. (685 mm) square. These ingots are *soaked* (heated below their melting point until they are at an even temperature). These almost white-hot ingots are then sent through a blooming or slabbing mill and, through a series of heavy rolls, reduced to blooms, billets, or slabs.

Each furnace load of from 50 to 500 tons [45 to 450 metric tons (mass)] consists of one alloy of steel. Thus, it will be formed into *one* of the three basic shapes. The steel used for I beams is not of the same composition as that used for hot-rolled steel (HRS) plate or cold-rolled steel (CRS) bars. Thus, if you order a special alloy you will have to wait until your lot can be run through all the several processes.

ROLLING MILLS

The blooms, slabs, or billets are reheated to about 2200°F (1200°C) and run through a series of rollers to produce the shape, size, and finish which is specified.

Whether for rolling flat, round, or shaped parts, rolling mills are made up of a series of *stands* (Fig. 25-2). Each stand has a set of rolls which is set closer together than the previous one. There may be six or more stands of rolls needed in one roughing or finishing line to reduce the stock to the required size.

Rolling Shapes

When forms such as H beams (see Fig. 25-3), channels, bars, or angles are to be hot-rolled, a set of rolls is made so that each pair changes the shape somewhat. This is partially illustrated in Figs. 25-4 and 25-5.

As shown in Fig. 25-3, many sizes of round, square, hexagonal, and rectangular shapes are hot-rolled from billets 3 to 6 in. (75 to 150 mm) square in the same kind of equipment.

Hot-rolling Plate, Sheet, and Strip

These wide flat pieces of steel or other metal are rolled much the same as the shapes. However, some important changes must be made in the rolls.

Plate is a sheet of steel $\frac{1}{4}$ in. (6.35 mm) or over in thickness. Plate 6 in. (150 mm) thick and over is fre-

Flowchart of Steelmaking

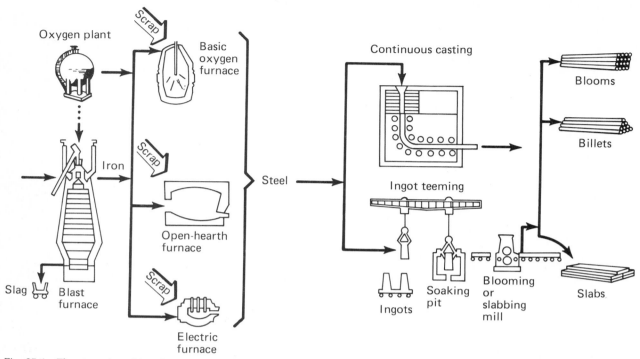

Fig. 25-1   The steps from blast furnace to blooms, billets, and slabs of steel. (*American Iron and Steel Institute.*)

Fig. 25-2   A six-stand rolling mill producing steel plate. (*Blaw-Knox Foundry & Mill Machinery, Inc., a subsidiary of White Consolidated Industries, Inc.*)

quently rolled. It is made in varying alloys, often 4 to 8 ft (1200 to 2400 mm) wide and 20 ft (6000 mm) long.

Sheet is $\frac{3}{16}$ in. (4.8 mm) or less in thickness. This is specified in USS or Manufacturer's gage, from 28 ga. (0.0149 in.) to 7 ga. ($\frac{3}{16}$ in.) thick, though thinner and heavier gages are rolled. This corresponds to approximately 0.4 to 5.0 mm thickness.

Sheet is usually rolled into large coils (Fig. 25-6) and later cut to the widths and lengths needed for automobile bodies and fenders, refrigerator panels, and similar work. It is most often made in a grade of low-carbon steel.

Strip is a narrow plate or sheet of carbon steel, stainless steel, or alloy steel for use in such things as garden tools, furniture, hand saws, and many other small products. It is supplied in many widths, usually 30 in. (760 mm) or less, and in coils or cut lengths.

The *rolls* used for making plate and sheet are often over 10 ft (3000 mm) long and are subjected to very high pressures. This pressure will cause them to bend in the middle unless they are reinforced. This is why four- and six-high rolling stands are used as shown in previous illustrations.

The smaller roll, in contact with the metal, has the "bite" needed to pull the hot steel through the pinched down space. The larger rolls prevent the deflection which would otherwise make a sheet or plate which was too thick in the center.

Uses of Hot-rolled Steel

The beams, angles, and channels are used for buildings and bridges. They are also used for the framework of many machines because they are inexpensive, strong, and easy to weld or bolt together.

The plate is used for large and small tanks for oil refineries, paper mills, and many other containers. The round, square, and hexagonal bars are used as the raw stock for thousands of different machined parts.

One of the largest uses of the sheet and bars is for the stock used for making cold-rolled, drawn, or pierced products, which will be described later in this chapter.

Many different alloys, including tool steel, are made as hot-rolled shapes which are then machined into shafts, cams, latches, cutting tools, and hundreds of other parts.

CAUTION: Hot-rolled bars, beams, etc., are not made to close tolerances. Round bars have ±0.005 in. (0.125 mm) or greater tolerance. Straightness is allowed to be 3° or more out of line. The makers of these products often produce stock much closer

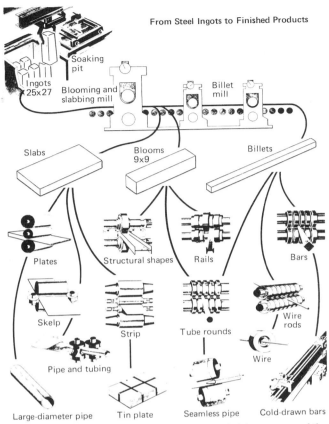

From Steel Ingots to Finished Products

Fig. 25-3 The steps from blooms, billets, and slabs to some of the many finished steel products. (*American Iron and Steel Institute.*)

than the maximum allowed, but the designer should be aware of the possible variability in hot-rolled-steel products.

COLD-ROLLING

The product called *cold-rolled steel (CRS)* is widely used in many fields. However, the round, square, and hexagonal bars are not rolled, they are drawn. Thus they are actually *cold-drawn steel (CDS)*. However, the term CRS has been used for so many years that it is unlikely to be changed. Cold drawing will be described later in this chapter.

Cold-rolled sheet and strip are made on continuous roll stands similar to those previously described. However, the rolls are ground and polished so that the surface of the steel (or any other metal) is smooth and shining as shown in Fig. 25-6.

To make this cold-rolled sheet, the hot-rolled sheet steel is passed through *pickling tanks* containing sulfuric or hydrochloric acid, then it is passed through cold- and hot-water rinses and a dryer. The clean steel, which is now "cold," that is, at from room

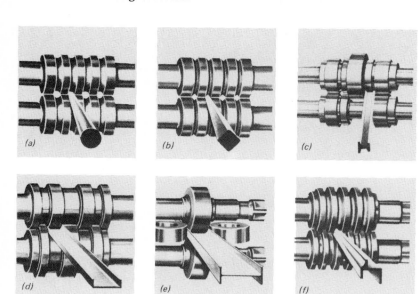

Fig. 25-4 The types of rolls used to make steel shapes. (a) Round bars. (b) Square bars. (c) Railroad rails. (d) Channels. (e) H beams. (f) Special sections. (*American Iron and Steel Institute.*)

temperature to 300°F (150°C), is passed through a series of successively smaller rolls and may be reduced to as little as one-fifth its original thickness. This also lengthens it a corresponding amount, so the final coil may be 7000 ft (2130 m) long.

This drastic reduction in thickness *cold-works* the steel (or other metals), and it may now be too hard for most uses. In this case, the coil is put into an annealing furnace to soften it. After annealing, the sheet or strip is passed through another set of rolls to bring it to exact size, finish, and temper. It may be shipped in coils or cut into convenient lengths.

Coatings of porcelain, enamel, copper, tin, nickel, or plastics may be applied before the steel is used.

Fig. 25-5 A 36-in.-wide (914-mm) flange beam partially shaped. It is white hot. (*Bethlehem Steel Corp.*)

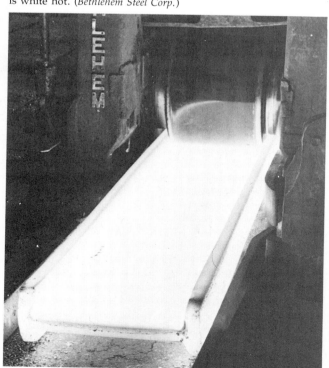

Fig. 25-6 Cold-rolled sheet coming through the last sizing roll and being coiled for shipping. (*Bethlehem Steel Corp.*)

Coating with zinc is known as *galvanizing*. Galvanized sheets are used in roofing, siding, automobiles, containers, etc.

The thickness of cold-rolled products is very closely controlled by automatic gaging devices on the roll stands. Tolerances of $+0$, -0.002 in. ($+0$, -0.05 mm) are frequently maintained.

Slitting Sheet Steel

The coil of steel or other metals may be up to 90 in. (2300 mm) wide, and this is too wide for many uses. Thus, at the steel mill, at the wholesale warehouse, or at the user's plant, the rolls are cut into narrower strips. This process is called *slitting*.

COLD-DRAWING

Thousands of tons of shiny round, square, and hexagonal bars of many alloys of steel are used each year. They are customarily ordered as cold-*rolled* steel (CRS), even though almost all this material is cold-*drawn* steel (CDS).

Uses of this material are endless. In building machinery, CRS may be used as purchased or turned, milled, and cut to many shapes. The round CRS bars may be used without machining as shafts in packaging, spinning, and similar machinery.

The reasons for its great usefulness are:

1. Accuracy. Round bars up to $1\frac{1}{2}$-in. (38-mm) diameter are made to $+0$, -0.002 in. ($+0$, -0.05 mm) tolerance.

2. Finish. The original finish is below 16 μin. (0.4 μm) and, if carefully handled, it is free of nicks and scratches.

3. Tensile and yield strength are higher than those for the corresponding grade of HRS.

4. Machinability is excellent. The chips produced, in some alloys, break up easily and have less tendency to form a BUE, and an excellent surface finish is usually possible.

CAUTION: The surface of cold-drawn bars is under tension. Thus when flat bars are machined, they will "move" or "walk" as shown in Fig. 25-7. This is less likely to happen with bars $\frac{1}{2}$ in. (13 mm) or thicker but it should always be considered when designing with CRS.

Making cold-drawn products starts with a bar or coil of hot-rolled steel, or other metals, which is $\frac{1}{32}$ to $\frac{1}{16}$ in. (0.8 to 1.6 mm) larger than the final size wanted. The hot-rolled bar is descaled, pickled in acid, washed, and coated with lime and sometimes other lubricants

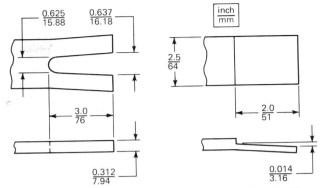

Fig. 25-7 Actual results of milling cuts on cold-"rolled" steel. "Movement" or distortion caused by removing stressed areas.

such as soap or molydisulphide. The front end of the bar or coil is then tapered by cutting or squeezing.

The *die* is made of tungsten carbide and set in a tool-steel holder as shown in Fig. 25-8. The carbide will resist wear due to friction, and the steel helps take the pressure, which is often over 100,000 psi (690 MPa). For simple forms, these dies cost only $200 to $500 and will last for years. Dies are made for special shapes also, using P/M and electrical discharge machining (EDM) (see Chap. 29).

Drawing is done by pushing the tapered end of the stock through the die, grasping it in puller tongs, and with an endless chain or hydraulic-type drawbench, pulling the bar through the die. Frequently, the hot-rolled bar is cut to a length which will result in a 36-ft (11,000-mm) length when drawn. This is then cut into 12-ft (3650-mm) lengths for sale.

Wire of all kinds and materials is usually cold-drawn in much the same way. However, the wire usually passes through a series of dies, each smaller than the preceding one. The dies are made from tungsten carbide, and sometimes of diamonds, with the holes cut by lasers or by an electron beam. Figure

Fig. 25-8 Drawblock (or bull-block) with carbide die for cold-drawing steel or nonferrous metals.

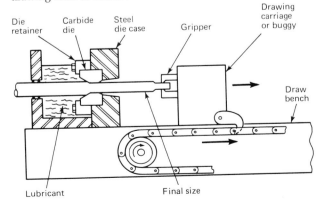

Fig. 25-9   A nine-block wire-drawing machine. Input $\frac{3}{16}$-in.-diameter (4.8-mm) steel wire. Output up to 6000 fpm (1800 m/min). Die box shown in front of each reel. (*Fenn Mfg. Co.*)

25-9 shows a typical high-speed wire-drawing machine.

The forces involved in cold-drawing can be quite large. To "squeeze" the wire, bar, or tubing at the die, the radial force must be greater than the yield point of the metal. However, the pulling force of the drawbench must be lower than the yield point or the finished bar or wire will stretch or break.

Thus, when drawing material at 50 to 150 fpm (15 to 45 m/min), the amount of reduction must not be great enough to require a pulling force greater than the final product can withstand.

PRESS-BRAKE BENDING

A *press brake* is a relatively simple machine used for forming sheet metal into a wide variety of shapes. It is used mostly for long, not very wide, parts such as architectural trim, support frames for machines and farm equipment, and box shapes. These are shown in Fig. 25-10.

Much of the work is done on metal $\frac{1}{4}$ in. (6.35 mm) or thinner, though 1500-ton-capacity (13.3-MN or 1460-t) press brakes are capable of bending $1\frac{1}{2}$-in.-thick (38-mm) mild steel plate. A press brake capable of bending long, $\frac{1}{4}$-in.-thick (6.35-mm) steel may also be used to bend shorter $\frac{3}{8}$-in.-thick (9.6 mm) plates. However, the power needed varies in the ratio 1:1 with the length, but it varies as the square of the thickness. One type of brake can bend $\frac{3}{16}$ in. × 16 ft (4.8 × 4900 mm) or $\frac{3}{8}$ in. × $6\frac{1}{2}$ ft (9.5 × 1980 mm) steel.

The Press Brake

The press brake, as shown in Fig. 25-11, has a ram which moves vertically to press against the bed. Punches of various shapes and lengths are bolted onto the ram, and they mate with a matching die fastened to the bed.

Press brakes are made with from 4- to 24-ft-long (1200- to 7300-mm) working space and from 15- to 2500-ton (133-kN to 22.3-MN) capacity, using motors of 1 to 75 hp (0.75 to 56 kW). They cost from $4000 to $50,000 or over, depending on the equipment used.

Both mechanically and hydraulically operated machines are made. The mechanical press brakes are somewhat faster, but the hydraulic brake has a longer, more easily adjustable stroke length.

Automation of the setting of backstops and stroke

Fig. 25-10   Examples of the shapes which can be made with a press brake.

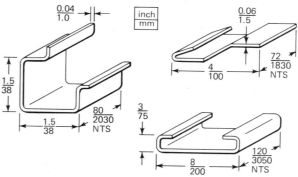

Fig. 25-11   A 17-ton (151-kN), 48-in. (1200-mm) hydromechanical press brake. (*Di-Acro Division, Houdaille Industries Inc.*)

lengths has increased the production and accuracy of press brakes (Fig. 25-12). The metal sheets or preforms are still fed in and removed by the machine operator.

The die sets, punch and die, are inexpensive and are made from steel or heat-treated cast iron. They cost from $200 to $800 for the standard shapes depending on their length. Some of the most used shapes and the use of two of these shapes to form a box are shown in Fig. 25-13. Note that several die sets may be mounted in one press brake. Setup time is usually $\frac{1}{2}$ to 3 h.

Urethane is an elastomer which behaves like a "solid fluid," that is, under pressure it will change its shape but not its volume. Since it has a "memory," it always returns to its original shape.

Urethane can be poured into molds (cast), and some formulations can be machined in lathes, milling machines, etc. Today, it is often used to replace the die in a press brake as shown in Fig. 25-14. A solid block of urethane may be used as the male (die) half with several different female (punch) styles. This cuts both die cost and setup time almost in half. The life of the urethane inserts may be from 5000 to over 50,000 impressions.

The *accuracy* of press-brake work is about ±0.016 in. (0.4 mm) for stock up to 0.125 in. (3.18 mm) thick, though larger tolerances are needed for thicker stock.

Fig. 25-12 Select-A-Form (semiautomatic) 135-ton (1.2MN) hydraulic press brake. Back gage, stroke, and pressing cycle can be preselected from the console. (*Pacific Press & Shear Co., a division of Conron Inc.*)

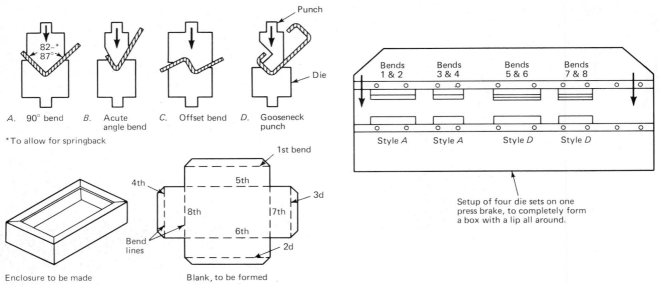

Fig. 25-13 Four of the most used punch and die sets for press brake use. A single brake with four die setups.

With the new semiautomatic brakes, repeatability is excellent, and tolerances may be one-half the above.

Economical quantities are low. From 1 to 100 is a frequent lot size, though the modern semiautomatic press brakes may be economical for lot sizes up to 1000 parts. Over this quantity roll forming (discussed in this chapter) may be less expensive.

CAUTION: A 90° angle in mild steel is made with an 85 to 88° die set, since metal will "spring back" after bending. The amount of spring-back increases as the yield strength of the metal increases.

Safety precautions must be strict, as the operator usually hand feeds and removes the stock, and sometimes two operators must work together. Work supports, covered foot pedals, nonrepeating mechanisms, two-hand palm switches, and light beam guards are some of the safety features used.

SHEARING

One of the fastest ways to cut sheets and plates is with a power-operated shear. The machine closely resembles a press brake but has upper and lower knife edges instead of dies.

The most used shears will cut mild steel plate up to $\frac{3}{8}$ in. (9.5 mm) thick × 20 ft (6000 mm) long, though capacities up to $1\frac{1}{2}$ in. × 12 ft (38 × 3600 mm) are listed, and smaller sizes are often all that is needed. Small, foot-operated shears cost about $600, and a $\frac{1}{2}$ in. × 12 ft (12.8 × 3600 mm) shear will cost about $40,000 with some automatic equipment. Figure 25-15 shows two frequently used sizes. With this modern

Fig. 25-14 Some of the ways in which urethane die pads are used. (*Di-Acro Division, Houdaille Industries Inc.*)

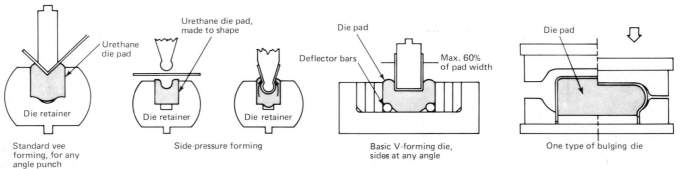

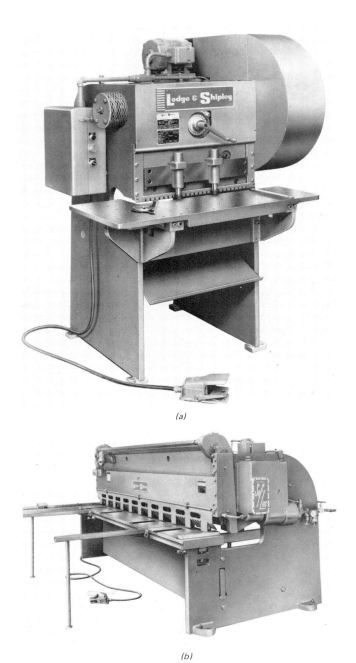

(a)

(b)

Fig. 25-15 (a) Speed Shear 40 in. (1000 mm) wide, 12-gage (0.1046-in.) capacity. (b) A 14-ft (4300-mm) mechanical shear. Models for ¼- to ½-in. thick (6.35 to 12.7-mm) mild steel. (*Lodge & Shipley Co.*)

equipment sheets, plates, and galvanized or painted metal can be cut rapidly to within ±0.010 in. (0.25 mm) of exact size for use in press brakes, roll-forming machines, punch presses, etc. Special tables and knives are made for cutting the double or single 30° and 45° vee notches on plates for welding.

Two improvements in metal-shearing equipment

are shown in Fig. 25-16. The automatic stacker (Fig. 25-16a) collects cut pieces and stacks them so they can be picked up by crane or forklift. The micrometer-adjusted, pin-located shear (Fig. 25-16b) allows previously punched sheets to be cut to ±0.002 in. (0.05 mm) accuracy, with only a very short setup time needed.

COLD-ROLL FORMING

Cold-roll forming (the word "cold" is often omitted) is a process in which a flat strip of metal is passed through a series of contoured rolls which gradually form the metal into the desired shape. The process is sometimes called *contour-roll forming*.

Any metal may be roll-formed, though carbon steel, aluminum, and stainless steel are the most used. Rolling is done at room temperature, except for a few metals such as magnesium and titanium, which are preheated to 400 to 500°F (200 to 260°C). Unless coolant is used, the temperature of a steel part may rise 50 to 100°F (28 to 56°C) due to the cold working of the metal. A few of the shapes rolled are shown in Fig. 25-17.

The sheets rolled are most often 0.187 in. thick × 20 in. wide (4.8 × 510 mm) or less, though sheets 60 in. (1500 mm) wide and narrow plate ¾ in. (19 mm) thick have been roll-formed. The stock is usually in coils, so any desired length can be made. Chopper or flying cutoff saws are used to cut the finished product to length.

Roll-forming Machines

Roll-forming machines commonly have 6 to 20 roll stands. Each stand holds two rolls which are shaped to progressively form the part. A 12-station (or 12-stand) cold-roll-forming machine is shown in Fig. 25-18.

The machine shown is an "inboard" type, with bearing supports on both ends of the rolls. For light gage metals, up to 0.030 in. (0.75 mm) thick, "outboard" machines with the rolls supported only on one end are used. Spindle diameters must be large enough to stand the pressure without bending. Spindles 1.5- to 3.5-in. (38- to 90-mm) diameter are common, though much heavier spindles have been used. A small outboard cold-roll-forming machine may cost $5000. Heavier, longer machines cost $15,000 and up.

Accuracy of roll-formed products is excellent. As noted in Fig. 25-17, overall tolerances as low as 0.005 in. (0.125 mm) are often held. Repeatability is excellent. If the stock is a uniform thickness, the dimensions will be maintained within 0.001 to 0.002 in. (0.025 to 0.050 mm) repeatability.

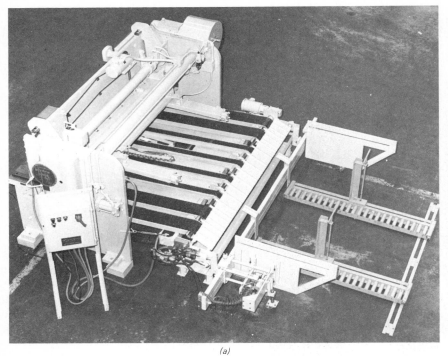

(a)   *(b)*

Fig. 25-16   (*a*) Rear view of a shear, showing automatic stacking attachment. (*b*) Second operation, accurately shearing prepunched plates for instrument panels. (*Lodge & Shipley Co.*)

The *advantages* of cold-roll forming are:

1.   High production capacity of 50 to 150 fpm (15 to 45 m/min).

2.   Finish is not improved, but it is maintained. Coated, painted, galvanized, and polished materials can be roll-formed successfully.

3.   Additional operations, such as notching and perforating, can sometimes be done while rolling.

4.   The strain hardening of the metal is sometimes an advantage.

The *disadvantages* of cold-roll forming are:

1.   The tool-steel forming rolls cost $300 to $600 a pair. This initial cost is high when up to 20 pairs are needed; thus design changes definitely should be avoided.

2.   The strain hardening may require annealing before using.

3.   For quantities under 5000 to 10,000 ft (1500 to 3000 m) or for prototype production, press-brake bending is usually less expensive.

TUBE BENDING

Round tubing, made from many materials, is used for fuel lines, many plumbing supplies, cooling and heating coils, and automobile mufflers. Both round and square tubing are used in making furniture, light machine supports, lawnmower frames, etc.

To form this tubing, in diameters from $\frac{1}{8}$ to 10 in. (3.2 to 250 mm), special machines are available. One of these tube-bending machines is shown in Fig. 25-19. Small, hand-operated machines are also available.

N/C tube benders will, at the command of the punched tape, measure the length, bend to the

Fig. 25-17   A few of the shapes which have been roll-formed. (*Roll Forming Corp.*)

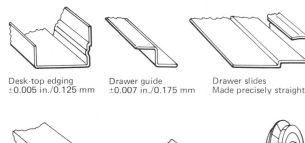

Desk-top edging   Drawer guide   Drawer slides
±0.005 in./0.125 mm   ±0.007 in./0.175 mm   Made precisely straight

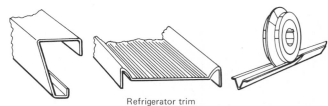

Conveyor track   Refrigerator trim   Trademark, and
length: 38 ft/11,600 mm   Pattern applied while   typical forming roll
rolling

Fig. 25-18 A 12-stand roll-forming machine with cutoff and stacking table. (*Yoder Co.*)

proper angle, rotate the part if necessary, clamp, measure and make the next bend, and cut off automatically. Bends will be accurate within 0.10 to 0.50°, and length may be within 0.015 to 0.050 in. (0.375 to 1.25 mm) of the drawing dimension. Production can run as high as 500 bends per hour for $\frac{1}{2}$-in. (13-mm) tubing.

The tooling needed is shown in Fig. 25-20. While the tooling must be accurately made, it costs only $250 to $500 a set for $1\frac{1}{4}$-in.-diameter (32-mm) tooling. Setup takes only 0.5 to 1.0 h; thus, short runs of 5 to 10 pieces can be economical.

CAUTION: Most metals have a ''spring-back'' of from 3 to 12°. This requires a corresponding over-

bend in order to achieve accurate angles. This overbend is often determined by trial and error on the first parts made.

HOT EXTRUSIONS

Sometimes the hot extrusion process competes with cold-roll forming. This process makes long, regular or irregularly shaped parts such as those shown in Fig. 25-21. Lengths are, as in roll forming, limited by the space available. Most parts are made in 12- to 40-ft (3.7- to 12-m) lengths, though 105-ft (32-m) lengths have been made.

The cross-sectional size depends on the material. The part must fit within a circumscribing circle of

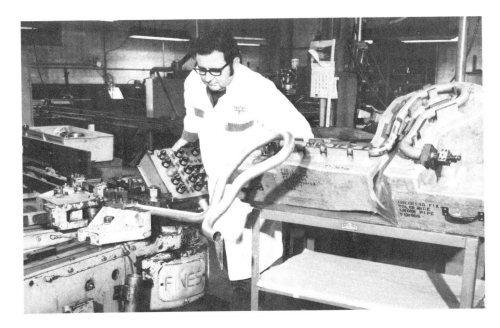

Fig. 25-19 A $1\frac{1}{4}$-in.-capacity (32-mm) hydraulic rotary bending machine. Note checking fixture on the right. (*Teledyne Pines, a division of Teldyne, Inc.*)

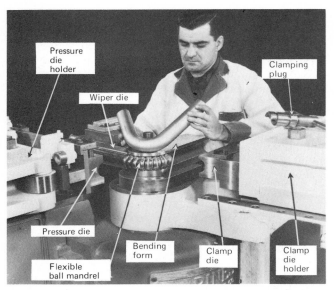

Fig. 25-20   One type of tooling used for rotary bending of tubing. (*Teledyne Pines, a division of Teledyne, Inc.*)

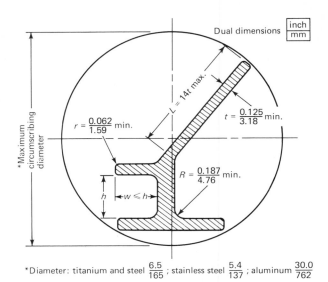

Fig. 25-22   Suggested dimensional limits for hot extrusions. (*From ITT Harper.*)

$5\frac{3}{8}$-in. (136-mm) diameter for stainless steel, $6\frac{1}{2}$-in. (165-mm) diameter for carbon steel and titanium, and 30-in. (760-mm) diameter for aluminum (see Fig. 25-22).

The Process

The desired alloy is melted, then hot-rolled into bars from 6- to 10-in. (150- to 250-mm) diameter. These bars are cut into billets or slugs of the desired length (about 150 to 600 mm), center drilled, and then lathe peeled to elimate all scale and surface defects.

Fig. 25-21   Examples of hot-extruded shapes which have been made in stainless steel, carbon steel, titanium, etc. Part (*a*) is cut into $\frac{1}{2}$-in. (13-mm) pieces and made into a catch. Part (*b*) is cut into gears. (*ITT Harper.*)

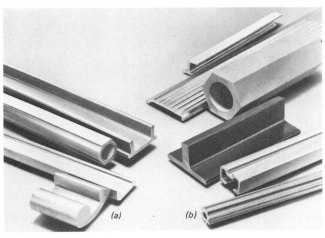

The billets are then induction heated above the critical temperature. Approximate temperatures are: steel and stainless steel, 2200°F (1204°C); aluminum, 800°F (427°C); copper alloys, 1500°F (816°C).

The billets are then placed in a press (usually hydraulic) and squeezed out through a die into which the shape has been cut.

The die is made of tool steel. The approach and exit angles will vary considerably, as the die design is still more of an art than a science. Figure 25-23 shows a typical arrangement. Dies cost $150 to $850 each.

Lubrication of the sides of the billet and the die is critical. If metal-to-metal contact is made at any point, the billet or the die will be scratched. The scratches may show up on the finished part. Lithium, graphite, and other high-ressure lubricants are used. For steel extrusions, powdered glass is used as the lubricant. This is called the *Ugine-Sejournet process*.

Pressures used are very high. To hot extrude steel takes 125,000 to 180,000 psi (860 to 1240 MPa). The hydraulic presses have capacities up to 14,000 tons (115 MN) and over.

The *rate of production* is high, since extrusion may be at 25 to 600 fpm (7.6 to 180 m/min). Complete extrusion of a billet may take only 4 to 10 s. Setting up for each shot will take 5 to 15 min, and changing a die takes about the same time. As die costs are low, hot extrusion may be economical for fairly small lot sizes.

Tolerances are good, but increase as the part gets larger. One company suggests tolerances of 0.020 to 0.046 in. (0.5 to 1.17 mm) for steel, but 0.003 to 0.012 in. (0.075 to 0.3 mm) can be held on some aluminum extrusions.

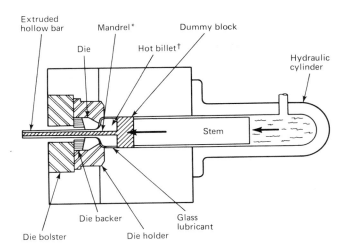

Extruded
hollow bar

Die

Mandrel*

Hot billet†

Dummy block

Hydraulic
cylinder

Stem

Die backer

Die holder

Glass
lubricant

Die bolster

*Omit mandrel for shapes without a center hole.
†Billet is gundrilled so that mandrel will fit through.

Fig. 25-23   Schematic diagram of a typical hot-extrusion press (*Adapted from ITT Harper.*)

The *finish* will usually be 125 to 250 μin. (3.2 to 6.4 μm) AA and will have some scale, especially on the ID. Most extruded shapes are stretch straightened to $\frac{1}{8}$ in. (3.18 mm) in 5 ft (1500 mm) or better in both twist and camber.

Review Questions and Problems

25-1.   Describe two methods of producing blooms, billets, and slabs.

25-2.   Why are rolling stands usually made four and six rolls high?

25-3.   How is hot-rolled sheet stock prepared for cold-rolling?

24-4.   Why is it necessary to anneal cold-rolled stock?

25-5.   How is slitting done? Why is it sometimes necessary?

25-6.   What are the advantages gained by cold-rolling, or cold drawing?

25-7.   What must be considered when considering a design using cold-rolled stock?

25-8.   Why are tungsten carbide dies used in cold-drawing operations?

25-9.   Describe the cold-drawing process, and explain what limits the amount of reduction possible in one draw.

25-10.   What are the advantages of the mechanically operated brake and the hydraulically operated brake?

25-11.   How is urethane used in some press brakes?

25-12.   What is "spring-back"?

25-13.   Discuss cold-roll forming and the reasons for having both inboard and outboard types.

25-14.   What are the advantages and disadvantages of cold-roll forming?

25-15.   What is tube bending? What are the capabilities of an N/C tube bender?

25-16.   Describe the hot extrusion process.

25-17.   A 1 × 36 in. slab enters a hot-rolling mill at 250 fpm. After passing through several stands, it has been reduced to $\frac{1}{4}$ × 36 in. At what speed will the slab be traveling as it leaves the rolling mill?

25-18.   A hot steel bar $4\frac{1}{2}$ in. × 16 in. × 36 ft long is to be reduced to $3\frac{3}{4}$ in. in one pass. If the width does not change, what will be the length of the bar after this reduction?

25-19.   A 0.063-in.-diameter carbon-steel wire is to be reduced 20 percent in diameter by cold-drawing. If the yield strength is 80,000 psi, what is the maximum drawing force which can be used?

25-20.   If the wire in Prob. 25-19 is being drawn at an input speed of 150 fpm, what will be the speed as it leaves the die?

25-21.   A manufacturer of aluminum gutters for do-it-yourself homeowners wishes to make them in 24-ft lengths. Consider manufacture using a brake, extrusion, or cold-roll forming. Which method would be most economical for orders of 1 to 100? In lots of 500 to 1000? In lots over 1000 pieces? Make your decisions on the basis of the economical lot sizes for each process as mentioned in this chapter, as well as other factors.

References

1.   *Cold Finished Steel Bar Handbook*, American Iron and Steel Institute, 1968.

2.   *Cold Roll Forming Practice in the United States*, American Iron and Steel Institute, 1966.

3.   Fiorentino, R. J.: *Comparison of Cold, Warm and Hot Extrusion by Conventional and Hydrostatic Methods*, Society of Manufacturing Engineers, Detroit, Mich., MF-159, 1972.

4.   Kervick, Richard J., and R. K. Springborn (eds.): *Cold Bending and Forming Tube and Other Sections*, Society of Manufacturing Engineers, Detroit, Mich., 1966.

5.   *Metalforming with K-Prene® Urethane*, Di-Acro Div. Houdaille Industries, Inc., 1971.

6.   *Metals Handbook*, vol. 4, *Forming*, American Society for Metals, 1969.

7.   *Picture Story of Steel*, America Iron and Steel Institute, 1000 16th St. N.W., Washington D.C. 20036, 1969.

8.   "Steel-Making Flow Charts," American Iron and Steel Institute, 1970.

presswork 26

Presswork of many kinds is used in making many of the things we use every day. Automobile bodies, hub caps, mounting brackets, and dashboards are produced in presses. Toasters, stoves, refrigerators, and metal furniture are a few items for the home which are dependent on punch-press work.

It is estimated that there are about 400,000 presses of various sizes and styles in the United States.

English	Metric Conversions
1 ton (2000 lb) = 8.896 kN (force)	
1 ton (2000 lb) = 0.907 t (metric tons)	
1000 psi = 6.9 MN/m² = 6.9 MPa	

TYPES OF WORK DONE IN PRESSES

A variety of work is done in presses, and one must know the vocabulary in order to discuss their types and uses. The following operations on presses are listed approximately in the order of their frequency of use, based on feeding a *strip* of stock from a coil through the press.

Piercing, sometimes called *punching* or *perforating*, makes a hole (often round, but any shape is possible) in the metal. The punched-out piece (the *slug*) is scrap (see Fig. 26-1a).

Blanking means that the punch cuts out a complete outline of the part with a single press stroke (Fig. 26-1b). This leaves a scrap strip, sometimes called the *skeleton*.

Notching (Fig. 26-1b) removes a piece of metal from the edge of the strip. It may be done to free some metal for forming or to cut out a part of the workpiece so that blanking will be more easily done.

Cutoff means to cut all the way across the strip, thus separating a piece (Fig. 26-1c). No metal is removed, and the cut may be of any shape.

Parting is like cutoff in that it goes all the way across the strip. However, as in Fig. 26-1d, it removes metal (makes scrap) and is used when the blanks (pieces cut off) do not have mating shapes.

Lancing is cutting a slit of any shape part way across the strip or blank. Figure 26-2a shows that no metal is removed. Lancing usually is done to free a section of the piece so that it can be bent.

Forming (Fig. 26-2b) is bending one or more tabs on the part or shaping the part in a relatively shallow shape. This may be done either before or after the piece has been cut off or parted from the strip.

All the above operations can be done either on the strip or on separate pieces, called *blanks*. The next five operations are performed only on individual pieces, which may have been pierced and blanked in a previous operation. The parts are hand fed or, if possible, fed by automatic equipment.

Drawing is a stretching operation which drastically re-forms the blank into various cuplike shapes such as in Fig. 26-2c. Very deep drawing is usually done as

Fig. 26-1 Some of the operations done on a punch press.

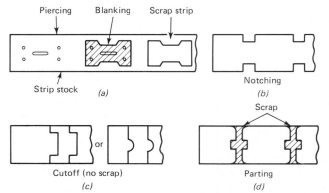

348

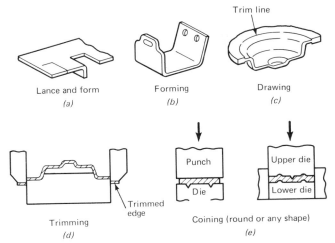

Fig. 26-2 Some of the operations done on presses.

a separate operation and will be considered later in this chapter.

Trimming often follows a forming operation. It removes surplus metal around the edges, which was necessary in order to hold the work for drawing. This may be done in a simple die such as that shown in Fig. 26-2d. Larger parts can be trimmed in a *shimmy* or *Brehn die*, which moves a die around the edge, cutting on all four sides.

Coining, or embossing, uses a closed die and is a squeezing operation used to impress a pattern or a slightly depressed area on a part. Our coins and tableware have their designs impressed this way. Figure 26-2e shows other uses.

Embossing, the same as coining, uses the same setup

as *sizing*, which means to press a part into a more accurate shape.

Shaving is sometimes used to trim off the rough edges and burrs usually left by the blanking operation. Only a few thousandths of an inch are removed by dies shaped to the contour of the part.

THE PRESS

Punch presses are made with power supplied either mechanically or hydraulically. They range in size from a small 16-ton (142-kN or 14.5-t) press with over 600 strokes per minute to 4000-ton (35.6-MN or 3600-t) giants with a maximum of 8 strokes per minute. Larger presses are in use, but most work is done on presses of 600 tons (5.34 MN) or less. The small press costs about $5000 and a 200-ton (1.78-kN) press costs about $50,000.

Mechanical Presses

Mechanical presses are basically simple in construction. The principal parts are labeled in Fig. 26-3. The flywheel supplies most of the energy used. It runs continuously and is connected to the shaft by an air- or mechanically operated clutch, which engages when a stroke is wanted, either intermittently or continuously. A brake is attached to the drive shaft to stop the stroke.

As a crank or eccentric is usually used to bring the punch down, its energy is greatest at the bottom of the stroke. Thus, mechanical presses are rated by the tonnage available at $\frac{1}{8}$ to $\frac{1}{2}$ in. (3.18 to 12.7 mm) from the bottom.

The mechanical press is faster and less expensive

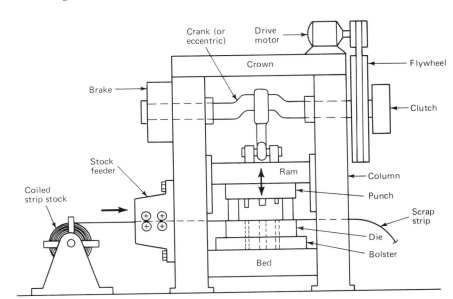

Fig. 26-3 Schematic of a crank drive, straight-side press.

than the hydraulic and uses a much smaller motor. However, it has a fixed stroke length and a fixed cycling speed. It is used for most piercing and blanking work.

Hydraulic Presses

The hydraulic press is built much the same as the mechanical press, except that one or more hydraulic cylinders are used to drive the punch. Hydraulic presses exert a constant force during the stroke, and their strokes per minute and stroke length can be controled by valving arrangements. They are made with higher capacities, occasionally up to 15,000 tons (133.5 MN), and long strokes. Thus, they are especially useful for drawing and extruding (described later in this chapter) and for piercing and blanking stock over $\frac{1}{8}$ in. (3.18 mm) thick.

Types of Presses

There are only two basic types of presses. They are the *C-frame press* and the *straight-sided press*. The C-frame is usually an open back, inclinable (OBI) design.

The *OBI* (Fig. 26-4) is made up to about a 200-ton (178-kN) capacity for light work. The entire press can be tilted back, usually 15 to 20°, so that gravity helps eject the work. The press is open on all sides, so strip feeding and die handling is simplified. The C-frame press (or *gap-frame*, as it is often called) will "open up" slightly under heavy loads, and this can cause die wear due to angular misalignment.

The *straight-sided press* (Fig. 26-5) is the basic design used for most mechanical and hydraulic presses. Very heavy support or spring or hydraulic cushions can be placed under the die, and the side frames and tie rods tie the press solidly together.

Mechanical Press Drives

The simplest and most frequently used drive is the direct, nongeared drive, often called *flywheel drive*. The crank or eccentric and the flywheel are directly connected, through the clutch, to the punch platen,

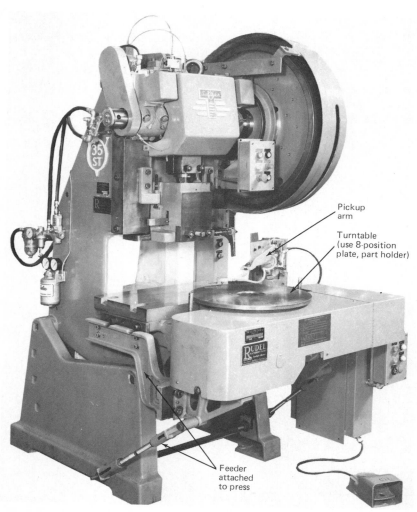

Pickup arm

Turntable (use 8-position plate, part holder)

Feeder attached to press

Fig. 26-4 A 35-ton open-back-inclinable (OBI) press. Shown with a dial feed attachment to position single pieces under the punch, thus "no hands in the die." (*V & O Press Co., Inc.*)

Fig. 26-5 A 150-ton two-point straight-side, eccentric-drive mechanical press. (*Verson Allsteel Press Co.*)

as shown in Fig. 26-6*a* and *b*. The rpm of the flywheel becomes the strokes per minute of the press.

Single-gear and *double-gear* drives are used for slower speeds and a greater mechanical advantage.

The *knuckle-joint press* (Fig. 26-7*a*) has a short stroke and can dwell at the bottom. It is used for coining, embossing, extruding, and sizing.

Toggle-action presses (Fig. 26-7*b*) are used extensively for drawing items like kitchen pans, electrical parts, and small home appliances. Large models are used to draw automobile panels, steel bathtubs and sinks, and other large forms.

A *single-point press* means that the force is applied only at the center of the die.

Two- and four-point presses are used when the die area gets large. A die area of 60 × 72 in. (1500 × 1800 mm) needs pressure applied at four points, the four corners, to prevent bending.

A *single-action press* is the usual punch-press action and is used for most pierce and blank work. However, when drawing some types of parts, it is necessary to hold the outside of the blank, and then press the form in the center. This requires a *double-action*

press. Even triple-action presses are occasionally used, with the third slide often up through the bed of the press.

THE PRESS DIES

A complete die or *die set* consists of a punch, the top section (male), and the die, the bottom section, which has cavities closely fitting the punches and the stripper plate. The parts of a die set are shown in Fig. 26-8.

The *shoes* are fastened to the ram and bed of the press, and the dies are accurately located and bolted to the shoes. The stripper plate holds the stock down so that it is not pulled up by the friction of the punch on its return stroke.

A great variety of dies are used to pierce, blank, form, etc. An experienced die designer is very versatile and resourceful in finding methods of making stamped parts economically. The dies listed below do not include those for drawing or extruding, since these will be discussed separately.

Materials for dies are high-carbon tool steels (such as A2, W1, O1, and D2) and M2 high-speed steel. Tungsten carbide is frequently used today for inserts in press dies, especially if a million or more pieces are to be made or if tough materials like stainless steel are being used. By using electrical discharge machinery

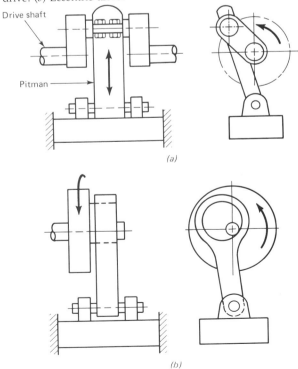

Fig. 26-6 Two types of press drives. Gear drives can be combined with eccentric drive, as well as the crank drive shown. (*a*) Crank drive. (*b*) Eccentric drive.

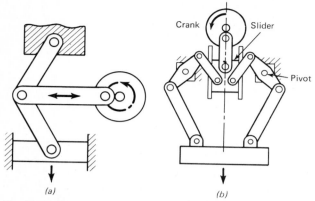

Fig. 26-7 (*a*) Knuckle and (*b*) toggle drives for punch presses (both in raised position).

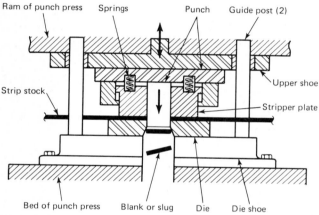

Fig. 26-8 Basic parts of a punch and die set shown assembled on a press.

(EDM) (Chap. 29), most of the diamond grinding of the die shape is eliminated, and carbide dies, properly handled, give four to ten times as many parts per sharpening.

Die life is gaged by the amount of burr produced on the punched-out parts. Often not over 0.005-in. (1.25-mm) burr is allowed. The number of parts produced depends on the die material and the work, but at least 20,000 to 50,000 parts can be expected for most situations, before regrinding, though carbide dies may run ten times this many.

After the burr gets too heavy, the tops of the punch and die are ground down 0.006 to 0.012 in. (0.15 to 0.3 mm), which "sharpens" the die set. From 5 to 10 regrinds are often possible.

Types of Dies

Only four major classifications of dies are used in presswork, but many variations and combinations are used.

Single-operation dies are used when there are only piercing or simple trimming and parting operations. A simple forming die might also be considered a single-operation die.

Compound dies are the most frequently used, as many punched metal parts are not very complex. The compound die pierces and blanks in the same stroke and is used to produce parts like those shown in Fig. 26-9. These dies make close tolerance and concentric parts, as all work is done in one stroke.

Progressive dies can perform very complex work, doing piercing, blanking, forming, lancing, and notching. The die is made up of a number of *stations*, each station having a punch and die set which performs one operation. There may be 2 to 15 of these stations in a press with a bed wide enough to hold them. Figure 26-10*a* shows a *strip*, showing the work performed at each station. While accurate work is done on progressive dies, the motion between stations makes it more difficult; however, production is high and though the dies are very expensive (from $1000 to $20,000 per set) their maintenance is low.

Transfer dies must be used if the workpiece is separated from the strip of stock while going through the press. These multistation dies are similar to progressive dies, but a mechanism must be added to pick up the unattached pieces and move them accurately to each station. These mechanisms are semistandard but expensive, costing $2000 and more. One type of transfer mechanism is shown in Fig. 26-11.

Forming dies may be included in either progressive or transfer dies. For a detailed description of several types, the *Metals Handbook*, Volume 4 is one of the best references.

SELECTING A PRESS

The selection of a mechanical, hydraulic, crank, ec-

Fig. 26-9 Some typical parts pierced and blanked in one stroke with a compound die.

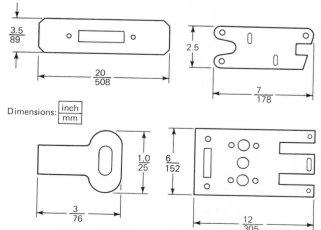

(a)

(b)

Fig. 26-10 (a) A progressive die for making pencil clips. Notice the "strip" at the bottom. (b) A 7-ft-long (2100-mm) progressive die which makes rotor and stator motor laminations. Both die sets are made with carbide die inserts. (*Oberg Mfg. Co.*)

centric, or toggle press depends on the type of work, as explained in the descriptions above.

The size of the press opening needed is governed either by the space needed for the die [progressive dies may be 60 in. (1500 mm) long] or by the size of the workpiece, such as large sheet metal panels.

The *tonnage* required for a pierce and blank job is governed by the area of metal sheared and the shearing strength of the metal. The formula is

$$P = LtS$$

Where P = total pressure, lb
L = total length of all piercing and blanking cuts, in.
t = thickness of the metal
S = shear strength of the metal, psi

Shear strength of aluminum is 8,000 to 14,000 psi, low-carbon steel 48,000 psi, and stainless steel about 60,000 psi.

In the metric system:

P is in kilonewtons (kN) or meganewtons (MN)

L and t are in meters

S is in newtons per meter squared (N/m²), or pascals (Pa)

Shear strengths are: aluminum, 76 MPa; steel, 331 MPa; stainless steel, 414 MPa.

EXAMPLE 26-1
An 8 × 12 in. (203 × 305 mm) rectangle of steel is to be blanked out of ¼-in.-thick (6.35-mm) plate.

$$P = 2(8 + 12)(\tfrac{1}{4})(48,000) = 480,000 \text{ lb}$$

Tonnage = 480,000 ÷ 2000 = 240 tons needed

Fig. 26-11 Two of the gripping fingers of a 12-station transfer press. Note rack-and-pinion movement. Forming die shown between the blank and the "bubble." (*Verson Allsteel Press Co.*)

In *metric*,

$$P = \frac{2(203 + 305)}{1000} \times \frac{6.35}{1000} \times 331 \text{ MPa}$$

$$= 1.016 \times 0.00635 \times 331 = 2.14 \text{ MN needed}$$

Conversion $= 240$ tons $\times 8.896 = 2135$ kN
$$= 2.14 \text{ MN (check)}$$

Accessories for Presswork

Buying machine tools is something like buying a car. There is the base price plus some "extras." In addition to the press, if coils of strip stock are used, a reel and an adjustable stock feeder and often a straightener (to remove the bend from the coil) are needed on the input side. For single-part feeding, a hopper and chute, sliding die, or robot feeder may be used.

For removing parts, an air jet, sliding die, or mechanical pickup may be added. Chutes or conveyors then send the product on to storage or the next operation. Sometimes a cutoff, or chopper, is added to cut the scrap strip into small pieces because a long coil is hard to dispose of.

Safety

Today, safety is strictly regulated by the Occupational Safety and Health Act (OSHA), and all presses *may* have to be made to a "no hands in the die" specification.

Many devices, from cables that pull the operator's hands back, to automated feeds, two-hand pushbuttons, and "light shields" are added to the press. The object is, of course, to avoid the loss of fingers and hands, which has occasionally happened in almost every shop. These safety precautions are often quite expensive, but by federal law and for humanitarian considerations they are necessary.

MULTISLIDE MACHINES

These medium-sized machines do many of the jobs that could also be done on a regular press with progressive and transfer dies. However, within their capacity, they are less expensive, cheaper to "tool up," and often faster. Figure 26-12 shows one model of a multislide press and some of the work done. Items such as hose clamps, pipe clamps, brush holders, and mounting brackets come from these machines completely formed and ready to use, except possibly for deburring.

These machines are fed with coil stock on edge (width dimension vertical), and the short-stroke ($\frac{3}{8}$ to $1\frac{1}{2}$ in., 9.5 to 38 mm) die assemblies (rams) move the punches horizontally. The rams are cam-actuated for both punch and retract stroke. They can be spaced at any point on the shafts on either the front or the back of the machine. From one to four rams may be used, depending on the size of the machine, and for light work a *split slide* can be installed, giving up to double the number of dies. Piercing, blanking, coining, embossing, bending in any plane, and seaming may be done on a workpiece in one pass through the machine.

By proper timing and arrangement of the cams, a 20-ton (178-kN) multislide machine can make 20-ton strokes with *each* of the rams in succession. Thus, work which would require a 40 to 60-ton (356 to 534-kN) conventional press can be done on the lighter, less expensive multislide machine.

Stock up to $\frac{3}{16}$ in. (4.8 mm) thick and 8 in. (200 mm) wide can be used in the larger multislide machines. Parts are produced at rates from 50 to 290 per minute or faster on light work. Setup, die changes, and readjustments take from 3 to 10 h, so these machines are usually used where at least 20,000 parts are needed. Multislide machines cost $30,000 to $50,000, and tooling will add $8000 to $15,000.

FINE BLANKING

All piercing and blanking operations leave a burr on the workpiece because of the way the metal separates under the pressure. Figure 26-13 shows what happens when a punch and die close.

The length of the smoothly cut edge in conventional blanking will be about 40 percent of the thickness of mild steel. The rest of the edge is rough, as it is not cut but broken off (failed in shear).

A process called *fine blanking* (or straight edge, fine-edge, smooth edge, or fine flow blanking) makes punched-out parts which have almost no burr and no rough edges.

Advantages of Fine Blanking

1. If a smooth edge is specified on a conventional stamped-out part, an added shaving operation is necessary. This requires an additional accurately made shaving die, which is not necessary with fine-blanked parts.

2. No shaving or reaming is needed to make smooth, accurately sized holes with no taper.

3. Holes smaller in diameter than the stock thickness can be punched with fine blanking. This is difficult or impossible with conventional presses.

4. The sheared edge often has a 32-μin. (0.8-μm) finish.

5. Tolerances in diameter and location of 0.0005 to 0.002 in. (0.0125 to 0.05 mm) are being held in commercial production.

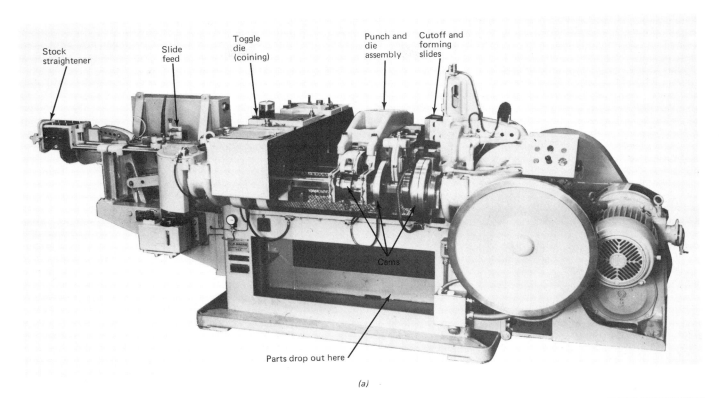

Stock straightener

Slide feed

Toggle die (coining)

Punch and die assembly

Cutoff and forming slides

Cams

Parts drop out here

(a)

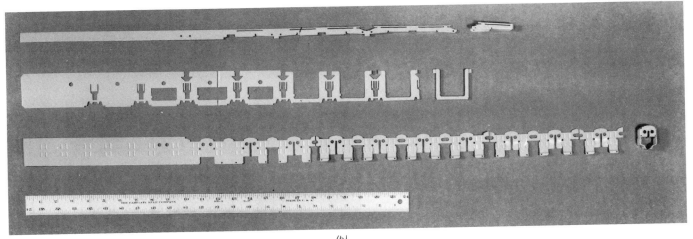

(b)

Fig. 26-12   *(a)* A 160 strokes per minute multislide machine handling up to $\frac{3}{32} \times 3$ in. (2.4 × 76 mm) stock. *(b)* Some parts made in multislide machines. *(U.S. Baird Corp.)*

The combined effect is to eliminate several expensive, time-consuming secondary operations with their necessary tooling, handling, and paperwork.

Disadvantages of Fine Blanking

1.   A triple-action hydraulic or mechanical press is required, and the press must be in good condition. These presses are expensive. A 200-ton press costs about $70,000. A typical press is shown in Fig. 26-14*a*.

2.   The total die assembly will cost 10 to 50 percent more than a conventional die, though about the same or less than the blanking plus the shaving die. Fine blanking dies are made to closer tolerances than conventional dies, often with clearances of only 0.0002 to 0.0004 in. (0.005 to 0.001 mm).

3.   Cycle time is longer, though up to 150 parts per minute has been attained.

4.   Die life between sharpening tends to be shorter than for conventional dies. A life of 10,000 to 30,000 parts per grind is typical.

5.   Total press tonnage cannot be used for piercing

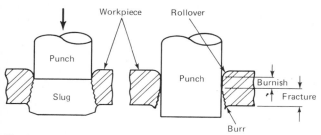

Fig. 26-13   The edge condition of the part in conventional piercing and blanking operations.

and blanking since considerable force is used for the blankholder and the special impingement ring (stinger plate) shown in Fig. 26-15. About two-thirds of the total rated press capacity is used for piercing, blanking, and forming.

The Fine-blanking Process

The main steps in fine blanking are shown in Fig.

26-15. The strip of metal is first gripped by the *stinger plate* or impingement ring, then held tightly between the punch and pressure pad, and forced through the die; thus the metal never breaks loose. The punch and pressure pad then separate, and the part and any slugs are forcefully blown or pushed out into chutes or receptacles. This, incidentally, is much quieter than conventional press blanking, and noise is one of the factors now controlled by OSHA.

Size of Work Done

Fine-blanking presses are made from 40 to 300 tons capacity (356 kN to 2.67 MN) and a few up to 800 tons (7.1 MN) total capacity. Parts up to $\frac{1}{2}$ in. (12.7 mm) thick and 10-in. (250-mm) diameter are frequently made, and 16 in. (400 mm) square and $\frac{3}{4}$ in. (19 mm) thick can be made. However, most of the great variety of parts made are under $\frac{1}{4}$ in. × 6 in. (6.35 × 150 mm).

Fig. 26-14   (*a*) A 175-ton (1.56-MN) fine-blanking press. (*b*) The flow of material during fine blanking. Note small radius and burr, and smooth walls. (*American Feintool, Inc.*)

(a)

(b)

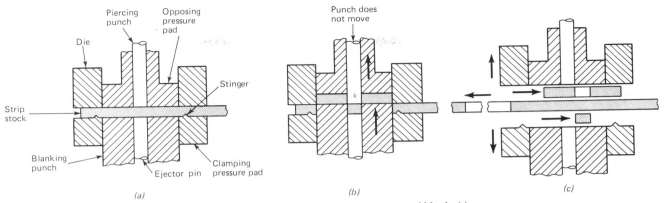

Fig. 26-15   The principal action of a fine-blanking press. (*a*) Press closed. (*b*) Pierce and blank. (*c*) Tool opens, part and slug are ejected, strip advances, and part and slug are blown or swept aside. (*Adapted from American Feintool, Inc.*)

STEEL RULE DIES

When piercing and blanking and moderate forming are needed for 50 to 500 parts, it would not be economical to spend the money for a regular tool-steel die set.

A process used for years for cutting fabric, leather, and cardboard cartons also works on metal. It is called a *steel rule die*. Many parts of the Apollo command ship and the Saturn rocket were made by this process.

Making the Die

To make a steel rule die, a full-size layout of the shape is made on special resin-bonded, laminated, compressed wood. This may be done with a vernier height gage for very accurate work, or it may be scribed around a template, or sometimes an accurately made drawing is pasted to the block. This shape is carefully jigsawed with special blades which are made to an exact width.

The steel rule is carefully bent to fit the slot and hammered into place. A flat ground steel plate, often $\frac{1}{2}$ in. (12.7 mm) thick, is then cut and ground to fit exactly within the steel rule. Figure 26-16 shows the complete setup. Notice that the upper wood block has a $\frac{1}{4}$-in. (6.35-mm) steel backup plate which supports the steel rule.

Cutting can be done on light gage metals using just the upper punch against plywood or urethane, especially if only 10 to 20 parts are needed. *Any size* can be made, from jigsaw puzzle pieces to 48 × 72 in. (1200 × 1800 mm) panels. The punch and die are mounted in any press of sufficient area and tonnage.

THE HOLE-PUNCHING MACHINES

When sheets of metal from 24 to 96 in. (600 to 2400 mm) long and up to $\frac{1}{4}$ in. (6.35 mm) thick need many

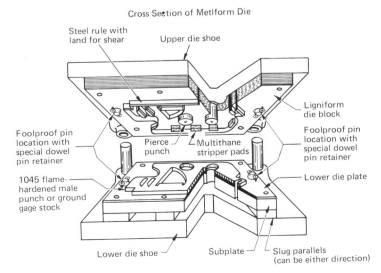

Cross Section of Metlform Die

Steel rule with land for shear

Upper die shoe

Ligniform die block

Foolproof pin location with special dowel pin retainer

Pierce punch

Multithane stripper pads

Foolproof pin location with special dowel pin retainer

1045 flame-hardened male punch or ground gage stock

Lower die plate

Lower die shoe

Subplate

Slug parallels (can be either direction)

Fig. 26-16   A complete steel rule die. Notice the many urethane stripper pads. (*J. A. Richards Company.*)

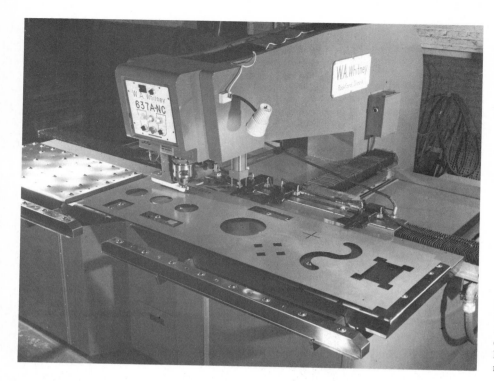

Fig. 26-17 A 30-ton (267-kN) single-tool N/C punching machine made for English or metric dimensions. Note edge clamps on far side. Capacity is 90 hits per minute, maximum 5-in. (125-mm) diameter. *(W. A. Whitney Corp.)*

openings cut in them, the usual punch press often does not have the space or capacity. If these parts are also needed in lots of 1 to 10 pieces, it certainly is not economical to make conventional dies for this work.

For this work an economical method is provided by hole punching machines, more aptly described by trademarks such as Fabricator (Strippit Division, Houdaille Industries, Inc.), Panelmaster (W. A. Whitney Corp.), or Perf-O-Mator (Producto Machine Co.). These machines have been used for many years but when numerical control was added, their productivity increased up to ten times, and their use has increased considerably.

The Machine

These punching machines (Figs. 26-17 and 26-18) consist of a large flat table to hold the sheet metal and 1 to 30 punching stations. The table moves freely along the X and Y axes so that any part of the sheet can be brought under the punch. Hydraulic hold-downs and edge clamps hold the work in place. The punches consist of a punch, die, hold-down, and stripper plate in a compact assembly. In the single-tool machines, the die set can be changed in a minute or less. In the multitool machines, the upper punch and lower die index together on N/C or manual command.

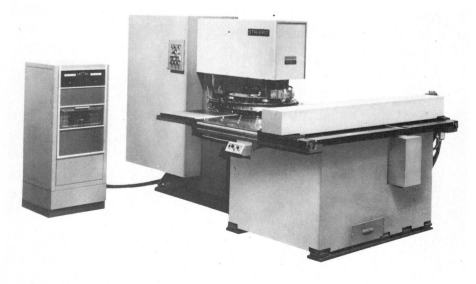

Fig. 26-18 A 30-ton N/C hole-punching machine with 20 tools in automatic tool changer. Capacity is 3½-in. (89-mm) diameter, 65 hits per minute. *(Strippit-Houdaille, a division of Houdaille Industries Inc.)*

The machines are made with capacities from 15 to 120 tons (133 to 1068 kN), with the most used machines around 30 tons (267 kN) capacity.

The Dies

Figure 26-19a shows, in the center, the 20 punch and die shapes which were set up in one machine. Many other sizes and shapes can be bought "off-the-shelf" or made to order. As each die is a separate unit, a broken punch is quickly and inexpensively replaced.

Uses of Punching Machines

As shown in Fig. 26-19b, panels for any type of work can be made in sizes up to 48 × 96 in. (1200 × 2400 mm) or larger. Thickness is most frequently under $\frac{1}{4}$ in. (6.35 mm), but punches are made which will punch through 1-in. steel plate.

Speed is excellent. Typical times are 10.5 min for 89 operations, with six tool changes, in 1-in. (25.4-mm) steel plate, or 588 holes in $\frac{1}{2}$-in. (12.7-mm) steel plate in 12 min. All holes were $\frac{13}{16}$-in. (20.6-mm) diameter. Slots and large square and round areas are cut by repeated, overlapping strokes of round, square, or rectangular punches. This is sometimes called *nibbling*.

Accuracy of location is ±0.005 in. (0.125 mm) or better. Hole sizes are more accurate, depending on punch and die wear.

Multiple patterns may be punched out on one large

Fig. 26-20 First operation in shearing apart repeated patterns made in punching machine. Second operation was shown in Chap. 25. (*Lodge & Shipley Co.*)

sheet, using numerical control, and then accurately sheared in one of the new semiautomatic shears as shown in Fig. 26-20. These patterns may be calculated with a computer-aided numerical control (CNC) unit, which can be purchased with some hole-punching machines for about $10,000.

Lot sizes can economically be very small, but long runs with speeds of up to 100 *hits per minute* and over can often be justified.

DEEP DRAWING

Deep drawing is used to produce a cup, cone, box, or similar shape directly from sheet metal, usually at room temperature.

Drawing parts from sheet metal is done in compound and progressive die operations, combined with piercing and blanking. However, when the depth of the draw is greater than half the diameter, the process is called *deep drawing*. The metal must be in the annealed (softened) condition, and for very deep drawing (over $1\frac{1}{2}$ times the diameter), the part may have to be drawn and annealed and drawn again several times, since the drawing operation causes work hardening.

The presses used may be mechanical or hydraulic.

Fig. 26-19 (a) Demonstration panel showing capabilities of turret-type punching machine at maximum of 175 hits per minute. (b) Two typical panels made on punching machines. Bending was done afterwards in a press brake. (*Warner & Swasey Co., Wiedemann Division.*)

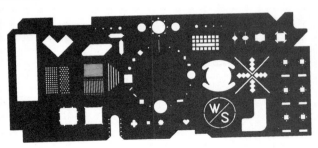

(a)

(b)

They are usually the double-action type with a stroke over twice the depth of the part.

The drawing operation is basically simple, but often trial and error is needed to achieve die shapes and a method which will prevent wrinkles and scoring and achieve accurate size. Figure 26-21 shows the basic process. The blank may be cut from a strip, and is often circular in shape, though rectangular and irregularly shaped blocks are used for noncircular parts.

When forming a cup shape from a circular plate, the bottom circle remains its original thickness. However, the walls are formed by compressing a pie-shaped section of the blank into a vertical section of the cup's wall. This may cause a thickening of the wall. This compression also tries to form wrinkles in the drawn part, especially when using metal less than 0.060 in. (1.5 mm) thick. The die edge radius and the blankholder control this.

The edge at the open top end of the deep-drawn part is seldom even; thus, a trimming operation is usually necessary. This may be a separate setup, or may be done in one station of a transfer die.

Materials which can be deep drawn at room temperature are low-carbon steels, aluminum, copper alloys, and stainless steels, though not from all alloys of

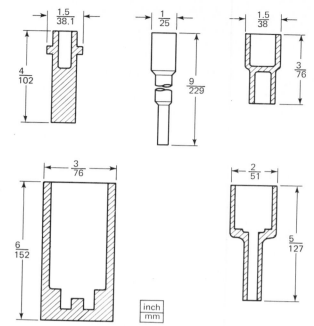

Fig. 26-22   Some parts which can be made by cold extrusion in punch presses.

these metals. Steel is often a *select grade* without external or internal defects.

Lubrication must be used as the sliding friction is great and galling (pressure welding) is likely. Lubricants may be oil, soap solution, grease, wax, molydisulfide, or water-soluble oil. The lubricant can be brushed or sprayed on, or the stock may be prelubricated.

COLD EXTRUSIONS

Cold extrusion, also called *impact extrusion* and *cold-forming*, is similar to cold forging, which is described in Chap. 27.

Cold extrusion uses a piece of metal (the slug) at room temperature. The slug may be cut from bar stock or partially formed in a previous operation. The process "upsets" or displaces the material at stresses above the elastic limit. The form is created by one or more punch and die sets. There is no waste (unless a hole is punched out), so the slug must be accurately sized. This is a squeezing or flowing of metal.

Products Made

Some of the shapes which are made are shown in Fig. 26-22. They may be gear blanks, pistons, bearing races, housings or caps, small shafts, copper fittings, etc.

Fig. 26-21   Two methods of deep drawing and some typical parts which can be drawn.

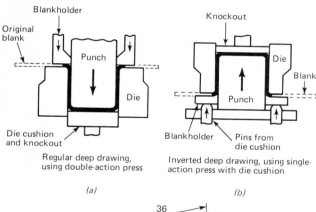

Regular deep drawing, using double-action press
(a)

Inverted deep drawing, using single-action press with die cushion
(b)

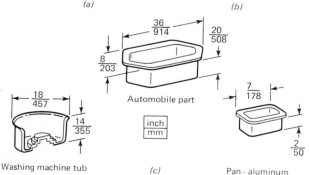

Washing machine tub

Automobile part

Pan– aluminum

(c)

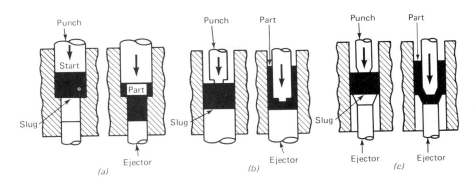

Fig. 26-23 Three methods of cold extrusion which are used with presses. (*a*) Forward extrusion. (*b*) Backward extrusion. (*c*) Combined backward and forward extrusion.

Methods of Cold Extruding

Cold extrusion is often done on the same presses which do piercing, blanking, etc. Mechanical presses are most frequently used, though hydraulic presses are preferred if long strokes are needed. A form of cold extrusion is also done with the special cold heading machines described in Chap. 27.

As shown in Fig. 26-23, three methods are used: forward, backward, and combination. In backward extrusion, the metal flows in a direction opposite to the motion of the punch.

The *extrusion ratio* is the ratio between the original area and the final area of the cross section of the part. As the section is not always constant, the ratios of the lengths of the part may be used.

$$R = \frac{A_1}{A_2} = \frac{L_1}{L_2}$$

where R = the extrusion ratio

A_1 and L_1 = the original area or extruded length

A_2 and L_2 = the extruded area or original length

Commonly used maximum extrusion ratios are: steel = 5, stainless steel = 3.5, pure aluminum = 40. Because of these limitations, it may be necessary to strike a part several times to complete an extrusion. This may be done in a large press using transfer dies.

Two things happen. First the temperature of the part may, in multiple strikes, go up a few hundred degrees. Second, the part is work hardened. The rise in temperature helps the forming a little but may shorten die life, and it does make the part difficult to handle. The work hardening raises the strength of the part, but also may make annealing (and subsequent descaling and cleaning) necessary if it gets too hard to form in the dies.

The Dies Used

The punch and die (Fig. 26-23) are shaped to the partial or final form of the work. Wall thickness of hollow parts is the distance between the punch and the die. These tools are made from several high-carbon tool steels or high-speed steels (such as H21, M2, M4) and, for long runs, from tungsten carbide.

Advantages and Disadvantages

Cold extrusion may be an economical way of making parts from ½ to 4 in. (12.7 to 100 mm) in diameter by 1 to 12 in. (25 to 300 mm) long. Larger parts have been made. *Advantages* of cold extrusion are:

Fig. 26-24 A large transfer press made specifically for cold-forming millions of spark plug bodies. Picture taken during tryout; thus wiring has not been enclosed. (*Verson Allsteel Press Co.*)

1. There is almost no waste material.

2. Finish can be about 63 μin. (1.6 μm). Tolerances from 0.001 to 0.005 (0.025 to 0.125 mm) can be held. Repeatability is excellent.

3. Accurate extrusion of a part may eliminate or minimize any further machining.

4. Both tensile strength and yield point are raised, sometimes quite dramatically.

However, there are also these *disadvantages*:

1. The tooling (punches and dies) is often expensive (from $1000 to $20,000), so cold extrusion may not be practical for production runs under 10,000 to 100,000 pieces.

2. For best results, the slugs or preforms should be free of either internal or external defects, which makes the stock somewhat more expensive.

Figure 26-24 shows a large transfer press ready for use in cold-forming sparkplug bodies.

There are several other methods of cold- and hot-forming which will be described in Chap. 27.

Review Questions and Problems

26-1. Name the following common sheet metal operations.

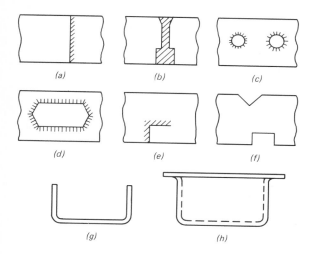

(a)   *(b)*   *(c)*

(d)   *(e)*   *(f)*

(g)   *(h)*

26-2. Describe a mechanical press used in presswork. What are the advantages and disadvantages as compared with a hydraulic press?

26-3. What is an OBI press and what are the advantages of this type of press?

26-4. Describe four ways that a mechanical press may be driven.

26-5. What materials are used to make dies used in presswork?

26-6. What determines die life?

26-7. How are a punch and die sharpened?

26-8. What are the four basic types of dies? Describe each.

26-9. Name some of the accessories required for presswork.

26-10. Describe a multislide machine. What are the advantages?

26-11. What are steel rule dies and how are they made?

26-12. A manufacturer wishes to make a cam $\frac{1}{8}$ in. thick (3.18 mm), approximately 3 in. (76 mm) in diameter, with a $\frac{1}{2}$-in. (12.7-mm) hole. What process would be used if the monthly quantity required is 150 pieces? If the quantity required is 2000 pieces? How would the required tolerance affect the process chosen?

26-13. Describe the fine-blanking process. What are the advantages and disadvantages?

26-14. A manufacturer of washers wishes to make a washer 1$\frac{3}{4}$-in. (44.45-mm) in diameter having a $\frac{5}{8}$-in.-diameter (15.88-mm) hole, in material 0.062 (1.58 mm) thick. If the shear strength of the material used is 30,000 psi (207 MPa), what pressure expressed in tons will be required? What press capacity is needed, in meganewtons (MN)?

26-15. What three methods are used to perform cold extrusions?

26-16. What are the advantages and disadvantages of cold extrusion?

26-17. Describe a hole-punching machine. To what type of work is it best suited?

References

1. *Computations for Metal Working in Presses*, Press Div., E. W. Bliss Company, Bulletin 38-E, 1972.
2. Conn, Harry: *Ipso Facto, Fabrication Economics*, W. A. Whitney Corp., Rockford, Ill.
3. Dallas, Daniel B.: "Pressworking: The Punching Machines Have Arrived," *Manufacturing Engineering and Management*, February 1973.
4. *Fine Blanking, an In-Depth Study; Stamping/Diemaking*, available from American Feintool, Inc.
5. Fox, William: "Some Helpful Diemaking Techniques (Steel Rule Dies)," *Modern Machine Shop*, November 1969.
6. *Forming Magnesium*, parts 2, 3, 4, booklets from the Dow Chemical Co., Metal Products Dept., Midland, Mich. 48640.
7. *Metals Handbook*, 8th ed., vol. 4, American Society for Metals.
8. Smith, Harold B.: "More on Hole Punching," *Manufacturing Engineering and Management*, July 1973.

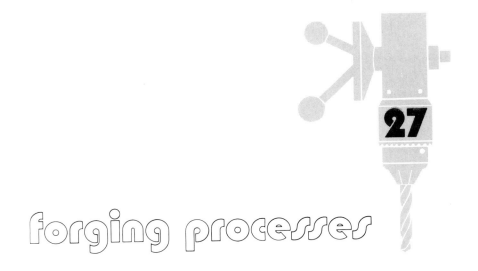

forging processes

Forging is a method of shaping a metal part by controlled plastic deformation of the metal, using hammers or presses, most often with dies. Forging is done *hot* (above the recrystallization temperature), *warm* (just below the recrystallization temperature), and *cold* (usually at or near room temperature). Forging, like casting, is used to reduce metal to a size and shape close to that of the finished part. In both processes, some surfaces are left as forged or as cast and some surfaces are machined.

Parts made by forging may weigh from ½ lb [227 g (mass)] to over 2 tons [1.8 t (mass)], though the greatest number of forgings made weigh from 5 to 100 lb [2.3 to 45 kg (mass)]. Some of the variety of forgings made are shown in Fig. 27-1. Many much larger forgings are made, especially from aluminum. Forged parts are used in the small tools, railroad equipment, automobile and truck, and aerospace industries, as well as in many smaller industries.

Practically all metals can be forged, though low-carbon and low-alloy steels and aluminum alloys account for over half the total tonnage. Stainless steel, nickel-based superalloys, and titanium are also forged, especially for aerospace uses.

Forgings are used when they cost less or have other advantages over castings or machined wrought stock.

Advantages of Forging

1. The grain flow follows the shape of the part, and the grain is refined (Fig. 27-2). This makes a stronger, tougher, more ductile part.
2. Forged parts have good uniformity in size and shape from part to part.
3. Metals can be machined and welded after forging, the same as similar wrought metals.
4. The forging process can, in many cases, be automated.

5. Production rates are good, as shown in Fig. 27-3, which was compiled from actual production records of board hammers. Other methods will give higher production.

Disadvantages of Forging

1. Tolerances for economical production are quite large, though tolerances as close as $\pm\frac{1}{64}$ in. (0.4 mm) can be achieved. Figures 27-4 and 27-5 show typical values.
2. Forging presses for large work are very expensive, and a set of dies for average-size work will cost $500 to $4000.
3. Heating furnaces (for hot forging) and trimming presses must be used, at added expense.

THE HOT FORGING PROCESS

It would seem to be a rather simple process to heat a piece of metal and then, with one or two very heavy blows, force it to fill a die. However, the metal is far from being fluid, and it just does not "flow" into the die cavities that easily. Thus, some or all of the following steps are needed in order to forge all except the simplest shapes.

1. *Cut to length* a slug or blank or preform a piece of stock. Cutting is done by sawing or by cutting with powerful shears. Preforming may be done hot in a roll former (described later) or in a separate forging machine. Today a growing number of preforms are made by powder metallurgy (P/M).
2. Heat the blank in furnaces (described later) or by resistance or induction heating. Steel and stainless steel are heated to about 2250°F (1230°C), and aluminum is heated to about 800°F (425°C).
3. The hot metal is brought to the roll former (if

Fig. 27-1   Some of the forged parts which have been made. (*Bethlehem Steel Corp.*)

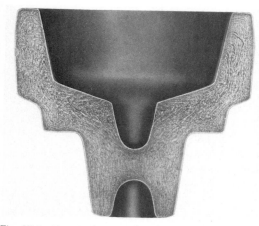

Fig. 27-2   Two examples of the characteristic grain flow in forged parts. These were made by impression-(closed) die forging. (*Bethlehem Steel Corp.*)

preforming is needed) or to forging presses, where it may be subjected to busting or upsetting (flattened or pancaked) to shape the metal and knock off much of the scale. Next may be *fullering* (making the center thinner) or edging and gathering (making lumps at the right places to help proper metal flow). Then the part is brought to the forging dies for blocking and rough and finish forging.

These steps are illustrated in Fig. 27-6 starting from the bottom of the picture.

4. After forging comes trimming, heat treatment (if specified), cleaning by sandblast or tumbling, and finally, machining to the finished size.

FORGING MACHINES

There are three principal types of forging machines and a number of variations of each type. The three types are the *drop hammer*, the *forging press*, and the *upsetter*. High-energy-rate forming (HERF) machines might be considered a fourth type.

Hammers (Drop Hammers)

Hammers for forging consist of a heavy frame, a falling weight (the hammer or ram), and an exceedingly heavy anvil. Hammers are used for *flat-die forging*

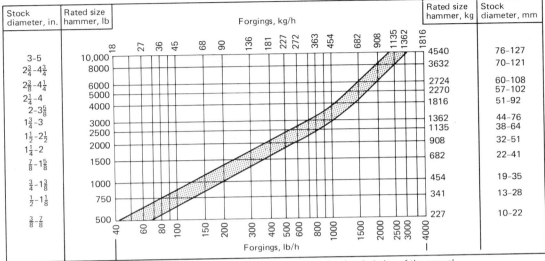

Fig. 27-3   Production rates for gravity drop hammers. (*Chambersburg Engineering Co.*)

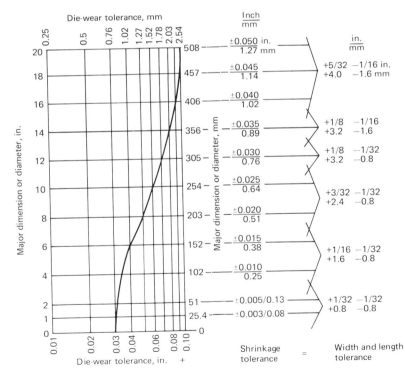

Fig. 27-4   Width and length tolerances for impression-die forgings. (*Adapted from Bethlehem Steel Corp.*)

(also called *open-die forging, hand* or *smith forging*). In this type of work, a part is shaped by manipulating the workpiece between simple flat, vee, or rounded dies. Drop hammers can also be used to make *impression-die forgings* (also called *closed-die forgings*), though the dies are never quite closed together. These are more or less complex forgings made between dies shaped to the form required. A fairly large

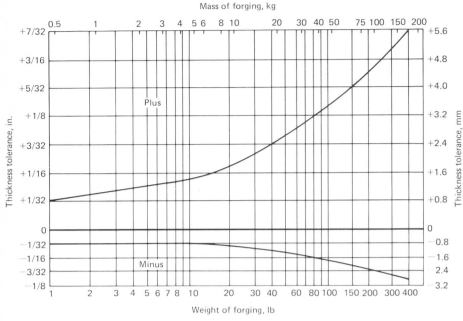

Mass of forging, kg

Bethlehem commercial-close thickness tolerances are normally two-thirds the ± values shown.

Fig. 27-5 Thickness tolerances for impression-die forgings. (*Adapted from Bethlehem Steel Corp.*)

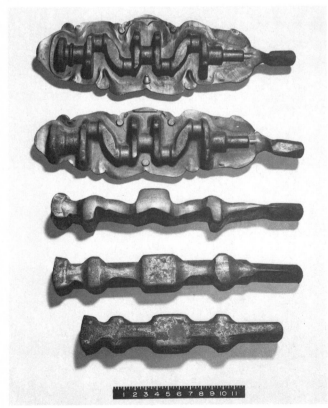

Fig. 27-6 The steps in changing a round bar to a forged workpiece. The first two steps (at bottom) were done by roll forming. (*National Machinery Co.*)

flash is always formed around the forged shapes. So-called *flashless forgings* are made in forging presses, which will be described later. Many of the parts shown in Fig. 27-1 were made as impression-(closed) die forgings.

Board hammers (Fig. 27-7) actually do use wooden boards held between gripping rolls to lift the heavy ram. When the grip is released, the hammer and the boards drop through a distance of 3 to 6 ft (0.9 to 1.8 m), and the force of gravity creates the forging force.

The most used sizes are 500 to 6000 lb [227 to 2720 kg (mass)] ram weight. The anvil, which must absorb the blow, is usually 20 times as heavy as the hammer. Wooden beams or special pads under the anvil help absorb the force. Board drop hammers can do a wide variety of work. However, the boards must be replaced about every 44 h at a cost of $120 plus labor and down time; thus other hammers are more used today.

Air-lift hammers (Fig. 27-8a) are modern drop hammers, which also derive their force from a free-falling ram. However, the ram is raised by an air cylinder. This makes a faster stroke possible, necessitates maintenance of fewer machine parts, and permits a selection of short or long (full height) blows. These hammers can make up to 50 to 75 strokes per minute, depending on their size, and are made in the same capacities as the board hammers. A 3000-lb [1360-kg (mass)] airlift hammer costs about $45,000 plus dies.

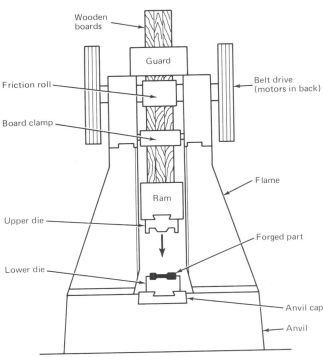

Fig. 27-7   Schematic drawing of a board drop hammer.

Labels in figure: Wooden boards, Friction roll, Board clamp, Guard, Belt drive (motors in back), Flame, Ram, Upper die, Forged part, Lower die, Anvil cap, Anvil

Due to lower maintenance costs, higher speeds, and easier operation, air-lift hammers are replacing board hammers. Figure 27-8b shows an advanced version of a forging machine with air lift and air downward drive. Program control is preset to control the force and the number of blows at each die station. This automation can materially increase the production rate.

Power drop hammers, better known as air or steam hammers, use air or steam at about 100 lb pressure added to the force of gravity to increase the force of the forging blow. A power drop hammer will supply two to two and a half times as much energy as the same rating of gravity drop hammer.

Steam (or air) hammers are the largest of the forging hammers and are made from 1000 lb [454 kg (mass)] up to a 50,000-lb [22,700-kg (mass)] falling weight hammer with an anvil weighing 1,000,000 lb (45,300 kg).

Forging Presses

Forging presses (Fig. 27-9) are similar to punch presses in construction, but they have heavier side

(a)   (b)

Fig. 27-8   (a) An air-lift, gravity drop forging hammer. (b) A modern "die forger" with air lift and air drive, plus program control. (*Chambersburg Engineering Co.*)

frames, and operate at higher velocities. The most used sizes are 300 to 6000 tons (272 to 5443 t) though larger presses are made. Like punch presses they may be driven either mechanically or hydraulically. The mechanical presses are usually crank or scotch-yoke drive.

Forging presses may cost more than twice what an equal capacity hammer costs. However, they are up to twice as fast, often use fewer forging strokes, and can be operated by less skilled labor. Forging presses are used for impression- or closed-die work only. They can do anything a hammer can and sometimes, due to their squeezing action, do better with difficult-to-forge metals.

The two largest hydraulic presses in the United States are 50,000-ton [45,360-t (metric ton)] presses one of which is shown in Fig. 27-10. A 75,000-t press is being used in the Soviet Union. These cost several million dollars and are used principally for aerospace forgings.

for quick estimates

To compare the capacity of a forging press to that of a gravity drop hammer, multiply the forging press tonnage rating by 2.5 to get the equivalent pounds rating

Fig. 27-9   A 4000-ton forging press, using a 250-hp (186-kW) motor. (*Ajax Manufacturing Co.*)

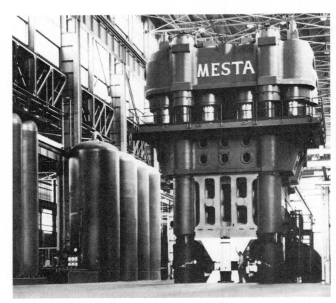

Fig. 27-10   A 50,000-ton (43,350-t) hydraulic forging press, at an Alcoa plant, for very large closed-die forgings. (*Mesta Machine Co.*)

of a drop hammer. Thus, a 2000-ton press does work equivalent to 2000 × 2.5 = a 5000-lb drop hammer. In metric, the metric ton rating times 1.25 equals the kg (mass) of the equivalent gravity drop hammer.

To estimate the capacity of press needed, the following equivalents give approximate results:

A = total cross-sectional area of the forging, in square inches, including $\frac{1}{2}$ in. of flash all around the part.

1.   Forging press capacity needed, in tons = A × 25 for steel; double this for stainless steel.

2.   Drop hammer capacity needed, in pounds = A × 65 for mild steel.

Forging Dies

Forging dies are made from several grades of tool steel. They must be hard to stand the abrasion, but tough in order to stand the pounding. Most dies have more than one "impression" for blocking, rough forging, and finish forging, as mentioned earlier. Typical die sets are shown in Fig. 27-11.

Dies are *sunk*, that is, the cavities are formed, by milling, electrical discharge machining, or sometimes the ceramic casting processes. Machining and shrinkage allowances and draft must be built in (Figs. 27-4 and 27-5) just as in patterns for casting. Draft may be from 3° to 10° depending on part size. Additionally, flash gutters are cut into the die to handle extra metal.

Die life varies widely due to many factors such as the metal being forged (steel, aluminum, etc.), the amount of scale on the part, the depth of the impressions, and the temperature and uniformity of the slug or preform. The type and proper application of lubrication can also be important.

In general, impression (closed) dies last 15,000 to 30,000 *platters* before they need reworking. A platter is the forging with its surrounding flash, as shown in Fig. 27-11. Each platter may contain from one to six parts.

The dies are often made in sections, called *inserts*, fitted into a die block. This is economical, as the blocker section may last several times as long as the finish forging section.

Dies must be heated before the first forging is made, and often gas or electric heaters are used to keep the dies hot so that the forgings will not cool too fast, shrink, and be difficult to eject. Many mechanical forging presses have lower die ejectors, and upper die strippers can be furnished.

Heating the Stock

It is important that the stock placed in the forge be at the correct temperature. Too cool a temperature may require a higher pressure, which could crack the dies. If the temperature is too high, the added heat from the forging process may actually cause melting of a portion of the surface of the forging. One of the types of furnaces used is shown in Fig. 27-12.

The *slot* or *box-type furnace* is the least expensive. It is merely a rectangular steel shell maybe 8 ft (2400 mm) wide by 4 ft (1200 mm) deep, with a refractory lining. These furnaces are heated by oil or gas. The bars or preforms are placed side by side inside a low "slot" through which the furnace operator reaches with tongs. Usually two people tend all types of furnaces, one feeding in the cold stock and the other bringing heated stock to the forge operator.

Rotary-hearth furnaces are doughnut shaped and

Fig. 27-11   Two examples of dies used for closed- (impression) die forging. (*a*) Upper and lower forging dies for an 8000-ton press. (*b*) Die making two parts on each platter. Flash is removed with a separate trim press. (*National Machinery Co.*)

(a)

(b)

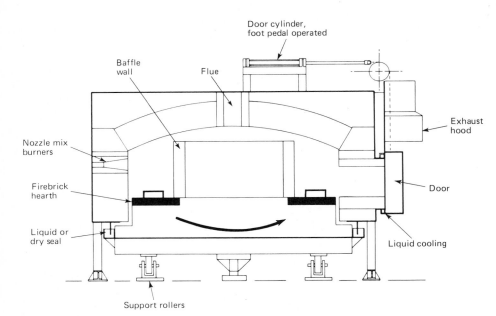

Fig. 27-12   A gas-fired rotary hearth furnace. Hearth diameters 4 to 7 ft (1200 to 2100 mm). Similar furnaces are also heated electrically. (*Sunbeam Equipment Corp., a subsidiary of Sunbeam Corp.*)

are set to rotate slowly so that the stock is heated to the correct temperature during one rotation. These also are heated by gas or oil.

Conveyor furnaces of several types can be used if only one end of the work must be heated, though they also will heat complete bars. Especially for larger stock, a pusher furnace may be used. This has an air- or oil-operated cylinder to push stock end-to-end through a narrow furnace.

The formation of scale, due to the heating process, especially on steel, creates problems in forging. New styles of gas-fired furnaces have been developed to reduce scaling to a minimum. Electric heating is the most modern answer to scaling, and it also heats the blanks more uniformly.

Induction heating (passing the stock through induction coils) greatly decreases scale, can often be operated by one person, requires less maintenance than furnaces, and is faster. Delivery to the forging machine operator can be by slides or automatic handling equipment.

Resistance heating, in which the stock is connected into the circuit of a step-down transformer, is even faster than induction heating and often is automated. Fixtures must be made for holding each different length, shape, and diameter of stock. However, the fixtures are often quite simple, and some can be adjusted to handle a "family" of parts.

NO-DRAFT (PRECISION) FORGING

Most forgings must have considerable draft and flash. This wastes material and requires a trimming opera-

tion and extensive machining. No-draft forgings (Fig. 27-13) are closer to the finished size, yet retain the excellent strength and grain-flow characteristics of a forged part.

Practically all no-draft forged parts are made from aluminum alloys, though some experimental work is being done with titanium.

Since the forging is totally enclosed in the die, the volume of the blank, or preform, must be controlled within $\frac{1}{2}$ to 8 oz (14 to 227 g), depending on the part size. When quite complex parts are made by this process, the blank is often preformed in conventional closed dies so that metal will not have to flow too far.

Dies for no-draft precision forging are considerably more complex (and more expensive) than regular forging dies. The no-draft dies must have very close-fitting (0.004 in., 0.1 mm) clearances in punch and die, knockout pins, and a stripper mechanism to aid in removing the part. Die life may be as low as 200 pieces before some maintenance is needed.

Tolerances are sometimes as low as ±0.015 in. (0.38 mm) but more often are about ±0.030 in. (0.76 mm) in aluminum. Production rates are from one-third to two-thirds those for conventional forgings.

Thus, close tolerance, no-draft (precision) forging is considered only for making parts with complex contours which are difficult to machine, especially if the part must meet high-strength requirements.

ROLL FORGING

Roll forging is a relatively simple and inexpensive way of doing certain forging jobs. The machine, shown in Fig. 27-14, is operated by one person. Heated stock is usually bars 6 to 36 in. (150 to 900 mm) long and from $\frac{1}{2}$ to 5 in. (13 to 75 mm) in diameter.

The *rolls* (from 10 to 38 in., 250 to 960 mm in diameter) have the shape machined into one-half to two-thirds their diameter. As shown in Fig. 27-15, the rest of the roll is cut off so that the heated rod can be pushed through them to a backstop.

When the foot treadle is pressed, the rolls grip the heated stock and push it toward the operator, while partially forging it to shape. The operator then places the rod in each succeeding groove until the final form is rolled.

Forging rolls may be used to make large reductions in the cross section and distribution of the metal of a billet, thus saving considerable work in the forging hammer or press. These rolls also may be used to elongate the shanks of hammer-forged parts.

HORIZONTAL-ACTION FORGING MACHINES

This group of forging machines includes three types, but all are basically similar.

Fig. 27-13 Comparison of no-draft (precision) forging with three other methods of forming a part. (*Wyman-Gordon Co.*)

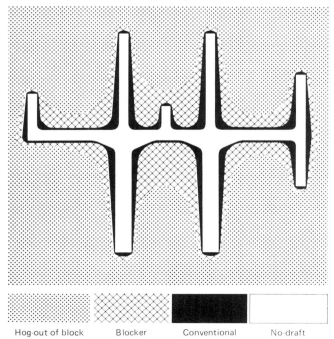

| Hog-out of block | Blocker | Conventional | No-draft |

No-draft forging gives much improved weight ratios of forging to finished part. Typical factors are: Hog-out 9X, blocker forging 3-4X, conventional forging 1.5X, no-draft 1.2X.

Fig. 27-14 A 40-hp (30-kW) roll-forging machine in operation. (*Ajax Manufacturing Co.*)

Forging machine, or *hot upsetter,* is the name given to the machine shown in Fig. 27-16a. This machine uses heated round or square rod or tubing stock and one to six cavities in the dies. The gripping dies are near the front of the machine, and the operator holds the part with tongs and places it in one cavity and then in the other. Forming of the part is done principally by the heading tool but can also be done by the clamping die. Figure 27-16b shows a set of dies used in a 3-in. (75-mm) forging machine.

Figure 27-17a shows a set of forging dies, and Fig. 27-17b shows the steps in making the gear blank. These are sometimes called *hot-upset forgings*. Gear blanks, pistons, socket wrenches, bearing races, and garden tools are some of the products made on these machines. Steel, some aluminum, and stainless steel are the most used metals.

Hot forging machines are fast, 20 to 80 strokes per minute, have no wasted metal, give excellent grain flow, and produce parts which can be used "as forged" or with a minimum of machining.

Machines for hot-upset forging are rated by the largest diameter of mild steel round bar which they can forge. The most used sizes are 1 to 6 in. (25 to 150 mm) with rated capacities of 100 to 1200 tons (91 to 1090 t)

Cold formers, or *cold headers,* are most often used to

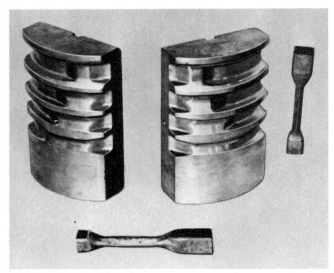

Fig. 27-15 Flat-backed segmented dies (four passes) used for short parts. (*Ajax Manufacturing Co.*)

make bolts and other fasteners of all sizes and shapes. Hundreds of these machines turn out hexagonal head, round head, and specially shaped bolts and screws. However, nuts, washers, shafts, couplings and special shapes are also cold forged. Figure 27-18 shows some of these parts. The principal material used is steel, with stainless steel, copper, and aluminum also used. Lot sizes of 50,000 to 1,000,000 pieces are commonly run, since dies and setup are expensive.

Cold formers have the same advantages as hot forging, plus speeds up to 600 per minute on small fasteners, and no scale from a heating cycle. Cold working also increases the tensile and yield strength.

Cold-forming machines (Fig. 27-19) differ from hot formers in that they use coils of "wire" from $\frac{1}{4}$- to $1\frac{3}{4}$-in. (6.4- to 44-mm) diameter as the stock. A cutoff station is included in the machine. Machines are ordered to handle a limited range of diameters. An automatic roll-threading station is added to the machine when it is making bolts.

Some cold formers have only one or two gripping stations, and the punches or rams are moved from one station to the other. The larger four- to seven-station machines have grippers which, with the aid of ejectors, move the stock from die to die. At the last die, the stock is dropped into a chute or onto a conveyor. One type of gripper and transfer mechanism is seen in Fig. 27-20; the punches are not shown.

HIGH-ENERGY-RATE FORGING (HERF)

High-energy-rate forming or forging (HERF), sometimes called *high-velocity forming* (HVF), machines are

(a)

(b)

Fig. 27-16   (*a*) A 2 $\frac{1}{2}$-in. (63-mm) hot-forging machine. (*b*) Closeup of a die section, showing three steps in forming. (*Hill Acme Co.*)

vertical counterblow machines used principally for hot forging, although they can also be used for cold forming, making powdered metal parts, and deep forming sheet metal parts.

HERF machines (Fig. 27-21), except for one which uses exploded gas, are operated by the sudden re-

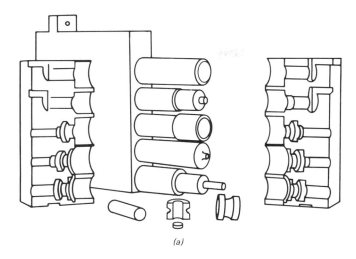

(a)

Multiple-operation Step Gear Dies

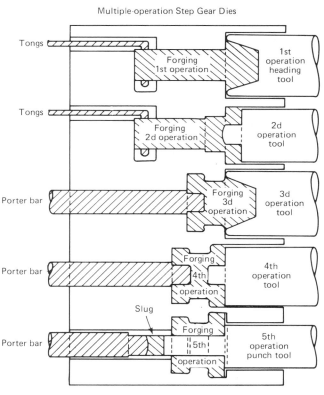

Tongs — Forging 1st operation / 1st operation heading tool

Tongs — Forging 2d operation / 2d operation tool

Porter bar — Forging 3d operation / 3d operation tool

Porter bar — Forging 4th operation / 4th operation tool

Slug

Porter bar — Forging 5th operation / 5th operation punch tool

(b)

Fig. 27-17 (*a*) A die set for making a gear blank in a forging (upsetting) machine. (*b*) Diagram of the steps in forging. The part is turned end-for-end between steps 2 and 3. (*Hill Acme Co.*)

lease of compressed nitrogen gas. This gas drives the ram and bolster toward each other at rates from 30 to 65 ft/s (9 to 20 m/s). In the machine shown, the ram moves a maximum of 12 in. (305 mm) and the bolster about 1.38 in. (35 mm). Most parts are formed with a single blow.

At these high velocities, the metal flows plastically;

thus deep, straight-sided parts can be made with zero or very small draft, which means much less metal to be machined off later. Much thinner walls can be forged: sections as thin as 0.15 in. (3.8 mm) have been made in experiments. Tolerances of ±0.010 in. (0.25 mm) and sometimes less have been achieved. Materials successfully forged in HERF machines included low-alloy steels, aluminum, and nickel steels.

High-energy-rate forming machines are listed with capacities of 15,000 to 1,808,000 ft-lb (20,000 to 2,450,000 N·m) of energy. This is adjustable over a wide range on each machine. The HERF machines cost from $90,000 to over $500,000. However, when compared to an equal capacity of other forging machines, they actually are less expensive, and they do not require as heavy foundations as other forgers.

Dies for HERF machines are made in the usual ways and of certain grades of tool steel. They are held in die chambers which strengthen the die. The die chambers are fastened to the ram and bolster of the machine.

The *disadvantages* of HERF machines are:

1.   A relatively slow cycle time. The actual forging takes only 0.005 s. However, dwell time and the raising of the ram are slower. Thus 5 to 12 parts per minute is average. However, these rates will be improved considerably by development work now going on.

2.   As the forging is done in confined (totally closed) dies, the stock must be very accurately sized. Preforms from previous forging or P/M preforms may be used, as may accurately cut lengths of stock.

3.   Hand loading is common at present though, as noted previously, automation could be used.

4.   Die wear can be short. In deep, straight-walled parts, die life may be as low as 2000 pieces. However, die life to 20,000 parts is achieved with other shapes.

Proper lubrication is very important to achieving good die life.

Fig. 27-18 Typical parts made on cold-heading machines. Some two-blow and some six-blow parts. (*National Machinery Co.*)

Fig. 27-19 Typical setup of a cold-heading process. Coiled wire, straightener, redraw to size, and enter the machine for forming. (*Ajax Manufacturing Co.*)

It seems as though the major uses of HERF will be for parts requiring amounts of energy beyond the range of standard forging machines, for parts for which the economies achieved from better tolerance and less draft are fairly large, or for parts which otherwise simply cannot be forged.

AUTOMATION OF FORGING

Some automatic transfer mechanisms and hammer blow regulating systems have already been mentioned in this chapter. In a continuing attempt to make forging less of an art and more of a science, two machines seem worth noting.

The **Impacter** from Chambersburg Engineering Company was first sold about 1963. It uses the counterblow principle of having two moving dies meet at the center, with the workpiece in between (Fig. 27-22). Because the workpiece is held between the dies, it has been called *midair forging*. This is a closed-die operation similar to all hot forging presses.

A control panel regulates the blow's force, the number of blows per station, and the travel of the automatic tongs which take one platter at a time through the press.

The counterblow Impacter has been used to make wrenches, pliers, scissors, pipe fittings, grab hooks, tractor parts, etc., from all forging metals.

Fig. 27-20   The transfer mechanism in a six-station cold-forming machine producing an automative wrist pin. (*National Machinery Co.*)

Positioning grippers

Part produced

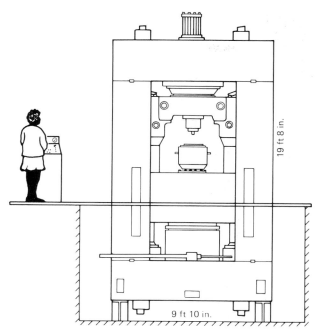

Fig. 27-21 A medium-size CEFF (controlled-energy-flow forming) HERF high-velocity forging machine. Energy range 181,000 to 400,000 ft-lb (245,000 to 542,000 N·m). (*Mesta Machine Co.*)

Input is cut lengths of stock. These can be resistance heated and automatically fed to the automatic tongs. The completed platters are dropped onto a short conveyor belt (see Fig. 27-22) which delivers them to a large container. Delivery can be to another conveyor which feeds stock to the trimming die press. One of the relatively new robots could also be used to transfer the platters.

These machines are expensive, but possible speeds are 90 to 170 blows per minute, accuracy is improved, and several parts may be forged in each platter. Because the scale tends to fall off, dies may last longer. As the hot workpiece is only in brief contact with the dies, the dies stay cooler and the work stays hotter.

A **Unimate** or other robot teamed with a forging press and an automatic electric stock heater comes close to complete automation, as shown in Fig. 27-23. The setup uses standard press forging dies.

The robot picks up the heated stock and positions it correctly on each die just as the forge operator would do by hand. Travel of the robot is electrically interlocked with the timing of the press blows. All operations for the die forger are set up and timed on a control panel. The robot is "taught" the sequence, direction, and distances to move, and then repeats these motions automatically.

The only skilled labor is the setup operator. Once the job is set up, only an attendant is needed, and the machine does not get tired and is not as affected by

the noise and heat. This certainly is one direction in which automation of forging may be achieved.

SAFETY IN FORGING

Standard safety precautions, such as glasses and safety shoes, must be used. In addition, the law now requires the use of *ear* protection, usually the "ear muff" type, because the noise in a forging shop can be over 100 decibels (dB).

Protection must be provided against occasional flying pieces of flash. Proper maintenance and inspection of the dies and the machine is especially important, as loose keys, guards, or cables can be further loosened by the vibration.

Review Questions and Problems

27-1. What is forging? List the advantages and disadvantages.
27-2. Describe the hot forging process.
27-3. List several products or parts which might be originally made by forging, either hot or cold.
27-4. Describe the three types of drop hammer forging machines.
27-5. Describe press forging. How does it differ from drop forging?
27-6. How large a forging press would be needed to forge a part with a projected size of 3.5 × 8.6 in.
Note: Add ½-in. flash all around.
27-7. What material is used to make forging dies? What physical properties must the dies have?

Fig. 27-22 An Impacter showing the part suspended and both dies open. (*Chambersburg Engineering Co.*)

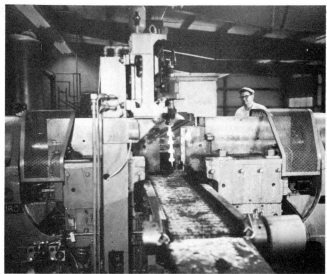

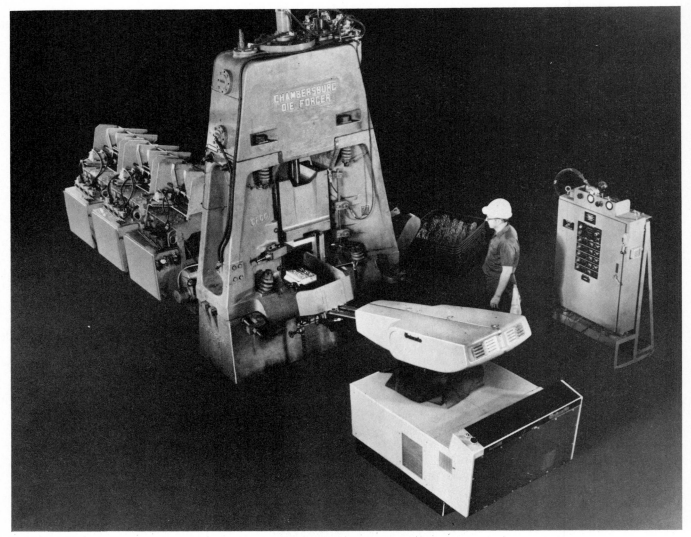

Fig. 27-23   A completely automated forging setup. Bar stock is heated, cut to length, forged, and put aside. Operator monitors and loads bar stock. (*Chambersburg Engineering Co.*)

27-8.   What determines die life?

27-9.   Why is proper heating of the work to be forged important? How is it done?

27-10.   Describe roll forging and its principal uses.

27-11.   What is hot-upset forging and how is it done? What are the advantages?

27-12.   How are hot-upset forging machines rated?

27-13.   How do cold-forming machines differ from hot forging machines? What are the advantages of each?

27-14.   Describe two machines which were designed to add automation to forging. What are the advantages of each?

27-15.   Describe a machine designed for high-energy-rate forging. For what type of work would you buy a HERF machine?

27-16.   What safety precautions are required when working around forging machines?

27-17.   What type of forging machine or other process would you use to make the following (all made of steel):

a.   Bolts $\frac{3}{4}$ −10 (thread) × 4 in.

b.   Machine base 24 × 60 × 15 in.

c.   Connecting rod (two holes) 15 × 3 × 1 in.

d.   Headed rods 1-in. diameter × 48 in. long with $1\frac{1}{2}$-in.-diameter head.

e.   Retaining pin $\frac{1}{2}$-in. diameter × $1\frac{1}{2}$ in. long with $\frac{5}{8}$-in. diameter × $\frac{1}{4}$-in. head.

References

1. Altan, Taylan: *Characteristics and Application of Various Types of Forging Equipment*, Society of Manufacturing Engineers, Detroit, Mich., MFR-72-02, 1972.
2. Bruno, E. J.: *High Velocity Forming of Metals*, Society of Manufacturing Engineers, Detroit, Mich., 1968.
3. *Closed Die Forgings*, Bethlehem Steel Company, Bethlehem, Pa.
4. *Forge Shop Modernization*, Chambersburg Engineering Company, Chambersburg, Pa.
5. *Forging Industry Handbook*, Forging Industry Association, Cleveland, Ohio, 1970.
6. Guillemot, L. F.: *High Energy Rate Forging*, Society of Manufacturing Engineers, Detroit, Mich., MF71-20, 1971.
7. Jenson, Jon E.: *Development of Useful Tolerances for Impression Die Forgings*, Society of Manufacturing Engineers, Detroit, Mich., MF70-602.
8. *Metals Handbook*, 8th ed., vol. 5, American Society for Metals, 1970.
9. *Parts from Wire* and *Upsetting*, booklets from National Machinery Company, Tiffin, Ohio, 1971.

gear making

Gears, sprockets, splines, escapement mechanisms, and racks and pinions are seldom seen except by the repair crew. However, they are a vital part of automobiles, watches, machine tools, many kitchen appliances, gear reducers, farm equipment, and many other items. They may be only a fraction of an inch or 25 ft (7600 mm) in diameter, from 0.010 to 72 in. (0.25 to 1800 mm) wide, and made from many different materials.

The manufacturing of gears is a very specialized business. Several companies make only gears and similar items. Many small- and medium-sized companies buy most of their gears from these companies because they actually get better quality at a lower price than if they made them themselves.

This chapter will not attempt to go into great detail about the manufacturing and mathematics of gears but will discuss the most used methods.

GEAR TERMINOLOGY

A discussion of gear design is beyond the scope of this book. However, some students may not yet have studied gears, so a brief review of terminology is presented here.

Figure 28-1 illustrates several of the terms used in describing gears and their dimensions.

The *involute tooth* shape (Fig. 28-1a) is the most common form of gear tooth. It is the curve generated by the end of a taut (tightly stretched) line as it is unwound from the circumference of a circle.

The *pitch diameter* (PD), or *pitch circle* (Fig. 28-1b), is the center of the working depth. It is the point at which the driving and the driven gear, when in contact, have equal thickness. This is the *mating* diameter which is used in calculating required center-to-center (c-c) distances of gears, racks, and pinions.

The *outside diameter* (OD) is seldom specified, as it is related to the standard American Gear Manufacturer Association (AGMA) standards based on PD and pitch.

$$OD = PD + 2 \times Addendum$$

Addendum is the radial distance from the PD to the OD. The addendum plus root clearance equals the *dedendum* in standard types of gear teeth.

Clearance, at the root fillet, is to allow for possible inaccuracies so that mating teeth will not hit bottom.

Backlash is a clearance made by cutting gear teeth slightly thinner than theoretical width to prevent tooth interference from allowable tolerances and to prevent expansion from temperature increase while running. Backlash can also be regulated by moving a pair of gears slightly closer together or farther apart.

The *pitch of teeth* is defined in two ways. *Circular pitch* (CP) (Fig. 28-2a) is the actual arc length between identical points on two adjoining teeth at the PD. The same distance on a flat gear (a rack) is simply called the *pitch*.

Diametral pitch (DP) is the pitch specified for most gears today. It is π divided by circular pitch, or the number of teeth in 3.1416 in. (79.8 mm); thus,

$$DP \times pitch\ diameter = total\ number\ of\ teeth\ in$$
the gear

The actual size of some of the gear teeth is shown in Fig. 28-2b. DP may be from 2.5 to 400, though 5 to 30 DP are the most common.

The *metric system* uses a module m instead of a diametral pitch. A module of 1 has a circular pitch of 3.142 mm, and a module of 10 has a CP of 31.416 mm. Thus,

PD = module × number of teeth

A 5-DP gear is about equal to a 5 module, a 20-DP gear is close to a 1.25 module. The conversion is

$$\text{Diametral pitch (DP)} = \frac{25.4}{m}$$

$$\text{or module } (m) = \frac{25.4}{\text{DP}}$$

Tooth thickness t is the thickness or width of the tooth at the pitch diameter (see Fig. 28-2*a*).

Table 28-1 shows the calculations for basic dimensions of English and metric gears. A comparison table of English and metric gears is in Appendix K.

The *width* of a gear tooth is the distance across the teeth parallel to the axis of the gear. This may be only 0.010 in. (0.25 mm) for small gears and $\frac{1}{2}$ to 3 in. (13 to 75 mm) for most gears. For large, heavy gears, widths of several feet have been made (Fig. 28-8).

Pressure angle refers to the central angle of two tangent involute curves. The standard pressure angles

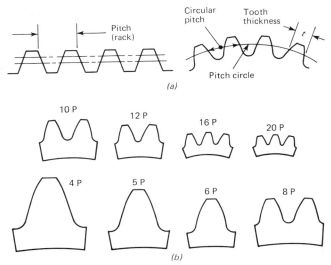

Fig. 28-2 (*a*) Circular and rack pitch. (*b*) Actual sizes of some diametral pitch teeth. (*Adapted from National Broach & Machine Division, Lear Siegler, Inc.*)

are $14\frac{1}{2}°$ and 20°. The 20° angle is the most used today, as it is a thicker, stronger tooth shape. A 20° stub tooth (slightly shorter) gear is popular because it avoids undercutting at the base of the teeth.

TYPES OF GEARS

There are six major types of gears as well as several special types. This book will deal only with the major types, since the others are not as widely used and knowledge of them can be gained on the job when necessary.

Fig. 28-1 (*a*) Generating an involute curve with a taut line. (*b*) The major elements of gear terminology. (*Adapted from National Broach & Machine Division, Lear Siegler, Inc.*)

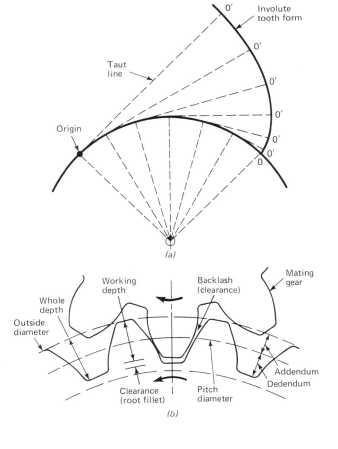

TABLE 28-1   BASIC DIMENSIONS FOR SPUR GEARS

To Find	English (in.)	Metric (mm)
Pitch diameter (PD)	$\dfrac{N}{\text{DP}}$	mN
Addendum (a)	$\dfrac{1}{\text{DP}}$	Module (m) in mm or (OD–PD) ÷ 2
Dedendum (d)	a + clearance	
Circular pitch (P)	$\dfrac{\pi}{\text{DP}}$	πm
Standard outside diameter (D_o)	D + 2a	D + 2m
Tooth depth (ht) whole depth	2a + clearance	
Average backlash per pair (B)	$\dfrac{0.040}{\text{DP}}$	0.040m
Tooth thickness (t)	$\dfrac{1.5708}{\text{DP}}$	0.0618m

N = Number of teeth on the gear
DP = diametral pitch
m = module (metric system only)

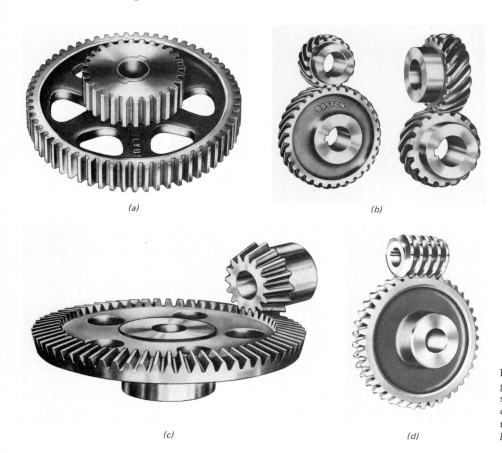

(a)

(b)

(c)

(d)

Fig. 28-3　(a) Spur gears. (b) Helical gears with parallel and right-angle shafts. (c) Bevel gears, 16:64 = 1:4 reduction. (d) Worm and worm gear, 40:1 reduction. (*Boston Gear Division, Rockwell International Corp.*)

Spur Gears

These are the most widely used and least expensive gears (Fig. 28-3a). They are made in a great variety of sizes, from less than 1 in. (25.4 mm) to several feet in diameter. They are made of steel, brass, and other metals, as well as of several kinds of plastic. Spur gears are made with both $14\frac{1}{2}°$ and 20° pressure angles.

Helical Gears

These gears (Fig. 28-3b) are like spur gears with their teeth at an angle. As more teeth are in contact at one time, the helical gear is quieter and stronger. It is widely used in machines today and is not much more expensive to make than a spur gear. One disadvantage is that helical gears exert an end thrust which requires the addition of thrust bearings. They are made in sizes and types similar to spur gears.

Herringbone Gears

A right- and lefthand helical gear side by side form the herringbone gear (Fig. 28-4). This neutralizes the thrust and makes a strong, quiet gear. Cutting these

gears is slower and more difficult, so they are most frequently used in the larger sizes, especially for heavy machinery.

Bevel Gears

Sometimes called *miter gears*, bevel gears (Fig. 28-3c) are used when shafts are at an angle with each other. A 90° angle is the most used, though almost any angle can be made. As the teeth are not parallel, it is more expensive to cut bevel gears. While bevel gears can be made in any size, the most usual sizes are 12-in. (300-mm) diameter and under. They are used 1:1 for angular drive or as reducing gears as illustrated.

Rack and Pinion Gears

These consist of a flat gear, called the *rack*, and a mating spur gear, called the *pinion*. The rack is easily cut with equipment much the same as that used for spur gears. These gears are used for creating straight-line motion from rotary motion or the reverse. Very fine racks are used in gages and small equipment, and coarser racks are used on lathes and other equipment.

Worm Gears

Worm gears and worms (Fig. 28-3d) are used when a large speed reduction ratio is wanted without incurring a large expense. Ratios as high as 100:1 can be made in diametrical pitches of 3 to 48 (8.5 to 0.5 module, metric). With a single pitch worm, the speed reduction ratio is the number of teeth in the worm gear. Worm gears lose considerable power due to friction. They require thrust bearings on both the worm and the worm gear shafts. Single pitch worms are self-locking; that is, the gear cannot turn the worm.

MAKING GEARS

Gears are made from plain and alloy steels, brass, bronze, stainless steels, aluminum, zinc, cast iron, and several plastics. Sometimes the smaller sizes can be produced ready to use by *powder metallurgy* or die casting. Plastic gears are made by the injection-molding process, very similar to die casting (see Chap. 30).

Small- and medium-sized gears, especially if made of brass, may be *extruded* in 12-ft (3600-mm) lengths and sawed off the bar. Thin gears (such as those used in clocks, watches, gages, and adding machines) can be made by *blanking* and shaving or by fine blanking in a punch press.

Gears for large, slow-moving equipment may be sand-cast iron or flame-cut steel. Small gears are also

Fig. 28-4 A heavy-duty triple-reduction gear drive, about 40:1 reduction. Input is at the near side. (*Philadelphia Gear Corp.*)

made, complete with hubs, by upsetting or hot pressing in forging presses. Both internal and external gears may be broached. Broaching is relatively expensive and slow, but it gives accurate, smooth-finished shapes.

Gear Blanks

Most gears used in automotive and machine tools are made from *blanks*. The blank is cast, forged, or machined to the approximate diameter and width of the final gear, and then accurately machined to the desired OD, with parallel ends on the hub or face. An accurately sized hole is bored or reamed in the blank, and this locates the gear for all further machining. The rough shape of the teeth is not usually included on the gear blank.

Machining Gears

In quantities of 1 to 10 gears (usually for experimental work), gears can be cut on a *horizontal milling machine*. Teeth are cut one at a time by a milling cutter having the correct tooth form. The cutter is slowly fed to depth as it traverses across the face of the gear. When one tooth is finished, the gear is rotated to the next tooth location.

Most gears used in larger quantities are first rough cut by hobbing or shaping, and then finish cut by shaving, rolling, grinding, etc.

Gear Hobbing (Generating the Gear Form)

A *gear hob* (Fig. 28-5) can be described as a cylinder on the surface of which a *thread* has been cut. The thread has the shape of the involute gear tooth. Gashes are then cut across the spiral to create cutting edges at regular intervals. The hob is most frequently 3 to 6 in. (75 to 150 mm) in diameter. Double- and triple-thread hobs are made. These cut faster (though not proportionately) but are less accurate. Hobs cost from $75 to $500 each depending on their size and grade.

The hob and gear blank are fastened to shafts which are at right angles to each other. The axis of the hob may be either vertical or horizontal. The hob is positioned to full depth (clear of the gear blank) and is then fed across the face of the gear. Figure 28-5b and c show the cutting action and positions. The hob is geared to make one complete revolution for each tooth which passes by it (or one revolution for two teeth if it is a double-thread hob). Cross feed is 0.002 to 0.015 in. (0.05 to 0.375 mm) per revolution of the gear.

As the gear blank is rotating, each successive hob tooth cuts a little deeper, first down into one side of

(a)

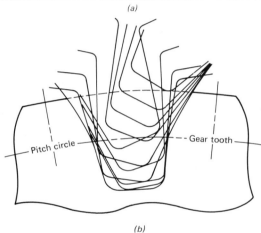

Pitch circle — Gear tooth

(b)

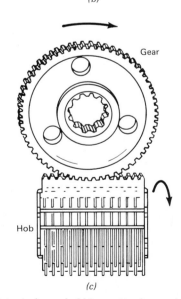

Gear

Hob

(c)

Fig. 28-5   (*a*) A typical gear hobbing cutter (a gear hob). (*b*) Diagram showing the action of the hob on a gear tooth. (*c*) The relationship of the gear and hob during the cutting. (*Barber-Colman Co.*)

the gear tooth and then up out of the other side. This process is called *generating* a gear tooth profile (Fig. 28-5*b*).

Hobbing is the most used method of making gears. It is fairly quick to set up; therefore, it is economical for short or long runs. The smooth cutting action makes it possible to generate quite accurate gears, and one hob can be used to cut spur and helical gears of different angles. Hobbing can also be used to make worm gears, sprockets, splines, and special gears and shapes.

Hobbing cannot be used to cut internal gears, or where flanges, etc., might interfere with the cutter. Gears from less than 1-in. (25-mm) up to 200-in. (5080-mm) pitch diameter and from 1 to 300 DP (25 to 0.085 module, metric) have been hobbed. Hobbing of gears of 2- to 4-in. (50- to 100-mm) PD is done at 35 to 75 gears per hour. The operation can be manual, or magazine- or hopper-feeding mechanisms can be added to fully automate the operation. One method of automation is shown in Fig. 28-6.

Hobbing machines (Figs. 28-7 and 28-8) are rated according to the maximum diameter of gear blank which they can machine. Sizes 4 to 15 in. (100 to 380 mm) are the most used. The hob's spindle may be mounted horizontally or vertically, whichever makes handling the blank easier. The gear's spindle is at right angles to that of the hob.

The machine must have rigid mountings for both hob and blank, and quite wide adjustments of

Fig. 28-6   One method of automating a gear hobber. (*Barber-Colman Co.*)

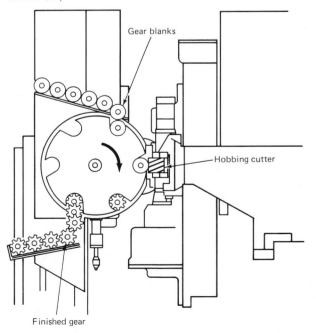

Gear blanks

Hobbing cutter

Finished gear

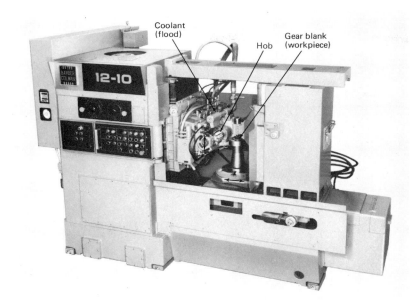

Coolant (flood)

Hob

Gear blank (workpiece)

12-10

Fig. 28-7 A medium-sized vertical-spindle gear hobbing machine. (Barber-Colman Co.)

center-to-center distances, speeds, and feeds. Rapid approach cycles and both rough and finish feedrates may also be included. A 10-in. hobbing machine will cost $30,000 or over, depending on its equipment. Other hobbing machines may cost from $17,000 to $100,000.

Gear Shaping

Gear-shaping machines are very versatile. Besides cutting gears, they can cut cams, odd-shaped holes, gear segments, ratchets, and many other regular and odd-shaped parts.

Shaping is used for spur, and herringbone gears, ratchets, splines, sprockets, etc. of almost any pitch and diameter, internal and external gears. Shaping can work closer to a shoulder than hobbing, and one shaping cutter can cut all spur gears of the same pitch. Gear face widths up to 12 in. (300 mm) can be made.

Shaping is also a *generating process*, but it differs from hobbing because the cutter and gear move on the same axis as shown in Fig. 28-9a and b. Both the hob and the gear rotate, and the cutter also reciprocates like the cutting tool in a standard shaper but at a rate of 50 to 450 strokes per minute.

The shaping cutter feeds in to full depth, and then continues cutting (generating) teeth. The gear is usually completed in one revolution.

A gear-shaping cutter is an accurately ground gear with a "dished" face which gives the teeth a positive rake. These, like hobbing cutters, must be sharpened carefully to maintain the original "geometry" of the cutter.

Gear-Shaping Machines

Gear-shaping machines (Figure 28-10) are rated according to the largest diameter gear they can handle. Sizes are from 6 to 40 in. (150 to 1000 mm) and larger pitch diameter. Smaller shaping machines are made for shaping small precise gears of 3-in. (75-mm) PD or less, and diametral pitches (DP) of 24 to 40 (1 to 0.64m, metric). Gear shapers cost from $25,000 to $80,000.

Full-Depth Gear Rolling

Full-depth rolling of bolt threads, worm gears, and similar items from round stock has been done in many shops for years. However, due to the deep

Fig. 28-8 A 20-ft-diameter (6100-mm) helical gear being hobbed. Rough and finish cutting required 800h total. (*Philadelphia Gear Corp.*)

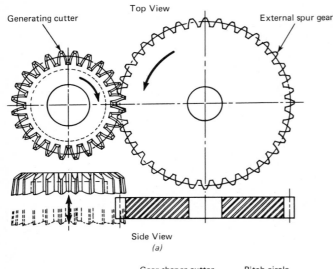

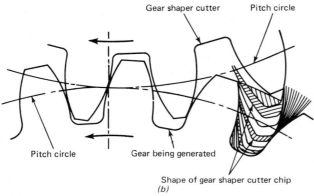

Fig. 28-9   (*a*) Diagram illustrating the vertical "shaving" motion, and the position of gear and shaping cutter. (*b*) The generating of the tooth form. (*Adapted from Fellows Corp.*)

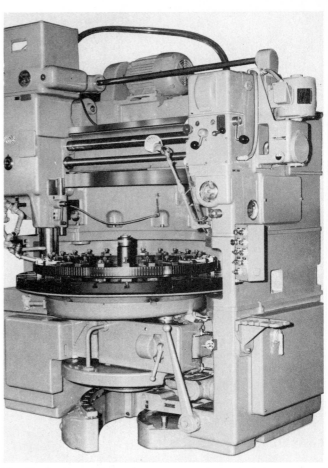

Fig. 28-10   A gear shaper for up to 40-in. (1000-mm) diameter and 6-in. (150-mm) face width gears. Cutter spindle is back of the gear, at the left. (*Fellows Corp.*)

shape of the teeth, it has been difficult to start with a round blank and "squeeze out" the shape of a spur or helical gear.

Full-depth rolling of 28 to 36 DP (0.9 to 0.7 m), rather fine pitch, gears is being done. However, tool life is short, so at the time this book is written, the process is not used commercially.

FINISHING GEARS

Hobbed, P/M, extruded, and die-cast gears may be used without further finishing or heat-treating for many relatively light machines such as some packaging machinery, low-power gear reducers, and some appliances. However, for high speeds or heavy loads, further work is necessary.

Hobbing and shaping with high speeds and fast feeds do not generate the accurate, smooth form needed for the more critical uses.

If gears must be heat-treated for hardness or toughness, there is almost always some distortion. Thus, a finishing operation is often necessary.

Grinding

Grinding the form is the only way, if the gear is harder than HRC 45 and more than 0.002 in. (0.05 mm) stock has to be removed. This was discussed in Chap. 15. Finishes to 16 μin. (0.4 μm) or better are made.

Honing

Gear honing, like the honing discussed in Chap. 16, is sometimes used as a final finishing operation to remove nicks and improve surface finish. Less than 0.002-in. (0.05-mm) stock should be left for honing.

The honing tool is an accurately shaped gear with a plastic- or metal-bonded abrasive coating. The abrasive "gear" is positioned at an angle with the metal gear, and the two are run together at high speeds under light pressure for 12 to 30 s. Finish is often 10 μin. (0.25 μm) or better, and honing can materially increase gear life and load-carrying ability.

Shaving

Gear shaving, since 1932, is the most frequently used finishing method for spur and helical gears with a hardness less than HRC 40. Shaving will remove

0.002 to 0.010 in. (0.05 to 0.25 mm) of metal and will produce a surface finish averaging about 32 μin. (0.8 μm). It improves the accuracy of tooth shape and spacing. Production may be 240 gears per hour for a 12-tooth, 1.125-in. (28.6-mm) PD steel gear, or 5 h each for a 400-tooth, 72-in. (1830-mm) steel bull gear.

The *shaving cutter* is basically a hardened and ground helical gear with two modifications. The sides of the teeth have square-sided cuts, gashes, or serrations, which form cutting edges, and (as shown in Fig. 28-11) a large relief is cut at the root for chip clearance. Shaving cutters will produce up to 20,000 gears before they need sharpening.

The shaving cutter is mounted in a *crossed axis position* with the gear as shown in Fig. 28-12. As the two rotate together, either the gear, on the smaller machine, or the cutter, on the larger machine, moves back and forth (traverses) so that the entire gear width is machined. Rotation is reversed so that both sides of the gear are shaved.

The *gear-shaving machine* is, like other gear-cutting machines, rated according to the size gear it can handle. As shown in Fig. 28-12, the gear is fixed in position (though it can rotate and traverse), and the cutter is mounted on a head which can be rotated to create the crossed axis position. A 12-in. (300-mm) gear shaver costs about $40,000.

Fig. 28-11 An assortment of gear-shaving cutters. (*National Broach & Machine Division, Lear Siegler, Inc.*)

Fig. 28-12 Gear-shaving machine showing cam operation and adjustments. (*National Broach & Machine Division, Lear Siegler, Inc.*)

Gear Rolling

Gear rolling (chipless machining) has been used for many years as a way of finishing worm gears, splines, and ball screws. In 1968 the Ford Motor Company and Teledyne Landis Machine developed a gear-rolling machine for finishing *helical* gears up to about 4-in. (100-mm) pitch diameter. Since then other companies have made similar machines, and the types and sizes of gears which can be roll finished are increasing.

As an example of the speed of the process, Ford Motor Company, in 1968, used only three roll-finishing machines to handle the output of 25 hobbing machines making 27,000 pinions (1.05-in. PD) per day. The gears were held to closer tolerances than had been possible with shaving, and finishes of better than 10 μin. (0.25 μm) were maintained.

The process is a combination of crushing and sliding or flowing the metal on the sides of the gear teeth by pressing the gear between two circular "dies," as shown in Fig. 28-13. The dies are hardened gears mounted on die heads, which can be moved in to create the necessary pressure. The life of a pair of roll dies is over 1,200,000 pieces before die failure. To speed production, several types of automation can be developed. One method now in use is shown in Fig. 28-13.

Machines for finish-rolling helical gears must be

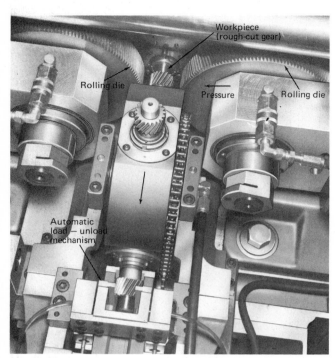

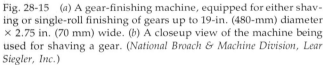

Fig. 28-13   Closeup of one make of gear-rolling machines. "Dies" at top of picture. Semiautomatic loading fixture shown. (*Teledyne Landis Machine.*)

Fig. 28-14   One type of double-die machine for finish-rolling helical gears. Notice automatic loading slide at the right. (*National Broach & Machine Division, Lear Siegler, Inc.*)

sturdy enough to withstand the 30,000- to 100,000-lb (133- to 445-kN) pressures needed. One type of double-die machine is shown in Fig. 28-14.

There are also *single-roll* gear-rolling machines which are less expensive and have the advantage of not displacing metal in two directions on the teeth. Direction of rotation is reversed to finish both sides of the teeth. One type of machine is shown in Fig. 28-15. This machine can also be used to shave gears up to a 19-in. (480-mm) diameter.

The necessary high forces are provided by an air

Fig. 28-15   (*a*) A gear-finishing machine, equipped for either shaving or single-roll finishing of gears up to 19-in. (480-mm) diameter × 2.75 in. (70 mm) wide. (*b*) A closeup view of the machine being used for shaving a gear. (*National Broach & Machine Division, Lear Siegler, Inc.*)

(a)

(b)

cylinder–toggle mechanism which raises the entire knee. Loading can be manual, or, by using a magazine feed, fully automatic operation is possible.

Gear-rolling machines cost from slightly over $50,000 for a small single-roll machine without tooling up to over $100,000 for a double-die machine with tooling. They are made by several companies, and because of their speed and long die life, their use is increasing.

At this time (1974) *spur* gears have not been successfully rough or finish rolled in production.

Lapping

Lapping of hardened gears is sometimes used after shaving. A slurry of fine grit and oil is poured onto a pair of gears rotating together or a gear running with a cast-iron lapping gear. The process is fast and much less expensive than grinding, and gives a very fine finish.

CHECKING AND INSPECTING GEARS

Checking gears requires special equipment, since the gear shape is complex and since the variety of shapes, including special modification, is quite large. Gears are checked for finish and tooth shape, the accuracy of tooth spacing, root diameter, radial position, tooth thickness, and depth. They are also tested for noise. An experienced inspector can listen to a gear running with a master gear and diagnose the gear errors very closely.

A frequently used checking machine is the Red Liner gear checker made by The Fellows Shaper Company. On this machine, the gear to be checked is run with a master gear, and any movement between the centers is measured and recorded by a red line on a moving chart.

OTHER TYPES OF GEARS

Bevel gears, spiral bevel, hypoid, and Zerol gears are cut with special equipment. As these gears are much less frequently used than spur and helical gears, this equipment is not described in this book. The references at the end of this chapter or information from the manufacturers will supply this information if it is needed.

Review Questions and Problems

28-1. What shape tooth is most common in gear tooth design, and how is it generated?

28-2. Define pitch diameter, addendum, dedendum, and backlash as used in gear design.

28-3. What is the difference between circular pitch and diametral pitch?

28-4. When dimensioning gears in the metric system, what is a module?

28-5. Define pressure angle. What two angles are most used today?

28-6. Name the six major types of gears and describe each. What are the advantages and the disadvantages of each?

28-7. What are the requirements for high-quality gears?

28-8. Compare gear hobbing and shaping. What are the advantages and disadvantages of each?

28-9. Describe full-depth gear rolling. What are the disadvantages?

28-10. Why are gear-finishing methods required after gear cutting?

28-11. Name five methods used to finish gears and briefly describe each.

28-12. What factors are usually checked in gear inspection?

28-13. Draw a sketch of a gear and indicate the pitch circle, addendum, dedendum, circular pitch, working depth, and whole depth.

28-14. A spur gear with 70 teeth has a pitch diameter of 6 in.
   a. What is the diametral pitch?
   b. What is the addendum?
   c. What is the tooth thickness?
   d. What is the whole depth?

28-15. A 36-tooth spur gear is to be hobbed with a 4-in.-diameter hob at 90 fpm.
   a. What is the work speed in rpm if the hob has a single thread?
   b. If it has a double thread?

References

1. *The Art of Generating with a Reciprocating Tool*, Fellows Corporation, Springfield, Vt., 1963.
2. Buckingham, Earle: *Analytical Mechanics of Gears*, Dover, New York, 1963.
3. *The Involute Curve and Involute Gearing*, The Fellows Gear Shaper Company, 1969.
4. *Metal Cutting Tool Handbook*, Metal Cutting Tool Institute, 1965.
5. Michalec, G. W.: *Precision Gearing, Theory and Practice*, Wiley, New York, 1966.
6. *Modern Methods of Gear Manufacture*, National Broach and Machine Div., Lear Siegler, Inc., 1972.
7. *Roll Forming of Gears at Ford Motor Company*, American Gear Manufacturers Association, AGMA-109.19, October 1968.
8. Samuelson, Marshall: *Gear Manufacturing Processes*, Cutting Tool Div., Barber-Colman Company, 1965.

less conventional machining processes

The processes described in this chapter are less frequently used, but they solve certain manufacturing problems which otherwise would be difficult or impossible.

Each process is described in sufficient detail so that the basic parameters are clear, and the principal uses, advantages, and limitations are shown. No attempt is made to go into technical details, as the literature is quite extensive and readily available if the student needs further information.

ELECTRICAL DISCHARGE MACHINING

Electrical discharge machining (*EDM*) is a process of machining by "melting" large numbers of very small particles of metal by bombarding the metal with sparks from a shaped electrode. The voltage used is low, from 1 to 45 V (volts), and the amperage increases as the area which is covered increases. From 5 to 60 A (amperes) is commonly used, though machines are made with capacities of several hundred amperes.

EDM has been used for small work for many years. Then about 1964 the power supplies were greatly improved, and EDM machines began to be used for making the punches and dies for presswork and the cavities for forging dies. Today a die-making shop could not compete without the use of electrical discharge machines.

About 1970 the aerospace industries began to use EDM for certain difficult jobs in machining the tough metals they use. Since then the use of EDM in production lines for such jobs as drilling the holes in carburators and many other jobs has been rapidly increasing.

Basic Principles of EDM

The basis for electrical discharge machining is that when two conductors (a shaped electrode and a workpiece) in an electric circuit are brought close together, a spark will go between them. As shown schematically in Fig. 29-1a, this spark (estimated to be at about 10,000°F, 5500°C) melts and vaporizes a spot in the metal. The direct-current (dc) circuit is caused to cycle on and off (pulsed) from 250 to 500,000 times per second (250 Hz to 500 kHz), creating the condition shown in Fig. 29-1b.

Because of the way electrons flow, the work is normally the positive, and the electrode the negative side of the circuit. Reverse polarity is used for some work.

The tank in which the *cutting* is done (or *burning* as some people call it) is filled with an *electrolyte*. This is a nonconducting liquid, often a petroleum distillate with additives. The vaporized metal, coming immediately in contact with the liquid, solidifies into a tiny hollow ball. Thus, the wet *chips* from EDM look like a fine black sludge.

The Electrical Discharge Machine

The electrical discharge machine itself is basically a vertical-spindle milling machine with a rectangular tank on the worktable, as shown in Fig. 29-2. The machine table may be moved along both the X and Y axis by handwheels, or occasionally by numerical control.

The *tank*, which is fastened to the table, may be 13 × 20 in. (330 × 500 mm) on a small EDM machine

Fig. 29-1 Comparison of low- and high-frequency sparks in EDM.

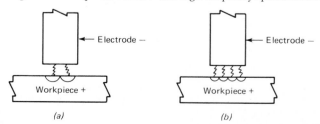

(a) (b)

Fig. 29-2   A 100-A dual supply, electrical discharge machine. Front of tank is removed so fixture can be seen. Note X and Y axis read-out on the right to assist in accurate table positioning. (*Elox Division, Colt Industries Inc.*)

costing about $8500 or up to about 48 × 96 in. (1200 × 2400 mm) on larger sizes costing $75,000 and over, plus the electrical unit. A typical electrical discharge machine is shown in Fig. 29-2, and a schematic of the entire setup is shown in Fig. 29-3.

The spindle does not usually rotate but is fed downward by a very sensitive servomechanism which keeps the electrode properly positioned as it cuts. The spindle is equipped with collets, vee clamps, or a flat plate for holding various shapes of electrodes.

A *flushing system* is a part of the electrical discharge machine. A tank, often part of the machine base, holds 30 to 75 gallons (114 to 284 litres). This is used to fill the work tank to about 25 mm above the work. A pump circulates this dielectric (nonconducting fluid) through a fine filter to remove the chips. The electrolyte is also used to flush the chips away from the electrode as will be described later.

The *electrical system*, except in the small 10- to 15-A machines, is sold separately. These are rated at 25 to 400 A and cost from $6000 to $50,000 each. Most units today are made so that split circuits can be used; that is, a 50-A service may be used to feed two electrodes using 25 A each or 10 small electrodes using 5 A each.

Controls are used to vary the frequency, amperage, voltage, capacitance, and the off-on time cycle.

the electrode

The electrodes are the cutting tools. Today they are usually made of fine grain carbon (graphite) or a carbon-copper mixture. These are easily machined to shape, are fairly inexpensive, will cut accurately, and can give good finish. A 1 × 1 × 6 in. (25 × 25 × 150 mm) electrode will cost about $5. Copper, brass, copper tungsten, and silver tungsten are also used. The last two give superb finishes but are difficult to cut to shape. The electrode and the workpiece should not be made of the same material.

The shape of the electrode is the shape of the cut. Thus, as shown in Fig. 29-4, any shape can be cut. Electrical discharge machines are made which will cut 0.005-in.-diameter (0.125-mm) holes in hardened steel using copper-tubing electrodes. The large machines will finish cut large forging, blanking, or press dies.

Thus, the real skill is in cutting the electrodes to precise size and shape. Milling machines, tracer-millers, and special machines have been developed to cut electrodes to two- and three-dimensional shapes.

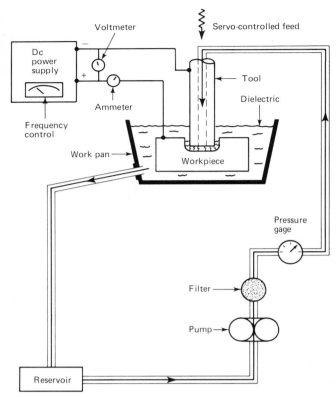

Fig. 29-3   Schematic drawing of an EDM system. The "tool" is the electrode. (*Adapted from Chemform Division, KMS Industries, Inc.*)

Wear Ratio

As the spark bombards the work, the electrode is also subjected to electron bombardment. Thus, the electrode is eroded (worn away) during the cut. This can be an important factor in EDM work.

$$\text{Wear ratio} = \frac{\text{volume of work metal removed}}{\text{volume of electrode consumed}}$$

This is often simplified to

$$\text{Wear ratio} = \frac{\text{depth of cut}}{\text{decrease in usable length of electrode}}$$

The wear ratio for carbon electrodes is up to 100:1. Other wear ratios (for cutting steel) are copper, 2:1; brass, 1:1; and copper tungsten, 8:1.

Thus, a piece of copper cutting 25 mm deep into steel will wear 12.5 mm. Because of this, the spindle must travel 37.5 mm to make the cut. These ratios are approximate and will vary considerably.

Reverse polarity (electrode positive) practically eliminates electrode wear. The workpiece metal will "plate" onto the electrode so that it will actually grow very slightly larger.

This is called *no wear EDM*. It is used mostly for

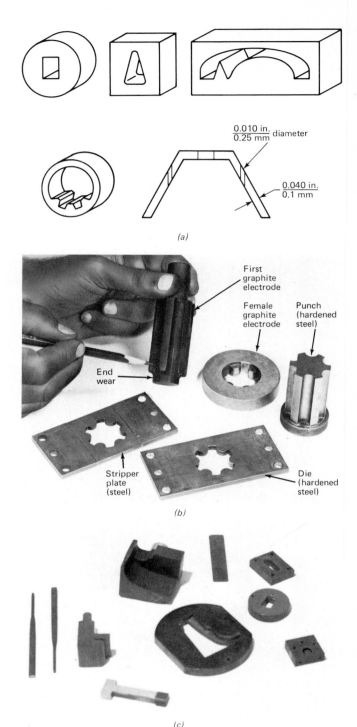

Fig. 29-4   (*a*) A few of the shapes which can be cut into any metal by EDM. (*b*) The two electrodes and the punch and die set made with them. (*Union Poco.*) (*c*) Some graphite electrodes, both male and female cutting.

roughing cuts and is slower than conventional EDM. However, it does save the expense of making a new electrode for each part made.

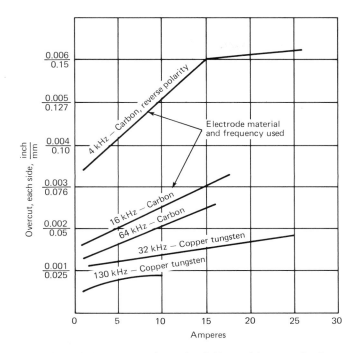

Note: Higher amperages give rougher finishes, and shorter cutting time.

Fig. 29-5 Overcut versus amperage for several combinations of electrode and frequencies.

Overcut

All electrodes will cut a hole or cavity larger than their own size. This overcut will be 0.001 to 0.008 in. (0.025 to 0.2 mm) on *all* surfaces.

This overcut *increases* with higher current and *decreases* with higher frequency. The amount of overcut must be known if final work size is to be held. All manufacturers today can supply charts like those shown in Fig. 29-5. These are quite accurate, so electrode sizes can be figured ahead of time.

Metal-Removal Rate (MRR)

Electrical discharge machining is *not* a fast way to remove metal. The maximum rate is 0.045 in.³ (0.74 cm³)/h/A. Thus a 25-A EDM machine will remove a maximum of 1.13 in.³ (18.5 cm³) of material per hour. Fine finishing cuts may be made at one-tenth this rate.

Thus, if a large cavity (such as a large forging or forming die) is to be made, it is often more economical to rough out most of the metal, in the annealed condition, by conventional milling. The die may then be hardened, and EDM will economically finish the shape to size, though some filing and polishing may still be required.

The *material being cut* will affect the MRR. Experi-

ments indicate that the metal-removal rate varies *inversely* as the melting point of the metal. The approximate value is

$$MRR = \frac{2.4}{(\text{melting point, °C})^{1.25}}$$

Thus EDM will cut aluminum much faster than steel.

Feeds and speeds for EDM cannot easily be taken from published tables as they can be for turning, drilling, and milling. The first run of a new job often requires some experimenting with control settings. However, there are controls, and they do affect the metal-removal rate. Some guidelines are:

Frequency—Lower frequencies give higher metal-removal rates and poorer finishes. Top amperage is limited by the area of the electrode and by the gap between the electrode and the work. The current control knob is sometimes labeled the "removal rate selector." Finish gets poorer as amperage is increased. The maximum amperage with carbon electrodes is 50 A/in.² (7.75 A/cm²).

Duty cycle—This is the relative off-on time of each pulse of electricity. On most EDM machines this can be hand controlled. A longer duty cycle tends to increase MRR but causes a rougher finish.

Voltage—This should be merely enough to ionize the dielectric fluid and cause the spark to cross the gap. The working voltage is seldom over 50. It is a result of the characteristics of the dielectric, the work material, and the electrode.

Flushing the Electrode

Basically an electrical discharge machine works with very little trouble, and down time is small. The most troublesome job is getting the sludge chips out from under and around the electrode. This job requires a smooth, constant flow of fluid across *all* sparking surfaces. Only 5- to 20-psi (34- to 138-kPa) pressure is needed. Some of the methods used are shown in Fig. 29-6.

In *burning out* deep complex molds (such as one for a plastic camera body), the work is sometimes put upside down on the spindle with the electrode on the table. Considerable ingenuity and planning is sometimes needed to get efficient flushing around the electrode.

Special EDM Equipment

Rotating spindles may be used to cut internal threads or similar shapes.

A *traveling wire EDM*, with an endless wire elec-

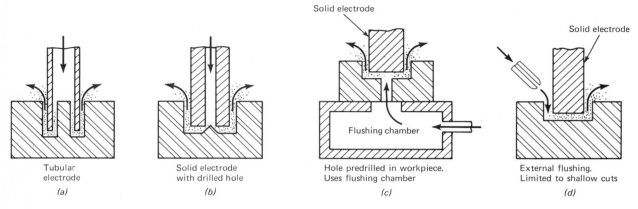

Tubular electrode
(a)

Solid electrode with drilled hole
(b)

Solid electrode

Flushing chamber

Hole predrilled in workpiece. Uses flushing chamber
(c)

Solid electrode

External flushing. Limited to shallow cuts
(d)

Note: In *a, b,* and *c,* a vacuum may be used instead of pressure.

Fig. 29-6   Examples of methods of flushing the chips from parts while cutting by EDM.

trode, is being used to cut shapes in hard metal. Numerical control is used to guide the table movement.

Electrode material attached to the end of a punch can be used to cut a die. The electrode material is glued to the die with special adhesives. A copper, female electrode is then used to finish form the punch for near-perfect fit with uniform clearance.

Advantages of EDM

Electrical discharge machining can be used to cut any shape into almost any conductive material (including the carbides), regardless of how hard it is. No burrs are formed, and no forces are exerted on the workpiece. Dies can be finished *after* hardening, thus avoiding the warpage which occurs if they are hardened after finish machining.

Sizes of cuts are from 0.010-in.-diameter (0.25-mm) holes to press dies over 48 in. (1200 mm) long, though most electrodes are from 2 to 12 in. (50 to 300 mm) on their largest side.

Finishes to 10 μin. (0.25 μm) can be obtained, and tolerances of $\pm$0.002 in. (0.05 mm) and better can be held. The EDM machine, on long duration cuts, needs very little attention. One operator can tend two or three machines.

Disadvantages of EDM

The major disadvantage of EDM is its slowness in removing metal. However, in die work and for narrow slots and small-diameter holes, the time still may be less than half of that required by any other method.

The electrode must be accurately machined, which takes skilled help. However, cutting carbon is a lot faster and easier than cutting steel.

The EDM process, due to the heat involved, leaves 0.001 to 0.005 in. (0.025 to 0.125 mm) of resolidified metal around the surface of the cut as well as an annealed layer under this. In die work this sometimes seems to be an advantage. However, if the work is used under stress, the hairline cracks in the resolidified areas may cause early failure.

Because of the wearing of the electrode, two or more electrodes may have to be made to complete one job. As many as six electrodes have been needed to accurately complete complex molding dies.

ELECTROCHEMICAL MACHINING (ECM)

Electrochemical machining will do many of the same jobs as electrical discharge machining. It will do them very much faster, but the equipment and the tooling is more expensive, uses more electricity, and takes more floor space. Thus, if a lot of metal must be removed or if a large number of parts are to be "machined," ECM may prove to be the most economical method.

The ECM Process

Electrochemical machining uses an electrolyte fluid and electric current to ionize and remove metal from the workpiece. This is similar to electroplating, the difference being that the freed positive metal ions combine with the electrolyte to form an insoluble precipitate which is washed away by the rapidly flowing electrolyte.

The most frequently used electrolytes are NaCl, that is, salt water (1 to 3 lb salt per gallon), and $NaNO_3$, sodium nitrate. These form oxides and hydroxides with the metals, and also hydrogen gas, which must be vented from the machine.

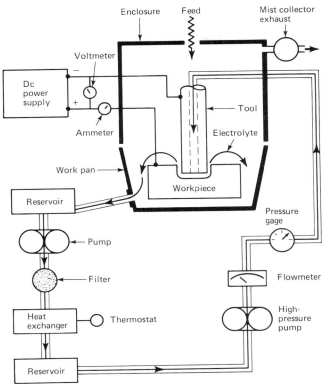

Fig. 29-7 Schematic drawing of an ECM system. (*Adapted from Chemform Division, KMS Industries, Inc.*)

The Electrochemical Machine

The principal parts of an electrochemical machine are shown in Figs. 29-7 and 29-8. The electrode holder is controlled much the same as in electrical discharge machines. The *worktable* is moved along the X and Y axes by handwheels or automatic controls. The work-holding tank is usually made of stainless steel or plastic to resist the corrosive action of the electrolyte.

The *power supply* furnishes only 5 to 24 V but 500 to 25,000 A of direct current (dc), though most work can be done with 1000 to 5000 A. The power supply is separate from the machine, to avoid corrosion from the electrolyte.

The *electrolyte* is circulated under pressure, usually 50 to 150 psi (345 to 1035 kN/m² or kPa) and occasionally higher. A 2500-A machine is supplied with a 200-gallon (750-l) tank, and a 5000-A machine may need an 800-gallon (3000-l) storage tank. Stainless-steel circulating pumps supply up to 120 gallons per minute (450 l/min) at the pressures noted above.

Filters, or *settling tanks*, or centrifuges are used to clean the electrolyte, and cooling coils may be needed for the larger machines when they are used near full capacity. Temperature should be about 120°F (49°C) maximum.

The *electrode* is usually copper, stainless steel, or tungsten copper. As shown in Fig. 29-9, it is insulated except at the cutting tip. Insulation today is usually

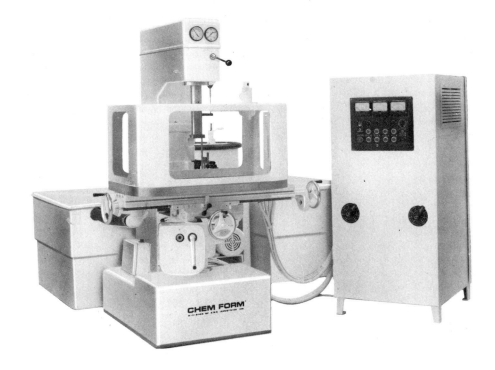

Fig. 29-8 A small ECM machine with a 200-gallon (750 l) fiber-glass tank, 0.006 to 0.30 ipm (0.15 to 7.62 mm/min) feed-rate, and 2500-A capacity. (*Chemform Division, KMS Industries, Inc.*)

plastic, though baked-on ceramic has been used. There is almost no wear on the electrodes, so they last indefinitely on most types of ECM cuts.

These electrodes may be made by powder metallurgy or tubing, or they may be machined to shape, as in EDM.

The *worktable* and enclosing tank must be sized to hold one or several workpieces. One company lists a 41 × 24 × 19 in. (1040 × 610 × 480 mm) enclosure as standard. One of the largest ECM machines has 25,000-A capacity with an 84 × 48 × 25 in. (2100 × 1200 × 630 mm) table enclosure.

Typical work which has been done with electrochemical machines is shown in Fig. 29-10.

Control of Electrochemical Machining

The *feedrate* of an ECM spindle may be set from 0.050 to 0.500 in./min (1.25 to 12.7 mm/min), but most work is done at about 0.150 in./min (3.8 mm/min). This is limited by the capacity of the machine, turbulence of the electrolyte flow, and the accumulation of hydrogen gas bubbles.

The *finish* produced will sometimes be as low as 5

Fig. 29-9   A typical ECM setup using reverse flushing. Electrode can be any shape. (*Adapted from Chemform Division, KMS Industries, Inc.*)

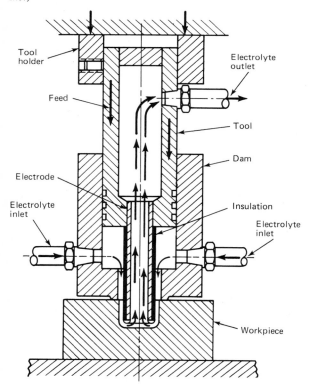

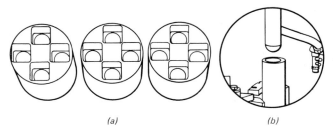

Fig. 29-10   Two types of work which were done by ECM. (*a*) It took less than 2 min to ECM the four holes in these rock boring bits. (*b*) A 3½-in.-diameter (89 mm) × 9-in.-deep (229-mm) hole in solid tungsten—cut by ECM in 3½ h; by EDM in 120 h. (*Cincinnati Milacron Inc.*)

μin. AA (0.125 μm) though more often 20 to 150 μin. (0.5 to 3.75 μm). Better finish results when *high* feedrates and high amperages are used. In general, hardened metals machine to better finishes. Different metals also react differently to ECM. Cobalt may get a polished surface, while aluminum will always have a matte finish when machined by ECM.

Tolerances as close as ±0.001 in. (0.025 mm) can be held, though ±0.005 to 0.010 in. (0.125 to 0.25 mm) are more realistic. The more complex the cut, the more difficult it is to hold uniform tolerances.

Repeatability (uniform results on many pieces) is excellent.

Advantages of ECM

The greatest advantages are that ECM does not produce burrs, does not have any appreciable "heat-affected" area, and does not cause machining stresses.

ECM is especially useful in cutting slots, etc., in hardened steel or difficult-to-cut alloys. On tough alloys ECM may cut faster than any other method.

The metal-removal rate can run as high as 6 in.³/h/1000 A (98 cm³/h/1000 A), or it can be a great deal less when ECM is used to cut small-diameter holes (sometimes 25 at a time) through tough metals.

Disadvantages of ECM

The initial cost is the principal item slowing down the use of ECM. A 500-A electrochemical machine costs about $19,000, and a 5000-A machine may cost $45,000 plus accessories and tooling for each job.

The tooling (electrodes, flushing arrangement, etc.) is also more complex, and thus more expensive. The use of a lot of electricity and the corrosive nature of the electrolyte must also be considered. Parts machined by ECM must be thoroughly washed immediately after machining.

Summary

Electrochemical machining (ECM) has proven to be a time- and money-saver for cutting burr-free slot shapes and holes in hard-to-machine metals, and in cutting deep holes of small diameter and in fast removal of material of complex shapes. The rather high cost of ECM could be justified on the basis of direct time saved in machining large quantities, or by savings of secondary operations, such as deburring, or merely because there was no other way in which certain parts can be made economically.

ELECTROCHEMICAL DEBURRING (ECD)

A great many person-hours are spent in removing burrs from machined parts. Much of the burring, more properly called *deburring*, is still done by hand with files and scrapers and many people. Deburring by hand may take less than a minute or larger parts may require 10 to 60 min to deburr.

Electrochemical deburring is a very simple application of ECM. The ECD machines are smaller, use less electricity and are easier to operate than a cavity-making ECM machine.

The process can be made semi- or fully automatic, and will do a better job than can possibly be done by hand, often in 5 to 20 s.

ECD Machines

The ECD machine is similar to the electrochemical machine, without the spindle for the electrode. Instead, one to four or more stations (Fig. 29-11b) each have an insulated fixture to hold the work and simple electrodes which move into or around the area to be deburred. The electrode movement may be by cams, levers, or air or hydraulic cylinders. The electrode, once in place, is not moved until deburring is completed. Figure 29-11a shows the basic system.

The electrolyte is the same as for ECM, but needs only 10- to 50-psi (70- to 345-kPa) pressure, and often only 25 to 200 A of electricity per station will do the job. The electrolyte is stored in the reinforced fiberglass tank in the base of the machine.

ECD Operation

As shown in Fig. 29-12, the copper or stainless-steel electrode is positioned with a 0.005- to 0.025-in. (0.125- to 0.64-mm) clearance from the burr. It thus often cuts the burr off at the root. The length of deburring time and the amount of current used regulate the amount of radius or chamfer on the corners of the

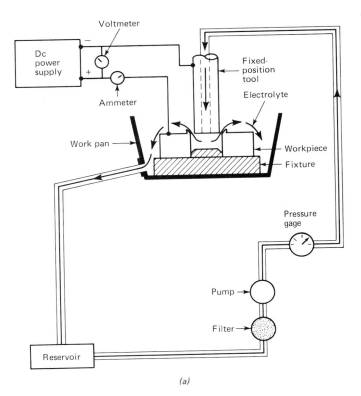

(a)

(b)

Fig. 29-11 (a) Schematic drawing of an electrochemical deburring machine. (b) A four-station ECD machine, 1000-A capacity. (*Chemform Division, KMS Industries, Inc.*)

work. Parts are transferred to a washing tank immediately after deburring.

Doughnut-shaped electrodes are used for external burrs and gear shapes, and any desired irregularly shaped electrode can be made quite easily. A 2000-A, four-station ECD machine will cost about $25,000, and machines as small as 250 A are made, costing about $8000; tooling for both machines is extra.

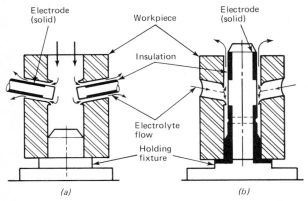

Fig. 29-12   Two ECD operations. (*a*) Removing burrs when side holes were drilled last. (*b*) Removing burrs when the center hole was drilled last. This is the less expensive setup. (*Adapted from Chemform Division, KMS Industries, Inc.*)

To be economical in simple jobs, yearly production of 20,000 to 50,000 parts may be needed. However, the need to deburr fragile parts or tough metals, or the need for increased quality, can make ECD a profitable process for smaller annual quantities. Completely automated ECD machines are now being used in production lines.

ELECTROCHEMICAL GRINDING (ECG)

Electrochemical grinding is being used to save many hours, and greatly cut the cost of grinding wheels for grinding multiple-tooth milling cutters, carbide-tipped tools and tool bits. Also, because like all electrochemical processes it puts very little pressure on the work, it is used to grind slots, flats, and forms in thin-walled or fragile workpieces. Some typical work is shown in Fig. 29-13.

ECG Machines

An ECG machine looks much like a standard grinding machine. In fact, standard grinders can be converted to electrochemical grinders. The most used

styles are surface grinders, tool grinders, and form grinders, though OD and ID grinders are also being used if quantity production makes them economical.

The total package (Fig. 29-14*a*) uses the grinding machine, a 25- to 40-gallon (95- to 150-l) tank, and a 250- to 1000-A power supply at 4 to 15 V. The electrolytes used are the same as in other electrochemical machining processes. The enclosure is usually made of plastic to resist the corrosive action of the salt.

ECG machines must have the spindle insulated from the rest of the machine. Brushes, or the newer mercury bearings, are used to conduct electricity to the grinding wheel. The grinding wheel is the cathode (negative) side of the dc circuit.

grinding wheels

The grinding wheels are standard round, cup, or formed shapes, but they must have a *conductive bond* and a nonconductive abrasive. The *abrasives* most often used are diamonds, Borazon, and aluminum oxide, often about 100 grit.

The bond is often copper, a copper carbon, or copper-plastic mix. The wheel is trued or shaped with a diamond, the same as conventional grinding wheels. However, the abrasive must project slightly beyond the bond, so *dressing* is done by reversing the current and deplating a few thousandths of an inch (or hundredths of a millimeter) of the metal bond.

The ECG Process

Figure 29-14*b* shows the typical ECG process applied to surface grinding. The flow of the electrolyte between the wheel (cathode) and the work (anode) dissolves the metal. The grits projecting from the wheel are set to just barely touch the work. Thus, they act as spacers and also "wipe" away the oxides formed on the workpiece. As the oxides are soft, there is very little grinding wheel wear.

Advantages and Disadvantages of ECG

Reports indicate that an ECG machine requires only normal maintenance. Wheel wear is greatly reduced.

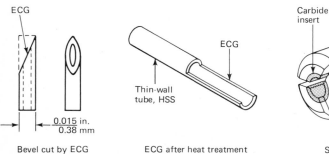

Bevel cut by ECG   ECG after heat treatment   Slots done by ECG

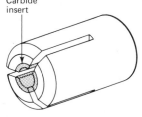

Fig. 29-13   A few of the cuts possible with electrochemical grinding.

An automotive company grinding multiple-tooth face mills reports using only one diamond wheel per year (previous use was over 50 per year, at $500 per wheel). The entire machine should be thoroughly washed occasionally, as the electrolyte is corrosive.

The greatest advantages are that all work is completely free of burrs; no heat is developed, so no heat cracks or distortions are developed; and very little pressure is exerted on the work.

ECG, when grinding cutting tools, takes the full cut in one pass (no rough and finish cuts). In fact, ECG works better the larger the wheel "contact" area is.

The major disadvantage is the cost of the ECG system. A 1000-A machine, with tank, pumps, and electrical controls, may cost $50,000. Because of this, ECG is most often used when long runs are made between setups, or when conventional grinding or milling of slots and forms creates excessive burrs or (especially on tough aerospace materials) takes too long.

OTHER CHEMICAL PROCESSES

Two other chemical machining processes are used for special types of work, usually in quantities of from 100 to 500.

Chemical Milling

This process was originally used on large panels for aerospace work, to make them a few pounds lighter by *milling* large areas a few thousandths of an inch below the normal surface. The panels may measure from 2 to 10 ft (600 to 3000 mm) long. The panel is very thoroughly cleaned, then covered with a *mask* of a rubbery plastic by spraying or dipping.

The mask is carefully cut and removed from the areas which are to be etched (milled). The plate is then dipped completely into a tank of chemicals which will dissolve (etch) away the exposed metal.

By controlling the time of immersion and the strength of the chemical bath, depth can be controlled to ±0.002 to 0.005 in. (0.05 to 0.125 mm). By using the proper electrolyte, almost any metal can be chemically milled.

Chemical Machining

Also sometimes called *chem-milling*, chemical machining is photoforming, photofabrication, etching, and chemical blanking. This process, like chemical milling, uses a chemical to dissolve the metal, which may be from 0.0005 to 0.125 in. (0.013 to 3.2 mm) thick. Material as thin as 0.005 in. (0.13 mm) is almost impossible to run on a punch press. However, in chemi-

Fig. 29-14 (a) An electrochemical surface grinder for flat or contour work. (b) Schematic drawing of the ECG surface grinder. (*Chemform Division, KMS Industries, Inc.*)

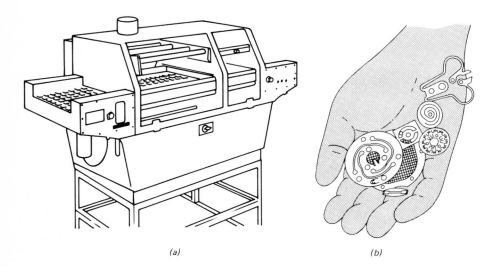

(a)   *(b)*

Fig. 29-15 (*a*) A small, conveyorized spray etcher, speed 1 to 30 ipm (25 to 760 mm/min) with top and bottom spray banks, and 15-in.-wide (380-mm) conveyor. (*b*) A few typical parts which have been made by chemical machinery (etching). (*Chemcut Corp.*)

cal machining, the etching is allowed to eat completely through the metal, so the result looks like a stamped-out part.

The chemical-machining process starts with an accurately made drawing or artwork of the part. To reduce errors, this is usually made 10 or 20 times actual size. The drawing is photographically reduced to the actual part size, and a negative or positive is made.

In the meantime the metal (aluminum, steel, titanium, or super alloy) is coated with a *photoresist* which is light-sensitive. This surface is now exposed to the light through the negative, just as in developing pictures. The exposed metal is then *developed* to remove unneeded portions of the photoresist. This may be done on either one or both sides of the metal.

The treated metal is next put into a machine (Fig. 29-15*a*) which sprays it with a chemical etchant, or it may be dipped into the solution. The etching chemical may be sodium hydroxide (for aluminum), hydrofluoric acid (for titanium), or one of several other chemicals. After 1 to 15 min, the unwanted metal has been eaten away, and the finished part is ready for immediate rinsing to remove the etchant. Figure 29-15*b* shows some of the work which has been done.

Chemical machining is an excellent method of getting complex parts from very thin metals without the cost of a punch-press die. The advantages are that this process does not distort the workpiece, does not produce burrs, and can easily be used on the most difficult-to-machine materials. However, the process is slow, and thus it is not usually used to produce large quantities or to machine metal over $\frac{1}{8}$ in. (3.2 mm) thick. Some small parts are made 10 to 100 at a time on a single plate, which speeds up production.

LASERS

Lasers are devices which are capable of generating a very intense beam of optical radiation. They have advanced from a laboratory curiosity in 1958 to a wide variety of industrial uses today. The word "laser" is an acronym for *l*ight *a*mplification by the *s*timulated *e*mission of *r*adiation.

Basic Construction

A laser operates on the principle that electrons in certain atoms oscillate when energy is supplied. A basic laser circuit (Fig. 29-16) consists of three parts: a pair of mirrors, a source of energy, and an *optical amplifier*. This amplifier is popularly called the *laser*. To these basic parts must be added a control system and a cooling system.

The *two mirrors* are flat or slightly concave. Sometimes prisms are used and are placed facing each other with the amplifier between them. One of the mirrors is an almost perfect reflector, and the other is only partially reflective, that is, some light can go through it.

Energy for some lasers is supplied by a source of light. This may be a flashlamp filled with xenon, argon, or krypton gas. The lamp is placed close to the amplifier inside a highly reflective cylinder so that as much energy as possible can be absorbed by the laser material. Other lasers, mentioned later, are supplied with energy by pulses of dc electricity.

The *amplifier* (laser) is a solid or gaseous material or, recently, a liquid material which contains a small percentage of a particular kind of atom or molecule. This amplifier may be a rod smaller than your finger or a gas-filled tube several meters long.

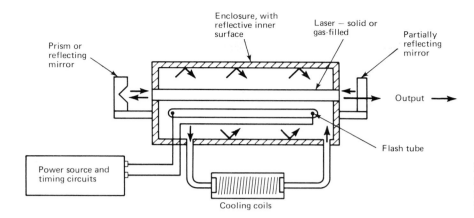

Fig. 29-16 Simplified schematic of a solid-rod laser system. Gas-tube systems are similar but do not have a flash tube.

Laser Operation

When energy is pumped into the amplifier, light begins to bounce back and forth between the mirrors. One color (wavelength) of the light is amplified each time it oscillates. This light rapidly becomes very intense. At some point in this buildup of energy, light goes out through the partial reflector. This is the useful output. All of this, of course, happens in 0.001 to 0.000001 s each time a shot of energy (pumping) is supplied.

This pulse or beam of light can be focused to a small diameter by lenses and reflected around corners by mirrors. It can also weld parts which are inside transparent glass or plastic enclosures.

Within limits set by the materials involved, the frequency and the amount of energy supplied by each "pumping" can be varied.

Cooling of the laser and the lamp is necessary because lasers are very inefficient. Often less than 1 percent of the pumped energy is sent out as useful laser energy, although efficiencies range from 0.02 to 20 percent.

Thus, 99.98 to 80 percent of the input energy is converted to heat, so all but the smallest lasers must have a cooling system. As in gasoline engines, both air and water cooling are used. Often cooling water is circulated around both the laser and the lamp (when lamps are used), and the water is cooled in a radiator just as in an automobile. The laser radiator may be only 4 in. (100 mm) square, or larger than the one your car uses.

Types of Lasers

A *ruby laser* (amplifier) is a solid synthetic ruby (Al_2O_3) rod containing chromium, which gives it the red color. This rod, and the neodymium (Nd) rods, are polished at both ends. Ruby is the most famous laser, but neodymium is becoming the most used.

Nd glass amplifiers are made of glass to which a small percent of neodymium has been added. This is less expensive than the ruby rod.

Nd:YAG neodymium-doped (added to) yttrium-aluminum-garnet crystals are in some ways superior to both of the above, though they are expensive. A $\frac{1}{4}$-in.-diameter × 3-in.-long (0.25 × 75 mm) Nd:YAG rod costs over $1000. The cheapest laser using such a rod costs about $4000. Figure 29-17 shows a commercial Nd:YAG laser with control by N/C tape. A complete outfit costs $30,000 to $50,000.

Neodymium lasers can emit continuous beams, but most commercial solid lasers operate in bursts of energy. These pulses can (when a device called a *Q switch* is used) be as short as a few billionths of a second (nanoseconds, ns) or about $\frac{1}{100}$ s (10 ms) long.

Depending on the use, pulse rates commonly range from 1 to 20 per second, though higher and lower frequencies are used.

CO_2 gas lasers consist of a glass tube filled with carbon dioxide. The tubes vary in length from 10 in. (250 mm) to 100 ft (30.5 m). The necessity for long lengths is being avoided today by methods such as circulating the gas, and by using transverse excited atmospheric (TEA) higher-pressure gas in the tubes with many discharge points along the length of the tube. CO_2 lasers usually cost $5000 and more.

Gas lasers are *pumped* by a direct-current (dc) electrical discharge of the kind used for neon sign tubes. They usually operate continuously (CW), but they can be pulsed or flashed by pulsing the input energy. CO_2 lasers may have a beam diameter from 1 to 100 mm, and produce power levels from less than 1 mW to over 1 kW.

Helium-neon (HeNe) gas lasers are made in small,

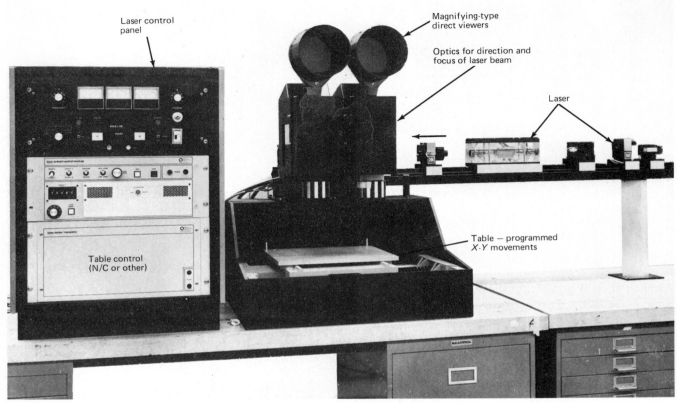

Fig. 29-17   A Q-switched Nd:YAG laser with dual outfit beams and 50:1 magnifying viewers and N/C tape control. Used for drilling, cutting, or scribing, (*Holobeam, Inc.*)

inexpensive sizes that are used in virtually every industry in the country.

They are safe, emitting only 0.001 W (1 mW) or so in a visible continuous orange beam of light (though 50-mW HeNe lasers are made, which are not safe). Their most important use is in alignment and measurement of machinery. Modern surveyor's transits now use HeNe lasers, and some new machine safety devices use them.

Uses of Lasers

CO_2 lasers are used extensively for cutting and welding almost any material from cloth to steel slabs. They have been used to cut 3-in. (75-mm) steel slabs, and a clothing manufacturer is using an numerically controlled CO_2 laser for cutting men's suits from cloth. It is faster and more accurate, and the edges of synthetic materials are sealed by the heat of the beam.

However, even though higher-powered lasers are being made and present tantalizing possibilities, the commercial uses of lasers for drilling, cutting, welding, and scribing are still usually associated with small work.

drilling

Drilling of almost any material is possible with lasers. However, the holes often have a 1° to 10° taper (larger at the top) and are not always smooth inside.

Advantages are that there is no drill breakage or dulling, no pressure on the part, and depth to diameter ratios of up to 40:1 can be drilled. Beam diameter and power can be varied to meet nearly any condition.

Swiss watchmakers drill 0.030-mm-diameter (0.0012-in.) holes in 0.25-mm-thick (0.010-in.) rubies using Nd:YAG lasers. It takes three to six pulses per hole. Also, lasers, using about 250 pulses per side, are used for drilling the diamond dies used for fine wire drawing.

A 0.05-mm (0.002-in.) hole is "drilled" through 0.05-mm-thick (0.002-in.) nickel foil (using an Nd:glass laser) for use in an electron gun.

welding

Welding requires close-fitting joints, and penetration of most commercial welding done is not over 1.0 mm (0.040 in.), though high-power CO_2 lasers can penetrate deeper. One automobile company is using up to 10-kW lasers to weld body seams.

Stainless steel 0.18 mm (0.007 in.) thick is welded, using overlapping spot welds, with a ruby laser. An Nd:YAG laser is used to weld two 0.05-mm (0.002-in.) wires to 0.38-mm-thick (0.015-in.) terminals, and 0.012-mm-thick (0.0047-in.) leads are welded to a thin layer of gold on a silicon chip. Figure 29-18 shows a pulsed YAG weld made on a small stainless-steel angle.

cutting

Cutting, especially of thin materials, can often be done economically with lasers. When metals or ceramics are being cut, the melted metal sometimes gets back and remelts on the work. Thus, a jet of oxygen, nitrogen, or argon is often used with the laser. CO_2 lasers are most frequently used for cutting.

Titanium is cut with oxygen assist at rates of 2.5 to 15.2 m/min (8 to 50 fpm). Plastics can be cut with lasers using nitrogen gas, which clears the area and cools the work. Numerical control is also used by an aerospace company to guide a laser for cutting any shape in 9.65-mm-thick (0.380-in.) titanium. A 50 percent saving in time is reported. An N/C-guided 6-kW CO_2 laser is being used to cut $\frac{3}{8}$-in. (9.6-mm) steel at 45 ipm.

other uses

Other uses of lasers include using N/C and Nd:YAG lasers to generate the complex patterns on film which are used to make integrated circuits (IC), to repair or cut patterns in thin gold film on electronic equipment, and to "trim" carbon resistors to accurate readings.

Safety with Lasers

Special care must be taken to protect the eyes, as laser beams are completely or partially reflected from many surfaces. Snug-fitting safety glasses with special lenses should be worn any time a laser of over 2 mW capacity is operating, though lasers below 10 mW may not be totally damaging.

Of course, one's body should never be in front of any high-power laser. It is easier to cut flesh than steel, and the laser beam does not care what is in front of it.

Many lasers use high-voltage, high-power circuits, so it is wise not to touch or operate this equipment unless you are trained to use it.

When a laser is being used, locked doors, warning signs, or barricades should prevent nonoperating

Fig. 29-18 A small, 1-in.-long (25-mm) stainless-steel angle welded in overlapping pulses by an Nd:YAG laser. (*Holobeam, Inc.*)

people from getting too close. In spite of some of the pictures, you cannot see most laser beams.

Review Questions and Problems

29-1. Describe EDM. What are the advantages and disadvantages?

29-2. When cutting a $\frac{1}{4}$-in.-deep (6.35-mm) cavity in steel by the EDM process using a copper electrode, how far must the spindle travel? When using a copper tungsten electrode?

29-3. What average current would be required to remove metal by EDM at the rate of 0.15 in.^{3}/min? Assume maximum rate.

29-4. When is reverse polarity used? What are the advantages and disadvantages of reverse polarity?

29-5. Why is proper flushing so important in EDM?

29-6. Compare electrochemical machining (ECM) with electrical discarge machining. What are the advantages and disadvantages of ECM?

29-7. Describe electrochemical deburring. What are the advantages?

29-8. For what type of work would you specify electrochemical grinding?

29-9. In what ways are the grinding wheels for ECG different from standard grinding wheels? Why?

29-10. Compare chemical milling and chemical machining. For what types and sizes of work would you specify each process?

29-11. From what materials are lasers (amplifiers) made?

29-12. For what types of work are lasers now used? Can you describe some possible future uses for lasers?

29-13.   Why is it necessary to have a cooling system with most lasers?

References

1. Charschan, S. S.: *Lasers in Industry*, Van Nostrand Reinhold, New York, 1972.
2. *ECM, ECD, ECG Simplified,* Chemform Division of KMS Industries, Inc., Pompano Beach, Fla., 1970.
3. Filshie, Ian S.: *EDM Electrodes—Design and Use*, Society of Manufacturing Engineers, Detroit, Mich., MR72-190, 1972.
4. Gent, William E.: "Tooling for ECD," *Manufacturing Engineering and Management*, May 1972.
5. Hach, Ralph J.: *Electrochemical Machining Electrolytes*, Society of Manufacturing Engineers, Detroit, Mich., MR69-136, 1969.
6. Hughes, F., and Notter: *Evaluation of the Electrolytic Grinding Process*, Diamond Div., DeBeers Industrial, London, England.
7. *Machining Data Handbook*, 2d ed., p. 691ff, Machinability Data Center, Metcut Associates, Inc., 1972.
8. "N/C—EDM Travelling Wire Processes," *N/C World*, August 1969.
9. Springborn, R. K. (ed.): *Non-Traditional Machining Processes*, Society of Manufacturing Engineers, Detroit, Mich., 1967.

plastics

30

There are today several hundred different plastics produced by many companies. Many of these are slight variations of basic plastics which have been "tailored" to do specific jobs. The changes may be made by adding reinforcing, such as glass fibers, or by chemical treatment or additives.

WHAT ARE PLASTICS?

Plastics are *polymers*. *Poly* means many, and *mer* means a unit or part. The mers are often single molecules, called *monomers*, such as the ethylene gas monomer (C_2H_4) shown in Fig. 30-1a.

These monomers, through the use of heat, pressure, and chemical processes and additives, become linked together into long chainlike formations like those shown in Fig. 30-1b and c. These chain, or giant, molecules have become heavy enough to become solid plastic. This description sounds simple, but actually very complex chemical plants are necessary to produce some of the plastics.

Thermoplastics are made from the chains of molecules described, sometimes with *branches* attached as shown in Fig. 30-2b. These plastics can be softened by heat. They can be reshaped while in the softened state, and they will reharden; thus the scrap can be used again. Most of the plastics in use today are thermoplastics. They may become too soft to use at temperatures from 150 to 600°F (66 to 315°C).

Thermosetting plastics are made from chains which have been linked together, referred to as *cross-linked*, shown in Fig. 30-2c. These have a three-dimensional network of molecules and will not soften when heated. They are practically insoluble, fireproof, and usually hard and brittle. These plastics cannot be reused.

Elastomers are made of mers which are less tightly bound together. An elastomer can be repeatedly bound together. An elastomer can be repeatedly stretched to at least twice its original length and must return with force to its original length when it is released. Examples of these are rubber, silicone, urethane, and chlorinated polyethylene. These are used for gaskets, molds, foam mattresses (polyurethane), and insulation.

In addition to the usual properties of tensile strength, coefficient of expansion, dielectric strength, etc., there are two properties which are especially important in plastics: heat-distortion temperature and recommended service temperature.

Heat-Distortion Temperature

Note: 100 psi = 689.5 kN/m² or 690 kPa.

This is the temperature at which an arbitrary amount of deflection in bending will take place with a load of 264 psi (1820 kPa). This value varies from 100 to 500°F (38 to 260°C). A similar test, with a load of only 66 psi (455 kPa), may also be used. A large difference between the two is a warning that stresses must be limited if the articles are subjected to heat. These tests are specified in ASTM D-648.

Recommended Service Temperature

This is a temperature, based on the general experience of manufacturers and users, at which a plastic

Fig. 30-1   A monomer and two of the polymers which are made from it.

Ethylene monomer
(a)

Polyethylene
(b)

Polypropylene
(c)

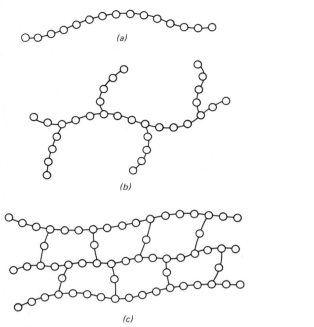

Fig. 30-2 (a) A linear molecule. (b) A branched molecule. (c) A cross-linked polymer as in thermosetting plastics.

may be used continuously under zero stress. It is not necessarily related to the heat-distortion temperature. This temperature may be from 120 to 600°F (49 to 315°C). Use of a plastic product above this temperature would certainly be unwise. The values given in this chapter should be used only as a guide as these temperatures are only approximate.

Fillers

Often other materials are added to the basic plastic. The best known are glass fibers. Other fillers are fibers of chopped cotton, asbestos, wood flour, clay, and sometimes metal powders. These fillers are used to increase the strength, working temperature, and hardness of the plastic.

Color can also be added to most plastics. This is usually added to the resin, but coatings of color or metallizing can sometimes be added after molding.

Entire books are written describing the 30 or so basic plastics, and the manufacturers' catalogs list many variations on these. This chapter will list a few of the most used plastics and their outstanding characteristics and uses.

THERMOSETTING PLASTICS

Epoxy Resins

The epoxy resins are cured or cross-linked by the addition of a hardener. They have excellent chemical

resistance and electrical insulating properties. Their working temperature is from 300°F (150°C) up to 500°F (260°C) with fillers and additives.

While epoxy adhesives are well known, these plastics also are used as castings for pipe fittings, electrical equipment, and other equipment. As laminates they are used to make printed-circuit boards, boat bodies, etc. Some trademarks are Devcon, Epocast, Epolite, Epoxylite, and Hysol.

Melamine-Formaldehyde and Urea-Formaldehyde

These two thermosetting plastics are called *amino resins*. The melamine can be molded into very hard, scratch-resistant dinnerware, business machine housings, electric switch cover plates, radio cabinets, etc. The heat-distortion temperature for filled aminos may be as high as 400°F (205°C), and their working temperature is about 325°F (160°C). The urea compounds are less water resistant but better electrical insulators than the melamine.

Both are used as adhesives in making plywood, and melamine is laminated with cloth to make table and counter tops.

Some trademarks are Beetle, Cymel, Permelite, and Plaskon.

Phenol Formaldehyde

This very hard plastic is the original Bakelite, one of a group of resins called the *phenolics*. These are hard, brittle, heat-resistant thermosets. They are inexpensive (about 22 cents per pound), excellent insulators, and have heat-distortion temperatures up to 350°F (180°C) and working temperatures up to 500°F (260°C). They are often used with a wide variety of fillers.

These may be molded or cast, and they are also used as coatings and in adhesive applications. Phenolics are used to make distributor caps, tool housings, coffee pot handles, gears, etc.

Some trademarks are Bakelite, Plyocite, Durez, and Resinox.

THERMOPLASTICS

This chapter will not attempt to list all the thermoplastics or their trademarks. The *1973 to 1974 Modern Plastics Encyclopedia* lists over 2700 trademarks for plastics and accessories such as presses, catalysts, and controls. The following are some of the plastics most frequently used.

ABS (Acrylonitrile-Butadiene-Styrene)

This combination of three monomers makes a plastic

which is fairly strong and can be made to have nearly the top impact strength of the plastics. Its heat-deflection temperature averages about 200°F (93°C), and it is opaque in natural ivory or darker colors.

ABS can be molded and extruded. It is often used for articles formed from large sheets, such as golf carts and lawnmower housings. It remains tough at −40°F (−40°C), is abrasion and scratch resistant, but is flammable.

Items such as gears, tool handles, power tool housings, and even canoes and small boats are made from ABS (Fig. 30-3). Some trademarks are Abson, Cycolac, Dylel, Kralastic, and Royalite. Cost averages about 45 cents per pound ($1.70 per cubic inch).

Acrylic

The full name is polymethyl methacrylate, and it is better known as Acrylite, Plexiglas, or Lucite. This beautifully clear, easily shaped plastic is most widely used in sheet form for signs of all kinds. Rods and tubes are cast in glass or metal cylinders or extruded. Acrylics can also be injection molded.

Its heat-distortion temperature is 190°F (88°C), so it is easily softened in boiling water if even an amateur wishes to form it. Acrylic has a "memory," so that if reheated (within a few weeks), it will return to its original shape. This makes errors easy to correct.

Acrylics will "pipe" light and are used in fiber optics and edge-lighted dials. This plastic is dimensionally stable and can withstand outdoor conditions better than any other material. It is tough but easily scratched.

Cellulosics

The original cellulose nitrate was called *celluloid*. It is transparent and tough but highly flammable. It is still made today and used for table tennis balls and "pearl" toilet seat covers, and can be cast into various handles and frames. It is not injection molded.

Cellulose acetate is also clear, tough, and scratch resistant, and not as flammable. It can be molded, extruded, and made in films. The heat-distortion temperature is only 100°F (38°C) or slightly higher, and tensile strength is among the lowest, about 2000 to 4000 psi (13.8 to 27.6 MPa). Cost is about 50 cents per pound ($2 per cubic inch).

Cellulose acetate butyrate (*CAB*) is equally clear and tough. It has a distortion temperature which may be as high as 150°F (66°C), greater weather resistance, and tensile strengths up to 6000 psi (41.4 MPa).

Trademarks are Ampol CAB, Tenite CAB, and Uvex. CAB is widely used for ballpoint pens, typewriter keys, steering wheels, toys, etc. It can be injection molded, extruded, and made into sheet and

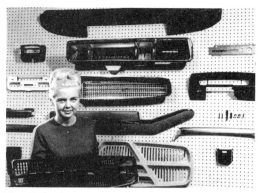

Fig. 30-3   Some automobile parts made from ABS. (*The Society of the Plastics Industry, Inc.*)

formed. Cost averages 65 cents per pound ($2.80 per cubic inch).

Cellulose acetate propionate is very similar to CAB, but is rated as slow burning. Trademarks are Forticel and Tenite C/P.

Fluorocarbon Plastics (Teflon)

Teflon is the most famous of this group. Teflon's complete name is polytetrafluoroethylene, sometimes abbreviated as *TFE* or *PTFE*. Other trademarks are Halon, Tetran, and Fluorosint. The outstanding characteristics of TFE are its extreme slipperiness, its ability to be used at temperatures from −450 to +500°F (−268 to +260°C), and its zero water absorption. It is the heaviest of the plastics, with a specific gravity of 2.15.

TFE cannot be processed by regular molding methods. Some forms can be made by a method similar to powdered metals. The Teflon powder is pressed, and then sintered at 620°F (330°C) to make gaskets, O rings, bearings, etc.

Rods and tubes are made by molding (Fig. 30-4). TFE is expensive, costing over $5 per pound.

Other fluorocarbon plastics similar to TFE can be extruded and injection molded. One, polychlorotrifluoroethylene (CTFE), can be expanded when made as tubing and then shrunk by means of heat to grip a wire tightly. However, it costs over $6 per pound.

Two other compounds, FEP (fluorinated ethylene propylene) and PVF (polyvinal fluoride), are similar, though somewhat less expensive, and can be cast or extruded.

Nylon (Polyamide)

Nylon, originally patented in 1931 by E. I. du Pont de Nemours Company, is now a generic (general, not limited to one company) name. Polyamide resins

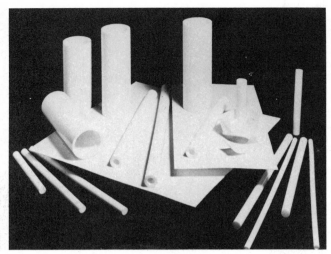

Fig. 30-4 TFE (Teflon) stock shapes: these can be machined into a wide variety of products. (*Polymer Corp., a division of ACF Industries, Inc.*)

are also sold as Nylatron, Nylafil, Plaskon, and Zytel.

There are several standard grades of nylon, and one company makes 21 variations for increased transparency, water resistance, wear resistance, and stability, and even uses a molybdenum disulfide filler for decreased friction.

Nylon costs from $.70 to $1.60 per pound depending on the type. It retains its properties from -40 to $+350°F$ (-40 to $+180°C$) and has a tensile strength up to 14,000 psi (96.5 MPa).

The nylons (polymides) are self-extinguishing, resist many chemicals, and are tough and wear resistant. They are not recommended for exposure to ultraviolet light, hot water, or alcohols. Nylon absorbs water up to 2 percent at high humidity, with some expansion resulting.

Nylon can be molded, extruded, formed into sheet and film, and, in recent years, has been very successfully cast (Fig. 30-5). It is used for bearings, gears, drawer slides, machine slides, and rollers besides the many filament uses in clothing, fishing lines, and rope.

Polycarbonates

This outstanding group of plastics came into use shortly after 1960. It is a medium-priced ($.75 to $1.30 per pound) competitor to even some metal products, as it is easy to handle for molding, extrusion, and machining. It can even be nailed and riveted without cracking.

Trademarks are Lexan, Merlon, and Polycarbafil. These plastics, due to their toughness, make excellent safety glass for street lamps, windows, and machine guards. They keep their strength and toughness at temperatures from -215 to $+250°F$ (-138 to $+120°C$). In thicknesses over 0.050 in. (1.25 mm), the polycarbonates do not support combustion.

These plastics are used in food processing because they resist all food stains. However, they turn yellow when exposed to ultraviolet light, and they are attacked by some detergents, ammonia compounds, and high-octane gasoline.

Polycarbonates are crystal clear or can be colored to any hue. They are used for housings for shavers, power tools, household equipment, and blow-molded bottles.

Acetal (Polyacetal)

Less expensive plastics (about 65 cents per pound) which compete with nylon are the formaldehyde-based acetals. These are known as Delrin, Celcon, and Formaldefil. They are easily molded at fairly low

Fig. 30-5 (*a*) Extruded Nylatron G.S. (nylon with molybdenum disulfide) shapes. (*b*) Injection-molded guide shoe. (*c*) Injection-molded nylon gears, etc. (*Polymer Corp., a division of ACF Industries Inc.*)

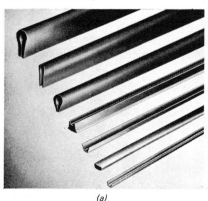
(a)

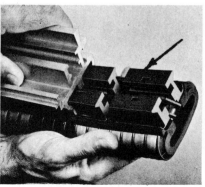
(b)

(c)

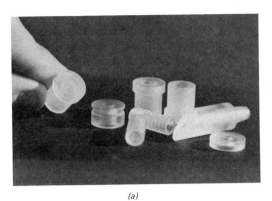 

Fig. 30-6   (a) Cross-linked polystyrene, machined into insulators and coil forms. (b) Many engineering plastics can be extruded to form seals, wear surfaces, edge protection, etc. (*Polymer Corp., a division of ACF Industries, Inc.*)

temperatures, retain their strength even at 240°F (115°C), and have very low moisture absorption.

The acetals are slow burning and lose strength and appearance if exposed for long periods to ultraviolet light. They can be machined and are tough enough to stand riveting. They have a low coefficient of friction, high fatigue endurance, and tensile strengths up to 10,000 psi (69 MPa).

Uses are very wide, including shower heads, stereo tape cartridges, toys, furniture casters, and cigarette lighter cases.

Polyethylene

Polyethylene leads all plastics in the volume of resin used each year. It is not a high-strength material and does not withstand ultraviolet light or temperatures over 200°F (93°C). However, it costs as little as 15 cents per pound, can be made in several densities, and can easily be molded, formed, and colored.

Several billion pounds per year are used in very thin sheet form for garment bags, trash can liners, wrapping for clothing, etc. It is injection molded into buckets, pipe fittings, flashlights, and trash containers.

Polyethlene's milky white color and waxy feeling is especially noticeable in the hundreds of blow-molded bottles, both the squeeze type and the more rigid kind. Thin films are colorless.

Although its shrinkage is high, it resists most chemicals except oil and gasoline. The trademarks are not well known to the public, but some are Alaton, Ameripol, Fortiflex, and Marlex.

Polypropylene

Polypropylene is in the same olefin group (extracted from petroleum or natural gas) as polyethylene. It costs a few cents more, but is very lightweight, can be

used up to about 250°F (121°C), and has excellent flex life. This latter property has led to its use as a molded-in hinge on several products.

Polypropylene is an excellent insulator and is stiffer than polyethylene. It can be molded or extruded into sheet, film, or pipe. It is used for automobile accelerator pedals, luggage, and hospital equipment. Some trademarks are Avisun, Marlex, and Pro-Fax.

Polystyrene

Polystyrene is number two in the total volume of resins used. It is a crystal clear, odorless, tasteless plastic with a high gloss. It costs as little as 15 cents per pound and is made in many grades. Most grades are quite brittle and (unless modified) will hold static electricity.

Polystyrene has a medium tensile strength, but it can be used only to 150 to 210°F (66 to 99°C). It is easily produced in any form, by most processes, and is easily joined by cementing.

Typical uses are for bottles, low-cost picnic utensils, model kits, signs, toys, etc., in a wide choice of colors (Fig. 30-6).

Expanded foam in slabs or beads is a leading use of polystyrene. Hot-drink disposable cups made of the beads are used by the millions. Slabs of the foam are used for insulation, children's floating toys, and packaging of fragile articles.

A few of the trademarks for polystyrene are Dylene, Kralon, Lustrex, and Styron.

Vinyl Plastics

This is the general name of a large group of plastics usually called *polyvinyl* _____. They are compounded to give almost any degree of flexibility, by adding fillers, plasticizers, and stabilizers.

Polyvinyl chloride (PVC) is the most frequently used of this group. This is a very clear, transparent plastic, easily colored, resistant to most chemicals, and very water-repellant. Cost is about 55 cents per pound.

PVC is self-extinguishing but, except in certain formulations, should not be used in temperatures over 160°F (71°C). It can stand outdoor exposure and is quite abrasion resistant, but it has a low tensile strength.

PVC can be extruded as wire insulation, chemical tubing, and refrigerator door gaskets. Items such as handlebar grips and pipe fittings can be injection molded. Blow-molded containers are tough and resist cracking. Gutters and siding for houses can also be formed.

Coating of fabrics of all kinds for industrial uses (such as drapery, tents, and tarpaulin-type covers) is a major use of PVC.

Some of the trademarks of the vinyls are Exon, Geon, Pliovic, Saran (slightly different plastic), Karoseal, and Naugahyde.

HOW TO SELECT A PLASTIC

Selecting the right plastic for a particular application certainly cannot be done by using the brief information provided here. One of the best solution methods is to call in a sales engineer from a company manufacturing resins or plastic parts. Another way is to use a reference book such as the latest edition of *Modern Plastics Encyclopedia*, which has over 150 pages of detailed specifications on the properties of all varieties of plastics.

Here is a checklist of what to look for, in the approximate order of importance.

1. At what temperature will it be used?
2. Must it resist gasoline, oils, or other chemicals?
3. What physical properties are needed: hardness, antifriction, toughness, elasticity?
4. Any requirements for electrical properties?
5. Can the plastic be molded, extruded, cast, or made in sheet or rod stock? Which is needed?
6. Can a less expensive plastic with similar properties be used?
7. Is the quantity you need large enough for economical use of high production equipment or can less expensive methods be used?
8. Can the product be made more easily in metal, that is, machined, die cast, press formed, etc.?

At each step in the checklist, some plastics will be eliminated, and finally, one or a few will be chosen for final consideration.

PROCESSING PLASTICS

Much of the machinery used in processing plastics is similar to that used for processing metals. However, certain special qualities of plastics must be considered when handling them.

Moisture, even in small amounts, if absorbed by the resin while in storage, can cause porosity, blisters, or surface defects. Thus, all resins must be kept dry.

Softening and melting temperatures are, in some plastics, very close together. Thus, accurate control of temperature is required during all processing.

Today many control systems have solid-state transistorized circuits. This eliminates several relays, and increases the speed and accuracy of response.

Plastics are relatively *low in strength*, so knockout pins and handling equipment must be arranged to avoid damage to plastic parts, especially while they are still hot.

The *bulk factor* (compression ratio), which was important in powdered metals, is also important with plastics, especially thermosets. The bulk factor in thermosets varies from about 3:1 up to 10:1 for glass-filled Melamine.

Shrinkage of plastics varies over a very wide range. In general, shrinkage from molding temperature to room temperature is about 0.003 to 0.009 in./in. or mm/mm, though some polystyrenes have only half as much shrinkage. Teflon and polyethylene may shrink 0.015 to 0.060 in./in. Molds and dies must be sized accordingly. This shrinkage helps to free parts from molds, but can cause trouble by binding on cores.

Compression Molding

Note:

    1 ton (2000 lb) = 0.907 t (metric tons)
    1 ton (2000 lb) = 8.896 kN
    1000 psi = 6.89 MN/m² = 6.89 MPa

Compression molding is used most often for the thermosetting plastics; other methods are faster for the thermoplastic resins. The basic process is shown in Fig. 30-7. The molding process is basically simple. A measured amount of powder, or a compressed preform, is placed in the lower, female cavity. This cavity is continuously heated by steam or electricity. The mold is immediately raised and contacts the male die (called the *force, plug, or core*), which also is heated. The combined pressure and temperature cause the plastic to flow into the mold. The press stays closed while the heat cures (hardens or sets) the plastic part.

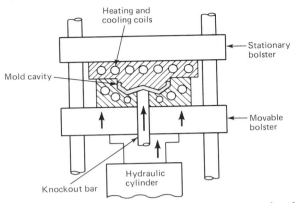

Fig. 30-7 Major parts of compression-molding press for plastics.

The complete cycle may take from 10 s for small parts under 0.100 in. (2.5 mm) thick to 5 or 10 min for large, thicker parts.

Molding pressures may be as low as 500 psi (3.4 MPa) for polyester and epoxy, but most thermosets require 2000 to 6000 psi (13.8 to 41.4 MPa), depending largely on the filler or plasticizer used. Molding temperatures are from 2000 to 4000°F (1093 to 2204°C), depending on the plastic, filler, etc.

A molding press is rated according to the tons of clamping pressure it can exert. Presses are made from 25 to 300 tons (222 kN to 2.67 MN) capacity in up-stroke presses, and up to 1500 tons (13.3 MN) and over in downstroke presses for large molds and laminated products. These presses can deep draw a complete television cabinet.

Transfer Molding

When intricate or quite accurate parts are to be molded from thermosetting plastics, compression molding is often not satisfactory. A modification called *transfer molding* gives the added quality and at a faster rate.

As shown in Fig. 30-8, transfer molding uses closed dies very similar to those used in die casting. The entire die set has three pieces, the male and female die halves, and the transfer pot into which the transfer ram plunges to fill the mold.

The material, often a preheated preform, is placed in the heated pot. As soon as the material is sufficiently softened, the plunger forces the almost fluid plastic into the closed mold. Pressures used are 50 to 100 percent higher than those used for compression molding; thus better details and higher strengths are possible.

Molds are vented by small passages along the die, as in die casting. A method using a spiral screw to pass the resin through a heated tube and then into the mold (called *screw transfer molding*) is especially easy to automate. Figure 30-9 shows a 300-ton (2670-kN) transfer-molding press without the dies.

Casting

Casting, that is, pouring a liquid into a mold, can be done with several plastics. Most of the thermosets (such as epoxy, phenolics, and polyesters) can be cast. The thermoplastics, acrylic, polyethylene, PVC, silicone, and urethane can also be cast.

Open molds into which the liquid resin, mixed with a catalyst or hardener, is poured can be used for most plastics.

Centrifugal casting, using the same methods discussed in Chap. 23, can be used with several plastics, such as the polyesters.

Rods and *tubes* of acrylic are cast in metal or glass tubes, then cured in a temperature-controlled liquid

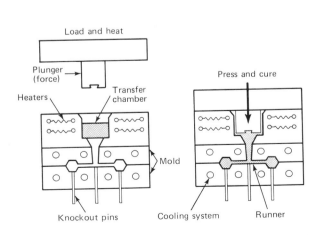
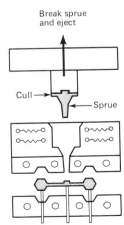

Fig. 30-8 Simplified diagram of the action of a transfer-molding press cycle for plastics.

Fig. 30-9   A 300-ton transfer press. Either steam or electric heating and three-speed controlled closing. The same press can be used for compression molding. (*Stokes Div., Pennewalt Corp.*)

bath. Sheets of acrylic may be poured between two parallel plates of glass, but a newer process continuously polymerizes the plastic between two highly polished continuous steel belts. Acrylic is exothermic (gives off heat while it cures); thus cooling can be a problem.

Nylon is cast into parts weighing from less than 1 up to 1500 lb [0.45 to 680 kg (mass)]. Nylon can be gravity cast, the same as metals, or made with automated pressure castings.

The above give an idea of the principal casting methods. Other methods, and many variations of those described, are being used today.

Injection Molding

Injection molding is the most widely used method of producing thermoplastic parts, from thimbles to entire front panels for home entertainment centers. Basically, injection molding is very similar to zinc-alloy die casting. The principal difference is that control of the temperature of the plastic all along its travel is quite critical. Several zones along the tube, *plus* nozzle and die temperatures, are closely controlled.

Figure 30-10a shows the basic process, starting usually with powder of $\frac{1}{8}$-in. (3.2-mm) cube-shaped pellets of resin poured into the machine's hopper. The resin falls into and is pushed along the heated tube by a 2- to 6-in.-diameter (50- to 150-mm) screw feeder until a sufficient volume of melted plastic is available at the injection nozzle end. This may take from 10 s to 5 or 6 min per shot. The entire screw is then plunged forward to force the plastic into the mold. Each *shot* may produce one or several parts, depending on the die used.

The ram is held under pressure for a few seconds so that the molded part can solidify (cool). It then retracts slightly, and the mold opens. Knockout pins eject the molded piece. The sprue and runners are then trimmed off, usually in a separate trimming press.

Pressures used will vary from 8000 to 25,000 psi (55 to 170 MPa) on the projected die area, depending on the plastic used and the thickness of the part.

Injection-molding machines are rated by the number of ounces of polystyrene that can be displaced by one forward stroke of the plunger. This is called the *shot size.* Capacities of 5 to 30 oz [140 to 850 g (mass)] are most frequently used, though machines are made with shot sizes from $\frac{1}{4}$ to 300 oz [7 g to 8.5 kg (mass)] or more. The necessary clamping pressure needed to hold the dies closed increases with the shot size. Injection-molding machines have from 10 to 3000 tons (90 kN to 27 MN) clamping pressure. Injection rates from 1 to 120 in.³/min (16 to 1970 cm³/min) are available.

A small, 1.2-oz (34-g) injection-molding machine may cost $9000; a 15-oz (425-g) machine costs about $35,000. Larger machines will cost over $100,000. Dies will cost from $800 to $10,000. Shrinkage allowance and some draft must be built into the dies.

Early injection-molding machines used a plunger which forced the pellets around a "torpedo" to heat them uniformly. This type is largely being replaced by the reciprocating screw machine described above and illustrated in Fig. 30-10b.

Automation of injection-molding machines is fairly common for production runs of 50,000 parts or over. Automatic unloading equipment is fairly easily designed for most parts.

Control of temperature in several zones and timing

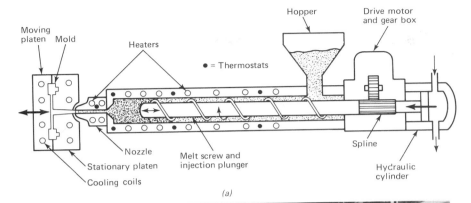

(a)

(b)

Fig. 30-10 (*a*) Simplified schematic of a single-stage, in-line reciprocating-screw injection-molding machine. A forward movement of the screw will inject the heated plastic into the mold. (*b*) A 75-ton, 4-oz. (113-g) shot injection-molding machine for thermoplastic or thermoset plastics. (*Gloucester Engineering Co.*)

of each part of the molding cycle are now being done by solid-state controls. These are more reliable and have longer life than the standard relay-type controls. *Cycle time* can be controlled by punched cards, punched tape, or computer. Thus, changing from one job to another can be done quite easily by simply "plugging in" the previously specified time and temperature information for the particular job.

Rotary-turntable injection-molding machines, both horizontal and vertical types, with up to 10 stations are used when high production justifies the higher cost.

Extrusion of Plastics

Extrusion means the continuous flow of material through a die. The extrusion machine in Fig. 30-11*a* can be used to make solid rods, pipe or tubing, and shapes such as those shown in Fig. 30-12. By using a wide, flat extrusion head, the same machine can extrude sheet plastic. Thin plastic film is similarly extruded through a circular die, but the process is different from then on, as will be explained later.

The pellets of thermoplastic material are made from powdered resins by extruding many strands through a special die head and chopping these strands into short pieces with rotating knives.

Plastic extrusion machines (Fig. 30-11) consist of a hopper into which the powder or pellets are fed, the screw feeding mechanisms, a nozzle, and the die assembly. The description is simple, but the machine is fairly complex. Up to 10 heating and cooling zones are used, which require heaters, oil and water circulation, and the equipment to control these.

Many plastic extrusion machines use a single-screw feeder. Today, however, more twin-screw machines are being made, as they have higher capacities and produce more uniform products with less down time. The twin screw (Fig. 30-11*b*) is designed with several zones of action for greatest efficiency.

These machines are rated according to the number

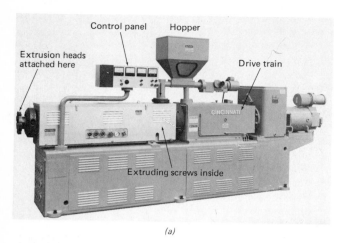

(a)

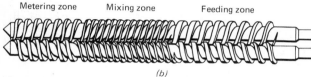

(b)

Fig. 30-11 *(a)* A twin-screw, 1000 lb/h (454 kg/h) plastic extruder. Can be equipped to make pipe, pellets, film, or shaped extrusions. *(b)* The twin intermeshing tapered screws used in some extruders. (*Plastic Machinery Division, Cincinnati Milacron Inc.*)

of pounds per hour (lb/h) or kilograms per hour mass [kg/h (mass)] of PVC pipe, rod sheet, etc., which they can produce. Capacities range from 100 to 1000 lb/h [45 to 450 kg/h (mass)]. Pipe can be made up to 22 in. (560 mm) OD, though diameters up to 8 in. (200 mm) are more often produced. Some of the shapes produced are shown in Fig. 30-12.

The total process includes running the pipe, tub-

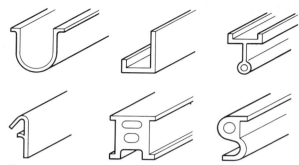

Fig. 30-12   A few of the plastic shapes made by extrusion.

ing, form, or sheet through a cooling cycle after it is extruded. This may be done by passing the product through 10 to 60 ft (3 to 18 m) of water in a trough or, for smaller work, by air cooling.

The product is pulled through the cooling cycle by a *haul-off*, which may be a flat or formed caterpillar type, or by wheels which lightly grip the cooled plastic. From the haul-off, the product goes to a *traveling saw* which cuts it to length, and then to storage, often by means of an automatic *tipping trough*.

thin-film extrusion

The making of the thousands of kilometres of thin film, often only 0.001 in. (0.025 mm) thick, starts with the extrusion machine just described. This is combined with other equipment, as shown in Fig. 30-13, to make a complete system. The die used is a circular ring with an air jet at the center and air-cooling jets around it as shown in Fig. 30-14*a*. Thicker film may be extruded through a flat die, as shown in Fig. 30-14*b*.

The "balloon" of plastic is pulled upward (sometimes downward) until thoroughly cool, and it is then rolled flat and cut into strips or bags or stocked on large rolls. The width and thickness of the film is

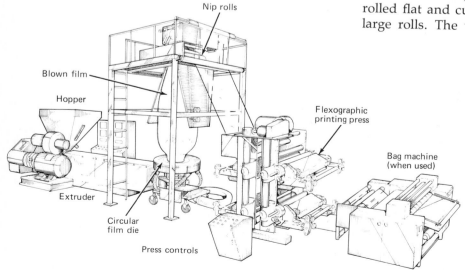

Fig. 30-13 A complete plastic film-making system. A sheeting winder may be used where printing and cutting is not desired. (*Gloucester Engineering Co.*)

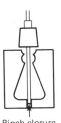
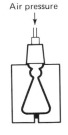

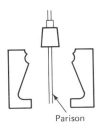

1. Parison in place

2. Mold closes

3. Parison expanded into mold

4. Mold opens, product released

5. Top and bottom "pinches" are trimmed off

(a)

(b)

Fig. 30-15   (*a*) The major steps in blow molding a bottle. (*b*) A complete thin-wall ceramic-cast beryllium copper blow mold. (*Unicast Development Corporation.*)

(a)

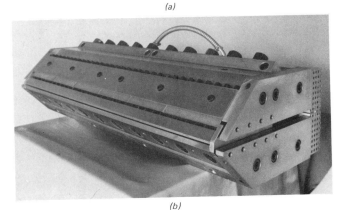

(b)

Fig. 30-14   (*a*) Center-fed blown film die, complete assembly. Sizes 4- to 36-in. (100 to 900-mm) diameter standard. (*Gloucester Engineering Co.*) (*b*) Sheet die, 24- to 96-in. (600- to 2400-mm) widths available. (*Extrusion Dies, Incorporated.*)

controlled by the circumference, OD and ID of the die. Films over 100 in. (2540 mm) wide can be made.

The blown film may be slit on the edges, thus making two sheets, or it may be shipped in the roll and later cut and sealed to make plastic bags, or it may be used in automatic machines for wrapping cigarettes, games, etc.

Blow Molding

Thousands of bottles of all sizes and shapes, toys, doll bodies, doll's legs and arms, tanks, and similar articles are made by blowing up a chunk of plastic until it fills a mold. The basic process is illustrated in Fig. 30-15*a*, though many refinements have been developed.

Today automatic multistation machines (Fig. 30-16) can turn out up to 1800 small bottles (or similar parts) per hour.

The heated *parison* (piece of plastic) may be solid or tubular. It is continuously extruded, cut to the proper length, gripped in the two-piece mold, and then air is injected to force the plastic against the walls of the water-cooled mold. Air pressure from 10 to 150 psi (70 kPa to 1.0 MPa) is used, so molds need only mod-

Fig. 30-16   A complete blow-molding machine which extrudes, cuts the parison, blows to shape, and loads to a conveyor. (*Plasti-Mac, Inc.*)

erate clamping pressures. The part cools, and then the mold is opened and the part is removed.

Larger items [such as 5-gallon (19-litre) containers, riding toys, waste baskets, etc.] are made in the same way except that the parisons are made intermittently. Shot sizes up to 100 lb [45 kg (mass)] have been used for large plastic items such as gas tanks and boat parts.

A complete blow-molding line will include an extruder and blow molder (often with two to eight molds), water-cooling system, air compressor, and finished part conveyors. Such a system may cost from $60,000 to over $300,000.

Thermoforming

Thermoforming, sometimes called *vacuum forming*, is a process used to make a wide variety of products from thin sheet plastic. These products include small jelly containers used in restaurants, luggage, refrigerator inner panels, and containers for packaging.

The tooling and equipment are relatively inexpensive. High production is possible with multiple molds, and several different plastics, such as polyethylene, polystyrene, ABS, and vinyl, may be used. However, scrap losses are high (though the scrap can be recycled). The maximum thickness is usually 0.125 in. (3.2 mm), and both sides of the part must be the same shape.

The basic process is quite simple. Figure 30-17 illustrates the steps. A piece of sheet plastic 0.005 to 0.125 in. (0.125 to 3.2 mm) thick is clamped into a frame. The plastic is heated, usually with electric heaters, until it begins to sag. Vacuum is then applied through small holes in the mold, and the plastic is rapidly pulled tightly against the mold. The frame is raised, the part is removed and then trimmed in a punch press. The mold may have several shapes, for the same part or for different parts.

The plastic used may be small [18-in.-square (460-mm)] sheets, or it may be continuously fed from rolls of the desired width.

There are several variations on the basic process, some of which use both an upper and lower forming die. The heating cycle may take over 30 s, but the vacuum cycle takes only 5 to 10 s. Therefore, rotary machines are made in which heating is done at two to five stations, a few seconds at each station, making the cycle time much shorter.

Foamed Plastics

Foamed plastics are today being used for coffee cups, insulation, flotation equipment, foundry patterns, packaging, and many other products. Over 100,000 tons of expandable *polystyrene* (EPS) alone are used.

There are several patented processes used in making foamed polystyrene products with densities from 2 to 30 lb/ft³ (32 to 480 kg/m³).

The second most used foam is *polyurethane*, usually referred to as *urethane*. This is used to make mattresses, automobile dashboards and seat cushions, and insulation for tank cars and refrigerated trucks. Some of the other plastics which are used in foam form are ABS, polypropylene, phenolic, silicone, and epoxy.

Extruded foam boards, sheets, and rods are made the same as other extrusions except that a chemical or gas *blowing agent* is added to the hot mix before it leaves the die. Thin, flexible sheets can be thermoformed into many shapes such as egg and meat packages. Rigid boards may be used "as is" for insulation or cut and cemented to form any desired shape.

Molded foam products are made by injection molding, usually at low pressures. While special machines have been designed, much foam molding is done with slightly modified, conventional reciprocating-screw machines. These use a resin blended or compounded with a blowing agent to partially fill the mold, which is then completely filled by the expansion of the gas in the plastic.

Structural Foam

If you want to make a one-piece typewriter cover, a carrying tray to hold 600 lb [270 kg (mass)], ducting, furniture parts, and many other items, the use of structural or engineering foam plastics may be the best method. The resins and processes developed since 1970 make it possible to make rigid, lightweight,

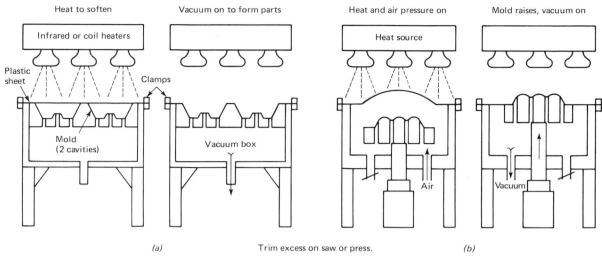

Fig. 30-17   Two of the several methods of thermoforming (vacuum forming) thin plastic parts.
(*a*) Simplest method of thermoforming. (*b*) Snap-back forming, for more consistent film thickness. An inverted-top plunger may also be used.

decibel-reducing, sturdy parts in one stroke of an injection-molding machine. Parts from 3 to 100 lb (1.4 to 45 kg) can now be made on special, or modified, injection-molding presses. Figure 30-18 shows some of the parts which have been made.

Structural or engineering foam plastics are specially formulated polycarbonate, polyester, and some other plastics. These have tensile strengths from 3500 to 12,000 psi (24 to 165 MPa), heat-distortion temperatures approaching 400°F (204°C) and are self-extinguishing. Their cost, in large quantities, is from $.70 to $1.00 per pound, higher if glass or mineral filling or "loading" is used.

The process is the same as injection molding except that low pressures (200 to 300 psi, 1.4 to 2.1 MPa) are used, and either a foaming agent is mixed with the plastic or nitrogen gas under pressure is injected with the plastic.

The pores in the plastic foam, when it presses against the walls of the mold, collapse and form a tough, smooth integral skin over the cellular core. Part thickness is usually between 0.100 and 0.375 in. (2.5 to 9.5 mm).

The structural foam part can be made from colored plastic, can be painted, or can be given a metallizing coating. Printing can also be done on it. Self-tapping screws, molded inserts, and ultrasonically inserted inserts may be used for fastening. Ribs and mounting posts can be included in the mold as integral parts of the final product.

Molding machines may be compact with simple controls, or they may be large machines with quite intricate controls for high-speed production or for making large slabs or formed parts.

TOLERANCES FOR MOLDED PLASTIC PRODUCTS

So many factors enter into tolerances of molded plastics that a designer should check with the supplier before specifying parts requiring special tolerances.

Shrinkage may vary due to variations in the cooling of different parts of a mold. Injection temperature will, of course, affect the shrinkage of a part. Small differences in the resin from one batch to the next, mold wear, humidity, and injection pressures all can affect the final size of a molded plastic part.

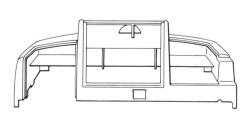

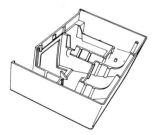

Fig. 30-18   Examples of parts made from engineering foam. (*General Electric Company, Plastics Business Division.*)

With ideal conditions, tolerances as low as 0.0005 in./in. (mm/mm) on diameters can be held on some plastics. However, 0.002 in./in. may be the minimum on plastics such as polyethylene and urea formaldehyde. Table 30-1 shows some practical tolerances for average conditions, according to a survey made by the Society of the Plastics Industry in 1966. A *fine tolerance* is about two-thirds to one-half that shown in the table.

FASTENING PLASTICS

Plastics may be fastened to plastics, metal, or wood by rivets, bolts, or screws. Plastics are often the base for adhesives and can be secured to themselves and to metal by "gluing" or cementing.

One frequently used method of fastening plastic parts together is by the use of localized heat. This can be done by hand with a torch or by more automated methods. One method often used is ultrasonic welding.

Fig. 30-19 Two styles of ultrasonic welding machines used for fastening plastics together. (*Branson Sonic Power Co.*)

TABLE 30-1   AVERAGE TOLERANCES FOR SOME MOLDED PLASTICS

Plastic	±Tolerance*	
	On a 3-in. Dimension†	On a 75-mm Dimension†
Alkyds (filled)	0.003	0.08
Melamine formaldehyde	0.0085	0.22
Nylon	0.0085	0.22
Phenolic (filled)	0.006	0.15
Phenolic (unfilled)	0.009	0.23
Polyethylene	0.0095	0.24
Polystyrene	0.005	0.13
Vinyl	0.005	0.13

*These will vary according to the specific composition of the plastic resin. Closer tolerances are often achieved.
†Additional inches or millimetres will be at one-third of the above, or less.

Ultrasonic welding uses very high frequencies, basically 20,000 Hz at very low amplitudes (length of stroke). These vibrations cause one plastic surface to move against the other, creating enough frictional heat to melt and weld the joining surface almost instantly without the use of solvents or adhesives.

The equipment, as shown in Fig. 30-19a and b, is quite simple. Pressures of 10 to 200 lb (44 to 890 N) are exerted, and a variety of "horns" (the part that contacts the work) can be used. Practically any plastic can be ultrasonically welded, seamed, or "sewed" to itself. Some plastics can be welded to other groups of plastics, but this varies considerably.

Modifications to the ultrasonic welding equipment enable it to be used for gang welding, staking, and assembling of sheet, cast, or molded plastic parts at rates to and over 300 ppm. Basic equipment costs are from $5000 and up.

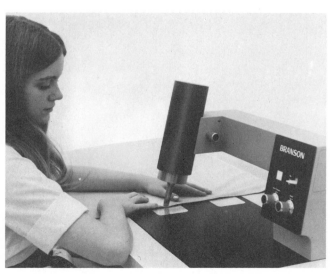

MACHINING PLASTICS

Many plastic parts are made which require little or no machining. However, prototype parts are sometimes more economically machined, and sometimes close tolerances are required which can only be maintained by machining. Cutting off, drilling, and tapping are the most frequently used processes.

General precautions include keeping the work cool to avoid sticking or excessive expansion or deflection. Many plastics are resilient (bounce back), so cuts must be made with very sharp tools. Sharp tools also avoid too much heat.

Cooling may be by air jet, vapor mist, or water and soluble oil, 5:1. Cutting tools should be high-speed steel, but for some thermosets, for filled plastics, and for long runs, carbide-tipped tools are preferred. No listing can give specific instructions for all the numerous plastic combinations. The following, gathered from several sources, will give a general idea of where to start.

Band Sawing

Much band sawing is done at 1000 to 3000 fpm (300 to 900 m/min), though polyester and polystyrene can be cut at 4000 fpm (1200 m/min). Use zero or positive rake blades with buttress or skip tooth form. Epoxy, formaldehydes, polyesters, and polystyrene are best cut with fairly fine spacing of 10 to 18 teeth per inch. A coarser 4 to 10 teeth per inch blade is best for other plastics, though the rule of finer teeth for thinner parts should be followed. Feeds should be as fast as possible without gumming up the blade.

Circular Sawing

Hollow ground circular saws, running at 3000 to 5000 fpm (900 to 1500 m/min), can be used for most plastics. Acetal, polycarbonate, polyethylene, polypropylene, and Teflon may be run at speeds up to 8000 fpm (2400 m/min). Carbide-tipped blades are best for epoxy, formaldehydes, and polyester. Feed steadily and rapidly, with the blade protruding well above the work. Blades for acrylic should have 10 to 15 teeth per inch, but 4 to 8 teeth per inch is satisfactory for most other plastics.

CAUTION: Brittle plastics (such as acrylic, polystyrene, and the brittle thermosets) will crack or chip quite easily when being sawed or drilled. Some experimenting may have to be done to obtain good results.

Drilling

Drills with polished, open flutes are best. The standard 118° point drill is used for acetal, ABS, epoxy, and PVC. A 140° point angle is suggested for drilling polycarbonates. A 90° point angle may work better on other plastics.

Feeds vary according to drill size but should be fairly heavy so that the heat is removed with the chips. An exception is polycarbonate, which should be drilled at low feeds.

Slow-spiral (low helix angle) drills are best for acrylic and polystyrene. These crack easily at breakthrough, so a backup board should be used.

Due to the resilience of plastics, it is often necessary to use 0.002- to 0.005-in. (0.05- to 0.125-mm) oversize drills if the drilled diameter is to be maintained.

Reaming

It is seldom necessary to ream plastics. When it is required, spiral-fluted reamers may work best. At least 0.005- to 0.010-in. (0.125- to 0.250-mm) stock should be left for the reamer to cut. If accurate inside diameters are required, it is often best to bore with a single-point tool.

Tapping

Only the coarse series of threads should be used in plastics, and oversize taps are needed for most materials. These are standard production with most manufacturers. Cutting speeds can be about the same as used with brass, except for the harder thermosets and glass- or asbestos-filled plastics. A 5° negative rake angle is suggested for polycarbonate.

Turning

Most plastics can be turned at 300 to 600 fpm (90 to 180 m/min). Sharp tool bits with 10° to 15° clearance angles and 10° to 20° positive rake angles work with most plastics. Negative rake angles (5° to 10°) are recommended when turning acrylics and polycarbonate. Fairly heavy feeds and depths can be used for roughing all except the harder plastics. Finish cuts should, of course, be at low feeds with a cutter having a nose radius.

Most plastics can be machined quite easily, though some experimenting may be needed. The references at the end of this chapter give the sources of a number of more detailed specifications, and the companies who make the plastics often have much information which they will furnish upon request.

Review Questions and Problems

30-1. Define "polymer."

30-2. What are the basic raw materials for the preparation of plastics?

30-3. Indicate an outstanding characteristic of the following:
   a. Thermoplastic polymer
   b. Thermosetting polymer
   c. Elastomer
   Give an example of each.

30-4. With a simple heating instrument, how would you differentiate between a thermoplastic and a thermoset?

30-5. When choosing a plastic for service:
   a. What thermal properties should be investigated?
   b. What mechanical properties?
   c. What chemical properties?
   d. What electrical properties?
   e. What physical properties?

30-6. How can the properties of plastics be modified?

30-7. What factors must be considered in the processing of plastics?

30-8. Name three ways in which transfer molding differs from compression molding.

39-9. What are the basic components of an injection-molding machine? Why is temperature control so critical in injection molding?

30-10. How does extrusion differ from injection molding?

30-11. How does extrusion of rods differ from thin-film extrusion?

30-12. Why is thermoforming a valuable method for the plastics manufacturer?

30-13. In fastening plastics to plastics, name:
   a. A mechanical method
   b. A chemical method
   c. An electrical method

30-14. Name five factors which must be considered in machining plastics.

30-15. If, for a specific application, you have to choose among a metal, a ceramic, or a plastic, what factors must you take into account in order to choose the proper material?

30-16. List in tabular form the following polymers, their most outstanding properties, and important uses: epoxy, Bakelite, Lexan, ABS, Lucite, cellulose acetate, Teflon, nylon, acetal, polyethylene, polypropylene, polystyrene, PVC.

References

1. Baird, R. J.: *Industrial Plastics*, Goodheart Wilcox Co., 1971.

2. Kobayashi, A.: *Machining of Plastics*, McGraw-Hill, New York, 1967.

3. "Machining the Engineering Plastics, Cadillac Plastic and Chemical Co.," *Plastics Design and Processing*, September 1968.

4. "Materials Selector," *Materials Engineering*, September issues.

5. Milby, R. V.: *Plastics Technology*, McGraw-Hill, New York, 1973.

6. *Modern Plastics Encyclopedia*, Modern Plastics, Hightstown, N. J., 1972–1973.

7. *Plastics Engineering Handbook*, 3d ed., The Society of the Plastics Industry, Inc., 1960.

8. Spur, G.: *Machining of Thermoplastic Synthetics*, Society of Manufacturing Engineers, Detroit, Mich., EM72-153.

The first step in making either iron or steel is to convert iron ore into molten iron. This is done in a *blast furnace* as shown schematically in Fig. A-1.

The furnace is charged from the top with alternating layers of coke, iron ore, and limestone, and tons of hot air from the *stoves* are blown through this charge. Gases from the burning coke remove oxygen from the ore, and the limestone causes other impurities in the ore to flow. The impurities combine into *slag* and float on top of the molten iron, which settles to the bottom. The slag is tapped off fairly frequently, and the molten metal is tapped off when about 150 tons (136,000 kg) have accumulated.

The molten iron is tapped into an insulated *hot metal car* which may go directly to a steel mill. Some of the iron is tapped into ladles and then poured into molds, making *pigs* for use in iron foundries and some small steel mills.

This process is, in the larger mills, continuous until the blast furnace needs repair or sales demand decreases.

CAST IRON

The product of the blast furnace is often remelted with some scrap and then processed and poured into molds. This iron has from 2 to 4 percent carbon, with a maximum of 1 percent phosphorus (P) and 0.2 percent sulfur (S). Varying amounts of silicon, and some manganese, will also be present. Four kinds of iron can be made.

White cast iron is produced if a casting with 1.5 percent maximum silicon is cooled very rapidly. This causes most of the carbon to remain as iron carbide. This material is so hard that it can be machined only by grinding. It is used, especially as a surface produced by chilling, where extreme resistance to wear is needed.

The most common type of iron used is the familiar *gray cast iron*. Gray iron contains up to 3 percent silicon and is cooled fairly slowly. As a result, most of the carbon is present in thin, angular-shaped flakes of graphite. A finer graphite flake size, and thus stronger casting, can be obtained by *inoculating* the iron with a small amount of ferrosilicon, aluminum, calcium silicate, zirconium, etc.

Nodular cast iron, also called *ductile iron*, is replacing gray cast iron in many castings. The addition of magnesium (Mg) or cerium (Ce) to the ladle just before pouring causes the graphite to form in spheroidal (ball-shaped) particles. These spheroids do not have the sharp *stress raiser shape* of the graphite flakes in gray cast iron. Thus, nodular iron is stronger and tougher, though somewhat more expensive.

Malleable iron, or malleable cast iron, is made from medium carbon, low silicon, white cast iron. The change is made by annealing the white iron, followed by a fast cooling to about 1400°F (760°C), and then slow cooling. The cooling cycle takes at least 10 h. Large furnace loads may need a cycle three times as long. This process forms *temper carbon*, which is similar to the nodules in ductile iron. The resulting product is very tough and easily machined, though more expensive than other types of iron castings.

Alloy cast irons might be called a fifth type. These have alloying metals added to increase resistance to corrosion, heat, or wear, and to improve mechanical properties.

The most common alloying elements are chromium, nickel, copper, molybdenum, and vanadium. There are many combinations of these alloys, both low and high alloys. Some of these are listed in Appendix B.

MAKING STEEL

A brief definition of steel is that it is iron with less than 2 percent carbon. In most steels the sulfur and phosphorus are also kept to a very low percentage. Steel is made from blast furnace iron plus 10 to 50 percent scrap steel. Steel is made by five basic processes and several variations of these. Only the basic processes will be briefly described here.

Bessemer furnace or *converter* is seldom used today because better processes have been developed. This is a pear-shaped, insulated vessel which holds about 25 tons (22, 700 kg). It is loaded with molten iron, and a blast of air is blown up through the liquid iron. The cycle only takes about 15 min, but the steel has more phosphorus and sulfur than open-hearth steel.

Open-Hearth Steel Refining

The open-hearth method of refining steel is so named because the metal is open to the sweep of flames across the hearth (see Fig. A-2).

A single furnace load (called a *heat*) may be 600 tons (545,000 kg) or more, though 300 tons (272,000 kg) is

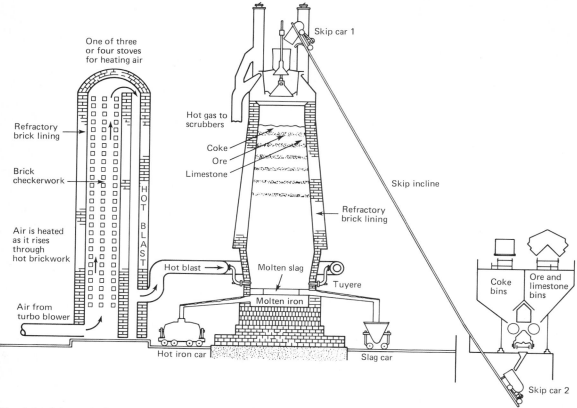

Fig. A-1   Schematic drawing of a blast furnace. [*American Iron and Steel Institute (AISI).*]

more common. The initial load, or *charge*, is made up of pig iron, 25 to 50 percent scrap metal, limestone, and some iron ore.

After the scrap metal charge has started to melt, molten iron from the blast furnace is added. The carbon in the iron is carried off by the air, and the melted limestone combines with other impurities to form a slag. As the entire process takes 5 to 8 h, samples can be taken and analyzed, so that the final alloy is quite closely controlled.

Today an *oxygen lance*, a water-cooled steel pipe, is used to blow a high-velocity stream of oxygen on the melt. This can shorten the time for a heat by 2 or 3 h.

Because of its ease of control, large batches, and moderate time per heat, the open-hearth method has been the most widely used process. However, today, the basic oxygen process is replacing it to quite an extent.

Basic Oxygen Process

This process, commonly referred to as *BOP*, is newer, faster, and more efficient than the open-hearth process, and is replacing it in most steel mills. The pear-

shaped furnace somewhat resembles the old Bessemer furnace. However, air has been replaced by a high-pressure stream of pure oxygen, which is blown in *above* the molten metal, as shown in Fig. A-3.

The *BOF* (*basic oxygen furnace*) is tilted sideways and charged with 25 to 40 percent scrap; the balance is molten iron. The furnace is then tilted up straight, the

Fig. A-2   Schematic drawing of an open-hearth furnace. [*American Iron and Steel Institute (AISI).*]

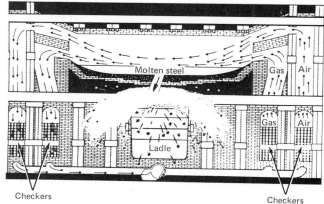

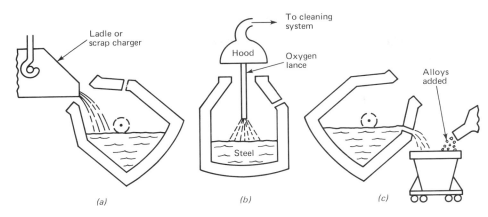

Fig. A-3 Sequence of load-refine-pour in a basic oxygen furnace. (*a*) Scrap and molten iron are loaded in. (*b*) Oxygen is blown in, and gases escape. (*c*) Steel is poured into a ladle. Alloys may be added.

water-cooled oxygen lance is lowered, and high-purity oxygen is blown onto the top of the metal at supersonic speed.

The oxygen combines with carbon and other impurities, and lime, added as a flux, also assists. BOP will refine a 300-ton (272,000-kg) batch of steel in about 45 min. Alloys are added both in the furnace and during final pouring.

Q-BOP

In 1971 the U.S. Steel Company and others completed development work on a process patented by Maxhutte of West Germany and in 1973, a 200-ton Q-BOP furnace was operating.

This process charges the furnaces (with up to 50 percent scrap) near the bottom of the furnace, and the oxygen is blown in through the bottom of the molten steel through specially designed tuyeres.

The Q-BOP process is claimed to be about 10 percent faster than BOP and can use more scrap and less of the more expensive molten iron. The capital costs are less, since there is no need for the high structures to handle loading the furnace and to handle the oxygen lance as in the basic oxygen furnace. The Q-BOP may, if production tests prove satisfactory, replace the present BOP furnaces in future refineries.

Electric Furnace Refining

Because the refining conditions can be regulated more closely than in the open-hearth process or BOP, electric furnaces are especially valuable for producing stainless steels, high-alloy steels, and tool steels, though some carbon steels are also refined this way.

Electric-arc furnace is the most used and will handle a charge of 25 to 50 tons of carefully selected steel scrap plus alloying elements as needed. The entire top, which holds the three or six large carbon electrodes, moves out of the way for charging (Fig. A-4).

The electrodes are lowered to within a few inches of the surface of the metal. The intensely hot arc plus the flow of current through the charge quickly melts the metal. In about 4 h refining is completed; the entire furnace is tipped forward on rockers, and the molten steel is poured.

Induction Furnace

For small lots, from a few hundred pounds up to about a ton, an induction furnace can be used. As shown in Fig. A-5*a*, an induction coil of copper tubing is wound around the furnace. After the furnace has been charged, a high-frequency current is turned on. Heat is generated first in the outer part of the charge, spreading gradually by conduction to the center.

Fig. A-4 Diagram of an electric-arc furnace.

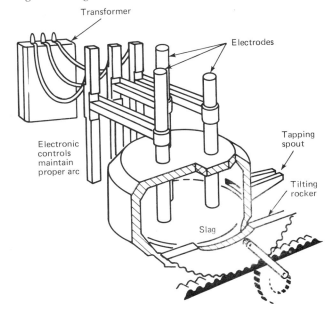

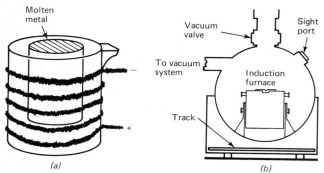

Fig. A-5   (a) Schematic of an induction furnace for melting steel. (b) Vacuum melting, with an induction furnace enclosed in a spherical shaped vacuum chamber. [*American Iron and Steel Institute (AISI).*]

ing. One system pours ingots from BOP or electric-arc furnaces in the shape of large electrodes. These electrodes, like a giant arc welder, become part of the electric circuit in a vacuum chamber (Fig. A-6).

The gaseous impurities are drawn off by the vacuum, and the molten steel drops into the water-cooled mold below. This remelted product is practically free of porosity and included gases.

The *electroslag* remelting process is similar, except that the electrode is lowered into a blanket of molten slag. As the fine droplets of metal pass through the slag, sulfur and other inclusions are reduced. Tests have shown that ESR tool steels have higher fatigue strengths and are less susceptible to cracking.

To make the highest quality alloy steel, *vacuum melting* is used (Fig. A-5b). Refining in a vacuum helps to rid the metal of dissolved gases such as hydrogen, nitrogen, and oxygen. These do no harm in ordinary uses, but under high temperature and stress the purest steel is needed.

Consumable Electrode Refining

To get the best quality of high-alloy and tool steels, the steel is often processed twice, referred to as *remelt-*

CONTROL OF GAS CONTENT OF STEEL

When steel is teemed (poured) into molds from the open-hearth or basic oxygen processes, considerable gas is still in the steel. This forms gas bubbles and oxide particles, which make the product unsuitable for some work.

Vacuum degassing can be done by pouring the steel into ingot molds which are enclosed. A vacuum is maintained which draws off gas from the hot metal as it flows past. This is called "stream degassing." De-

Fig. A-6   Consumable electrode remelt refining in a vacuum. (*Metallurgical Bulletin No. 21, Vanadian-Alloys Steel Company.*)

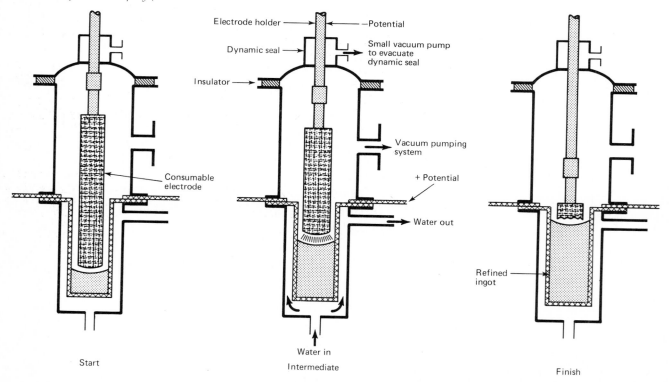

gassing can also be done in the ladle by drawing the steel up into a small vacuum chamber, and then returning the degassed steel to the ladle.

Rimmed steel is made by adding a small amount of aluminum to the steel as it is poured into the mold. This prevents excessive carbon monoxide formation, and the bubbles that do form are trapped below the surface layers. This is used mostly with low-carbon grades of steel.

Killed steel is so called because the steel lays quietly in the mold, without the bubbling action which occurs with rimmed steel. Killed steel is made by adding enough aluminum to the pour so that the steel is completely deoxidized. This process is used with high-carbon grades of steel and causes a large *pipe* or shrinkage hole to form at the top of the mold. This must be cropped off and is remelted. Killed steel is very "clean" for high-quality work. *Semikilled steel* has less aluminum added, so that the "pipe" is much smaller and more of the ingot can be used. Deoxidation is not complete, but the quality is good.

References

1. Avner, Sidney H.: *Introduction to Physical Metallurgy*, 2d ed., McGraw-Hill Book Company. 1974.
2. Heine, Loper, Rosenthal, Principles of Metal Casting, Second Edition, Chapters 15 and 16. McGraw-Hill Book Co. 1967.

Much information is available from such organizations as:

Gray Iron Founders' Society
Cleveland, Ohio 44114

Steel Founders' Society of America
21010 Center Ridge Road
Rocky River, Ohio 44116

American Iron and Steel Institute
1000 16th Street, N. W.
Washington, D.C. 20036

STEEL, WROUGHT

Wrought steel is steel that has been rolled, drawn, or extruded preparatory to direct use or machining to shape. Many of these steels can also be cast, in which case the same numbers are also to be used.

Steel, with very few exceptions, is identified with a four-number system which is, in general, standard with the American Iron and Steel Institute (AISI) and the Society of Automotive Engineers (SAE).

The AISI sometimes uses a letter prefix. B means a Bessemer furnace steel, though today it is usually actually made in the open hearth. C means an open-hearth steel, and E means a steel refined by electric arc or similar process.

Lead pellets are sometimes added to the steel during pouring, to make a more easily machinable alloy. These steels are identified by the letter L between the second and third number and are called *leaded steels*. Examples are 12L40, 41L40, etc. These are not listed separately in the following list. The addition of the small amount of lead does not change the physical properties of the steel.

CODING SYSTEM FOR STEEL

The first two digits of the four-digit number specify the general composition of the steel. These are listed in Table B-1.

The second two digits specify the average percent of carbon in the steel. A decimal point is assumed to be before the third digit. Thus 1008 steel has 0.08

TABLE B-1

10 XX	"Carbon steels" with no alloying metals
11 XX	Free machining, resulfurized carbon steel, originally a Bessemer furnace steel
12 XX	Similar to 11 XX, but an open-hearth steel
13 XX	Manganese steels
20 XX	Nickel steels, largely replaced by the 80 XX steels
31 XX	Nickel-chrome steel
40 XX	Plain molybdenum steel
41 XX	Chrome-molybdenum steels
43 XX	Nickel-chrome-molybdenum steels
46 XX	Nickel-molybdenum steels
48 XX	Nickel-molybdenum steels
50 XX, 51 XX, 52 XX	Plain chrome steel
61 XX	Chrome-vanadium steel
86 XX	Nickel-chrome-molybdenum but lower alloying percentage than 43 XX
92 XX	Silicon steels, especially for use as laminations in electric motors, etc.

percent carbon, and a 4140 steel has 0.40 percent carbon. The only common steel that has over 0.60 percent carbon is 52100, which has 1.00 percent carbon. This is the steel used for most ball bearings and bearing races.

A few of the most used steels are listed in Table B-2. Complete lists are available in various handbooks. Remember that not all the listed steels are regularly made or stocked. You can get many alloys only if you can order at least 20 tons or if you will pay a premium for a special melt.

STAINLESS STEEL

Stainless steel has a three-digit numbering system. The first number indicates the type of stainless, and the second two have no special standard meaning. The groups are:

2XX—Chromium-nickel-maganese, austenitic, nonmagnetic, not hardenable by heat treatment.

3XX—Chromium-nickel, austenitic, nonmagnetic, not hardenable by heat treatment.

4XX—Chromium alloy only. They are in two groups. Ferritic is magnetic but not hardenable. Martensitic is magnetic and hardenable by heat treatment. Ferritic alloys are 405, 430, 442, 446.

5XX—Low chromium, martensitic, magnetic, hardenable.

Research during World War II developed a precipitation-hardening, low-carbon, high-strength stainless steel. The label PH for precipitation hardening identifies these.

Most stainless steels are weldable, but carbide precipitation occurs when the 300 series is welded and the joint may start to leak under some conditions. Thus, three special stainless steels were developed for use in tanks and pressure vessels which require welding.

304L—extra low carbon (0.03 percent)

321—with Ti added

347—with Cb or Ta added

Most stainless steels are difficult to machine. Thus, sulfur or selenium has been added to some alloys to make *free machining stainless steel*. Sulfur is not noted in the number, but Se is added to the number if it is added.

TABLE B-2   DATA ON SOME FREQUENTLY USED STEELS

Number	Machinability	Tensile Strength* psi	Tensile Strength* MPa	Approximate Composition, %†
1010	66	60,000	414	0.45 Mn
1018	66	64,000	442	0.75 Mn
1045	55	91,000	628	0.75 Mn
B1112 C1212	100	78,000	538	0.85 Mn, 0.2 S, 0.09 P
B1113 C1213	137	75,000	518	0.85 Mn, 0.28 S, 0.09 P
1117	90	70,000	483	Same as C1213
1215	137	78,000	538	
3140	55	100,000	690	0.8 Mn, 1.25 Ni, 0.65 Cr
4140	60	108,000	745	0.85 Mn, 0.95 Cr, 0.2 Mo
4340	57	120,000	828	0.7 Mn, 1.85 Ni, 0.8 Cr, 0.25 Mo
50B46	70	80,000	552	0.9 Mn, 0.3 Cr, 0.0005 Bo
E52100	40	100,000	690	0.35 Mn, 1.45 Cr
8620	70	96,000	662	0.8 Mn, 0.55 Ni, 0.5 Cr, 0.2 Mo
8640	66	124,000	856	Same as 8620
9255	54	115,000	793	0.85 Mn, 2.0 Si

*Tensile strength varies with amount of cold working, heat treatment, and size of part.
†Almost all have approximately 0.05 S and 0.04 P.

Table B-3 gives data on several frequently used stainless steels.

CAST STEEL—CARBON AND LOW-ALLOY

Most steels can be used to make castings, and 4130 and 8620 are used to make high-strength valves. However, ASTM has a number of specifications of special casting alloys.

A few of the steels from each class are listed in Table B-4.

CAST STAINLESS STEEL

The Alloy Casting Institute (ACI) has established specifications for a number of stainless-steel alloys, several of which are similar to the wrought alloys. There are two general classifications:

The initial letter H indicates alloys generally used where high temperatures exceed 1200°F (649°C). The second letter represents an increase in the nickel contant from A, 1 percent to X, 66 percent.

Numerals following the C (corrosion resistant) al-

TABLE B-3   DATA ON SOME STAINLESS STEELS

AISI Number	Machin- ability*	Tensile Strength psi	Tensile Strength MPa	Approximate Composition, % C	Cr	Ni	Other
201	50	115,000	794	0.15	17	4.5	6.5 Mn
202	50	105,000	725	0.15	18	5	9 Mn
301	50	100,000	690	0.15	17	7	2 Mn
302	50	90,000	621	0.15	18	8	2 Mn
303	70	90,000	621	0.15	18	8	0.15 S min.
303 Se	65	80,000	552	0.15	18	9	0.15 Se min.
304	55	85,000	587	0.08	19	10	
316	50	90,000	621	0.08	17	12	2.5 Mo, 2.0 Mn
316 F	63	80,000	552	0.08	17	12	0.07 P or S min.
321	40	90,000	621	0.08	18	11	Ti = 5 × C min.
347	40	90,000	621	0.08	18	11	Cb or Ta = 10 × C
410	54	75,000†	518	0.15	13	. . .	1.0 Mn
416 Se	90	75,000†	518	0.15	13	. . .	1.25 Mn, 1.0 Se
420 F	75	90,000†	621	0.15	13	. . .	1.0 Mn, 0.4 S
440 C	40	110,000†	759	1.10	17	. . .	0.75 Mo max.
502	65	70,000	483	0.10	5	. . .	1.0 Mn, 0.55 Mo
17-4 PH	65	200,000	1380	0.04	17	4	4 Cu, 0.3 Cb + Ta
17-7 PH	50	215,000	1483	0.07	17	7	1.15 Al

*Compared to B1112 as 100 percent.
†Tensile strength can be more than double that shown, by proper heat treatment.

TABLE B-4   DATA ON THE MOST USED ASTM CAST STEELS

Class Number	Tensile Strength		Approximate Composition, %				
	psi	MPa	C	Cr	Ni	Mo	Other
65-35	65,000	449	0.30	0.40	0.50	0.20	0.7 Mn, 0.3 Cu
70-40	70,000	483	0.25	0.40	0.50	0.20	1.2 Mn, 0.3 Cu
WCA	60,000	414	0.25	0.40	0.50	0.25	0.7 Mn, 0.5 Cu, 0.03 V
WCC	70,000	483	0.25	0.40	0.50	0.25	1.2 Mn, 0.5 Cu, 0.03 V
WC5	70,000	483	0.20	0.65	0.80	1.10	0.55 Mn
WC6	70,000	483	0.20	1.25	. . .	0.55	0.65 Mn
WC9	70,000	483	0.18	2.50	. . .	1.10	0.55 Mn
C5	90,000	621	0.20	5.50	. . .	0.55	0.55 Mn
LC1	65,000	449	0.25	. . .	. . .	0.55	0.65 Mn
LC3	65,000	449	0.15	. . .	. . .	. . .	0.65 Mn, 3.5 Ni

All contain approximately 0.04 to 0.05 P, 0.05 to 0.06 S, 0.6 to 0.8 Si and 0.5 Ni.

loys specify the maximum carbon content, and letters following the numbers indicate additional alloying element. Table B-5 lists some of the most used alloys.

GRAY CAST IRON

ASTM Specification A48 lists seven classes of gray iron castings. The class number is the minimum tensile strength of test bars in thousands of pounds per square inch. The classes are as follows, with the equivalent megapascals in parenthesis: 20 (138 MPa), 25 (173), 30 (207), 35 (242), 40 (276), 50 (345), 60 (414).

The strength of cast iron varies with the relative proportion of free carbon (graphite) and combined carbon (cementite). The higher strength irons have lower carbon content. They are also more brittle.

Machinability of classes 20, 25, and 30 is about 100 percent. This decreases, and class 60 will have a machinability of 50 percent.

MALLEABLE CAST IRON

More commonly called *malleable iron*, this is identified by a five-number system. The entire number gives the approximate yield strength in pounds per square inch. The final one or two numbers give the percent elongation. The basic alloy is close to the same for all numbers, and the difference in strength is attained by variations in heat treatment. The five grades listed in Table B-6 are the most frequently used.

DUCTILE CAST IRON

This is commonly called *ductile iron,* or *nodular iron,* though it is a true cast iron. The numbering system consists of three pairs of numbers representing, in order, minimum tensile strength–minimum yield point–percent elongation. The 80-55-06, used "as cast," is one of the most used.

The top two strengths can be attained either by

TABLE B-5   DATA ON SOME HEAT AND CORROSION RESISTANT (STAINLESS STEEL) CASTING ALLOYS

ACI Number	Similar Wrought Alloy	Tensile Strength		Approximate Composition, %			
		psi	MPa	C max.	Cr	Ni	Other
CA-15	410	200,000*	1380	0.15	13.0	1.0	
CA-40	420	220,000*	1518	0.40	13.0	1.0	
CB-7Cu	17-4PH	190,000	1311	0.07	16.0	4.0	2.8 Cu
CD-4MCu	. . .	110,000	759	0.04	26.0	5.5	2 Mo, 3 Cu
CF-8	304	77,000	531	0.08	20.0	10.0	
CF-8M	316	80,000	552	0.08	20.0	10.0	2.5 Mo
CN-7M	. . .	69,000	476	0.07	20.0	26.0	Mo and Cu, vary
HD	327	87,000	600	0.50	28.0	5.5	
HF	302B	87,000	600	0.30	21.0	10.0	
HH	309	87,000	600	0.35	26.0	13.0	0.2 N
HK	310	75,000	518	0.40	26.0	20.0	
HT	330	70,000	483	0.55	15.0	35.0	
HW	. . .	68,000	469	0.55	12.0	60.0	

*Maximum strength after heat treating.

NOTE: 1.   All contain approximately 0.04 P, 0.04 S, 1.0 to 2.0 Si, and 1.0 to 2.0 Mn.
        2.   Machinability is from 15 to 60 percent of B1112.
        3.   See also ASTM specifications A296 and A297.

TABLE B-6   DATA ON THE MOST USED MALLEABLE (CAST) IRON ALLOYS

ASTM Number	Tensile Strength	
	psi	MPa
32510	40,000	276
48005	70,000	483
60003	80,000	552
70003	90,000	621
80002	100,000	690

All grades contain 2.0 to 2.65 C, 0.9 to 1.4 Si, and 0.25 to 0.55 Mn.

proper treatment or by the use of small amounts of alloys such as copper, nickel, and molybdenum. The percent of silicon and manganese also can affect the properties of ductile iron. Table B-7 lists all the alloys.

ALUMINUM ALLOYS

Aluminum alloys are specified by a four-digit number for wrought aluminum (rolled, drawn, forged, extruded), and a three-digit number for casting alloys.

The first number indicates the principal alloying metal or metals. The other numbers have no special significance today. They sometimes are the same as those used in an older numbering system.

Letters and numbers following the alloy number give additional information.

F—as fabricated or cast

O—soft annealed condition

H—strain-hardened. This is used with the nonheat-treatable wrought alloys in the 1000, 3000, and 5000 series.

H1X—strain-hardened only

H2X—strain-hardened and partially annealed

H3X—strain-hardened and stabilized, applies only to 5000 and 6000 series

The second digit is from 1 to 8, with 1 the softest and 8 very hard. Thus, H14 means a wrought product which has been strain-hardened to medium hardness.

The heat-treatable wrought and cast alloys are followed by a tempering code with a T prefix. There are more than a dozen of these. The following are the most used.

T4—solution heat-treated and naturally aged

T6—solution heat-treated and artificially aged. This is the most used temper

T651—T6—followed by "stress relieving by stretching the plate or rod" 1 to 3 percent

The basic numbering systems are:

Number	Alloyed with
Wrought Alloys	
1 XXX	Pure, no alloy
2 XXX	Copper
3 XXX	Manganese
4 XXX	Silicon
5 XXX	Magnesium
6 XXX	Magnesium and silicon
7 XXX	Zinc
8 XXX	Any other
Cast Alloys	
1 XX	Copper and silicon
2 XX	Magnesium
3 XX	Zinc
4 and 5	Not used
6 XX	Zinc
7 XX	Tin

Tables B-8 and B-9 list the most frequently used alloys.

MAGNESIUM

Magnesium alloys are identified by two letters and two numbers. The letters identify the two principal

TABLE B-7   THE DUCTILE (CAST) IRON ALLOYS

ASTM Grade	Tensile Strength		Microstructure	Heat Treatment
	psi	MPa		
60-40-18	60,000	414	Ferrite	Annealed
65-45-12	65,000	449	Ferrite and pearlite	Used as-cast
80-55-06	80,000	552	Pearlite and ferrite	Used as-cast
100-70-03	100,000	690	Pearlite	Normalized
120-90-02	120,000	828	Tempered martensite	Oil quench and temper to desired qualities

NOTE: 1.   All grades contain 3.0 to 4.0 C, 2.0 to 3.0 Si, and some Mn and Ni.
  2.   Also available are "high alloy" ductile irons (corrosion- and/or heat-resisting), such as International Nickel's "Ni-Resist" series. These are approximately 3.0 C, 22 Ni, 2.5 Si, 2 Cr, 2 Mn.

alloying elements. Some, like A for aluminum and Z for zinc, are easily translated; however, S is for silicon, so Q is for silver. ASTM B-275-61 lists these.

The two numbers are the approximate percent of the two elements, rounded to a single number. The original alloy is followed by the letter A, and subsequent variations are lettered B, C, etc.

For example, ZK51A is a zinc (Z) and zirconium (K) alloy with approximately 5 percent zinc and 1 percent zirconium, and it is the original alloy (A). Table B-10 lists some of the most used magnesium alloys.

COPPER ALLOYS (BRASS AND BRONZE)

Brass is basically copper plus zinc, and bronze is copper plus tin. However, this distinction does not apply to the names of many of today's copper alloys.

The Copper Development Association (CDA) has a numbering system for dozens of copper alloys, and these numbers are gradually becoming more known. However, the old names like *red brass* and *admiralty metal* are still widely used. Moreover, many companies have their own numbering system.

The listing in Table B-11 shows only a few of the more frequently used alloys. For complete information, see the *CDA Standards Handbooks on Wrought Mill Products and Cast Products*.

In the CDA system, alloys numbered from 100 to 190 are mostly copper, with less than 2 percent alloy. Numbers from 200 to 799 are other wrought alloys. The 800 and 900 series are all casting alloys.

Most copper alloys cannot be heat-treated. Wrought stock can be cold-worked by rolling and is called *quarter hard, half hard, spring hard,* etc. Beryllium copper will age-harden.

TITANIUM

Titanium is a strong, corrosion-resistant metal about 60 percent as heavy as steel. It is stronger than steel at temperatures up to 1000°F (538°C). The numbering system is almost self-explanatory. The principal alloys are listed (by their chemical symbols) with the approximate percent of each as a whole number. Some alloys such as Ti-6Al-4V (6 percent aluminum and 4 percent vanadium) are so well known that they are referred to without the chemical designations. Thus one would use a Ti-6-4 alloy. Table B-12 lists some of the more frequently used titanium alloys.

Machining is not too difficult if proper speeds, feeds, and tooling are used.

SUPERALLOYS

These are sometimes called *exotic* or *high-temperature alloys, high-nickel alloys,* or *cobalt-based alloys,* or *aerospace alloys.* Many of these have no iron in them, they are very expensive, and machinability is from 10 to 50

TABLE B-8   DATA ON THE MOST USED WROUGHT ALUMINUM ALLOYS

| Alloy Number | Tensile Strength | | Approximate Composition, % |
	psi	MPa	
EC-0	12,000	83	99.5 Al (for electrical wire)
EC-H18	24,000	166	99.5 Al (for electrical parts)
1100-0	13,000	90	99 Al (commercially pure)
1100-H18	24,000	166	as above
2011-T8	59,000	407	93 Al, 5.5 Cu, 0.7 Fe, 0.5 Si
2014-T6	70,000	483	93 Al, 4.4 Cu, 0.8 Si, 0.8 Mn
2017-T4	62,000	428	95 Al, 4 Cu, 0.5 Mg, 0.5 Mn
2024-T6*	70,000	483	93 Al, 4.5 Cu, 1.5 Mg, 0.6 Mn
3003-0	16,000	110	98 Al, 1.2 Mn
3004-H38	41,000	283	97 Al, 1.2 Mn, 1 Mg
3005-0	18,000	124	99 Al, 0.8 Mg
5050-H38	32,000	221	98 Al, 1.4 Mg
5052-H38	42,000	290	97 Al, 2.5 Mg, 0.25 Cr
6061-T6*	45,000	310	97 Al, 1 Mg, 0.25 Cr, 0.25 Cu, 0.6 Si
6063-T6	35,000	242	98 Al, 0.7 Mg, 0.4 Si
7075-T6*	83,000	573	90 Al, 5.6 Zn, 2.5 Mg, 1.6 Cu, 0.3 Cr

NOTE: 1. Machinability is 200 to 500 percent of B1112.
2. Silicon bearing alloys cause more rapid tool wear.
*Most used.

TABLE B-9   DATA ON THE MOST USED CAST ALUMINUM ALLOYS

| Alloy Number | Tensile Strength | | Approximate Composition, % | Type of Casting |
	psi	MPa		
13	43,000	297	87 Al, 12 Si, 1 Fe	Die cast
43-F	19,000	131	96 Al, 5.5 Si	All
195-T6	36,000	248	94 Al, 4.5 Cu, 1 Si	Sand, permanent mold
214-F	25,000	173	96 Al, 4 Mg	All
B214-F	20,000	138	94 Al, 4 Mg, 1.8 Si	Sand
356-T6	33,000	228	92 Al, 7 Si, 0.3 Mg	Sand, permanent mold
380	48,000	331	87 Al, 9 Si, 3.5 Cu	Die cast
A750-T5	20,000	138	90 Al, 6.5 Sn, 2.5 Si, 1 Cu	Sand, permanent mold

NOTE: 1. Machinability is 200 to 500 percent of B1112.
2. High-silicon alloys cause considerable tool wear, and may require the use of carbide cutters.

TABLE B-10   DATA ON SOME MAGNESIUM ALLOYS

ASTM Number	Tensile Strength		Approximate Composition, %	Uses
	psi	MPa		
Pure Mg	26,000	179	99.8 Mg	Sheet and Plate
AM100-T6	40,000	276	90 Mg, 10 Al	Sand and permanent mold
AZ31B	40,000	276	94 Mg, 3 Al, 1 Zn	F extrusions; H24 sheet and plate
AZ61A-F	45,000	310	93 Mg, 5.5 Al, 1 Zn	Extrusions
AZ63A-T6	40,000	276	90 Mg, 6 Al, 3 Zn	Sand and permanent mold
AZ91A, B, C	40,000	276	90 Mg, 9 Al, 0.7 Zn	F die casting; T6 sand and permanent mold
EK30A-T6	23,000	159	95 Mg, 4 rare earths	Sand and permanent mold
HK31A-H24	37,000	255	96 Mg, 3.25 Th, 0.7 Zr	Sheet and plate; T6 Sand and permanent mold
HM21A-T8	34,000	235	97 Mg, 2 Th, 0.5 Mn	Sheet, plate, extrusion
ZK60A-T5	53,000	366	94 Mg, 5.5 Zn, 0.5 Zr	Extrusion
ZK61A-T6	43,000	297	93 Mg, 6 Zn, 0.2 Zr	Sands and permanent mold

TABLE B-11   DATA ON SOME COPPER ALLOYS

CDA Number	Trade Name	Tensile Strength		Approximate Composition, %
		psi	MPa	
102	OF (oxygen-free)	38,000	262	99.95 Cu
110	ETP (electrolytic tough pitch)	38,000	262	99.90 Cu
172	Beryllium copper	80,000	552	98 Cu, 1.9 Be
220	Commercial bronze	45,000	310	90 Cu, 10 Zn
230	Red brass	50,000	345	85 Cu, 15 Zn
270	Yellow brass	54,000	373	65 Cu, 35 Zn
360	Free-cutting brass	55,000	380	62 Cu, 35 Zn, 3 Pb
464	Naval brass	70,000	483	59 Cu, 40 Zn, 0.8 Sn
521	Phosphor bronze	75,000	518	91 Cu, 8 Sn, 0.25 P
614	Aluminum bronze	80,000	552	90 Cu, 7 Al, 2 Fe, 1 Mn
752	Nickel silver	65,000	449	65 Cu, 18 Ni, 17 Zn

Some Cast Alloys

826	Beryllium copper	82,000	566	97 Cu, 2.3 Be, 0.5 Co
836	Red brass, leaded	37,000	255	85 Cu, 5 Sn, 5 Zn, 5 Pb
854	Yellow brass, leaded	34,000	235	67 Cu, 29 Zn, 10 Sn, 3 Pb
907	Bronze	44,000	304	89 Cu, 11 Sn
955	Aluminum bronze	100,000	690	77 Cu, 11 Al, 4 Ni, 4 Fe, 4 Mn
976	Nickel silver	40,000	276	64 Cu, 20 Ni, 8 Zn, 4 Sn, 4 Pb

Machinability varies widely; beryllium and aluminum alloys can be very difficult to machine. Others are from 150 to 500 percent of B1112.

TABLE B-12   DATA ON SOME OF THE MOST USED TITANIUM ALLOYS

Ti-35A, 50A, 65A, 75A, and 100A are pure titanium. The number is the approximate tensile strength in kpsi [except Ti-100A has 85,000 psi (587 MPa) T.S.]. The difference in strength is determined by the "interstitial" elements, chiefly oxygen (0.18 to 0.40) and iron (0.12 to 0.30).

Alloy	Tensile Strength*		Percent Elongation	Remarks
	psi	MPa		
Ti-5 Al-2.5 Sn	120,000	828	10	Alpha
Ti-8 Al-1 Mo-1 V†	130,000	897	10	Alpha-lean beta
Ti-6 Al-2 Sn-4 Zr-2 Mo	130,000	897	10	Alpha-lean beta
Ti-679	145,000	1000	10	Alpha-beta
Ti-6 Al-6 V-2 Sn	155,000	1070	8	Alpha-beta
Ti-6 Al-4 V†	130,000	897	10	Alpha-beta
Ti-6 Al-2 Sn-4 Zr-6 Mo	150,000	1035	10	Alpha-beta
Ti-8 Mo-8 V-2 Fe-3 Al	125,000	862	10	Beta

*Tensile strength at room temperature. Higher if heat treated.
†Also available in "ELI" (extra low interstitials) for temperatures to −423°F (−253°C).
NOTE: Ti-6 Al-4 V is the most used general purpose titanium alloy.
Data from TIMET, a division of Titanium Metals Corporation of America.

TABLE B-13   DATA ON SOME HIGH TEMPERATURE (SUPER) ALLOYS

| Alloy Number | Tensile Strength | | Approximate Composition, % |
	psi	MPa	
A286	146,000	1007	56 Fe, 26 Ni, 15 Cr, 2 Ti, 1.35 Mn
Astroloy	120,000	828	56 Ni, 15 Cr, 15 Co, 4.4 Al, 5.25 Mo, 3.5 Ti
Discaloy	145,000	1000	56 Fe, 26 Ni, 14 Cr, 2.7 Mo, 1.7 Ti
Hastelloy C	130,000	897	54 Ni, 16 Cr, 17 Mo, 5 Fe, 2.5 Co, 4 W
Hastelloy X	114,000	787	49 Ni, 22 Cr, 18.5 Fe, 9 Mo, 1.5 Co
Inconel 718	208,000	1435	55 Ni, 18.6 Cr, 18.5 Fe, 5 Cb, 3.1 Mo
Monel 400	90,000	621	65 Ni, 31.5 Cu, 1.35 Fe
Nimonic 90	179,000	1235	59 Ni, 19.5 Cr, 18 Co, 2.4 Ti, 1.4 Al
Rene 41	206,000	1421	57 Ni, 19 Cr, 11 Co, 10 Mo, 3 Ti
Udamet 500	176,000	1214	57 Ni, 18.5 Co, 18 Cr, 4 Mo, 2.9 Al
Waspaloy	185,000	1277	60 Ni, 19.5 Cr, 13.5 Co, 4.3 Mo, 3 Ti
M252	180,000	1242	65 Ni, 20 Cr, 10 Co, 2.5 Ti, 2 Fe
HS-31	108,000	745	56 Co, 25.5 Cr, 10.5 Ni, 7.5 W, 1 Mn
HS-152	90,000	621	64 Co, 21 Cr, 11 W, 2 Cb, 2 Fe

Machinability is 10 to 50 percent of B1112.

percent of B1112. Their principal use is in aerospace or other applications where strength is necessary at temperatures sometimes up to 2200°F. There are many of these. A few of the more frequently used superalloys are listed in Table B-13.

ZINC DIE-CASTING ALLOYS

The most used name for these alloys is Zamak. Two alloys, Zamak 3 and Zamak 5, are the die-casting alloys. The copper in Zamak 5 makes it somewhat harder and stronger. Machinability of both is 200 to 300 percent of B1112.

Tensile strength   Zamak 3   41,000 psi (283 MPa)
Zamak 5   47,600 psi (328 MPa)

Composition   Zamak 3   96 Zn, 3.5 to 4.3 Al
Zamak 5   94 Zn, 3.5 to 4.3 Al, 0.75 to 1.25 Cu

References

1. Avner, Sidney H.: *Introduction to Physical Metallurgy*, 2d ed., McGraw-Hill, New York, 1974.
2. *Machining Data Handbook*, 2d ed., Machinability Data Center, Metcut Research Associates, 1972.
3. Walton, Charles F.: *Gray and Ductile Iron Castings Handbook*, Gray and Ductile Iron Founders' Society, Inc., 1971.
4. Wilson, Frank W., and Cox, editors, *Machining the Space-Age Metals*, Society of Manufacturing Engineers, Detroit, Mich., 1965.

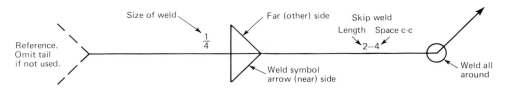

Basic Arc and Gas Weld Symbols

Type of weld							
		Plug or slot	Groove				
Bead	Fillet		Square	V	Bevel	U	J
⌢	◺	⏝	‖	V	V	Y	J

Supplementary Symbols

Weld all around	Field weld	Contour	
		Flush	Convex
○	●	—	⌢

Application of Dimensions to Intermittent Fillet Welding Symbols

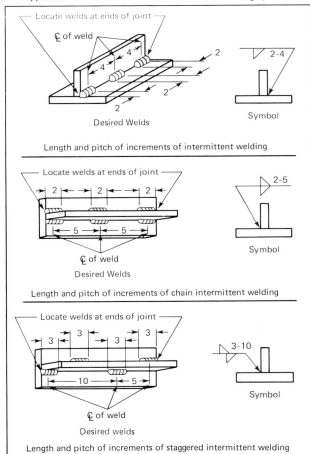

Designation of Size of Combined Welds with Specified Root Penetration

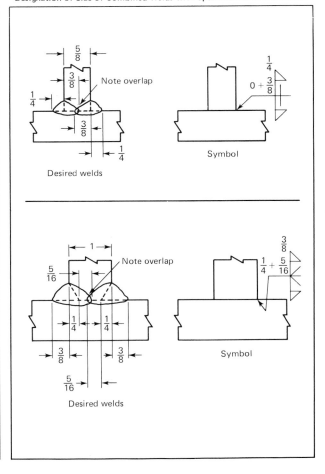

AWS symbols. Examples from The Lincoln Electric Company.

Application of Dimensions to Fillet Welding Symbols

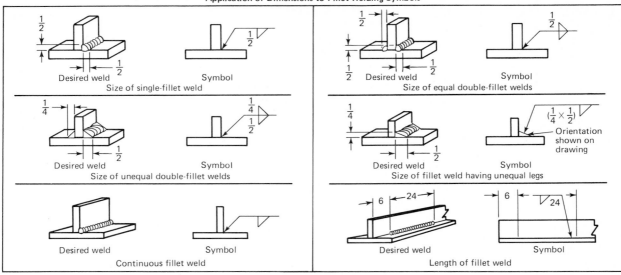

Application of Fillet Welding Symbols

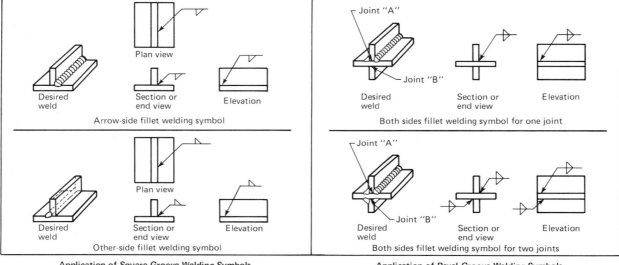

Application of Square-Groove Welding Symbols

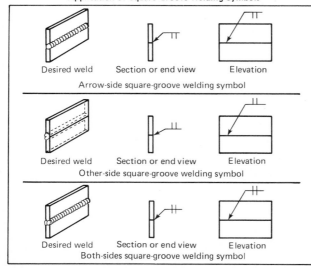

Application of Bevel-Groove Welding Symbols

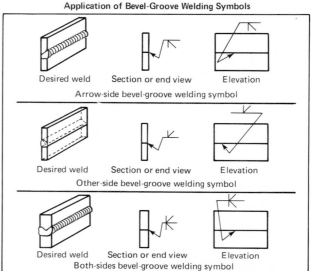

Designation of Size of Groove Welds with No Specified Root Penetration

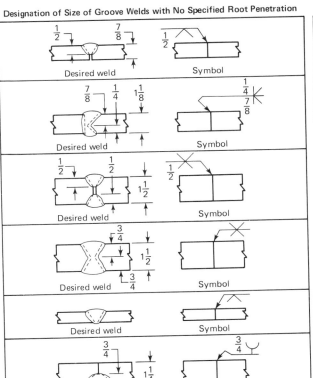

Designation of Size of Groove Welds with Specified Root Penetration

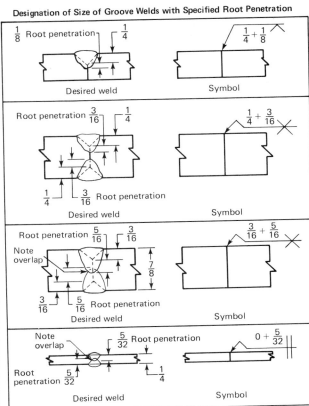

Designation of Location and Extent of Fillet Welds

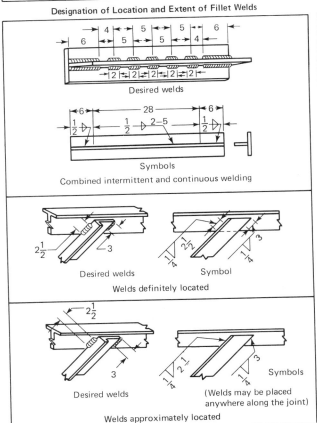

Designation of Extent of Welding

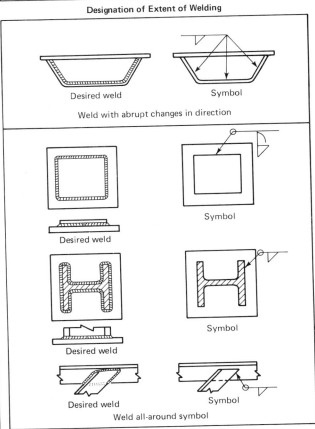

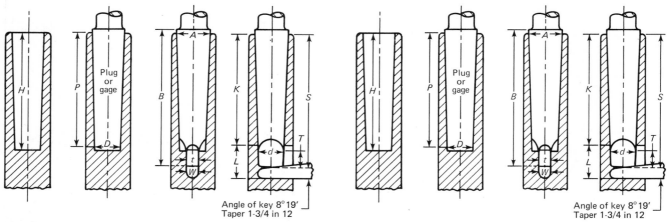

Fig. D-1 Morse tapers.

Fig. D-2 Brown & Sharpe tapers.

TABLE D-1 MORSE TAPERS

Number of Taper	Diameter of Plug at Small End	Diameter at End of Socket	Standard Plug Depth	Whole Length of Shank	Depth of Hole	End of Socket to Keyway	Length of Keyway	Width of Keyway	Length of Tang	Diameter of Tang	Thickness of Tang	Shank Depth	Taper Per Foot	Taper Per Inch
	D	A	P	B	H	K	L	W	T	d	t	S		
0	0.252	0.356	2	$2\frac{3}{8}$	$2\frac{1}{32}$	$1\frac{15}{16}$	$\frac{9}{16}$	0.160	$\frac{9}{32}$	0.24	$\frac{5}{32}$	$2\frac{1}{4}$	0.625	0.0521
1	0.369	0.475	$2\frac{1}{8}$	$2\frac{5}{8}$	$2\frac{5}{16}$	$2\frac{1}{16}$	$\frac{3}{4}$	0.213	$\frac{3}{8}$	0.35	$\frac{13}{64}$	$2\frac{7}{16}$	0.600	0.0499
2	0.572	0.700	$2\frac{9}{16}$	$3\frac{1}{8}$	$2\frac{5}{8}$	$2\frac{1}{2}$	$\frac{7}{8}$	0.260	$\frac{7}{16}$	0.55	$\frac{1}{4}$	$2\frac{15}{16}$	0.602	0.0500
3	0.778	0.938	$3\frac{3}{16}$	$3\frac{7}{8}$	$3\frac{1}{4}$	$3\frac{1}{16}$	$1\frac{1}{16}$	0.322	$\frac{9}{16}$	0.75	$\frac{5}{16}$	$3\frac{11}{16}$	0.602	0.0500
4	1.02	1.231	$4\frac{1}{16}$	$4\frac{7}{8}$	$4\frac{1}{8}$	$3\frac{7}{8}$	$1\frac{1}{4}$	0.478	$\frac{5}{8}$	0.98	$\frac{15}{32}$	$4\frac{5}{8}$	0.623	0.0519
5	1.475	1.748	$5\frac{3}{16}$	$6\frac{1}{8}$	$5\frac{1}{4}$	$4\frac{1}{16}$	$1\frac{1}{2}$	0.635	$\frac{3}{4}$	1.41	$\frac{5}{8}$	$5\frac{7}{8}$	0.630	0.0526
6	2.116	2.494	$7\frac{1}{4}$	$8\frac{5}{8}$	$7\frac{7}{8}$	7	$1\frac{3}{4}$	0.76	$1\frac{1}{8}$	2.00	$\frac{3}{4}$	$8\frac{1}{4}$	0.626	0.0521
7	2.75	3.27	10	$11\frac{3}{4}$	$10\frac{1}{8}$	$9\frac{1}{2}$	$2\frac{5}{8}$	1.135	$1\frac{1}{2}$	$2\frac{11}{16}$	$1\frac{1}{8}$	$11\frac{3}{8}$	0.625	0.0520

NOTE: Same dimensions, changed to millimetres, are used for metric Morse tapers.

TABLE D-2 BROWN & SHARPE TAPERS

Number of Taper	Diameter of Plug at Small End	Diameter at End of Socket	Standard Plug Depth	Whole Length of Shank	Depth of Hole	End of Socket to Keyway	Length of Keyway	Width of Keyway	Length of Tang	Diameter of Tang	Thickness of Tang	Shank Shank Depth	Taper Per Foot	Taper Per Inch
	D	A	P	B	H	K	L	W	T	d	t	S		
5	0.45	0.5229	$1\frac{3}{4}$	$1\frac{7}{8}$	2	$1\frac{27}{32}$	....	....	....	....	....	$1\frac{25}{32}$	0.500	0.0418
6	0.50	0.599	$2\frac{3}{8}$	$2\frac{31}{32}$	$2\frac{1}{2}$	$2\frac{19}{64}$	$\frac{7}{8}$	0.291	$\frac{7}{16}$	0.460	$\frac{9}{32}$	$2\frac{7}{8}$	0.500	0.0419
7	0.60	0.725	3	$3\frac{5}{8}$	$3\frac{1}{8}$	$2\frac{29}{32}$	$\frac{15}{16}$	0.322	$\frac{15}{32}$	0.560	$\frac{5}{16}$	$3\frac{17}{32}$	0.500	0.0418
8	0.75	0.8985	$3\frac{9}{16}$	$4\frac{1}{4}$	$3\frac{11}{16}$	$3\frac{29}{32}$	1	0.353	$\frac{1}{2}$	0.710	$\frac{11}{32}$	$4\frac{1}{4}$	0.500	0.0418
9	0.90	1.0667	4	$4\frac{3}{4}$	$4\frac{1}{8}$	$3\frac{7}{8}$	$1\frac{1}{8}$	0.385	$\frac{9}{16}$	0.860	$\frac{3}{8}$	$4\frac{5}{8}$	0.500	0.0417
10	1.0446	1.289	$5\frac{1}{16}$	$6\frac{1}{4}$	$5\frac{3}{16}$	$5\frac{13}{32}$	$1\frac{5}{16}$	0.447	$\frac{21}{32}$	1.01	$\frac{7}{16}$	$6\frac{3}{32}$	0.5161	0.0430
11	1.25	1.53	$6\frac{3}{4}$	$7\frac{1}{16}$	$6\frac{7}{8}$	$6\frac{13}{32}$	$1\frac{5}{16}$	0.447	$\frac{21}{32}$	1.21	$\frac{7}{16}$	$7\frac{3}{32}$	0.500	0.0418
12	1.50	1.797	$7\frac{1}{8}$	$8\frac{9}{32}$	$7\frac{1}{4}$	$6\frac{15}{16}$	$1\frac{1}{2}$	0.510	$\frac{3}{4}$	1.46	$\frac{1}{2}$	$7\frac{15}{16}$	0.500	0.0416

Brown & Sharpe tapers 1-4 and 13-18 are not listed, as they are not frequently used.

Fig. D-3   Jarno taper shanks.

TABLE D-3   JARNO TAPER SHANKS GENERAL DIMENSIONS (SAMPLE SIZES)

Number	A	B	Sample Sizes	Taper Per Foot
2	1	0.250	0.20	0.600
3	$1\frac{1}{2}$	0.375	0.30	0.600
4	2	0.500	0.40	0.600
17	$8\frac{1}{2}$	2.125	1.70	0.600
18	9	2.250	1.80	0.600
19	$9\frac{1}{2}$	2.375	1.90	0.600
20	10	2.500	2.00	0.600

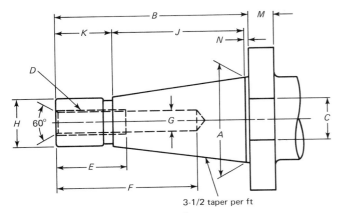

Fig. D-4   Milling machine taper arbors.

TABLE D-4   MILLING MACHINE TAPER ARBORS

M.M.*	A	B	C	D	E	F	G	H	J	K	M	N
30	1.250	$2\frac{3}{4}$	0.630 0.640	$\frac{1}{2}$–13	1	2	$\frac{27}{64}$	0.673 0.675	$1\frac{7}{8}$	$1\frac{13}{16}$	$\frac{3}{8}$	$\frac{1}{16}$
40	1.750	$3\frac{3}{4}$	0.630 0.640	$\frac{5}{8}$–11	$1\frac{1}{8}$	$2\frac{3}{8}$	$\frac{17}{32}$	0.985 0.987	$2\frac{11}{16}$	1	$\frac{1}{2}$	$\frac{1}{16}$
50	2.750	$5\frac{1}{8}$	1.008 1.018	1–8	$1\frac{3}{4}$	$3\frac{1}{2}$	$\frac{7}{8}$	1.547 1.549	4	1	$\frac{3}{4}$	$\frac{1}{8}$
60	4.250	$8\frac{5}{16}$	1.008 1.018	$1\frac{1}{4}$–7	$2\frac{1}{4}$	$4\frac{1}{4}$	$1\frac{7}{64}$	2.359 2.361	$6\frac{7}{16}$	$1\frac{3}{4}$	$\frac{3}{4}$	$\frac{1}{8}$

*MM = milling machine.

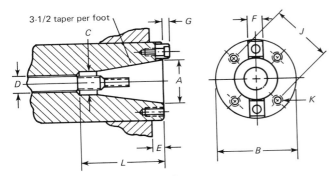

Fig. D-5   Milling machine spindle noses.

TABLE D-5   MILLING MACHINE SPINDLE NOSES

M.M.*	A	B Max.	B Min.	C	D	E	F	G	J	K	L
30	1.250	2.7423	2.7488	$1\frac{1}{16}$	$\frac{21}{32}$	$\frac{1}{2}$	$\frac{5}{8}$	$\frac{5}{16}$	$2\frac{1}{8}$	$\frac{3}{8}$–16	$2\frac{7}{8}$
40	1.750	3.4993	3.4988	1	$\frac{21}{32}$	$\frac{5}{8}$	$\frac{5}{8}$	$\frac{5}{16}$	$2\frac{5}{8}$	$\frac{1}{2}$–13	$3\frac{7}{8}$
50	2.750	5.0618	5.0613	$1\frac{9}{16}$	$1\frac{1}{16}$	$\frac{3}{4}$	1	$\frac{1}{2}$	4	$\frac{5}{8}$–11	$5\frac{1}{2}$
60	4.250	8.7180	8.7175	$2\frac{3}{8}$	$1\frac{3}{8}$	$1\frac{1}{2}$	1	$\frac{1}{2}$	7	$\frac{3}{4}$–10	$8\frac{5}{8}$

Speeds and feeds for machining are difficult to specify because one finds quite widely varying figures in different handbooks and catalogs. The most recent and reliable reference is the second edition of *Machining Data Handbook* from Machinability Data Center, Metcut Research Associates, 3980 Rosslyn Drive, Cincinnati, Ohio 45209. This book gives detailed information on most metals and cutting processes. The figures on feeds and speeds are on the high side, but many have been tested for optimum results.

Turning speeds must be slower when heavy feeds (over 0.020 in./rev or 0.50 mm/rev) and when deep cuts (over 0.200 in. or 5.0 mm) are made. Speed must also be reduced as the BHN and the carbon content of steel increase.

Speeds may be doubled when light finishing cuts are made, and tripled when the proper grade of carbide is used.

The following tables are representative of the recommendations in general and offer reasonable ranges for most work.

REAMING

Reaming can be done at drilling feedrates, or up to $1\frac{1}{2}$ to 3 times that rate. Finish required, accuracy needed, and tool life will affect the choice of freedrate.

Speeds for reaming are usually 40 to 60 percent of the speeds used for drilling, as the reamer's cutting edges are relatively weak.

TAPPING

Feeds

The feed in inches per revolution (in./rev) or millimetres per revolution (mm/rev) is fixed by the pitch of the thread. Every revolution of the tap (or die) will advance it one full thread. Thus feedrate is

$$\text{Tap feed, in./rev} = \frac{1}{\text{threads per inch}} = \text{pitch}$$

$$\text{Tap feed, mm/rev} = \text{pitch of thread, mm}$$

Feedrate *per minute*, ipm or mm/min:
$$\text{ipm} = \text{in./rev} \times \text{rpm} \quad \text{and} \quad \text{mm/min} = \text{mm/rev} \times \text{rpm}$$

Speeds

Tapping speeds should be one-quarter to one-third the speeds shown in Table E-1 for turning. However, on N/C machines, the manufacturer's recommended maximum rpm for tapping is from 200 to 400 rpm, even if the resulting fpm is very low.

TABLE E-1 SUGGESTED CUTTING SPEEDS, USING HIGH-SPEED (HSS) CUTTING TOOLS

Material	Drilling		Reaming		Turning	
	fpm	m/min	fpm	m/min	fpm	m/min
Aluminum	250–600	75–180	100–300	30–90	400–1000	120–300
Brass, free cutting	150–300	45–90	130–200	40–60	225–350	70–110
Bronze, soft	100–250	30–75	75–180	23–55	150–225	45–70
Cast iron:						
Soft	75–150	23–45	60–100	18–30	100–150	30–45
Medium	70–110	20–35	35–65	10–20	75–120	23–35
Hard	60–100	18–30	20–55	6–17	50–90	15–27
Copper	60–100	18–30	40–60	12–18	100–200	30–60
Magnesium	300–650	90–200	150–350	45–110	600–1200	180–360
Stainless steel:						
Free-machining	65–100	20–30	35–85	10–25	100–150	30–45
Other	15–50	5–15	15–30	5–9	40–85	12–25
Steel, free machining	100–145	30–45	60–100	18–30	125–200	40–60
Under 0.3 carbon	70–120	20–35	50–90	15–27	75–175	23–50
0.3 to 0.6 carbon	55–90	17–27	45–70	14–20	65–120	20–35
Over 0.6 carbon	40–60	12–18	40–50	12–15	60–80	18–25
Titanium	30–60	9–18	10–20	3–6	25–55	8–17
Zinc diecasting	200–400	60–120	125–300	40–90	300–1000	90–300

NOTE: Carbide tools may be run two to four times as fast as shown above.

TABLE E-2   END MILLS—SPEEDS AND FEEDS

| | Feed per Tooth | | | | Cutting Speed | | | |
| | $\frac{1}{4}$ in. (6.35 mm) Diameter End Mill | | 1 in. (25 mm) Diameter End Mill | | Roughing | | Finishing | |
Material	in./tooth	mm/tooth	in./tooth	mm/tooth	fpm	m/min*	fpm	m/min*
Aluminum	0.003	0.08	0.009	0.23	600	180	800	245
Bronze, medium	0.003	0.08	0.007	0.18	250	75	300	90
Bronze, hard.	0.002	0.05	0.005	0.13	125	40	150	45
Cast iron, soft	0.003	0.08	0.008	0.20	60	18	80	25
Cast iron, hard	0.002	0.05	0.005	0.13	50	15	70	20
Plastic, glass filled	0.003	0.08	0.012	0.30	150	45	160	50
Steel, low carbon	0.001	0.03	0.004	0.10	75	23	90	30
Steel, 4140	0.0005	0.013	0.003	0.08	50	15	70	20
Steel, 4340	0.0003	0.008	0.002	0.05	50	15	70	20
Stainless steel								
Type 304	0.001	0.03	0.004	0.10	55	17	75	23
Type 17-4PH	0.0005	0.013	0.003	0.08	35	10	50	15
Inconel	0.0002	0.005	0.003	0.08	30	9	40	12
Monel-K	0.0003	0.008	0.004	0.10	60	18	80	25
Ti-6Al-4V	0.001	0.03	0.004	0.10	25	8	40	12
Zinc die casting	0.002	0.05	0.010	0.25	800	240	1000	300

*Conversion rounded off.
NOTE: Face mills use up to twice the above feed rates. Carbide cutters, double the above cutting speeds.
Adapted from DoAll Company, for HSS cutters.

TABLE E-3   FEEDS FOR DRILLING—HSS AND CARBIDE DRILLS USE THE SAME

| Drill Diameter | | Drill Feed | |
in.	mm	in./rev	mm/rev
Under $\frac{1}{8}$	0–3.2	0.001–0.003	0.03–0.08
$\frac{1}{8}$–$\frac{1}{4}$	3.2–6.35	0.002–0.005	0.05–0.13
$\frac{1}{4}$–$\frac{1}{2}$	6.35–12.7	0.004–0.007	0.10–0.23
$\frac{1}{2}$–1	12.7–25.4	0.007–0.017	0.18–0.43
Over 1	25.5 up	0.015–0.030	0.38–0.76

NOTE: For finer finish use lower feeds. Horsepower used increases with feedrate, so feeds at the lower end of the suggested range ordinarily must be used for the difficult-to-machine metals, though too fine feeds can cause work hardening.

As the feedrate increases, the power required and the heat generated also increase. To compensate for these factors, rpm may have to be decreased.

Finishing Feeds (Turning)

These are usually from 0.0015 to 0.010 in./rev (0.04 to 0.25 mm/rev) depending on the finish desired. Studies have been made which prove that, with a

TABLE E-4   FEEDS FOR ROUGH TURNING—HSS OR CARBIDE CUTTERS

| | Feedrate | |
Material	in./rev	mm/rev
Aluminum	0.007–0.050	0.18–1.27
Cast iron	0.011–0.025	0.28–0.64
Copper alloys	0.005–0.022	0.13–0.56
Nickel alloys	0.005–0.018	0.13–0.46
Stainless steel	0.005–0.022	0.13–0.56
Steel	0.010–0.090	0.25–2.30
Titanium	0.007–0.018	0.18–0.46

NOTE: Cutting with single-point tools on a lathe is quite different from drilling, reaming, etc. Lathe tool bits can be made large in cross section so that very heavy cuts may be taken (up to the usable horsepower of the lathe) without breaking the cutting tool.

larger nose radius, fine finishes can be achieved with fairly high feedrates.

As mentioned before, with low feedrates and shallow depths of cut, the rpm may often be safely doubled, as very little heat is being generated during the finish cut.

Drill Sizes

TABLE F-1   NUMBER, LETTER AND FRACTIONAL INCH SIZES OF TWIST DRILLS

Drill Size	Decimal	Drill Size	Decimal	Drill Size	Decimal	Drill Size	Decimal	Drill Size	Decimal	Drill Size	Decimal	Drill Size	Decimal	Drill Size	Decimal
97	0.0059	72	0.0250	50	0.0700	29	0.1360	8	0.1990	L	0.2900	$\frac{29}{64}$	0.4531	$\frac{27}{32}$	0.8438
96	0.0063	71	0.0260	49	0.0730	28	0.1405	7	0.2010	M	0.2950	$\frac{15}{32}$	**0.4688**	$\frac{55}{64}$	0.8594
95	0.0067	70	0.0280	48	0.0760	$\frac{9}{64}$	**0.1406**	$\frac{13}{64}$	**0.2031**	$\frac{19}{64}$	**0.2969**	$\frac{31}{64}$	0.4844	$\frac{7}{8}$	**0.8750**
94	0.0071	69	0.0292	$\frac{5}{64}$	**0.0781**	27	0.1440	6	0.2040	N	0.3020	$\frac{1}{2}$	**0.5000**	$\frac{57}{64}$	0.8906
93	0.0075	68	0.0310	47	0.0785	26	0.1470	5	0.2055	$\frac{5}{16}$	**0.3125**	$\frac{33}{64}$	0.5156	$\frac{29}{32}$	0.9062
92	0.0079	$\frac{1}{32}$	**0.0313**	46	0.0810	25	0.1495	4	0.2090	O	0.3160	$\frac{17}{32}$	0.5313	$\frac{59}{64}$	0.9219
91	0.0083	67	0.0320	45	0.0820	24	0.1520	3	0.2130	P	0.3230	$\frac{35}{64}$	0.5469	$\frac{15}{16}$	**0.9375**
89	0.0091	66	0.0330	44	0.0860	23	0.1540	$\frac{7}{32}$	**0.2188**	$\frac{21}{64}$	**0.3281**	$\frac{9}{16}$	**0.5625**	$\frac{61}{64}$	0.9531
88	0.0095	65	0.0350	43	0.0890	$\frac{5}{32}$	**0.1562**	2	0.2210	Q	0.3320	$\frac{37}{64}$	0.5781	$\frac{31}{32}$	0.9688
87	0.010	64	0.0360	42	0.0935	22	0.1570	1	0.2280	R	0.3390	$\frac{19}{32}$	0.5938	$\frac{63}{64}$	0.9844
86	0.0105	63	0.0370	$\frac{3}{32}$	**0.0938**	21	0.1590	A	0.2340	$\frac{11}{32}$	**0.3438**	$\frac{39}{64}$	0.6094	1	1.000
85	0.011	62	0.0380	41	0.0960	20	0.1610	$\frac{15}{64}$	**0.2344**	S	0.3480	$\frac{5}{8}$	**0.6250**		
84	0.0115	61	0.0390	40	0.0980	19	0.1660	B	0.2380	T	0.3580	$\frac{41}{64}$	0.6406		
83	0.012	60	0.0400	39	0.0995	18	0.1695	C	0.2420	$\frac{23}{64}$	**0.3594**	$\frac{21}{32}$	0.6562		
82	0.0125	59	0.0410	38	0.1015	$\frac{11}{64}$	**0.1719**	D	0.2460	U	0.3680	$\frac{43}{64}$	0.6719		
81	0.013	58	0.0420	37	0.1040	17	0.1730	$\frac{1}{4}$	**0.2500**	$\frac{3}{8}$	**0.3750**	$\frac{11}{16}$	**0.6875**		
80	0.0135	57	0.0430	36	0.1065	16	0.1770	E	0.2500	V	0.3770	$\frac{45}{64}$	0.7031		
79	0.0145	56	0.0465	$\frac{7}{64}$	**0.1094**	15	0.1800	F	0.2570	W	0.3860	$\frac{23}{32}$	0.7188		
$\frac{1}{64}$	**0.0156**	$\frac{3}{64}$	**0.0469**	35	0.1100	14	0.1820	G	0.2610	$\frac{25}{64}$	**0.3906**	$\frac{47}{64}$	0.7344		
78	0.0160	55	0.0520	34	0.1110	13	0.1850	$\frac{17}{64}$	**0.2656**	X	0.3970	$\frac{3}{4}$	**0.7500**		
77	0.0180	54	0.0550	33	0.1130	$\frac{3}{16}$	**0.1875**	H	0.2660	Y	0.4040	$\frac{49}{64}$	0.7656		
76	0.0200	53	0.0595	32	0.1160	12	0.1890	I	0.2720	$\frac{13}{32}$	**0.4062**	$\frac{25}{32}$	0.7812		
75	0.0210	$\frac{1}{16}$	**0.0625**	31	0.1200	11	0.1910	J	0.2770	Z	0.4130	$\frac{51}{64}$	0.7969		
74	0.0225	52	0.0635	$\frac{1}{8}$	**0.1250**	10	0.1935	K	0.2810	$\frac{27}{64}$	0.4219	$\frac{13}{16}$	**0.8125**		
73	0.0240	51	0.0670	30	0.1285	9	0.1960	$\frac{9}{32}$	**0.2812**	$\frac{7}{16}$	**0.4375**	$\frac{53}{64}$	0.8281		

TABLE F-2 METRIC AND INCH TWIST DRILL EQUIVALENTS

mm	in.	mm	in.	mm	in.	mm	in.	mm	in.	mm	in.
0.04	0.0016	0.92	0.0362	2.60	0.1024	4.70	0.1850	7.60	0.2992	11.50	0.4528
0.06	0.0024	0.95	0.0374	2.65	0.1043	4.75	0.1870	7.70	0.3032	11.75	0.4626
0.08	0.0032	0.98	0.0386	2.70	0.1063	4.80	0.1890	7.75	0.3051	12.00	0.4727
0.10	0.0039	1.00	0.0394	2.75	0.1083	4.90	0.1929	7.80	0.3071	12.25	0.4823
0.12	0.0047			2.80	0.1102			7.90	0.3110		
		1.05	0.0413			5.00	0.1969			12.50	0.4921
0.15	0.0059	1.10	0.0433	2.85	0.1122	5.10	0.2008	8.00	0.3150	12.75	0.5020
0.18	0.0071	1.15	0.0453	2.90	0.1142	5.20	0.2047	8.10	0.3189	13.00	0.5118
0.20	0.0079	1.20	0.0473	2.95	0.1161	5.25	0.2067	8.20	0.3228	13.25	0.5217
0.22	0.0087	1.25	0.0492	3.00	0.1181	5.30	0.2087	8.25	0.3248	13.50	0.5315
0.25	0.0098			3.10	0.1221			8.30	0.3268		
0.28	0.0110	1.30	0.0512			5.40	0.2126			13.75	0.5413
0.30	0.0118	1.35	0.0532	3.15	0.1240	5.50	0.2165	8.40	0.3307	14.00	0.5512
0.32	0.0126	1.40	0.0551	3.20	0.1260	5.60	0.2205	8.50	0.3346	14.25	0.5610
0.35	0.0138	1.45	0.0571	3.25	0.1280	5.70	0.2244	8.60	0.3386	14.50	0.5709
0.38	0.0150	1.50	0.0591	3.30	0.1299	5.75	0.2264	8.70	0.3425	14.75	0.5807
				3.35	0.1319			8.75	0.3445		
0.40	0.0157	1.55	0.0611	3.40	0.1339	5.80	0.2284			15.00	0.5906
0.42	0.0165	1.60	0.0630	3.45	0.1358	5.90	0.2323	8.80	0.3465	15.25	0.6004
0.45	0.0177	1.65	0.0650	3.50	0.1378	6.00	0.2362	8.90	0.3504	15.50	0.6102
0.48	0.0189	1.70	0.0669	3.55	0.1398	6.10	0.2402	9.00	0.3543	15.75	0.6201
0.50	0.0197	1.75	0.0689	3.60	0.1417	6.20	0.2441	9.10	0.3583	16.00	0.6229
0.52	0.0205	1.80	0.0709			6.25	0.2461	9.20	0.3622	16.25	0.6398
0.55	0.0217	1.85	0.0728	3.65	0.1437	6.30	0.2480	9.25	0.3642	16.50	0.6496
0.58	0.0288	1.90	0.0748	3.70	0.1457	6.40	0.2520	9.30	0.3661	16.75	0.6595
0.60	0.0232	1.95	0.0768	3.75	0.1476	6.50	0.2559	9.40	0.3701	17.00	0.6693
0.62	0.0244	2.00	0.0787	3.80	0.1496	6.60	0.2598	9.50	0.3740	17.25	0.6791
				3.90	0.1535			9.60	0.3780		
0.65	0.0265	2.05	0.0807	3.95	0.1555	6.70	0.2638	9.70	0.3819	17.50	0.6890
0.68	0.0268	2.10	0.0827	4.00	0.1575	6.75	0.2658	9.75	0.3839	17.75	0.6988
0.70	0.0276	2.15	0.0847	4.10	0.1614	6.80	0.2677	9.80	0.3858	18.00	0.7087
0.72	0.0284	2.20	0.0866	4.15	0.1634	6.90	0.2717	9.90	0.3898	18.25	0.7185
0.75	0.0295	2.25	0.0886	4.20	0.1654	7.00	0.2756	10.00	0.3937	18.50	0.7283
0.78	0.0307	2.30	0.0906	4.25	0.1673	7.10	0.2795	10.10	0.3976	18.75	0.7382
0.80	0.0315	2.35	0.0925	4.30	0.1693	7.20	0.2835	10.25	0.4035	19.00	0.7489
0.82	0.0323	2.40	0.0945	4.40	0.1732	7.25	0.2854	10.50	0.4134	19.25	0.7579
0.85	0.0335	2.45	0.0965	4.50	0.1772	7.30	0.2874	10.75	0.4232	19.50	0.7577
0.88	0.0346	2.50	0.0984	4.60	0.1811	7.40	0.2913	11.00	0.4331	19.75	0.7776
		2.55	0.1004			7.50	0.2953			20.00	0.7874
0.90	0.0354			4.65	0.1831			11.25	0.4429		

NOTE: Not all listed sizes may be stocked.

TABLE G-1 SUGGESTED PERCENT OF THREAD FOR VARIOUS MATERIALS (BOTH INCH AND METRIC THREADS)

Material	% THD.	Material	% THD.	Material	% THD.
Steels:		Stainless steel:		Nickel base:	
B1112	70	302	50	Monel	55
C1020	65	303	60	Hastelloy	55
E4340	65	316	50	Inconel	50
52100	60	410	55	Irons:	
Aluminum:		Copper:		C. I. Soft	75
Wrought	65	Yellow Br.	65	C. I. Mild	70
Cast	75	OFC	55	C.I. Hard	65
BeCu	60	Everdur	55	Ni-Resist	60

NOTE: Above are actual percents. The normal oversize cutting of drills should be considered.
Adapted from Besly Products, The Bendix Corporation.

TABLE G-3 METRIC TAP DRILL SIZES

Metric Tap‡	Pitch mm	OD in.	Nearest Thread Per Inch	Metric Tap Drills (mm) and Percent Thread*		
M1.0	0.25	0.039	102	0.75(82)	0.78(72)	0.8(65)
M1.2	0.25	0.047	102	0.85(82)	0.88(72)	0.9(65)
M1.4	0.30	0.055	85	1.10(82)	1.15(68)	
M1.6†	0.35	0.063	73	1.25(82)	1.30(70)	1.35(58)
M2	0.4	0.079	63	1.60(82)	1.65(71)	1.70(61)
M2.5	0.45	0.099	56	2.05(82)	2.10(73)	2.15(64)
M3	0.5	0.118	51	2.50(82)	2.55(73)	2.60(65)
M3.5	0.6	0.138	42	2.90(82)	2.95(75)	3.00(68)
M4	0.7	0.158	36	3.30(82)	3.40(70)	3.50(58)
M5	0.8	0.197	32	4.20(82)	4.30(71)	4.40(61)
M6.3†	1.0	0.248	25	5.40(83)	5.50(74)	5.60(65)
M6	1.0	0.236	25	5.00(82)	5.10(73)	5.20(65)
M8	1.25	0.315	20	6.80(78)	6.90(72)	7.00(65)
M10	1.5	0.394	17	8.60(76)	8.80(65)	8.90(60)
M12	1.75	0.473	14.5	10.3(79)	10.5(70)	10.7(60)
M14	2.0	0.551	12.5	12.0(82)	12.2(74)	12.4(65)
M16	2.0	0.630	12.5	14.0(82)	14.2(74)	14.4(65)
M20	2.5	0.788	10	17.5(82)	17.75(73)	18.0(65)
M24	3.0	0.945	8.5	21.0(82)	21.25(75)	21.5(68)

*Because drills cut oversize, the *actual* percent thread may be 4 to 12 points less than shown in this table.
†Suggested by the OMFS Committee of ANSI.
‡Some countries use 1.7, 1.8, and 2.6; these are not ISO sizes.

TABLE G-2   TAP DRILL SELECTOR CHART—NUMBER AND INCH SIZES

Tap Size	75% Thread			65% Thread			55% Thread		
	Theoretical Hole Size Decimals	Standard Drills*		Theoretical Hole Size Decimals	Standard Drills*		Theoretical Hole Size Decimals	Standard Drills*	
		Size, No., or mm	Decimal Equiv.		Size, No., or mm	Decimal Equiv.		Size, No., or mm	Decimal Equiv.
Machine Screw Sizes									
0 80	0.0479	$\frac{3}{64}$	0.0469	0.0494	1.25	0.0492	0.0510	1.29	0.0508
1 64	0.0578	1.47	0.0579	0.0599	53	0.0595	0.0619		
72	0.0595	53	0.0595	0.0613	1.55	0.0611	0.0631	$\frac{1}{16}$	0.0625
2 56	0.0686	51	0.0670	0.0710	1.8	0.0709	0.0732	49	0.0730
64	0.0708	50	0.0700	0.0729	1.85	0.0729	0.0749	1.9	0.0748
3 48	0.0788	47	0.0785	0.0815	46	0.0810	0.0841	2.15	0.0846
56	0.0816	46	0.0810	0.0840	2.15	0.0846	0.0862	44	0.0860
4 40	0.0877	44	0.0860	0.0909	2.3	0.0906	0.0942	$\frac{3}{32}$	0.0938
48	0.0918	2.3	0.0906	0.0945	$\frac{3}{32}$	0.0938	0.0971	2.45	0.0965
5 40	0.1007	39	0.0995	0.1039	37	0.1040	0.1072	36	0.1065
44	0.1029	38	0.1015	0.1059	2.7	0.1063	0.1087	2.75	0.1083
6 32	0.1076	36	0.1065	0.1117	34	0.1110	0.1157	2.9	0.1142
40	0.1137	33	0.1130	0.1169	32	0.1160	0.1201	31	0.1200
8 32	0.1336	3.4	0.1339	0.1377	29	0.1360	0.1417	28	0.1405
36	0.1370	29	0.1360	0.1406	28	0.1405	0.1442	27	0.1440
10 24	0.1495	25	0.1495	0.1549	23	0.1540	0.1602	21	0.1590
32	0.1596	21	0.1590	0.1637			0.1677	19	0.1660
Fractional Sizes									
$\frac{1}{4}$ 20	0.2012	7	0.2010	0.2078	5	0.2055	0.2143	3	0.2130
28	0.2153	3	0.2130	0.2199	$\frac{7}{32}$	0.2188	0.2245		
$\frac{5}{16}$ 18	0.2584	F	0.2570	0.2656	$\frac{17}{64}$	0.2656	0.2728	I	0.2720
24	0.2720	I	0.2720	0.2774	J	0.2770	0.2827	K	0.2810
$\frac{3}{8}$ 16	0.3142	$\frac{5}{16}$	0.3125	0.3223	8.2	0.3328	0.3303	$\frac{21}{64}$	0.3281
24	0.3345	Q	0.3320	0.3399	R	0.3390	0.3452	$\frac{11}{32}$	0.3438
$\frac{7}{16}$ 14	0.3680	U	0.3680	0.3772	V	0.3770	0.3865	W	0.3860
20	0.3888	W	0.3860	0.3953	10.0	0.3937	0.4018		
$\frac{1}{2}$ 13	0.4251	$\frac{27}{64}$	0.4219	0.4351	11.0	0.4331	0.4450		
20	0.4513	11.5	0.4528	0.4578			0.4643		
$\frac{9}{16}$ 12	0.4814			0.4922	12.5	0.4921	0.5030	$\frac{1}{2}$	0.5000
18	0.5084			0.5156	$\frac{33}{64}$	0.5156	0.5228		
$\frac{5}{8}$ 11	0.5365	$\frac{17}{32}$	0.5312	0.5483	$\frac{35}{64}$	0.5469	0.5601		
18	0.5709	14.5	0.5709	0.5787	$\frac{37}{64}$	0.5781	0.5853		
$\frac{3}{4}$ 10	0.6526	16.5	0.6496	0.6656			0.6785		
16	0.6892	$\frac{11}{16}$	0.6875	0.6973			0.7053	$\frac{45}{64}$	0.7031
$\frac{7}{8}$ 9	0.7668	$\frac{49}{64}$	0.7656	0.7812	$\frac{25}{32}$	0.7812	0.7956	$\frac{51}{64}$	0.7969
14	0.8055			0.8147	$\frac{13}{16}$	0.8125	0.8240		
1 8	0.8783	$\frac{7}{8}$	0.8750	0.8945	$\frac{57}{64}$	0.8906	0.9107	$\frac{29}{32}$	0.9062
12	0.9188			0.9296			0.9404	$\frac{15}{16}$	0.9375

*For true hole size, add 0.001–0.005 in.
(0.025 to 0.125 mm) because drills will cut oversize.
Adapted from Besly Products, The Bendix Corporation.

TABLE H-1   CONSTANT MULTIPLIER FOR LENGTH
OF DRILL POINTS OF VARIOUS
ANGLES

Point Angle	K	Point Angle	K	Point Angle	K
60°	0.87	118°	0.30	135°	0.21
82°	0.58	120°	0.29	145°	0.16
90°	0.50	125°	0.26	150°	0.13

Vertical length of drill point = P
P = drill diameter x K
Point angle = total included angle between lips

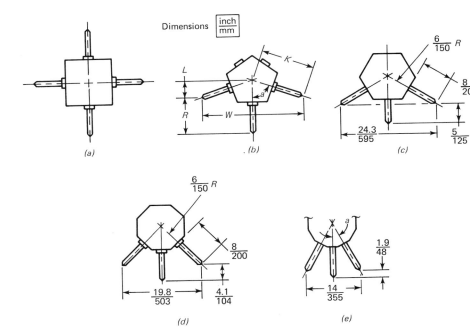

When ordering an N/C turret drill or other machines using turrets, it is sometimes economical to use up to 10 or 12 faces on the turret. However, it should be recognized that as the number of faces increases, the depth of cut, without interference from the adjacent tools, will decrease, as shown by the drawings in Fig. I-1, which are based on one assumed set of dimensions. The simple formulas are also given so that other conditions may be calculated.

Refer to Fig. I-1b.

Angle a = 360° ÷ number of sides

K = total distance, center line to tool tip

R = reach of tool before interference can occur

$L = K \cos a$

$R = K - L$

$W = 2(K \sin a)$

Fig. I-1 Numerical control turret drills. (a) Square turret—no interference. (b) Pentagonal turret—usually no interference. (c) Hexagonal turret—good clearance. (d) Octagonal turret—fair clearance. (e) Twelve-sided turret—small depth clearance.

Length Conversions

The following conversion tables (Appendixes J and K) are purposely quite brief. The first reason is that except for the temperature scale, interpolation can be accurately used; secondly conversion to smaller or larger numbers is easily accomplished by moving the decimal point. Finally, if one has need of more detailed conversion tables, these are readily available free from manufacturers or in a number of books and special slide rules which are now on the market.

The following tables will help to establish the new value system and can be used to obtain figures which are sufficiently accurate for most engineering and manufacturing jobs.

TABLE J-1 SUGGESTED EQUIVALENT MILLIMETRE TOLERANCES FOR USE IN DIMENSIONING DRAWINGS

For Inches	Use mm Equivalent	Actual in. Equivalent	Closer mm Equivalent	Actual in. Equivalent*
0.001	0.02	0.00079	0.025	0.00098
0.002	0.05	0.00197*		
0.003	0.08	0.00315	0.075	0.00295
0.004	0.10	0.00394*		
0.005	0.12	0.00472	0.125	0.00492
0.010	0.25	0.00984*		
$\frac{1}{64}$	0.40	0.01575		

*Almost identical tolerances.
For tolerances of 0.0001, etc., move the decimal point one place to the left.

TABLE J-2 INCH-MILLIMETRE CONVERSION*

in.	mm	in.	mm	in.	mm†	in.	mm†
0.0001	0.0025	0.001	0.025	1	25.4	10	254
0.0002	0.0050	0.002	0.050	2	50.8	20	508
0.0003	0.0075	0.003	0.075	3	76.2	30	762
0.0004	0.0100	0.004	0.102	4	101.6	40	1016
0.0005	0.0130	0.005	0.127	5	127.0	50	1270
0.0006	0.0150	0.006	0.152	6	152.4	60	1524
0.0007	0.0180	0.007	0.178	7	177.8	70	1778
0.0008	0.0200	0.008	0.203	8	203.2	80	2032
0.0009	0.0230	0.009	0.229	9	228.6	90	2286
				12	304.8	100	2540

*Rounding off error is 10 percent or less.
†Exact equivalent.

TABLE J-3 MILLIMETRE-INCH CONVERSION

mm	in.	mm	in.	mm	in.	mm	in.
0.001	0.00004	0.01	0.0004	1.0*	0.0394	100	3.937
0.002	0.00008	0.02	0.0008	2.0	0.0787	200	7.874
0.003	0.00012	0.03	0.0012	3.0	0.1181	300	11.811
0.004	0.00016	0.04	0.0016	4.0	0.1575	400	15.748
0.005	0.00020	0.05	0.0020	5.0	0.1969	500	19.685
0.006	0.00024	0.06	0.0024	6.0	0.2362	600	23.622
0.007	0.00028	0.07	0.0028	7.0	0.2756	700	27.559
0.008	0.00031	0.08	0.0031	8.0	0.3150	800	31.496
0.009	0.00035	0.09	0.0035	9.0	0.3543	900	35.433
						1000	39.370

*1 mm = 0.0393700 in.

TABLE J-4 FRACTIONAL INCHES TO MILLIMETRES

in.	mm	in.	mm	in.	mm	in.	mm
$\frac{1}{64}$	0.397	$\frac{17}{64}$	6.747	$\frac{33}{64}$	13.097	$\frac{49}{64}$	19.477
$\frac{1}{32}$	0.794	$\frac{9}{32}$	7.144	$\frac{17}{32}$	13.494	$\frac{25}{32}$	19.844
$\frac{3}{64}$	1.191	$\frac{19}{64}$	7.541	$\frac{35}{64}$	13.891	$\frac{51}{64}$	20.241
$\frac{1}{16}$	1.588	$\frac{5}{16}$	7.938	$\frac{9}{16}$	14.288	$\frac{13}{16}$	20.638
$\frac{5}{64}$	1.984	$\frac{21}{64}$	8.334	$\frac{37}{64}$	14.684	$\frac{53}{64}$	21.034
$\frac{3}{32}$	2.381	$\frac{11}{32}$	8.731	$\frac{19}{32}$	15.081	$\frac{27}{32}$	21.431
$\frac{7}{64}$	2.778	$\frac{23}{64}$	9.128	$\frac{39}{64}$	15.478	$\frac{55}{64}$	21.828
$\frac{1}{8}$	3.175*	$\frac{3}{8}$	9.525*	$\frac{5}{8}$	15.875*	$\frac{7}{8}$	22.225*
$\frac{9}{64}$	3.572	$\frac{25}{64}$	9.922	$\frac{41}{64}$	16.272	$\frac{57}{64}$	22.622
$\frac{5}{32}$	3.969	$\frac{13}{32}$	10.319	$\frac{21}{32}$	16.669	$\frac{29}{32}$	23.019
$\frac{11}{64}$	4.366	$\frac{27}{64}$	10.716	$\frac{43}{64}$	17.066	$\frac{59}{64}$	23.416
$\frac{3}{16}$	4.763	$\frac{7}{16}$	11.113	$\frac{11}{16}$	17.463	$\frac{15}{16}$	23.813
$\frac{13}{64}$	5.159	$\frac{29}{64}$	11.509	$\frac{45}{64}$	17.859	$\frac{61}{64}$	24.209
$\frac{7}{32}$	5.556	$\frac{15}{32}$	11.906	$\frac{23}{32}$	18.256	$\frac{31}{32}$	24.606
$\frac{15}{64}$	5.953	$\frac{31}{64}$	12.303	$\frac{47}{64}$	18.653	$\frac{63}{64}$	25.003
$\frac{1}{4}$	6.350*	$\frac{1}{2}$	12.700*	$\frac{3}{4}$	19.050*	1.0	25.4*

*Exact equivalent.

TABLE K-1   TEMPERATURE CONVERSION
FAHRENHEIT TO CELSIUS

°F	°C	°F	°C	°F	°C	°F	°C
−100	−73	0	−18	100	38	2000	1093
−90	−68	10	−12	200	93	3000	1649
−80	−62	20	−7	300	149	4000	2204
−70	−57	30	−1	400	204	5000	2760
−60	−51	40	4	500	260	6000	3316
−50	−46	50	10	600	316	7000	3871
−40	−40	60	16	700	371	8000	4427
−30	−34	70	21	800	427	9000	4982
−20	−29	80	27	900	482	10,000	5538
−10	−23	90	32	1000	538		

NOTE: To add °C equivalents of °F, add the °C and then 18°C more for
each added term. EXAMPLE: 1700°F = 1000°F   538°C

$$\begin{array}{rr} 700 & 371 \\ — & 18 \\ \hline 1700°F & 927°C \end{array}$$

TABLE K-2   SURFACE FINISH
MICROINCHES (0.000001 in.) to
MICROMETRES (0.001mm)
(APPROXIMATELY 40 μin. = μm)

μin.	μm	μin.	μm	μin.	μm	μin.	μm
1	0.03	20	0.51	125	3.18	300	7.62
2	0.05	32	0.81	130	3.30	400	10.16
3	0.08	40	1.02	140	3.56	500	12.70
4	0.10	50	1.27	150	3.80	600	15.24
5	0.13	63	1.60	160	4.06	700	17.78
6	0.15	70	1.78	170	4.32	800	20.32
7	0.18	80	2.03	180	4.57	900	22.86
8	0.20	90	2.29	190	4.83	1000	25.40
9	0.23	100	2.54	200	5.08	2000	50.80
10	0.25	110	2.79				

TABLE K-3   PRESSURE CONVERSION POUNDS PER
SQUARE INCH (psi) TO KILONEWTONS
PER SQUARE METRE (kN/m²) OR
KILOPASCALS (kPa)

psi	kN/m² kPa	psi	kN/m² kPa	psi	MN/m² MPa*
10	69	100	690	2000	13.8
15	103	200	1380	3000	20.7
20	138	300	2070	4000	27.6
30	207	400	2760	5000	34.5
40	276	500	3450	6000	41.4
50	345	600	4140	7000	48.3
60	414	700	4830	8000	55.2
70	483	800	5520	9000	62.1
80	552	900	6210†	10,000	68.9
90	621	1000	6.9		

*Meganewtons used to keep figures smaller.
†Or 6.21 MPa

TABLE K-4   POUNDS (MASS) TO KILOGRAMS (MASS)
(ACCURATE TO THREE SIGNIFICANT
FIGURES)

lb	kg (mass)	lb	kg (mass)	lb	kg (mass)
10	4.5	200	90.7	2000	907
20	9.1	300	136	3000	1360
30	13.6	400	181	4000	1810
40	18.1	500	227	5000	2270
50	22.7	600	272	6000	2720
60	27.2	700	318	7000	3180
70	31.8	800	363	8000	3630
80	36.3	900	408	9000	4080
90	40.8	1000	454	10,000	4536
100	45.4				

TABLE K-5   POUNDS FORCE TO NEWTONS
(J/m) or kg·m/s²

lb	N	lb	kN	lb	kN
10	44.5	200	0.890	2000	8.90
20	89.0	300	1.33	3000	13.3
30	133	400	1.78	4000	17.8
40	178	500	2.22	5000	22.2
50	222	600	2.67	6000	26.7
60	267	700	3.11	7000	31.1
70	311	800	3.56	8000	35.6
80	356	900	4.00	9000	40.0
90	400	1000	4.45	10,000	44.5
100	0.445				

TABLE K-6   CUTTING SPEED—FEET PER MINUTE (fpm) to METRES PER MINUTE (m/min)

fpm	m/min	fpm	m/min	fpm	m/min
10	3.05	200	61.0	2000	610
20	6.10	300	91.4	3000	914
30	9.14	400	122	4000	1220
40	12.2	500	152	5000	1520
50	15.2	600	183	6000	1830
60	18.3	700	213	7000	2130
70	21.3	800	244	8000	2440
80	24.4	900	274	9000	2740
90	27.4	1000	305	10,000	3048
100	30.5				

*NOTE: This is also a conversion of feet to metres, as the denominators (min) are the same.

TABLE K-7   HORSEPOWER (hp) to KILOWATTS (kW)

hp	kW	hp	kW	hp	kW
$\frac{1}{4}$	0.187	2	1.49	15	11.2
$\frac{1}{3}$	0.249	$2\frac{1}{2}$	1.87	20	14.9
$\frac{1}{2}$	0.373	5	3.73	50	37.3
$\frac{3}{4}$	0.560	$7\frac{1}{2}$	5.60	100	74.6
1	0.746	10	7.46	300	224
$1\frac{1}{2}$	1.12				